D0998667

The Origins of Agriculture and Settled Life

The Origins of Agriculture and Settled Life

by
Richard S. MacNeish

University of Oklahoma Press : Norman and London

Library of Congress Cataloging-in-Publication Data

MacNeish, Richard S.
The origins of agriculture and settled life / by Richard S. MacNeish.
p. cm.
Includes bibliographical references and index.
ISBN 0–8061–2364–8 (alk. paper)
1. Agriculture—Origin. 2. Agriculture, Prehistoric. 3. Indians—Agriculture. I. Title.
GN799.A4M34 1991
306.3′64—dc20 91–50304

The paper in this book meets the guidelines for permanence and durability of the Committee on Production Guidelines for Book Longevity of the Council on Library Resources, Inc. ∞

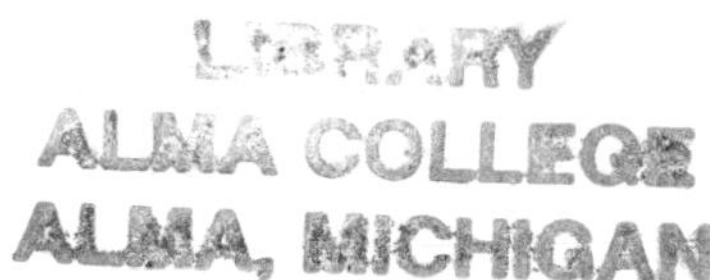

Contents

Illustrations

Tables

Preface

My first encounter with the problem of the origins of agriculture and settled life began one hot March afternoon in 1949 on a dusty "street" of Los Angeles, a tiny hamlet at the end of a rocky road in the Sierra de Tamaulipas, Mexico (MacNeish 1958). I had just walked in from Nogales Cave, some twenty miles away, after closing down our dig, and was at the head of five burros with all the equipment and a dozen workmen who looked like bandits. Two nights before, I had left La Perra Cave, telling my dig supervisor, Alberto Aguilar, to finish the final square, clean the wall profile, draw it, close up the camp, and then pack our supplies in to Los Angeles so that we could move on to preceramic excavations in other regions.

The night trip from La Perra Cave to Nogales Cave had been memorable. While walking the twenty kilometers down the Canyon Diablo, we happened on a *tigre* (jaguar). I trained the flashlight on it, but as we got ready for the kill, my guide dropped the rifle and I dropped the flashlight. Fortunately, the jaguar didn't jump. However, because we had to traverse the last five miles of the boulder-strewn canyon with no flashlight and very few matches, we didn't reach our Nogales Cave camp until three in the morning. Up at dawn, we spent that day frantically drawing profiles, tagging specimen bags, and filling trenches. Our last supper, of beef jerky, boiled beans, and tortillas, was washed down with stale water. Then, after a welcome night's sleep and a breakfast of stale, hard tortillas and coffee made with our stale water, we climbed the tortuous route out of the canyon. Shortly afterward, wading through the Los Angeles spring for my first bath in ten days, I pondered our season's work and thought that finding the first ceramic sequence for Mesoamerica made for a pretty good season. Tired but happy, I really perked up when I saw Alberto on the outskirts of town.

My first questions were: "Did you bring the equipment and specimens? Is the jeep packed?"

The answer was a deadpan, "No, señor."

"Why not?"

Alberto grinned. "Because we found what you said we might find."

"What's that?"

"It's hard to explain."

By now we had reached the unpacked jeep and the door of our hut. "Your explanation had better be good, or you're going to walk back to Pánuco!"

Alberto ran into the house and returned with a box labeled "NW corner of square N15W5, depth 14 inches." Opening it, he said, "Está muy bueno."

And he was right. It *was* good! Inside the box, on a woven mat, were three tiny corncobs tied with agave string. The depth told me it was 3,000 to 5,000 years old. It was the earliest corn ever found, and it was to change the course of my life.

The effect was immediate. We did not leave Sierra de Tamaulipas that March; instead, we returned to dig La Perra Cave. We peeled off each fine layer of preserved plant remains, bagged and labeled them, recorded the finds, and eventually, in May, took the specimens to the Instituto Nacional de Antropología e Historia (INAH), in Mexico City. We had analyzed only about one-half of the materials. We knew more had to be done, but we were out of money. (In fact, I had to borrow money from my sister to get us out of Mexico.) Fortunately, the late Helmut de Terra, of the then Viking Fund (now the Wenner Gren Foundation), offered to help. He said he would give our corncobs to Paul Mangelsdorf, a Harvard botanist then working on a corn project for the Rockefeller Foundation in Mexico, and that he would send some of our bottles of carbon and plant remains from various levels to a young University of Chicago physicist, Willard Libby, who had just invented a new dating technique, analyzing the carbon-14 isotope.

Then I was off to Ottawa, Canada, to do fieldwork for the National Museum of Canada. A week later I was busy in Yellowknife, in the Northwest Territories—a long way from my precious little corncobs. I heard nothing from Mangelsdorf. The one carbon specimen sent to Libby was dated 658 years old. (Later I learned that it had come from Level 1, not Level 14, where we had found our corncobs.) Most of the analysis of the stone artifacts had been accomplished, but the dates and botanical materials had not been analyzed. In 1951, having decided that something had to be done, I took my hard-earned savings and, on my winter vacation, was off to Mexico to try to find my lost carbon and corn samples. They were right where I had left them. With the help of Eduardo Noguera and José Luis Lorenzo of INAH, I got permission to take my specimens to Libby and Mangelsdorf.

After this "vacation," I went back to my duties in the Northwest Territories. When I flew out of Edmonton, Alberta, at the end of the next season and picked up my mail at the Bank of Commerce, I found two special letters. One was from Libby, the other from Mangelsdorf. Libby wrote that the bottle of carbon from Level 14 associated with the early corn was 4,445 ± 280 years old and therefore was older than any other corn yet uncovered—including that at Bat Cave, New Mexico, and at Swallow's Cave, Chihuahua. The letter from Mangelsdorf informed me that my corn was the most primitive yet found, and he asked when I could come to Cambridge to discuss the tiny cobs.

My reaction was, "Holy ----!," an expletive not often heard in the august halls of Canadian banks. Immediately I started making arrangements to travel. After an endless three-day train trip to Ottawa, I reached Cambridge and Mangelsdorf's laboratory in the Harvard Botanical Museum, where I began laying out the cobs by their stratigraphic levels. Thanks to Paul, his wife, Peggy, and Walton Galinat, I was introduced to the world of corn studies. It was fascinating! We were a compatible little group, and what is more, we all (but mainly Paul) decided that the best way to solve the problem of the origin of corn was to combine the skills of archaeologists with those of botanists. Although I had worked with a couple of geologists in Tamaulipas and had tried to take courses in other archaeologically useful disciplines at the University of Chicago, this was my first real introduction to the interdisciplinary approach. I had a lot to learn, not only about this approach but also about the problem of the origin of corn agriculture, which gradually grew into the problems of "the origins of agriculture and settled life."

Before I could begin fieldwork agriculture, I had to learn how to prepare grant proposals and plan research attacking the problems of a changing subsistence system that included the development of agriculture. The very first step was to bone up on agricultural studies. My first field class under the tutelage of Mangelsdorf was in the Ocampo regions of the Sierra Madre of southwest Tamaulipas in 1954–1955 (Kaplan and MacNeish 1960). This excavation, which was supported by my Guggenheim Fellowship, yielded abundant well-preserved materials in good stratigraphic contexts. The real learning came when we started our analysis. Of key importance was Mangelsdorf's two-week visit to our headquarters in Ciudad Victoria to separate our corn samples from other plant remains so that he could take the corn back with him. As we poured onto our lab tables the plant remains from various floors, levels, and squares of several caves, looking for the all-important corn, he separated out and showed us how to identify the remains of teosinte, tripsacum, beans, pumpkins, squash, and other domesticated plants as well as feces. Besides providing an informal lecture on the domestication of the plants, Mangelsdorf told us about the experts for each plant variety. This enabled me to establish long-term associations with Hugh Cutler and Tom Whitaker, experts in the domestication of pumpkins, gourds, squash, and other cucurbits; Larry Kaplan, in common, tepary, runner, lima, and other beans; Eric Callen, in setaria and human feces; and C. Earl Smith, in various New World domesticates. These contacts proved invaluable.

By the end of the season, we had concluded that Tamaulipas probably was not the birthplace of corn, beans, various kinds of squash, or teosinte; however, we suspected it might well have been the place where pumpkins were first domesticated. We were conceiving the idea of multiple places of origin of agriculture or domestication in pristine areas. In Mangelsdorf's course in economic botany, I learned of the hundreds of plants that had been domesticated and read the various theories of plant domestication. None of them seemed to fit the facts I was finding in Tamaulipas. Little by little, I was learning that plant domestication is a behavioral process, as is the development of agriculture and the concomitant evolution of village life. I began to realize that I had an opportunity to establish a generalization about an important culture change or process.

At that point, however, my theoretical contribution was relatively modest. I

Figure 1. Important sites in the search for the origins of corn

was simply too busy doing fieldwork, writing for the National Museum of Canada, and carrying on "the Great Corn Hunt" (see fig. 1) during my winter vacations. It was a nomadic way of life. By 1957, I was looking for evidence of the first corn along the Río Balsas in Guerrero; the next year I was in Honduras and Guatemala, which were too far south, as it turned out later. In 1959, I went first to Chiapas, digging Santa Marta Cave, which had good preceramics of the right age but pollen profiles showing no corn early enough (MacNeish and Peterson 1962). Then I went on to Oaxaca, looking at caves near Mitla, where the pollen profiles seemed to show that this region was north of the area where the first corn had been domesticated. By 1960, I had pretty well determined that the possible point of origin was south of the Valley of Mexico and north of Chiapas. Sure enough, that very year we found Ajuereado Cave, near the town of Coxcatlán in the Tehuacán Valley of Puebla, Mexico, and thus began the Tehuacán Archaeological-Botanical Project, under the auspices of the R. S. Peabody Foundation for Archaeology.

Because most of the factual results of the Tehuacán Project have been published elsewhere (Byers 1967a, 1967b; MacNeish 1970; F. Johnson 1972; MacNeish 1973), I shall simply provide an overview here. Initially (1961–1965), we excavated nine major stratified sites and tested eighteen more in all five of our ecozones. We uncovered more than a half million specimens from 160 stratified components (72 of which were preceramic occupations) and 531 surface sites (41 of which were preceramic). All these preceramic occupations were relevant to our investigation of the development of agriculture and the domestication of plants. The four preceramic phases initially uncovered—Ajuereado (more than 12,000–7600 B.C.), El Riego (7600–5000 B.C.), Coxcatlán (5000–3400 B.C.), and Abejas (3400–2300 B.C.)—were dated on the basis of 120 radiocarbon determinations, using about 10,000 stone artifacts and bones, as well as about 40,000 plant remains and feces, including many domesticated or cultivated plant samples, which were the most relevant to our investigation of the origins of agriculture. Using our reconstructed sequence (see chapter 3), by 1966 we could hypothesize about the origins of agriculture in the Tehuacán Valley and could even provide a testable theory regarding the development of village agriculture in other places at other times.

My next step, of course, was to test these hypotheses in another independent center of agriculture, and I chose the Andean area. Much work had already been done in this region. The coastal area had yielded a long preceramic sequence with abundant preserved plant remains (Towle 1961) leading to village agriculture (Lanning 1967b), but the theoretical framework was weak (Lathrap 1977) and much of the ecology was most unlike Tehuacán. Further, the Andean highlands, which we speculated might resemble Tehuacán in the preceramic record, were relatively unknown.

In 1966, therefore, I began a survey of the Andean region and, in the Ayacucho Valley, found the cave of Pikimachay, which showed that the region had excellent preceramic potential. Once again we organized an interdisciplinary expedition. In 1964, during the initial survey, Angel García Cook had found more than 450 sites, and in 1970 we began a major digging program. After its conclusion in

1973, the analysis, undertaken by scientists from many disciplines, continued into 1974. This time, our efforts revealed a 20,000-year preceramic sequence in the highland valley of Ayacucho, Peru: Pacaicasa (25,000–13,000 B.C.), Ayacucho (13,000–11,000 B.C.), Huanta (11,000–9000 B.C.), Puente (9000–7100 B.C.), Jaywa (7100–5800 B.C.), Piki (5800–4400 B.C.), Chihua (1400–3100 B.C.), and Cachi (3100–1750 B.C.). Even though we found fewer plant remains, the Ayacucho sequence was, if anything, better than the Tehuacán sequence. As chapter 2 will show, both highland and lowland sequences were basic to the development of my hypothesis explaining the lowland development of village agriculture. In the period from 1967 through 1974, however, I thought of the Ayacucho data as mainly confirming my Tehuacán hypothesis (MacNeish 1978).

The next step was to test my hypothesis about lowland village agricultural development, which seemed so different from the development in the highlands, at least in Peru. Was the lowland development in Mesoamerica different from the highland one that we knew so well? I had long suspected that it might be (MacNeish 1954) but had little evidence. Fortunately, in the late 1970s I had been in contact with Jeffrey Wilkerson, who was working in the Tecolutla drainage region of central Veracruz, where he had uncovered some late preceramic (3000 B.C.) materials, called Palo Hueco, at the bottom of his Santa Luisa site (Wilkerson 1975). The question now was whether we could get a preceramic sequence in this region that would allow us to test our lowland hypothesis derived from the Peruvian data. The practicality of interdisciplinary investigations in Veracruz was also questionable. Anti-American sentiment, local politics, and the difficulties of transportation in the local deltaic or coastal zones, which Wilkerson had determined were the most likely areas for early village developments, were all considerable obstacles. While we were trying to make up our minds, fate took a hand.

In the winter of 1979, Toni Nelken-Terner and I went to visit Norman Hammond's site near Orange Walk, Belize, to see if his Swazey pottery was as early as our Purrón pottery from Tehuacán. We were given a hospitable grand tour, and we saw not only Hammond's pottery but also a lot of Belizean countryside, which we realized resembled Wilkerson's Veracruz region in many ways. One afternoon we borrowed a truck to see if preceramic sites might be located in the same types of ecozones as in Veracruz. Sure enough, on the third stop, on a point in Progreso Lagoon, we found the first preceramic site in the Maya lowlands. This find raised the question whether we could get a long preceramic sequence in this Maya lowland area that would allow us to test our Peruvian lowland hypothesis. We decided to survey lowland Belize for its preceramic potential.

In 1979, Toni Nelken-Terner, Jeff Wilkerson, Viki Velasco, Tom Emerson, and I began an extremely successful hit-and-run reconnaissance of coastal Belize, finding 160 sites, approximately 100 of which were preceramic, and 5 or 6 of which held possibilities of stratigraphy. Our new interdisciplinary project took place from 1981 through 1983. Nine stratified sites yielded a tentative sequence—Lowe-ha (9000–7500 B.C.), Sand Hill (7500–6000 B.C.), Orange Walk (6000–5000 B.C.), Belize (5000–4000 B.C.), Melinda (4000–3000 B.C.), and Progreso (3000–2000 B.C.)—but the dates are poor (MacNeish, Wilkerson, and Nelken-Terner 1980). Our data are not yet completely analyzed, nor are they as complete as those from previous Mesoamerican and Andean investigations. The little information that we do have, however, can be used to test our lowland hypothesis further (MacNeish and Nelken-Terner 1983a).

As before, research brought me in contact with other projects focusing on the origins of village agriculture. I was particularly interested in the research being done in the tropical areas of Central and South America, where village life reputedly was based upon root crop (manioc) agriculture and had apparently developed in a distinct manner, promising possible new hypotheses.

Throughout the duration of the Belize Archaic Archaeological Reconnaissance project, I wrestled with these three hypotheses and attempted to test them with data from all over the world. Then, in the fall of 1982, I had an opportunity to organize and try out these hypotheses on a group of stimulating students (Keith Adams, Bill Barnett, Dennis Blanton, Pat Crawford, Mac Goodwin, Rick Kanaski, Bruno Marino, Steve Marozowski, and Barry Rowlett) and professors (K. C. Chang, Creighton Gable, Larry Kaplan, and Jim Wiseman). This took place during the first seminar I taught in the newly formed, innovative department of archaeology at Boston University. The seminar was held on Friday afternoons, was well organized, and was supplemented by a "beer break" at the pub across the street. As a result, the class and the arguments lasted not only the required two and a half hours but far into the night. We had a ball. Better than that, though, I was forced to alter, develop, and modify my hypotheses about the origins of agriculture and settled life.

However, the task of putting everything together in manuscript form was done under the auspices of the Andover Foundation for Archaeological Research, where the final draft of the manuscript was ably edited by Jane Libby. Colleen Barnitz, at New Mexico State University, typed the manuscript. From 1985 to 1988, while we were investigating origins of agriculture in the U.S. Southwest, both helped with the phrasing of the hypothesis and honed the first draft into this final version.

The results are new hypotheses—and this book. There is still much we don't know, but at least we've made a start. I hope you'll enjoy reading the results as much as we enjoyed creating them.

RICHARD S. MACNEISH

Andover, Massachusetts

The Origins of Agriculture and Settled Life

CHAPTER 1

Theories of the Origins of Agriculture

One of humanity's most important inventions is agriculture. This decisive step freed people from the quest for food and released energy for other pursuits. No civilization has existed without an agricultural base, either in the past or today. Truly, agriculture was the first great leap forward by human beings.

Yet, in spite of the importance of the process of plant domestication that ultimately led to agriculture, little is known about how, where, when, and why it happened. Many fields of study—geography, botany, anthropology, history, archaeology, and others—have attacked this knotty problem, but only recently have we begun to amass any relevant data. There has been much speculation on the origins of agriculture, yet many of the early speculations were diachronic hypotheses that could not be tested by hard historical or archaeological data.

In the last thirty or so years, however, interdisciplinary studies by botanists, geographers, archaeologists, and others have brought to light archaeological sequences of domesticated plant remains that not only are relevant to the problem of how plants became domesticated but also might help us understand why it happened. These hypotheses in turn might become the basis for yet other forms of research that would enable us to test the original hypotheses with independent data.

In fact, from these studies we are now deriving generalizations or laws about the development of agriculture and the concomitant evolution of settled life. Thus, our study of the origins of village agriculture not only may give us interesting historical information but also may provide us with scientific generalizations of value to the social sciences. Further, by providing the bases for laws of cultural change, these studies might open new avenues of investigation and encourage the development of new methodology in the social sciences.

Obviously, our new understanding of the cultural process that we call the development of agriculture needs more confirmation. The same also can be said of our methodology. The data we collected and described are marginally representative; our hypotheses are deduced from analyses of cultural-historical integrations; and the testing of these hypotheses is by the comparative method. In recent years we have defined some of the major problems, refined our methodology, and indicated the direction further research should take. We do not yet have final answers or laws, but we have made a significant step in that direction. Like the early humans who developed agriculture, people in the discipline of archaeology have taken a great scientific step forward.

This book presents a new theory about the origins of agriculture. In this chapter I set the stage by providing insight into the work of the people who have derived recent hypotheses—including my own theory—about the origins of agriculture and village life. Subsequent chapters examine the various areas where agriculture evolved: the Andean area, Mesoamerica, the Near East, the Far East, the eastern and southwestern United States, Europe, India, Southeast Asia, Oceania, the tropical lowlands of Central America and northeastern South America, and Africa (both tropical and northern). In each chapter I present raw sequential data and then proceed to test and modify my various hypotheses. I shall further confirm these hypotheses by examining similar developments that failed to culminate in the establishment of agriculture. Finally, I shall present my partially tested hypotheses and suggestions for future research.

THEORETICAL BACKGROUND

Just how far back to go in a consideration of theories about the origins of agriculture took some soul searching. Rather arbitrarily, I decided to consider only those theories, speculations, or hypotheses that have a direct bearing on the theories presented in my conclusions. I shall therefore emphasize research done since the 1920s.

Let us begin by examining the sequence of major theoretical contributions (see table 1.1). In the 1920s and 1930s and, to a lesser extent, in the 1940s, the major contributions were made by Edgar Anderson, Melvin Gilmore, and Ivan Vavilov, botanists; Carl Sauer, a geographer; and archaeologist V. Gordon Childe. Their publications, teaching, and research continued into later decades, of course; and their influence is still felt today.

During the 1940s and 1950s, Robert Braidwood and his colleagues—Linda Braidwood, archaeologist; Jack Harlan and Hans Helbaek, botanists; Bruce Howe, an archaeologist; Charles Reed, a zoologist; and others—led the field. Their era witnessed a subtle shift: the geographers and botanists began yielding the lead to teams of botanists and archaeologists—Paul Mangelsdorf, a botanist; Bob Lister and myself, archaeologists, in Mexico; Herbert Dick and Paul Martin, archaeologists; Hugh Cutler and Tom Whitaker, botanists; and others in the southwestern United States; Charles Heiser and Volney Jones, botanists; James Griffin, Clarence Webb, and Dick Yarnell, archaeologists; and others in the eastern United States; Hans Helbaek, Harold Godwin, and various archaeologists in Europe and the Near East; and various archaeologists as well as botanists Tom Whitaker and Margaret Towle in Peru. During this period the modern interdisciplinary studies were being born, and many new kinds of data collection were inaugurated.

This period continued well into the early 1960s. Interdisciplinary teams began intensive investigations into the origins of agriculture. Often directed by archaeologists, these teams included scientists from a variety of relevant disciplines:

Table 1.1. Diagram of the history of theories about the origins of agriculture

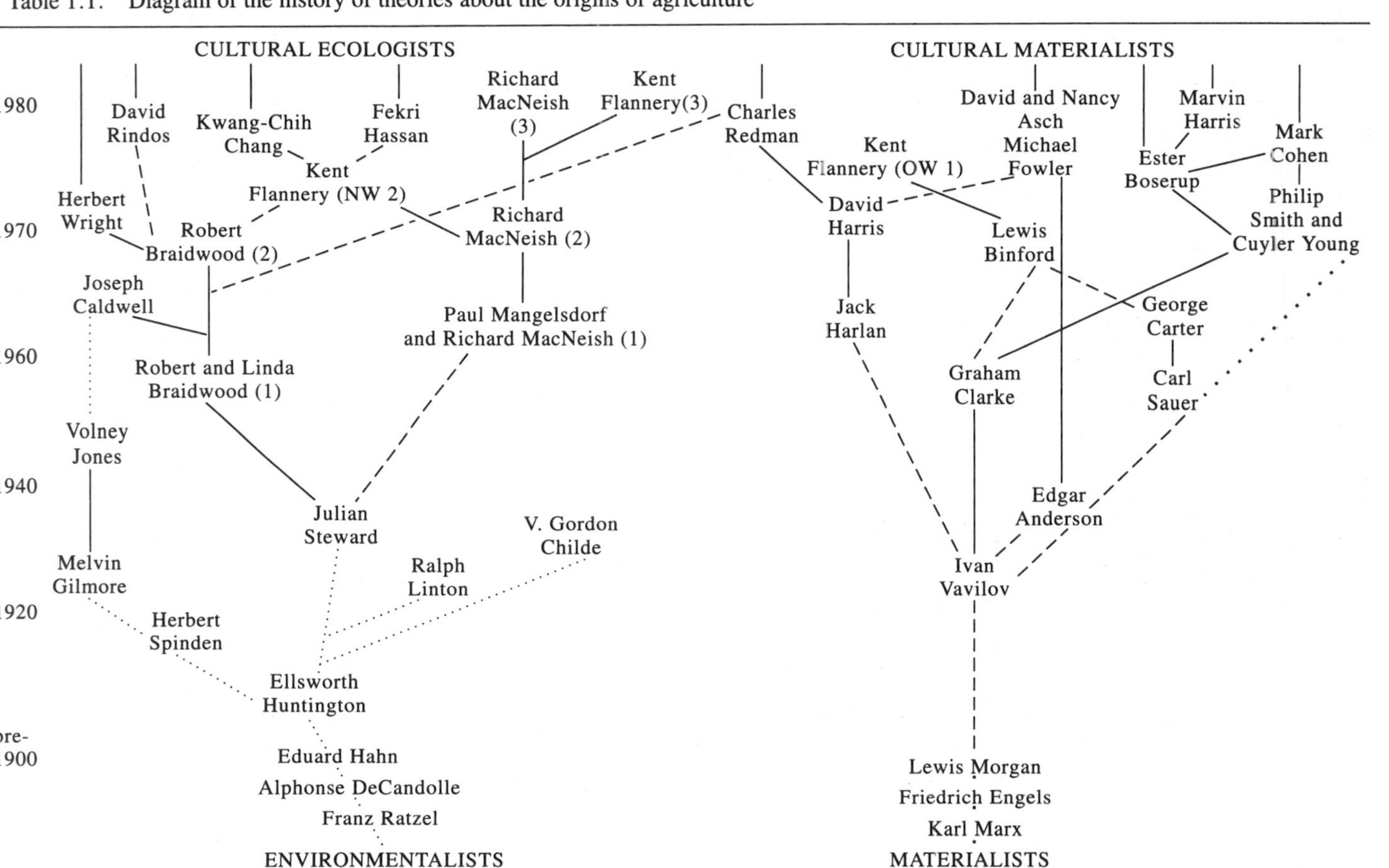

coprology, palynology, ethnobotany, taxonomic botany, phytolithic studies, isotopic studies, and so on. Out of this period came a host of new researchers who have given us new hypotheses from a number of fields: Ester Boserup, Karl Butzer, George Carter, and David Harris from geography; Herbert Wright from geology; David Rindos from palaeo-ethnobotany; George Beadle, Vorsila Bohrer, Vaughan Bryant, Eric Callen, John DeWet, Walton Galinat, Jack Harlan, Larry Kaplan, Barbara Pickerskill, and G. Yen from botany; and numerous archaeologists, including Lewis Binford, the late Joseph Caldwell, K. C. Chang, Mark Cohen, Kent Flannery, Fakri Hassan, Dick Ford, Melvin Fowler, Cynthia Irwin-Williams, David and Nancy Asch, Charles Redman, Philip Smith, Stuart Struever, and Patty Jo Watson. Undoubtedly, I have left out many important contributors from this period, for the ranks have been swelling at an ever-increasing rate since 1974. No doubt more new scientists will come, and some will develop new and better theories. Having the general picture in mind, let us look now at the different periods and their participants in more detail.

The first two scientists directly connected with modern thought about the origins of agriculture were Sir V. Gordon Childe, the British archaeologist, and Ivan Vavilov, the Soviet botanist. Both were avowed Marxists, believing that cultures changed by means of a series of economic revolutions brought about by the introduction of new means and modes of production. Their concepts about what caused the shift from food collection to food production (i.e., agricutlure) were, however, fundamentally dichotomous. In fact, their conceptualizations represent the major theoretical dialectic over the origins of agriculture. This basic conflict, which I refer to as "materialism" versus "environmentalism," has undergone several stages of development in its evolution to present-day cultural materialism and cultural ecology.

Let us begin by reviewing Childe's contribution. Although the following sketch of his theory might not do him justice, I shall endeavor to outline its salient features (Childe 1951). The fundamental event in the establishment of agriculture, Childe asserts, is the ending of the Pleistocene with its concomitant changing climate. In the Near East, desiccation expanded the deserts, causing a concentration of plants, humans, and animals around a series of oases. Recognizing the food potential of the associated plants and animals, humans began to experiment with them. Their experimentation led to a new means of food production—agriculture (i.e., the planting of seeds). Associated with this shift was the Neolithic Revolution, characterized by sedentary village life, expanding population, the use of ground-stone tools, the development of ceramics, and the emergence of a new type of social (and political?) organization.

Although this theory was most fully expressed in 1951, it was based upon years of research by Childe, starting with his work on the Danube in the 1920s. The germ of Childe's ideas had sprouted much earlier with nineteenth-century geographers who have been called environmental or geographical determinists. This large, literate group included Franz Ratzel (1882), Alphonse DeCandolle (1884), Eduard Hahn (1896), H. Ling Roth (who, in 1887, wrote a paper titled "On the Origins of Agriculture"), Ellsworth Huntington (1915), and Edward Semple (1911). These determinists believed that major environmental changes brought about fundamental culture changes, including civilization itself and, inferentially, agriculture. Thus, Childe's concepts had a rich heritage.

Somewhat comparable to this heritage is that of Ivan Vavilov, the other major theoretician of the 1920s. Vavilov's theories (1951) grew out of the dialectical materialism of Karl Marx, Friedrich Engels, and related thinkers. According to Marx, agriculture was invented during the stage of history called "barbarism." Explanations of the event, however, were vague. They merely set forth the proposition that people needed to produce more food for increased population. Vague as the thinking of these materialists was, it is apparent that they thought of population and a new means of food production as causes and not results, as Childe would have seen them. Thus the battle lines were drawn, and while Childe was collecting archaeological data to substantiate his hypothesis, Vavilov and his colleagues were collecting botanical evidence to back theirs.

In all fairness to Vavilov and his followers, their attempt was never to explain concretely how and why agriculture happened. They were much more interested in when and where it happened and therefore chose to classify and study the geographical distribution of varieties of domesticated plants. The Russian study, a tremendous worldwide undertaking by a team of dedicated Soviet scientists, brought together great masses of relevant data and made some fundamental botanical classifications of domesticated plants. The major contribution of this study, however, is the hypothesis that areas with the greatest concentration of domesticated plant varieties were the "hearths" of domestication, that is, the centers in which domestication was first practiced and from which all later use of domesticated plants and agriculture radiated and diffused.

Vavilov identified eight independent primary hearths of domestication, a ranking that still stands. In chronological order the eight hearths are: China (136 plants), India (117), the Near East (83), highland Guatemala (49), the Andes (46), the Sudanic-Abyssinian area (38), the South American tropics (35), and southeast Asia (24). Although Vavilov and his colleagues were never specific, the idea underlying their hypothesis was that humans first settled these areas because of the wide variety of edible plants available. These plants were domesticated because increasing populations needed more food. This hypothesis certainly followed the party line. (Ironically, however, Vavilov ended his career in a concentration camp because, according to Trofim Lysenko, official spokesman for Soviet biology at that time, he had deviated from the "true" party line.)

By 1930 the two major schools of thought—the environmentalists and the materialists—had evolved. Other studies that would influence later developments were also being undertaken. In anthropology, Alfred Radcliffe-Brown's functionalism and Julian Steward's cultural ecology marked the beginnings of a more systematic approach to problems of culture change. After anthropology's years of dominance by Franz Boas and his disciples, the older, materialistic ideas of cultural evolution under Alexander Lesser and Leslie White began to influence young archaeologists investigating the origins of village agriculture. Also gradually emerging was the field of ethnobotany—first under Ruth Underhill, Edward Castetter, and Melvin Gilmore, and later under Volney Jones—whereby archaeologists and botanists interacted informally. Also, Franz Boas's anthropology, although theoretically sterile, inspired such thinkers as Alfred Kroeber and his "California school" to investigate human geography and to cooperate with Carl Sauer's geographers.

These two fields—botany and geography—made the greatest theoretical contributions to the problem of the origin of agriculture in the 1940s and early 1950s. The two major contributors, both cultural materialists in my opinion, were Edgar Anderson, a botanist, and Carl Sauer, a geographer. Both influenced a host of students.

Anderson, for many years head of the Missouri Botanical Gardens, near Saint Louis, had special interests in plant geography, plant morphology, and human relationships to the plant kingdom (Anderson 1952). On the basis of both Vavilov's and his own observations, he concluded that most domesticates (often seed plants) are "open-habitat plants." Further, he noted that such plants readily mutate when grown in disturbed soils. Then, taking the meager relevant archaeological data of the 1940s, he suggested that the disturbed areas where mutants were most likely to form were the kitchen middens of sedentary peoples in open-habitat regions (Anderson 1956). From these middens humans collected seed plants, including some (presumably more beneficial) mutants and learned to plant these seeds. Because they needed food to feed their expanding population, our collecting villagers became Village Agriculturists. (To imply that population increases as well as successful sedentarism were causes of agriculture is cultural materialism.) Often called the "dump heap theory" of agricultural beginnings, this hypothesis has been employed continually in later periods with slight variations (Fowler 1973).

An even more popular hypothesis from the 1940s is remarkably similar even though it came from a geographer. While at the University of California at Berkeley, Carl Sauer had been influenced heavily by the ethnological distribution studies of Alfred Kroeber, and he also was familiar with the works of Vavilov. Even though his conclusions about hearths were different from Vavilov's, Sauer, like the Russian botanist, believed that the hearths of domestication were regions of marked plant and animal diversity, where humans could have easily obtained food. His geographical studies led him to conclude that agriculture most likely originated in such "wooded" tropical areas as Southeast Asia and northeastern South America (Sauer 1952). Here, he speculated, people could have settled down, had enough food to afford leisure time sufficient for developing various technical skills (some of which would predispose them to agriculture) and for experimenting with plants (often root crops) in a way that gradually led to planting and (with increasing population?) agriculture. According to Sauer's hypothesis, humans developed agriculture as a result not of hunger but of abundance. His scheme is similar to Vavilov's and Anderson's but emphasizes lusher regions with fewer tensions, or driving forces.

While Sauer and Anderson presented hypotheses that were materialistic, they were unlike the "true" Marxist hypotheses of the same period, which were best expressed by the British archaeologist J. Graham Clark. The following statement delineates Clark's materialism clearly: "Here it need only be emphasized that the rate and scope of social evolution have to a large extent been conditioned by the development of more effective means of obtaining food and most notably by its production through control of animal and plant breeding. More food leads to growth in the density of population, the creation of surplus, and a need for a new means of food production like agriculture" (J. G. Clark 1957, 11).

Much of Clark's work was on the Mesolithic of northern Europe, the general stage out of which village agriculture arose in that region. To Clark, better food production and increasing density of populations were basic prerequisites for change (J. G. Clark 1936). From many standpoints, Clark was an intellectual descendant of Childe, but because he rejected Childe's concept of environmental change as the motivating force behind the development of agriculture, he put himself squarely on the Marxist side.

Robert Braidwood, probably the best representative of the cultural ecologists, also rejected Childe's "oasis theory." Unfortunately, while rejecting the theory, Braidwood offered no strong theoretical alternative. While he established his theoretical position in the period from 1940 to 1960, we must depend heavily on his writing after that period. In terms of causation, Braidwood seems to see four general levels (Braidwood and Willey 1962):

1. During the late Pleistocene there was a trend toward regionality and intensified extraction of food resources from more localized environments (a band's territory?).
2. During the period from 20,000 to 8000 B.C., increasing manipulation of and experimentation with plants in their natural habitat led to plant domestication.
3. As groups of people settled outside the natural habitat zones, they gave up regional intensive collecting techniques and intensified their planting of domesticates.
4. As they used more plants, humans began to make more use of hybrids, did more planting, and increased group size to form agricultural villages.

Although Braidwood's hypothesis focuses on human adaptations to the environment (cultural ecology), it does not clearly state the cause of the development of agriculture. In fact, in his book *Prehistoric Men* (1951, 110–120) Braidwood states that "the groups became Agricultural Villagers because they were ready for it." This is hardly a theoretical statement about causation.

Braidwood's greatest contribution was not in theory but in methodology. His use of the interdisciplinary approach, in cooperation with the botanists Jack Harlan and Hans Helbaek, provided a model for others to follow. Robert Braidwood, Bruce Howe, and others found and dug some important sites—Jarmo, Palegawra Cave, Çayönü, and Karim Shahir—that yielded information pertinent to this problem. They point to the Hilly Flanks of the Zagros Mountains in Iran as a key area where village agriculture, or at least incipient agriculture, may have first occurred.

Data collection, methodology, and the interdisciplinary approach were also the emphasis of several people working on the origins of agriculture in the New World. In the U.S. Southwest, Paul Martin, an archaeologist, cooperated with the botanist Hugh Cutler, and Herbert Dick worked with C. Earle Smith and Paul Mangelsdorf, while George Carter, a Sauer student, worked alone; in the eastern United States, Volney Jones was identifying plant remains dug up in Kentucky caves by William S. Webb and William Funkhouser. Margaret Towle assisted the Virú Valley archaeologists in Peru, and Paul Mangelsdorf and I cooperated on projects in Tamaulipas, Mexico.

The researchers mentioned in the preceding paragraph could be classified as cultural ecologists because they believed that the causes of the development of agriculture were basically ecological, not merely population pressure and/or new means of production, as the materialists asserted. Although no new theories came out of the investigations of this group in the 1940s and 1950s, we did amass data that contradicted the hypotheses of such cultural materialists as Clark, Sauer, and Anderson. For example, it seemed to me that plant domestication and agriculture had taken a long time (MacNeish 1950) and that the evidence pointed to "a Neolithic evolution rather than a revolution" (MacNeish 1965). It appeared that "there were multiple origins of New World domesticates" and that there was "little evidence for the single hearth theory or even for a few well-defined hearths" (MacNeish 1965, 93). I also opposed the concept of a unilinear development in Mesoamerica, at least for the Formative period and probably much before it. Instead I argued that the coast developed one way and the highlands another (MacNeish 1954). (This assertion was somewhat in opposition to Herbert Spinden's earlier speculations that all agriculture and village life had begun in the harsh areas of highland Guatemala and Mexico because bands with increasing populations needed more food.)

Although my ideas opposed those of the materialists regarding the origins of village agriculture, my conclusions failed to specify why it happened. In fact, like Braidwood, we cultural ecologists were deductionists. Many of us felt we did not have enough data to justify setting up a new hypothesis, let alone a new theory. Nevertheless, we had high hopes that our methods would enable us to accumulate data that would lead to hypotheses that could be tested to yield a generalization.

While I came up with few definite conclusions from 1945 to 1955, I did begin to define some useful terms. Since these terms will be used throughout this volume, let me explain them here:

Domesticate. A plant or animal that differs from its wild ancestor because it has been changed genetically through human selection, either consciously or unconsciously.

Cultivar (or *tamed animal*). A plant (or animal) that differs morphologically but not genetically from its wild ancestor as a result of human alteration of the ecosystem, habitat, or environment.

Horticulture; incipient agriculture. Plant cultivation that emphasizes the planting of individual domesticates or cultivars in relatively limited plots (from *hortus,* garden).

Pasturing. An activity similar to horticulture in that it emphasizes the care of individual animals (from *pascere,* to feed or tend beasts).

Agriculture. The planting of multipropagators (i.e., seeds) of domesticates or cultivars in relatively large plots or fields (from *ager,* field).
Herding. The caring for a large number of tamed or domesticated animals.
Seasonality. Cyclic variation in economic and/or social patterns.
Scheduling. The hierarchy of priorities resulting in the temporal ordering of selected options.
Options. Alternatives for economic or social action.
Strategy. The conscious or unconscious adoption of a particular pattern of scheduled options.
Microband. A group of three to four nuclear families and linked individuals, usually associated in a temporary occupation.
Macroband. A group comprising more than four nuclear families and linked individuals who are temporarily associated with one another.
Task group; task force. One or more individuals acting together to carry out specific activities for a limited time.
Village; hamlet. Small community whose members occupy the same location throughout the year for a series of sequential years. Villages have central administrative, religious, or economic specialized areas; hamlets do not.

Of our team, Mangelsdorf was nearer to being a theorist than I. In a vague, unstated way, I felt that the development of agriculture had been slow and gradual and had been accomplished by seasonal, patrilineal bands—as defined by Julian Steward (Steward 1937)—who had collected plants susceptible to domestication and, by domesticating them, had maintained bigger groups (macrobands) that evolved into village agriculturists. Like Braidwood, we asked not *why*, but only *how,* it happened, and if Braidwood's theory was unsophisticated, ours was downright naive. The cultural materialists during this period had stated hypotheses about why agriculture had come into being, but we cultural ecologists were uncovering facts that contradicted their theories. More important, we were developing a method and an interdisciplinary approach that brought in new relevant hypotheses and data from a wider variety of fields.

Between 1956 and 1970, progress on the problems of the origins of agriculture accelerated as never before. It was exciting to be a part of this new period.

My first real connection with theories concerning the origins of agriculture began with the 1965 analysis of data from our interdisciplinary work at Tehuacán, Mexico (Byers, 1967a). At that time, following various logicians (Cohen and Nagel 1970), I began to use two terms for the various kinds of causes leading to changes in subsistence patterns. These were *necessary conditions,* prerequisites for cultural change or factors vital in allowing the changes to happen, and *sufficient conditions,* triggering causes or factors that directly brought about the change.

Our preliminary Tehuacán data suggested to me the following (MacNeish 1973, 502–503):

I. A culture system like the Ajuereado (hunting microbands) might evolve into an El Riego-type system if it experiences the following:

Necessary Conditions

A. Ecological diversity that yields food in different microenvironments or ecozones at different seasons.

B. A number of cultivable/domesticable plants and animals

Sufficient Conditions

A. A series of seasonally adaptive subsistence options, and the necessary eco-subsistence knowledge

B. An ecosystem change that reduces the faunal subsystems (biomass) to such an extent that a single subsistence option is no longer tenable

II. If the newly developed El Riego-type system, with a micro- macroband settlement pattern and a seasonally scheduled subsistence, occurs in an ecosystem that has the following Necessary and Sufficient Conditions, it will evolve into a macro- microband settlement pattern subsystem with a seasonally scheduled subsistence system that includes horticulture or animal domestication (Coxcatlán-type system):

Necessary Conditions

A. Cultivable and/or domesticable plants and animals

B. Ecological diversity, including marked seasonality

C. Easy communication and interaction with other areas of great ecological diversity or lush ecological uniformity

Sufficient Conditions

A. Population that has increased until the equilibrium in terms of energy flow between the ecosystem and subsistence is upset

B. Horticulture (and animal domestication?) coming into being as part of a scheduled subsistence system

III. If a Coxcatlán-type system (with a macro- microband settlement pattern that includes a seasonally scheduled subsistence system with horticulture) experiences the following Necessary and Sufficient Conditions, it probably will develop into an Abejas-type system (a central-based band with a seasonally scheduled subsistence system that includes both horticulture and agriculture of a wide variety of plants):

Necessary Conditions

A. Horticulture and cultivated/domesticated plants/animals

B. Great ecosystem diversity

C. Access to similar zones, where horticulture of other kinds of plants has developed

Sufficient Conditions

A stimulating interaction involving the exchange of cultivars and domesticates

IV. Under the following Necessary and Sufficient Conditions, the Abejas-type system (central-based bands with horticulture) may have evolved into an Ajalpan-type system (semipermanent hamlet communities with subsistence agriculture):

Necessary Conditions (i.e., in areas of lush uniformity)

Cultivars and/or domesticates susceptible to more productive hybridization or agriculture

Sufficient Conditions

An optimizing subsistence strategy, which results in a shift from horticulture to agriculture with village life

By 1974, I believed that investigations of incipient agriculture in Peru and Tehuacán suggested the following general statement (MacNeish 1977, 796).

> Pristine agriculture may evolve under the following conditions:
> *Necessary Conditions*
> A. In a Center with potential domesticates, considerable diversity of life zones, and possibilities of relatively easy interaction between those life zones
> B. After environmental changes, a realignment of man's subsistence options that coincides with the development of knowledge of subsistence
> *Sufficient Conditions*
> A. One of the following two ecosubsistence options:
> i. a seasonally scheduled subsistence system (with the result that plants and/or animals become domesticated and/or cultivated in one or more of the life zones of that Center)
> ii. a stimulating interaction between all or most life zones of that Center, including those with domesticates or cultivars as well as those with efficient wild-plant or animal subsistence strategies.
>
> As a result, man would be forced to settle down in villages. This event, in turn, might become one of the necessary conditions for the rise of civilization.
>
> I cannot help but think that these hypotheses are also worthy of consideration by scientists investigating the origins of agriculture in Old World Centers, such as the Near East and China.

This set of cultural ecological hypotheses, developed as new data emerged in the period from 1940 through 1960, directly opposed the theories of Anderson, Sauer, and the other materialists.

Perhaps the closest parallel to my theories were those of Kent Flannery. Although a student on Braidwood's digs in the Near East as well as on mine in Tehuacán, Flannery's concepts were new and differed from either of ours. In his judgment (Flannery 1966, 1986), based upon my data from Tehuacán, the end of the Pleistocene in highland Mesoamerica reduced the availability of staple resources, which caused a shift to new resource-procurement systems. This broad-spectrum resource utilization in turn led to a positive-feedback cycle between seasonal scheduling and the selection of various plants, which caused certain plants to become major food resources and ultimately led to seed planting and the shift to a new system—village agriculture.

Subsequently, Flannery and Michael Coe compared this hypothesis with data from the Pacific Coast of Guatemala, specifically the Ocós Formative development (Coe and Flannery 1964). Neither attempted to theorize about development on the coast, but later Flannery did excavate three caves and a single preceramic open site in Oaxaca, which resulted in a long preceramic sequence for that region. As reported in chapter 3, some of Flannery's archaeological phases did contain plant remains relevant to our basic problems. He also joined Frank Hole in the Deh Luran Valley of Iran to work on some of the earliest village materials of the Near East, which gave evidence of the use of domesticates in the Bus Mordeh and Ali Kosh phases, between 5000 and 7200 B.C. (Hole, Flannery, and Neely 1969). These materials relate to the story of the origin of village agriculture in the Near East, discussed in chapter 4. However, neither of his monographs on these materials or his more general article on the origin of agriculture indicated that Flannery had changed his basic Mesoamerican theory. Although he hinted that the development in the Near East differed from that in Mesoamerica, he never defined this other development (Flannery 1968). Flannery's theory, however, seemed to have moved closer to Lewis Binford's "margin theory" (Binford 1968)—to be mentioned shortly.

Flannery's basic ideas about the origins of agriculture for Mesoamerica were transferred to the Near East by Fekri Hassan, who was working in Palestine. Hassan's clearly stated theory is worth quoting here at length (Hassan 1977, 604–606):

> In summary, the emergence of agriculture in Palestine can be viewed as a culmination of certain adaptive trends which emerged during the Epipaleolithic. These trends were initiated by the transition from specialized hunting-gathering to a diversified subsistence base. The greater safety involved in a diversified subsistence base against periodic or seasonal shortage in any given resource could have rendered this mode of subsistence a successful alternative to specialized hunting-gathering if intensive microclimatic fluctuations occurred at the close of the Pleistocene. Field evidence usually focuses on macroclimatic changes, which tends to obscure seasonal and annual events which can directly affect any population that depends primarily on extracting natural resources. The inclusion of wild cereals among other resources that previously were either neglected or underutilized, was perhaps a major stimulus for further changes in the subsistence regime that led later to a mutual dependence between man and cereals. The inclusion of wild cereals could have been facilitated by their spread from their refuge areas after 14,000 B.P. (Van Zeist 1969:45). Once wild cereals were incorporated, a "positive feedback network" (cf. Flannery 1968) was established due to the great economic potential of cereals, which eventually led to a subsistence complex, a pattern of settlement and residence, and an intra- and intergroup organizational structure linked with intensive utilization of wild cereals.
>
> One of the major changes, in response to an increasing dependence on wild cereals, was the change toward sedentary habitation and larger size of the local group. . . . This change in residential pattern was positively reinforced by the development of nontransportable facilities for processing and storage, and by permanent dwellings, social integration and cohesion, trade and food exchange. The trend toward year-round sedentariness and large local population, however, was in conflict with two functional requirements. First, as the population continued to increase in size, either through ordinary growth or redistribution of population, and as the term of residency increased in response to improvements in processing, storage, harvesting, etc., the need to expand the catchment area from which additional resources were exploited had to grow accordingly. Animals and legumes were particularly in demand to provide protein of high quality. The expansion of the catchment area implied greater seasonal and periodic mobility. But higher mobility was in conflict with the trend toward sedentariness. The options, therefore, were either: (1) to disband, which is incompatible with the already established trends; (2) to split into mobile hunters and sedentary gatherers of wild cereals, which is incompatible with the trend toward cohesive social units but less incompatible than (1); (3) to establish greater symbiotic relations with already existing hunters, who did not join in the cereals-oriented subsistence system; or (4) to relocate to areas with a greater diversity of biotopes and/or more yield in high-quality food per unit-area. The second functional requirement was the need for a permanent and sufficiently abundant source of drinking water to meet the demands of a large and sedentary population. This obviously

> could be only solved by relocating toward areas where perennial springs were available. The survival of a large and sedentary group, dependent upon wild plants, is more vulnerable than that of a small and mobile group. The vulnerability can be reduced by increasing the area of the cereal stands and/or relocating the cereals to those areas with underground water reservoirs (near bountiful springs or river deltas and fans). A high premium would have been thus placed upon relocating cereals to such areas, where perennial drinking water, good pasturage, and diverse resources were present. Agriculture would seem to have been thus the more likely alternative, that is, most compatible with the pre-existing cultural trends. In addition, agriculture served to reinforce these trends. . . . The transition to agriculture, then, was grounded in a subsistence base heavily oriented toward the utilization of wild cereals and a settlement/residential pattern favoring large local group size and sedentary habitation. The transition occurred over a long span of time and involved a chain of mutual causal relationships between subsistence, settlement, group size, economy, and social organization.

The focus of Hassan's model is cultural dynamics. His theory is also good cultural ecology and stands in basic opposition to the idea that population pressures and changing means of production caused village agriculture to evolve.

While considerable progress was being made in the preceramic archaeology of the Levant and in the Tigris-Euphrates region—all in the lowlands of the Near East—Braidwood, the doyen of the archaeology of agriculture, continued significant investigation in the highlands of the Hilly Flanks. (As we shall see in chapter 4, Jean Perrot, Charles Reed, Philip Smith, Ralph Solecki, Cuyler Young, and others also were undertaking significant studies in the Near East and were finding hints of a long preceramic development.) Perhaps the most environmental-deterministic studies were those of Herbert Wright, whose conclusions were similar to those of Childe. Nevertheless, in spite of the accumulation of data crucial to our understanding of the development of village agriculture, little new significant theory was immediately forthcoming from Braidwood. He was, however, training students such as Kent Flannery, Frank Hole, Charles Redman, and Patty Jo Watson, who would develop new theories, most of which (as might be expected) were firmly on the cultural ecology side of the dialectic.

The cultural materialists, likewise, were numerous, vocal, and active. One whose theories were most compatible with those of the cultural ecologists was the geographer David Harris. Harris proposed four models for the development of village agriculture (D. Harris 1977, 190–91). In all four a crucial change after the Pleistocene is "reduction in availability of staple resources," which is good environmental determinism. Of his four models, however, the one leading to early food-producing systems cites "population pressures, resource specialization, and technical innovation and adaptation" as crucial sufficient conditions (good cultural materialism). In fact after the initial reduction in availability of staple resources, the sequence of events in this model is as follows:

1. Broader spectrum procurement of wild foods
2. Reduced mobility caused by population pressures
3. Increased sedentarism and immigration
4. Intensification of wild-food procurement

5. Improved seasonal scheduling
6. Resource specialization
7. Technical innovation and adaptation
8. Cultural selection of genetically responsive plants and animals
9. Development of food-producing systems (agriculture)

Harris also plots other routes that may follow the Pleistocene, but these develop in other ways for other regions and do not lead to village agriculture.

Harris is unlike the other cultural materialists not only in his theory but also in the sophistication of his model, which emphasizes early broad-spectrum food procurement, sedentarism, and population pressure. In this sense Harris's model is similar to Anderson's "dump heap theory" as revamped by Michael Fowler (Fowler 1973) and as apparently used by Stuart Struever and Kent Vickery (1973) in an attempt to explain the beginning of agriculture in the eastern United States.

Except for the emphasis on sedentarism and population pressures, Harris's theory bears little resemblance to what is perhaps the most popular general theory of the 1960–1970 period—Lewis Binford's "edge" or "margin" theory (Binford 1968). Binford's hypothesis was created to oppose Braidwood's theories, but in fact it opposed the theories of all the cultural ecologists and of the materialist David Harris. In simple terms, Binford's theory is as follows: When the availability of staple resources declined at the end of the Pleistocene, groups of people moved to the lush or coastal riverine regions, where they became sedentary foragers. Because they were sedentary, their populations increased, forcing certain groups to radiate inland to adjacent marginal regions that had low rainfall and diverse plant communities, including plants with drought-resistant domesticable seeds. The population pressures in the marginal zones caused strong selective pressure that favored the development of more effective means of food production, one of which was agriculture.

Binford initially developed his theory to explain the beginnings of agriculture in the Levant; however, it is now applied more generally. For example, Binford states that my Tehuacán data confirm his hypothesis. (They certainly do not.) Because he is the leader of the "New Archaeology," his theory has gained wide popularity—an amazing phenomenon when one considers that in area after area (as will be shown in chapters 2–11) relevant data fail to support his theory.

The remarkable studies of Ester Boserup, particularly her pocket-sized monograph, *The Conditions of Agricultural Growth* (1965), often are cited as theoretical confirmation of Binford's theory. They are actually more closely related to the studies of the classical cultural materialists, of whom Mark Cohen is perhaps our finest present-day example (Cohen 1977a). While Cohen's studies focused on the preceramic of the Peruvian coast, he stated that the data were "totally in keeping with the assumption that the technology of domestication developed in response to the pressure of population" (Cohen 1977b, 174). Philip Smith and Cuyler Young came to the same conclusion in their studies in the highlands of the Near East (Smith and Young 1972). These statements are in keeping with the dicta of another leader of the period—the "true" cultural materialist, Marvin Harris (M. Harris 1975).

Among the cultural materialists of this period, the botanist Jack Harlan (Harlan 1971) is considered a deviant. Harlan seems to agree with Cohen's assumptions about both the Near East and Africa. He has conducted major field investigations, many in conjunction with archaeologists, that have produced data pertaining to the problem of the origin of agriculture. His concept of Centers and non-Centers in plant domestication is a major contribution to solving the problem of the origins of domestications, if not of agriculture itself. Like Carl Sauer, however, he considers sedentarism, population, and lush environs to be the major causes of agricultural development. Yet in his investigations in Africa he recognized that the natural climatic change that desiccated the region from 7,000 to 4,000 years ago may have forced people to adopt or develop agriculture (Harlan 1977).

The last individual I shall consider in this final period fits in neither the cultural ecologist nor the cultural materialist school. In fact, he probably belongs to both. Charles Redman, a student of Braidwood, is more of a compromiser than an advocate of either category (Redman 1978). In concluding his studies of the Near East, he visualized two routes leading from the Pleistocene Hunters to Village Agriculturists. For the lowlands (e.g., the Levant) Redman views climatic change, dependence on cereal grasses, availability of potential domesticates, ecological diversity in the vicinity, and substantial, nonportable architecture as the causes of agricultural development. On the other hand, he sees climatic change, dependence upon migratory herd animals, ecologic uniformity in the vicinity, availability of potential domesticates in the area, and interpersonal stress within large, seasonally nomadic groups as the causes of highland village agriculture (e.g., in the Zagros Mountains).

Long before I began my research on the preceramic of the Maya lowlands, I too had hypotheses explaining both highland and lowland developments (MacNeish 1954). These I expressed more formally in 1978 (MacNeish 1978) and, in a slightly different format, in 1981 (MacNeish 1981a, 143–47) as follows:

> *Intensified food procurement and sedentarism.* From many standpoints the entire process, with its dichotomous implications, began at the end of the Pleistocene. Here the basic triggering causes were the changes in the biomass, specifically the diminution of herd animals and other megafauna that were hunted. This occurred at about the time that humans had developed an entirely new series of subsistence options, such as seed collection, seed storage, leaf and berry picking, trapping, and seafood collecting. These changes shifted the emphasis from a nomadic way of life, consisting of living in family groups and depending mainly upon hunting for subsistence, to a collecting way of life with a seasonally scheduled subsistence system and a settlement pattern with group size fluctuating from microband to macroband with each season. This latter is the way of life that in the New World is often called the Archaic.
>
> *Environmental factors.* Exactly what kind of Archaic way of life developed was in large part goverened by environmental factors, Necessary Conditions, if you will. Those who moved into harsh areas with steeply stratified ecozones and great seasonal variation formed seasonal collecting camps of "El Riego" types, whereas those who moved into the lusher areas with catchment-basin types of ecozones

formed semi-sedentary base camps, or "Conchita" types. Thus, by Developmental Period V, different culture systems or types were exploiting different environmental zones. Although these were interacting with each other, major evolutions in each sub-area continued to take place for rather different reasons.

The El Riego type, or seasonal collecting camp, perhaps represented by El Riego in Tehuacán, Piki in highland Peru, and Palegawra in hilly Iraq, experienced a positive feedback situation. Increasing seasonal seed selection led to the planting of domesticates and cultivars, and increased use of this new subsistence option and food storage led to increasingly longer stays and increasing numbers of macroband encampments, which in turn led to an increasing need for ceremonialism and shamanism as integrating forces to control these larger groups. These Sufficient Conditions occurring in the highland zones led to development of the "Coxcatlán" type, or seasonal horticultural (micro-macro-) band camp. The Necessary Conditions for change in these highland zones included the presence of more plants that could be domesticated than in the lowland zones and the distribution of these plants such that domesticates and cultivars were exchanged with other similar zones where people were using other kinds of domesticates. Here were the reasons for highland development from Period V to Period VI.

Because of the rather different environments, again a Necessary Condition, the semi-sedentary (base) camp, or "Conchita" type, of lowland culture developed into the collecting hamlet, or "Palma Sola" type, for rather different reasons than did the simultaneous highland shift. Here intensive collecting and the development of increasingly better collecting techniques, as well as seasonally scheduled exploitation from a base camp, led to a more sedentary way of life. This in turn led to increasing population in base camps, which caused more intensive collecting, and so on. Here were the Sufficient Conditions, very different from those of the highlands, that made a positive feedback situation leading to the change from Period V to VI. Perhaps the best example of this shift is the development of Canario to Encanto-Chilca in the central Peruvian coast. Furthermore, the shift from Kebaran A to early Natufian in the Levant and perhaps that from Conchita to Palma Sola on the Veracruz coast took place for much the same reasons.

Highlands. Now let us set up hypotheses about the conditions for, or causes of, change from Period VI to VII in both the dichotomous zones. In Tehuacán the development from Coxcatlán, with its horticulture (micro-macro-) band camps, to Abejas, with agricultural base camps, seems to have been caused by the increasing use of domesticates that diffused into the areas, as well as by the increasingly efficient production of food from the planting of these domesticates or cultivars. Those Sufficient Conditions started a positive feedback cycle characterized by a more sedentary way of life due to growing crops for storage, leading to population increases. These factors in turn led to more and better use of imported domesticates and/or cultivars, which led to a more sedentary way of life, population increase, and so on. However, I should point out that the exchange system and the increased horticultural food production were the causes or Sufficient Conditions in these highland situations: the population factors were the consequences of these and were in themselves, in turn, Necessary Conditions for change, along with such other factors as the potential of some of the domesticates or cultivars to give increasingly greater yields. This process seems also to have occurred in the development from Zarzi to Zawi Chemi in highland Iraq, from Chihua to Cachi in highland Ayacucho, and perhaps also in the development to Mito in nearby Huánuco.

> *Lowlands.* This is very different from what seems to have occurred in the contemporary dichotomous zones in the highlands. In Mexico, population pressure in the lowlands once again was a Sufficient Condition for change and was linked with the rise of sedentary life in the "Palma Sola" types or collecting (band) hamlet, which developed the "Palo Hueco" type, or sedentary tribal collecting village. Thus there was a shift from hamlets to villages, which led to increasingly more efficient collection and perhaps even to plant domestication via the dump-heap mechanism, as well as to diffusion of domesticates from the highlands. Both these factors acted in a positive-feedback loop along with a changing social organization of composite bands, which led to some sort of lineage type of social organization with strong ceremonial leadership. This allowed not only for better social integration but also for better planning of food production. Again, the Necessary Conditions were the environmental factors, which permitted these developments from Period VI to VII on the coast. Again, I feel that this was the sort of process that was occurring in the Levant between early and late Natufian, as well as the development from Encanto to Playa Hermosa, Gaviota, and the like, on the central Peruvian coast.
>
> The next shift in the lowlands, from the "Palo Hueco" type, or tribal collecting village, to the Santa Luisa type, or tribal horticultural village, at the transition from Period VII to VIII was paralleled by the shift from Playa Hermosa and Gaviota to cultures such as those of Paradiso and Río Seco of Peru. Again, a major Necessary Condition for change was population pressure, but now it was coupled with increased planting of domesticates and a ranked type of social organization with strong religious leadership.
>
> The causes of the contemporary shift in the highlands from "Abejas," with its agricultural base camps, to "Purrón," with its tribal agricultural hamlets, were similar to those in the lowlands. For the first time in the highlands, population was the major factor in the positive-feedback chain that included subsistence agriculture and led to greater sedentarism, which in turn led to population increases, the need for more and better food production, and so on. . . . Village agriculture in most pristine centers had developed in two dichotomous ways for a number of very different reasons.

Studies in regions such as the lowlands of tropical South America, Europe, the eastern woodlands of North America, and many other non-Centers—areas that did not have pristine development of agriculture—suggest that village agriculture developed there in ways than can be explained by other causes. Thus, to my mind, a tripartite or trilinear development, entailing a new theoretical framework, is required to explain all developments of village agriculture. Obviously, this theory is the main concern of this book. At this juncture, therefore, I shall outline the theory in the form of hypotheses to be tested against data provided in the chapters that follow.

THE TRILINEAR THEORY

I would like to begin by presenting my explanatory model—the trilinear theory—which shows how and why village agriculture happened everywhere. I shall try to present it in most general terms, but this is not a simple matter: there are three hypothetical models and three hypothetical sets of causes that seem to account for all occurrences of village agriculture. I call these my *primary, sec-*

ondary, and *tertiary* developmental theories. The chapters that follow discuss Centers of pristine development of agriculture and concomitant sedentary life. They will test and modify my primary and secondary developmental theories area by area, starting with the Andean region and going on to Mesoamerica, the Near East, and the Far East. The secondary developmental theory, and in addition the tertiary theory, also will be tested and modified by data from such non-Centers as the southwestern United States, Europe, the eastern United States, Southeast Asia and Oceania, the tropical lowlands in South and Central America as well as the Antilles, and, finally, Africa, both north and south of the Sahara. The final chapter will consider the validity of the theory and discuss problems to be studied in the future.

Fundamental to my hypotheses are the concepts of Center and non-Center. Since various scientists have used these terms in a slightly different manner, I should explain how I define them. Following Vavilov and many others, I consider that *Centers* are "those culture areas where a large number of plants were first domesticated." As we shall see, this means that initial domestications and cultivations came out of primary and secondary developments in dichotomous environments. Vavilov concluded that there were eight Centers, or hearths, of domestication. I will show that four of these—India, Southeast Asia, Africa, and northwestern tropical South America—were actually non-Centers that received their first domesticates from the real Centers—the Near East, the Far East, Mesoamerica, and the Andean area. These first four non-Centers domesticated their own plants. Other non-Centers, where there was secondary or tertiary development of domesticates, include the eastern and southwestern United States, Malaysia, Polynesia, and Europe. Some of the problem areas, such as the Aegean, Egypt, and the Indus Valley—all closely connected with the Near East Centers—can be classified as non-Centers for the simple reason that primary development of agriculture did not occur in them. It should be noted that this concept of Center and non-Center differs slightly from that of Jack Harlan (Harlan 1971). He defines three Centers—Mesoamerica, China, and the Near East—as those areas in which all domestication occurred in relatively confined areas, while their associated non-Centers—South America, Southeast Asia, and Africa—were geographical units where domestication took place over a wide spatial range.

Such definitions are of limited use, for they ignore the temporal aspects that are basic to any study of origins. Further, in the present state of research, it is difficult, if not impossible, to determine the exact place for domestication of most plants, let alone to locate clusters or Centers of domestication. Therefore, in this book, Centers will be considered those culture areas where a large number of plants were initially domesticated; and non-Centers, those areas where this did not occur.

In explaining how village agriculture developed, we must start with the assumption that village agriculture represents only one system within a temporal stage in cultural development that is preceded (and followed) by other stages. The number of stages leading up to village agriculture may vary considerably depending upon what criteria one uses to define the various stages (see table 1.2). I have utilized a three-stage scheme: (1) when most people produced no food and were food collectors; (2) a transitional stage when there was much food collection (albeit sometimes in a very specialized manner) and little or no real food production and many people were foragers; and (3) when food production had just come into being and there were food producers in Centers and some non-Centers. Stages 1 and 3 are clearly defined; the second stage, by contrast, is transitional. Obviously, these assumed stages are also hypotheses to be tested and modified by comparing the relevant sequences leading to agriculture in various parts of the world.

This three-stage scheme seems to explain best the evolution to village agriculture and all concurrent developments. Stage 1 consists of *Hunting-Collecting Bands* (system A in table 1.2). These bands subsisted on collecting, often with nonscheduled hunting, and their settlement pattern was nomadic and nonseasonally scheduled. Organized as microbands or family groups, these collectors had a band social organization (Steward 1955). Some representatives of stage 1 have survived into the present (via route 13). Examples include some Australian Aborigines and Tasmanians, perhaps Bushmen and Hottentots in Africa, the Semang of Malaysia, the Alacaluf of Patagonia, and the various Pygmy groups. Why these groups did not evolve but continued on route 13 (hunting and collecting or just collecting course) through stages 2 and 3 is not our concern here; it seems, though, that some of the causes of change were lacking in their cases. Many Paleolithic inhabitants of the Old World, as well as Early Paleoindians in the New World, also represent this arrested stage of development. The archaeological record contains many more examples of this first stage than does ethnography. In fact, the general rule is that changes in conditions usually caused Hunting-Collecting Bands to develop new cultural systems rather than to continue following the same route through a series of stages.

Perhaps the simplest and first development in many areas was the evolution (route 7) to *Efficient Foraging Bands,* system D. These bands had a scheduled subsistence system of collecting, a seasonally scheduled settlement pattern (often fluctuating between micro- and macrobands), and a band-exogamous social organization. Possible sufficient conditions for this development include worsening environmental conditions and decreasing food resources at the end of the Pleistocene, the development of new subsistence options, increased mobility, and greater ecosystem knowledge and seasonality; environments with varied ecosystems not exploitable from a base, seasonal food supplies, lack of domesticates, and distinct seasons are possible necessary conditions. The Indians of California and the Great Basin, many tropical forest tribes, and the tribes of the Boreal forest and the Arctic are good ethnographic examples of such groups. Numerous archaeological examples of these groups are found in the Epipaleolithic and Mesolithic of Europe, North Africa, and the Near East, in much of the Archaic of the New World, and in any non-Centers of agriculture.

System D may sometimes have been a brief transition between system A and other systems of stage 2, namely, *Foraging Villagers* (system C1, via route 16)

Table 1.2. Possible routes of development from Hunting–Collecting Bands—system A—to other systems

Stage 3
Food
Producers
in Centers
and some
non-Centers

Destitute Foraging Bands
Incipient Agricultural Bands
Agricultural Bands
Agricultural Villagers System E
Horticultural Villagers
Foraging Villagers
Affluent Foraging Bands
Semisedentary Bands with Domesticates
Efficient Foraging Bands
Hunting-Collecting Bands

route 2b
route 15
route 17
route 4
Agricultural Bands System B2
route 3
Horticultural Villagers System C2
route 11
Semisedentary Bands with Domesticates System D1
route 13
route 2a
route 10
route 9

Stage 2
Transitional
with many
Foragers

Incipient Agricultural Bands System B1
Foraging Villagers System C1
route 12
route 5
route 14
route 6
route 1b
route 16
route 8
Destitute Foraging Bands System B
Affluent Foraging Bands System C
Efficient Foraging Bands System D
Hunting-Collecting Bands
route 7
route 6
route 1a
route 13

Stage 1
Food
Collectors

Hunting-Collecting Bands System A

and/or *Incipient Agricultural Bands* (system B1, from Destitute Foragers—system B—via route 1a). While other conditions, whether purely environmental or not, might pertain to these developments, necessary conditions included circumscribed ecozones, some with abundant or lush food staples, and environments that had areal uniformity but regional diversity, which allowed exploitation from a single base of a number of foods from different ecozones. These factors, in conjunction with the development of a broad-spectrum subsistence pattern, the intensification of wild-food procurement techniques, resource specialization, and increased logistic mobility, often led directly from system A of stage 1 via route 6 to sedentarism and decreased residential mobility, system C1 (Foraging Villagers). Such developments occurred not only in early Centers of agriculture (e.g., the Natufian of the Near East and the Archaic of lowland South America and Mesoamerica) but also in such non-Centers as India, Egypt, the Sudan, and the southwestern United States. System C1 as well as System C, Affluent Foraging Bands, also came into being via route 10 among groups in non-Centers that never developed agriculture (e.g., the Northwest Coast tribes of North America, the Aleuts, various Siberian tribes, and some tropical forest tribes).

Another development, from system A via route 1a to the first Destitute Foraging Bands (system B) and on via route 1b to the Incipient Agricultural Bands (system B1), although significant, most often was only an initial step in the development of village agriculture. Only rarely have groups—such as the Owens Valley Indians of California and perhaps some New Guinea tribes—stayed in system B1 until the present and very few if any continue as system B—Destitute Foragers. Generally speaking, the transition to incipient agriculture is represented in specific parts of all Centers of agriculture: for example, from the Paleolithic to Zone 3 in Palegawra Cave to Zarzi and Zawi Chemi in the Near East; from Puente to Piki in the Andes; and from Ajuereado to Coxcatlán in highland Mesoamerica. In all these Centers, as well as in some non-Centers where the transition occurred, certain necessary conditions pertained. All had plants susceptible to domestication (King 1987, 58–59); that is, certain plants were genetically unstable, which made them susceptible to mutation and hybridization. Further, these plants all grew in characteristic environments that had distinct seasons, including one when food staples were scarce, and a number of ecozones with varied food resources available in different seasons. These ecozones were far enough apart that they could not be exploited from a single base, yet all were in areas readily accessible to each other. In other words, Centers that developed beyond system A were parts of interstimulating spheres of interaction.

These necessary conditions, however, did not cause development into system B unless certain sufficient conditions were also present. The archaeological record is full of cases in which band agriculture did not evolve (e.g., tribes of the Great Basin, California, and Africa) because the causative sufficient conditions were not present. In every case of evolution into system B, however, worsening environmental conditions and reduction of food staples led to a broad-spectrum subsistence system that included seed collecting and storage. Accompanying these changes was the development of a seasonally scheduled settlement pattern and a subsistence system that led to (1) an increase in mutants and hybrids of the domesticable seed plants, (2) the collecting and eventual cultivation of plants for storage, and (3) the occupying of sites for longer periods by larger groups (macrobands) in certain seasons. These various necessary and sufficient conditions may be conceived of as a positive-feedback model. According to this model, the feedback cycles in all of the above cases are as follows.

ROUTE 1A (see table 1.3)

Worsening environmental conditions that cause a reduced supply of basic food staples may force Hunting-Collecting Bands gradually to schedule their subsistence system. One of the bands' scheduled options may be seed collecting and

Table 1.3. Some necessary and sufficient conditions involved in the evolution from Hunting-Collecting Bands to Destitute Foraging Bands via route 1a—a primary developmental hypothesis to be tested and/or modified by comparative archaeologic data

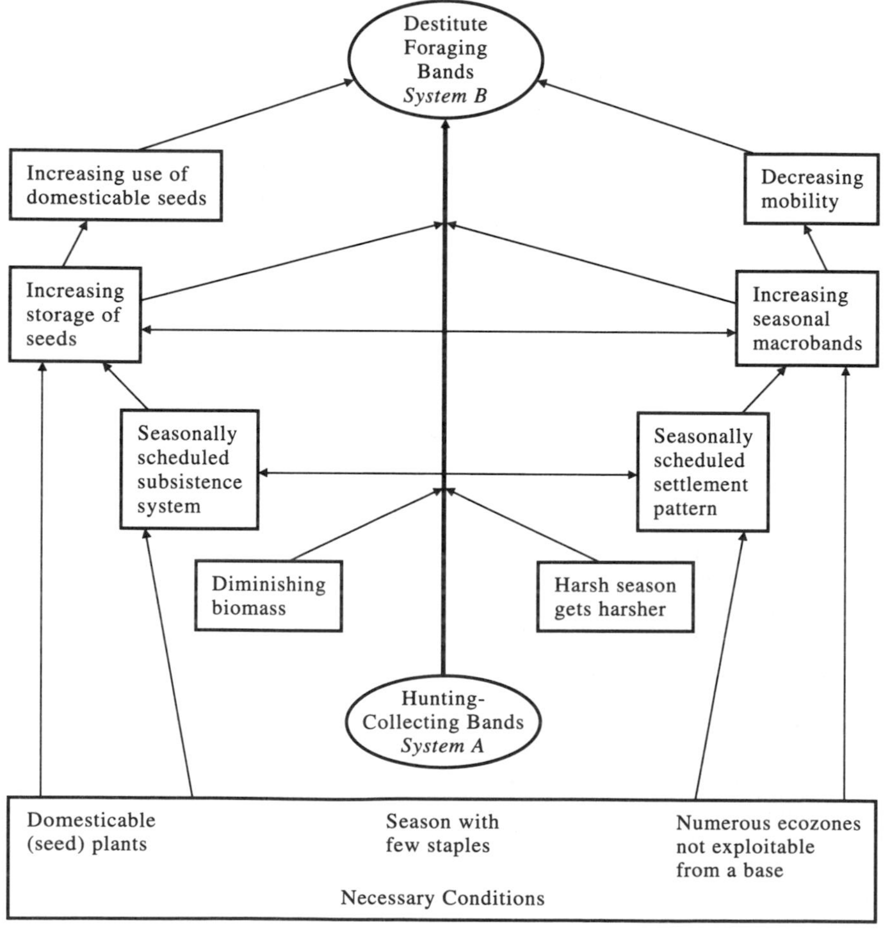

storage, provided their ecosystem has areas with seeds that fruit in certain seasons as well as harsh seasons with few foods. If the zones of their ecosystem are layered and diverse and cannot be exploited from a single base, the bands develop a seasonal settlement pattern. As their foraging becomes more efficient, there will be an accompanying tendency toward decreased residential mobility and seasonal population increases (the seasonal combination of microbands into macrobands). Thus a feedback system will develop involving increasingly long stays and bigger seasonal groups. The result will be more efficient seed collecting and storage, which in turn will lead to longer stays and bigger seasonal groups—requiring more collecting and intensification, and so on. In this way, *Destitute Foraging Bands* (system B) will evolve and will emphasize seed planting, collection, and storage.

ROUTE 1B (see table 1.4)

When Destitute Foraging Bands were growing in size, perhaps by increased foraging and decreased residential mobility, as suggested above, they may have randomly selected seeds that were genetically unstable or susceptible to change. Genetically unstable seeds may be susceptible to domestication. The foraging bands also may have randomly selected seeds that had genetic variations stamped into their genetic codes. It would follow then, that if these bands returned to the same ecozones from which they had collected their original seeds, and then dropped (or even purposely replanted) some of them in their camp areas, which were disturbed habitats, they would upset the ecological stability of that particular ecozone. Doing so might produce the chance of more mutants (DeWet, 1975) as a result of seedling vigor, increased germination rates, and more inflorescences with more seeds (King 1985, 1987).

As time passed, the demand for more food might become pressing as population increased and people stayed longer in one place. The answer would be to select larger and/or genetically different seeds to eat and store. Eventually, selecting larger seeds would lead to more inflorescences and greater seedling vigor and competitive advantage. By purposely planting the selected, larger seeds (which often were genetically changed) in disturbed habitats, these early people were cultivating and/or domesticating their food plants (King 1987). Thus the cycle was established: longer stays by larger groups needing more horticulture resulted in longer stays by still larger groups, and so on, resulting eventually in Foraging Bands with Incipient Agriculture.

The above description emphasizes the dynamics of the process of the first plant domestication and cultivation (table 1.4). The same generalization may be stated as a potential law, as follows:

If Hunting-Collecting Bands of system A have the following necessary conditions:

1. An environment with great ecological diversity
2. Potentially domesticable plants in one or more ecozones
3. Multiple resources inaccessible to or unexploitable from a single base
4. Natural seasonality, with a harsh season when few food resources are available

Table 1.4. Some necessary and sufficient conditions involved in the evolution from Destitute Foraging Bands to Incipient Agricultural Bands via route 1b—a primary developmental hypothesis to be tested and/or modified by comparative archaeologic data

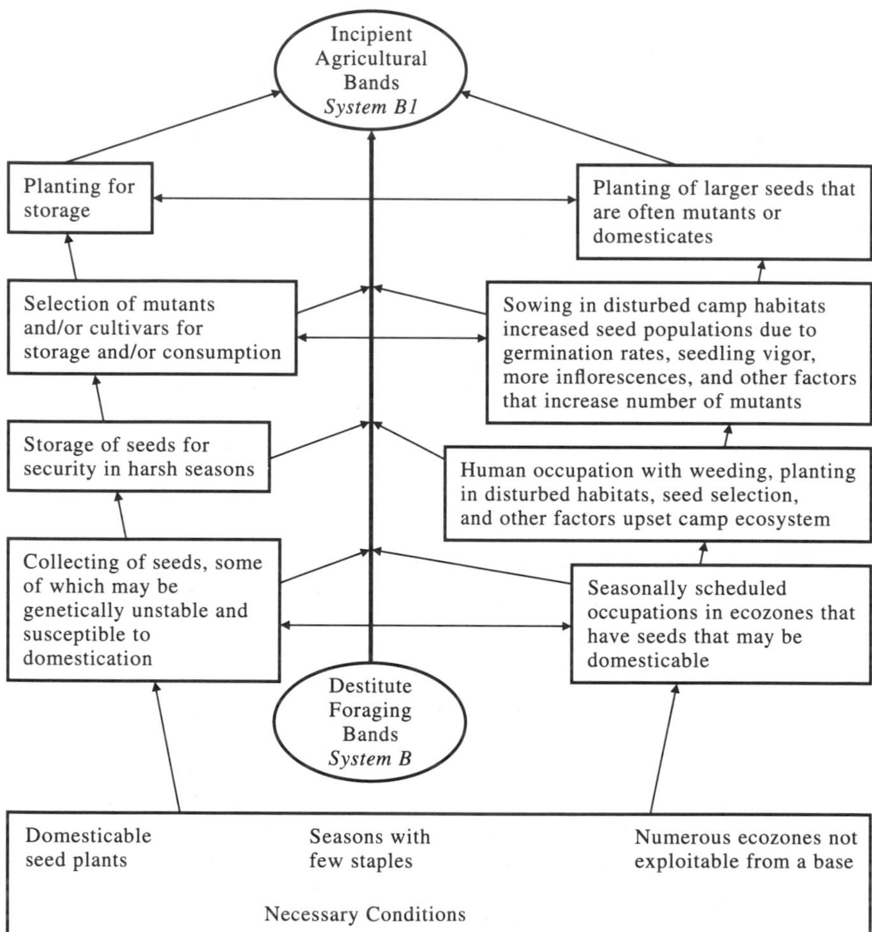

5. Human populations that are gradually increasing from microbands to macrobands, which further tax the environment

combined with the following sufficient conditions:

1. An environmental change that reduces the availability of food (faunal) staples and makes the harsh season harsher
2. Decreased residential mobility leading to further population pressures
3. An increase in logistic mobility leading to
   a. seasonal scheduling

b. development of broad-spectrum subsistence options
   i. seed collection
   ii. seed storage
4. An alteration in the ecosystem and/or in the genetics of some seeds being collected, reducing the energy expenditure per seed collected

then groups will start planting, probably in disturbed areas near their camp, the larger domesticated seeds for reasons of security and competition. In this way they evolve into Destitute Foraging Bands and later into Incipient Agricultural Bands (systems B and B1), the first step in the development of Agricultural Bands (B2) and then a Village Agricultural way of life (system E).

Now let us consider the transitions from stage 2 to stage 3. Some Hunting-Collecting Bands of system A failed to evolve further (route 13). Some that evolved via route 1 still exist today as Destitute Foraging Bands. Many Efficient Foraging Bands (system D) evolved (via route 7) into stage 3 and continue in that stage today, as did some Affluent Foragers (system C) that evolved via route 11 as well as Foraging Villagers (system C1) via route 10. However, three other evolutions from these three systems (A, D, and C) were brought about by what seem to be only three sets of causes.

Perhaps the primary developments occurred only in the pristine Centers or in subareas of certain specific regions, such as highland Mesoamerica, the highland Andean zone, and the Hilly Flanks of the Near East. Let us consider these developments as positive-feedback cycles.

ROUTE 2 OR 2A AND 2B (see table 1.5)

Changes in plant genetics and increases in planting to accommodate longer stays by larger foraging bands with incipient agriculture will lead to the use of more hybrid seed types for planting and the acceptance of other domesticated seeds. This process occurs where ecosystems with diversified areas and regions have gone through a similar horticultural evolution. These conditions lead to larger populations (macrobands) and longer stays, which eventually lead to bigger fields (agriculture). Bigger fields, in turn, increase the time populations remain in an area so that they become base-camp macrobands with multiseason stays. These base camps next turn into year-round hamlets. This sedentary living then leads to an increase in population, which must be further augmented by increased food production, until the procurement system is rescheduled to one that has *Agricultural Villages.*

This generalization may be phrased as a possible law, as follows:

If Incipient Agricultural Bands (system B1) or Agricultural Bands (system B2) have the following necessary conditions:

1. Domesticable plants
2. Relatively easy access to areas with and without domesticable plants
3. The practice of food storage for harsh seasons

combined with the following sufficient conditions:

1. Decreased residential mobility
2. More plants domesticated or extensive exchange that imports more plants for cultivation so there is more food grown for storage

Table 1.5. Some necessary and sufficient conditions involved in the evolution from Agricultural Bands and/or Incipient Agricultural Bands to Agricultural Villagers via route 2—a primary developmental hypothesis to be tested and/or modified by comparative archaeologic data

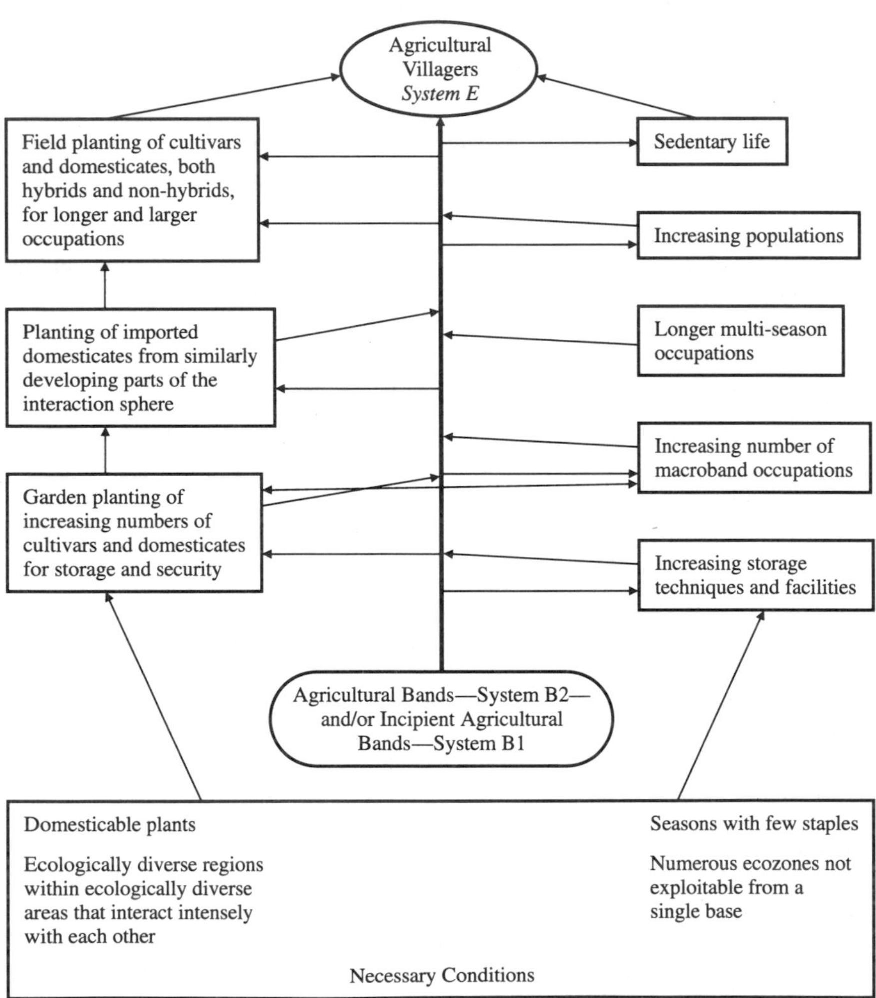

3. Increased populations residing longer in one place
4. Increased agriculture, often with crop specialization
5. Hybrids

then the inhabitants will cease to forage and will develop into Agricultural Villagers (system E).

The development of village agriculture seems to have occurred earliest in the

Centers; indeed, that is one of the reasons they are called Centers. The process described above seems to explain what happened in some of the highlands of Peru (see chapter 2), as well as what occurred in Tehuacán in the highlands of Mexico (chapter 3). Although the evidence is poorer, it is also a probable explanation of what happened in the Hilly Flanks of the Near East (chapter 4), and it might even be relevant to the earliest—as yet unfound—developments in the Far East (chapter 5).

The major, *primary development* of agriculture often seems connected with a *secondary,* slightly later development. This secondary development of village agriculture of systems C, C1, C2 and E occurred not only in certain parts of Centers but also in some non-Centers. In the Near East Center the secondary developments occurred in the Levant and in the Tigris and Euphrates valleys. In the Americas, secondary developments took place in the coastal region (and perhaps the montaña zone) of the Peruvian Andes, and perhaps in lowland Mesoamerica (see chapters 2–3). Secondary non-Center developments may have taken place in such regions as India and Egypt (chapters 9 and 11), the Aegean (chapter 7), some regions of the southwestern United States (chapter 6), and perhaps some of the coastal regions of the Far East (chapter 5). In such secondary developments, hypothetical necessary conditions could have created the following feedback cycles.

ROUTES 3 AND 12–15 (see table 1.6)

If the areas within the environment of Foraging Villagers have a positive-feedback cycle between population and food production as well as some sort of an exchange with an area in which domestication and/or horticulture and/or agriculture has occurred, then these groups will supplement or replace wild-resource specialization with horticulture from the outside areas that interact with the foragers. Horticulture leads relatively quickly to population increases. If the villages run out of territory or circumscribed lush ecozones into which they can expand, horticulture often destroys the carrying capacity of the environment. As food production from wild resources diminishes, more planting of the borrowed domesticates or cultivars as well as those domesticated locally becomes necessary, until the Foraging Villagers become Agricultural Villagers.

Despite apparent similarity, the causative factors of this secondary development are quite different from those of my primary developmental hypothesis. This difference becomes clear when the conditions of secondary developments are stated as a possible law:

If Foraging Villagers have the following necessary conditions:

1. Resource specialization in ecozones
2. Regional diversity but areal uniformity
3. An environment that
   a. can be exploited from a single base
   b. includes one or more "lush" circumscribed ecozones with easily exploitable, multiple resources
   c. has access to other major areas where primary domestications have occurred

Table 1.6. Some necessary and sufficient conditions involved in the evolution from Foraging Villagers and/or Affluent Foraging Bands to Horticultural Villagers and Agricultural Villagers via routes 6, 3, 12, and 15—a secondary developmental hypothesis to be tested and/or modified by comparative archaeologic data

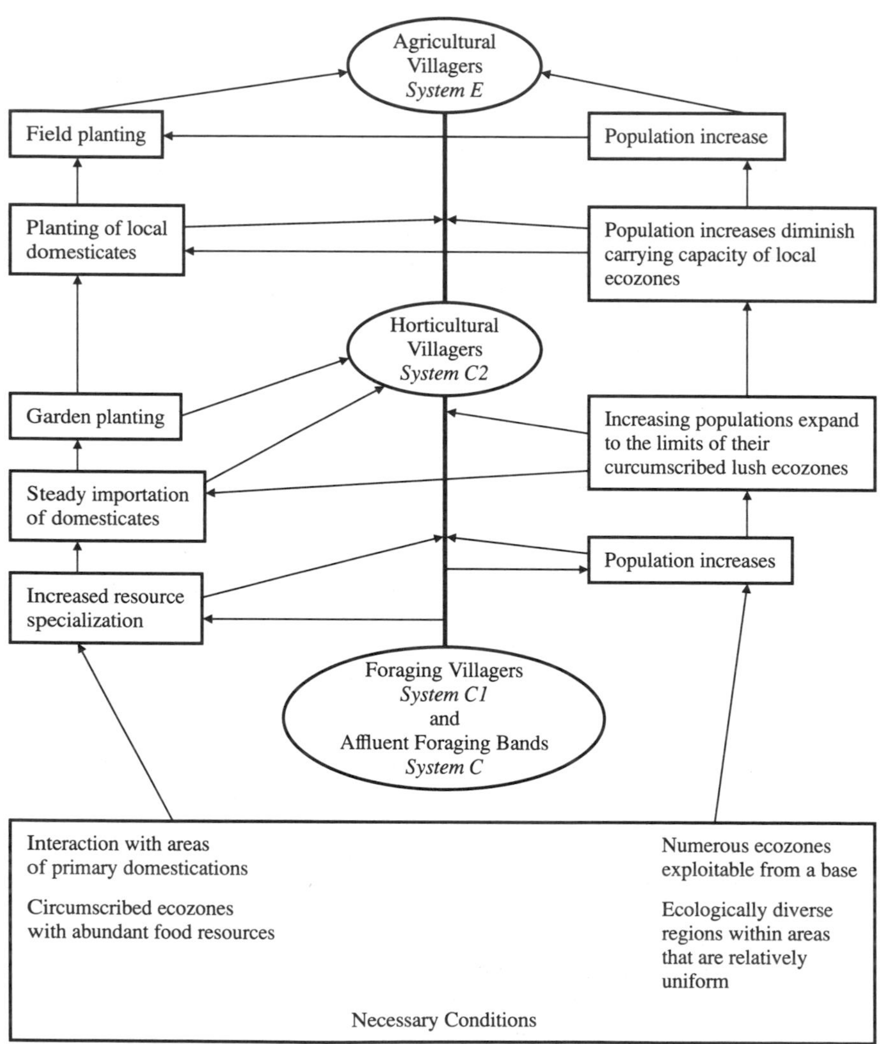

as well as the following sufficient conditions:

1. Increased populations, in large part the result of sedentarism
2. Decreased carrying capacity of one or more ecozones
3. Inability of resources of lush circumscribed ecozones to carry the population

4. Failure to develop or invent more resource specializations
5. Adoption of domesticates and Incipient Agriculture from neighbors in an interaction sphere

the Foragers will become Horticultural Villagers and/or Village Agriculturists relatively quickly.

The final pristine development of village agriculture seems to have occurred only in non-Centers. This *tertiary development* (routes 7, 5, 14, 4) not only took longer than the primary and secondary ones but came much later (table 1.7). Tertiary development of village agriculture seems to have occurred in such areas as western Europe (see chapter 7) and perhaps in some parts of the southwest United States (chapter 6), the eastern United States (chapter 8), Japan (chapter 5), tropical and western-Saharan Africa (chapter 11), Southeast Asia (chapter 9), and the tropics of South and Central America (chapter 10). Tertiary developments must be explained in terms of another set of laws of cultural change. The slow shift from Semisedentary Bands with Domesticates (system D1) through Horticultural Villagers (system C2) to Agricultural Villagers (system E) seems to have been caused by a unique set of necessary and sufficient conditions.

The positive feedback picture seems to be as follows.

ROUTES 7, 5, 14, 4, 15 (see table 1.7)

If bands have relatively efficient foraging systems (system D) in an easily exploitable and uncircumscribed environment, they will continue to improve this system until it leads to village life even if they have exchanges with an area in which domestication and/or horticulture and/or agriculture has developed. Only as their system begins to overextend its local ecozones and/or as that lush environment slowly changes for the worse will they begin to practice horticulture (system C2), at first adding just a few domesticates to the efficient foraging subsistence system (system D) that is the basis of their village life. Only slowly will they take on more and more horticulture as their population gradually increases and the environment's carrying capacity slowly diminishes or their exchange or redistribution system breaks down. At last, following this gradual pattern of development, they become Village Agriculturists.

Listing the necessary and sufficient conditions in this feedback system as a law allows one to discern the difference between primary and secondary developments, on the one hand, and tertiary development on the other.

If Efficient Foraging Bands (system D) and/or Semisedentary Bands with Domesticates (system D1) have the following necessary conditions:

1. Regional diversity but little areal uniformity
2. An area with some lush ecozones (often river valleys or coastal zones)
3. Indirect and/or casual exchanges with adjacent areas that possess cultivars and domesticates and that practice horticulture and/or agriculture

and if the following sufficient conditions occur:

1. Intensification of wild-food procurement
2. Resource specialization with energy-efficient foraging systems
3. Intensive foraging that eventually upsets the carrying capacity of the (lush) ecozones

Table 1.7. Some necessary and sufficient conditions involved in the evolution from Semisedentary Bands with Domesticates and/or Efficient Foraging Bands to Agricultural Villagers via routes 7, 5, 14, 4, and 15—a tertiary developmental hypothesis to be tested and/or modified by comparative archaeologic data

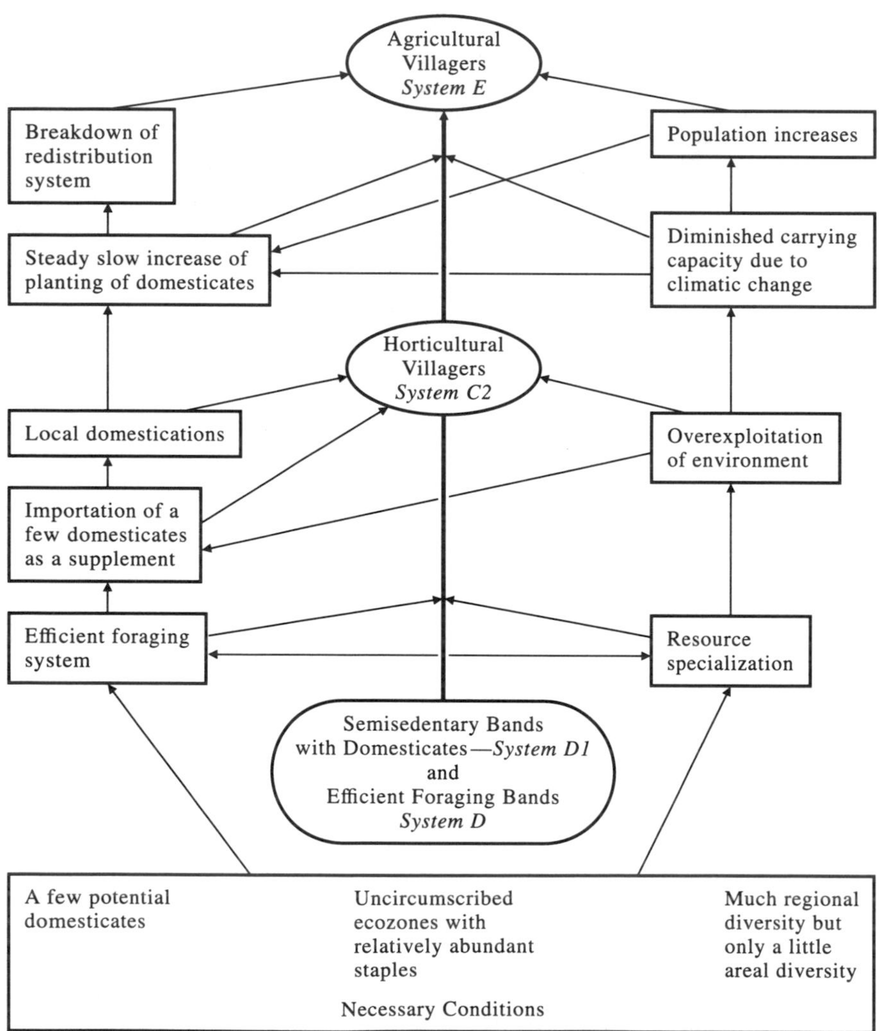

4. A natural reduction of the carrying capacity of the local foraging area
5. A gradual breakdown in the food-redistribution system

then they will slowly become Horticultural Villagers (system C2) and/or finally Village Agriculturists (system E)

The preceding statements about primary, secondary, and tertiary developments give us a "trilinear theory" about the evolution of village agriculture. These statements about the cultural process are in essence hypotheses. In the following chapters we shall test and modify these hypotheses, sequence by sequence, in each major area (Center or non-Center) in which village agriculture evolved. As will become evident, the data used to test and modify the hypotheses are uneven and sometimes unconvincing, and we have still to reach a final theory on this fascinating problem. Yet the very testing of the hypotheses may indicate directions for future research. I believe that we are moving in the right direction and that we have taken the first great leap toward a genuine theory on the origins of agriculture.

PART 1
The Centers

CHAPTER 2

The Andes

One of the major centers of domestication was the Andes. The foods that we owe to this area include over two hundred plants, among them potatoes, peanuts, and lima and common beans (see table 2.1). Shortly after the conquest of Peru, many of these plants were carried to Europe, Africa, and Asia, where they became staples. As time went on, their origins were forgotten. Indeed, there was little interest in their origin until this century.

A BRIEF HISTORY OF RESEARCH ON THE PROBLEM

Increased interest in plant origins coincided with the rise of anthropological and archaeological investigations in Peru in the 1940s. Certainly the early archaeologists such as Alfred Kroeber, Julio Tello, Max Uhle, and others were well aware that the Andean civilizations were based upon a solid agricultural subsistence system composed of many other plants in addition to the basic corn, beans, squash, and potato complex. However, although the general framework of the ceramic periods was being worked out by the 1930s, little or no information existed about the sequence of the first utilization or cultivation of these various plants, and even less was known about how or where they were first domesticated. Although identifications were made from archaeological materials by botanists and geographers such as Alphonse DeCandolle, Fortunato Herrera, Louis von Wittmack, and Edgar Yacovleff, rarely, if ever, were there speculations about origins.

All this began to change in the late 1930s when Peruvian archaeological investigations intensified and culminated in the famous Andean Institute (also known as the Virú Valley Program), with its concentration of well-known archaeologists—Wendell Bennett, Clifford Evans, James Ford, Julian Steward, Duncan Strong, Gordon Willey, and others—in the period from 1937 through 1949. Perhaps most important from the standpoint of the problem of the origins of agriculture was the discovery of many preceramic *huacas*—large middenlike "tells"—on the Peruvian coast. Sites such as Huaca Prieta (fig. 3, 30) in the Chicama

Valley, Huaca Negra (fig. 3, 28) in the Virú Valley, and Aspero (fig. 3, 23) in the Supe Valley began to provide insight into early Andean agriculture.

Excavations such as those of Junius Bird at Huaca Prieta yielded evidence of a whole new period called "Early Farmers" (ca. 3000–1000 B.C.). During this period people lived in sedentary villages and had some corn (*Zea mays*), gourds (*Lagenaria siceraria*), cotton (*Gossypium barbadense*), achira root (*Canna edulis*), beans (*Phaseolus sp.*), chile (*Capsicum sp.*), and lucuma fruits (*Lucuma bifora*), but not all the domesticates later used by Andeans. It was recognized that these coastal dwellers were basically living off maritime resources and that the domesticates, which were only a small portion of their diet, were probably imported from elsewhere, usually designated "the highlands" or "some region east of the Andes" (Bennett and Bird 1949, 28–32).

The next advance occurred following the Virú Valley endeavors, roughly in the period from 1949 to 1960. This time was one of cooperation between botanists (George Carter, Paul Mangelsdorf, Margaret Towle, and Thomas Whitaker) and archaeologists (Wendell Bennett, Junius Bird, Duncan Strong, and others). It culminated in Margaret Towle's scholarly compendium of prehistoric domesticated plants entitled *The Ethnobotany of Pre-Columbian Peru* (1961).

Another significant period lasted roughly from 1956 through 1966 and involved intensive investigations of the preceramic. This period had started with Frederic Engel's collecting and digging on the central coast of Peru. Following my lead in Tamaulipas, he showed a sincere interest in the origins of agriculture. His studies were followed by those of Edward Lanning, who worked out the first preceramic sequence—a six-thousand-year chronology—for the central coast (Lanning 1963). Although based on limited excavation, Lanning's six-period sequence of Red Zone, Oquendo (Period I), Chivateros (Period II), Arenal (Period III), Luz and Canario (Period IV), Encanto-Corbina (Period V), and Chuquitanta–Río Seco (Period VI) still stands today.

By 1966, Gene Hamel, Michael Moseley, Thomas Patterson, and others had supplemented these data with findings from more and better stratigraphic excavations. Also during this period the first highland preceramic cave—Lauricocha—was excavated by Augusto Cardich (Cardich 1964), providing some relevant data. Between 1966 and 1970 excavations of preceramic sites subtly changed and became more problem oriented. One of the main problems addressed was how and why village agriculture came into being.

It was about this time, 1966, that I entered the field with a definite problem-oriented approach, seeking to test my Tehuacán-derived hypotheses about the origins of agriculture. My first task was to find an area like Tehuacán, one situated in the highlands and having a wide variety of ecozones, potentially domesticable plants, a dry climate for plant preservation, and caves that might give a long sequence of the Archaic. Engel, Lanning, and Patterson welcomed me with open arms, and I began several months of survey. We finally decided that the Ayacucho Valley was the best place to do research to test my hypotheses.

In 1969, after a number of frustrating delays, we began our interdisciplinary project, which lasted until 1975 (MacNeish 1981a). Reconnaissance by Angel García Cook and others turned up about 600 sites. Of these, 15 stratified sites were dug (see fig. 2); they yielded a sequence of 22 phases, some with domesticated plant remains, covering a time period of about 25,000 years (see figs. 2 and 4). Of particular relevance to origins of agriculture were the late Early Man and Archaic phases—Puente (9000–7100 B.C.), Jaywa (7100–5800 B.C.), Piki (5800–4400 B.C.), Chihua (4400–3100 B.C.), and Cachi (3100–1750 B.C.).

Nearby, in the Junín area to the north, Peter Kaulicke, Danielle Lavallée, Ramiro Matos, John Rick, and others were digging other preceramic stratified caves. Although they worked out a closely related sequence, they found little or no preserved plant remains. Just north and west in the highlands, in the Callejón de Huaylas, Thomas Lynch, assisted by the botanist C. Earle Smith, was conducting an interdisciplinary project at Guitarrero Cave and others. Unlike the Junín area, the stratigraphy of this area was not clear-cut, but it did have preserved plant remains, including domesticated beans dating to the crucial period from 7,000 to 9,000 years ago. Also on the western slope of the Andes, Frederic Engel was digging Quiché and Tres Ventanas caves, which had early preceramic remains but few if any plant remains. All in all, however, in this brief period (1966–1975), extensive preceramic data were collected, and some relevant plant ecofacts emerged.

Plant remains were even more numerous in the coastal preceramic remains uncovered at Ancón by Michael Moseley, in the Lurín Valley by Thomas Patterson, and in the Chilca, Asia, and Paracas areas by Frederic Engel, although the stratigraphic data and artifacts found at those sites were much less numerous. The excavations mentioned above did not significantly change Lanning's basic six-period preceramic coastal sequence.

We now as a group had collected valuable data but did not have the distinctive characteristics of the earlier periods when plant domestication began. Perhaps most significant in the period from 1966 through 1975 was the rise of theory. Lynch spoke of transhumance as a cultural process (Lynch 1980), while Cohen (Cohen 1977a) wrote of deterministic population pressures. Moseley (Moseley 1978) and Patterson (Patterson 1973) both spoke of the influence of ecological factors. Still later, David Browman, Thomas Patterson, and I wrote about a dichotomous causative model (MacNeish, Patterson, and Browman 1975). Throughout this period new thoughts arose that may reach fruition in the future. Of course, more data are still being collected, particularly by Donald Lathrap on the coast of Ecuador, by Sheila and Tom Pozorski in the Trujillo area, by Deborah Pearsall and John Rick around Junín, by Duccio Bonavia at Gavilanes on the central coast of Peru, and by Robert Benfer in Frederic Engel's old Chilca region. In fact, more and more research programs are being proposed for this area to solve well-defined problems or test specific hypotheses rather than to collect new data.

THE ENVIRONMENT

Before discussing specific data from the major ecozones of the Andes, a brief description of those ecozones themselves is needed (see Figure 3). Two factors are paramount in determining the ecozones of the Andean area—hydrology (i.e., the oceans) and topography (specifically, the Andes). These factors divide the

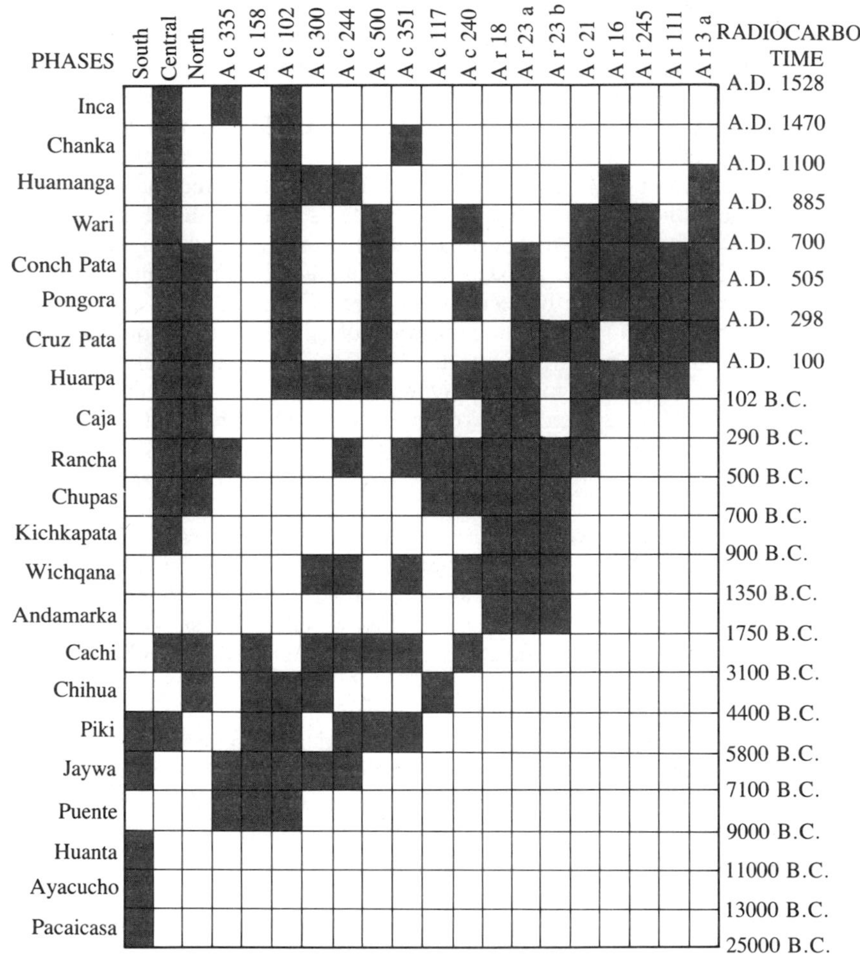

Figure 2. Complete cultural sequence of the Ayacucho Basin, with occupations of the various stratified sites

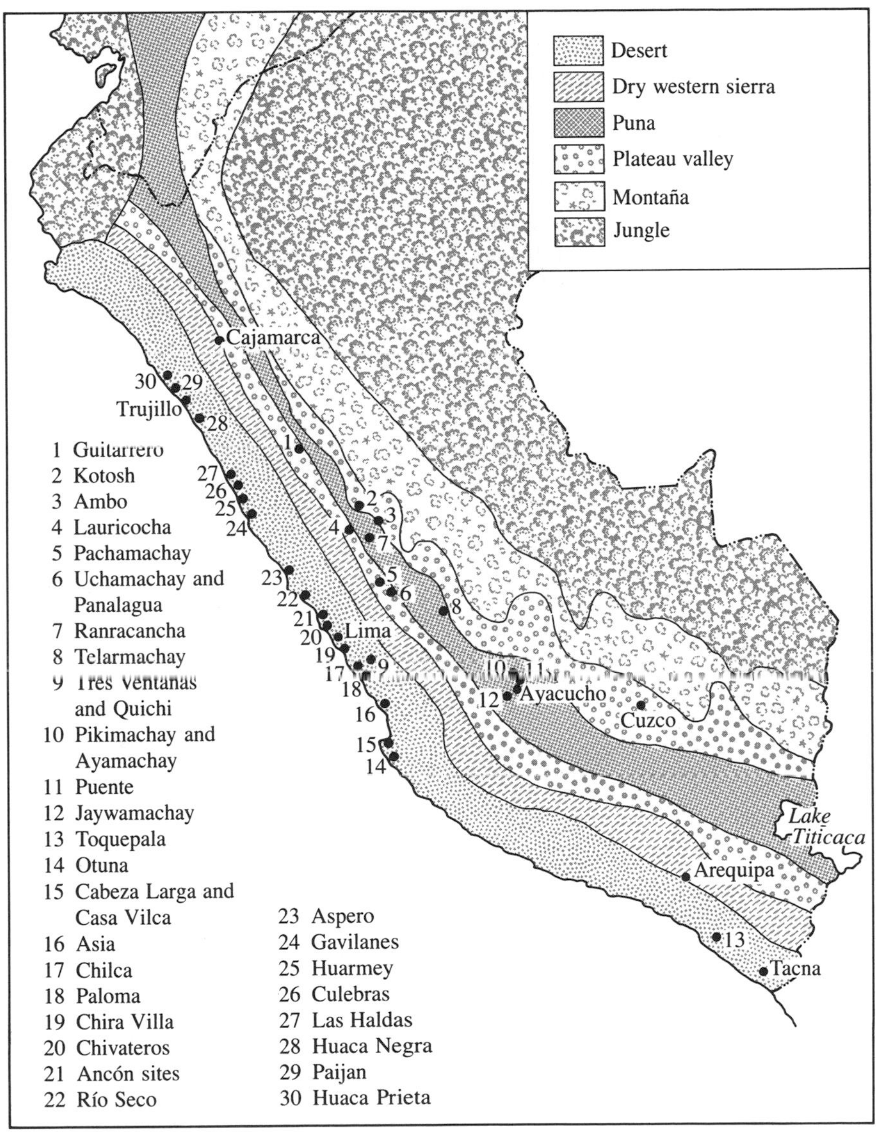

Figure 3. The environmental life zones and relevant sites in the Andean area

Andean area into three narrow topographic, vegetational, and climatic general geographical units that run from north to south: the west coast, the east slopes, and the Andes themselves, and each can be divided into life zones or subareas.

Because the cold waters of the South Pacific Coastal Current (Humboldt Current) run close to the western coast of South America, air masses are constantly being moved toward the east. The prevailing wind direction over the coastal lands south of Ecuador causes this air to be constantly warmed. This warming increases the capacity of the air to hold evaporated water, preventing rainfall on the coast west of the Andes and defines our desert life subarea or life zone. Rivers from the mountains cut across the coastal deserts, however, and in the spring-summer rainy reason (October–March), these rivers are often full of water, and their forested banks are home to much small game. The ocean also supplies abundant food, particularly anchovies and other fish.

In various combinations the above features create a variety of ecozones (often mislabeled microenvironments) within the coastal desert zone. Some of these, such as the delta, are teeming with life. Like an oasis, however, these fertile areas are limited in size and surrounded by less fruitful areas. In other words, their capacity for human exploitation is circumscribed, with the result that an expanding human population might quickly outgrow its resources in certain areas.

Ecuador and parts of the Pacific Coast of Colombia, the area north of this desert zone, are not affected by the Humboldt Current. Instead, the area is dominated by the Equatorial Counter Current, which literally dumps rain on this subarea. Here the coastal section is covered by lush tropical forest and mangrove swamps that supply some food resources. Because of the warmer ocean currents in this subarea, however, seafood resources are far scantier. Thus, this second coastal subarea has a different ecology that leads to different adaptations, because while it appears lush, its ability to support human populations does not even approach that of the rich, oasislike areas of the southern coastal deserts.

From many standpoints, the northern coastal tropical subarea is similar to the second major division—the eastern slopes of the Andes. The eastern slopes are basically affected by the currents from the South Atlantic Ocean that dump rains on the eastern flanks of the Andes. Within this major division are two subareas, the selva and the montaña.

The selva is the tropical rain forest at the relatively flat foot of the mountains. It is traversed by a series of eastwardly flowing streams whose waters come out of the mountains. Various riverine factors as well as topography affect the lush vegetation and food resources of this subarea, which can be divided into a series of exploitable ecozones. Like the northern Pacific Coast, however, the rain forest looks lusher than it really is.

The same may be said of the montaña, which is located on the steep, wet eastern slopes of the Andes and is cut by numerous gorges and deep, narrow river valleys. The montaña receives even more rainfall than the mountains, precipitation decreases as one ascends the slopes of the Andes. The montaña is divided into many ecozones, although none are particularly hospitable.

The third and final major division is the Andes themselves, which may be divided into a number of possible subdivisions. There are three parallel ranges—

the Cordillera Occidental, Cordillera Central, and Cordillera Oriental—and between these ranges are huge plateaus (altiplanos). These altiplanos are cut by a number of large rivers—the east-flowing Marañon, Montaro, Apurimac, and Urubamba, among others—and the west-flowing Callejón de Huaylas, Lurín, Chillón, and Guayas of Ecuador.

There are major differences in elevation in the Andes, the central Andean range being higher than the northern. The resulting different amounts of rainfall produce layered zones of vegetation with different congeries of fauna. This ecological variation is the source of many of our potentially domesticable plants (see table 2.1).

Although it is possible to subdivide the Andean zone in a number of ways, I use only three major subdivisions, both for convenience and because of the differences in their food resources. The first division is the west Andes, consisting of the dry western slopes of the Andes. This region has easy access to the lush coast and is dissected by steep, wooded canyons that are home to abundant large mammals (mainly deer, llama, and guanaco). Second is the high, treeless arca characterized by puna vegetation and rich faunal resources. Third is the plateau-valley, or basin, area that has both wooded vegetation and lusher riverine resources as well as access to puna resources at high elevations. Within each of

Table 2.1. Some native cultivated and/or domesticated plants of the Andean area

Scientific name	Popular name
Achras zapota	zapota beads
Acisanthera inundata	triana (fiber)
**Amaranthus caudatus*	amaranth
Andira inermis	quinillo colorado cabbage tree (medicine)
Aniba puchury	aniba (medicine)
**Annona reticulata*	anona (fruit)
**Annona squamosa*	sweetsop (fruit)
**Arachis hypogaea*	peanut
Aristotelia macqui	macqui (fruit)
Arrabidaean chica	bignonia chica (fiber)
**Arracacia xanthorrhiza*	arracacha root
Baccharis sp.	chilca (fiber)
Baccharis polyantha	chilca (dye)
Baccharis prostrata	chilca (dye)
Bombax ceiba	ceiba root fiber
**Bromus mango*	mango
Buddleia sp.	quisuare (wood)
**Bunchosia armeniaca*	armenica (fruit)
Caesalpina paipai	cactus (dye)
Caesalpina spinosa	cactus (dye)
**Campomanesia lineatifolia*	palillo
**Canavalia plagiosperma*	jack bean
**Capsicum sp.*	pepper
**Capsicum chinense*	chile pepper
**Capsicum frutescens*	chile pepper

Table 2.1. (continued)

Scientific name	Popular name
**Capsicum pubescens*	chile pepper
Carludovica palmata	bamboo palm, jipi-japa fiber
Caryocar amyadaliferum	almond-like seed
**Chenopodium pallidicaule*	cañihua
**Chenopodium quinoa*	quinoa (cereal)
Chusquea scandens	kurcas (fiber)
**Cucurbita andina*	squash
**Cucurbita ficifolia*	squash
**Cucurbita moschata*	squash
**Cyclanthera pedata*	achocha
Cyperus sp.	jurico root
**Cyphomandra betacea*	tree tomato
Cyphomandra splendens	tree tomato
**Erythroxylon novogranatense*	coca
Furcraea andina	chuchau (fiber)
Furcraea gigantea	penca (fiber)
Furcraea occidentalis	Peru coast fiber
**Indigofera suffruticosa*	añil (dye)
Inga endlicheri	pacae
**Lepidium meyenii*	maca
Lucuma bifora	lucuma (fruit)
**Lupinus mutabilis*	sweet chocho
Miconia chrysophylla	puca (dye or wood)
Myroxylon sp.	balsam (gum)
Myroxylon peruiferum	balsam (gum)
**Opuntia exaltata*	cactus
**Oxalis tuberosa*	oca
**Pachyrrizus ahipa*	jícama
**Pachyrrizus tuberosus*	jícama
**Passiflora sp.*	granadilla (fruit)
Pavonia paniculata	pavon
**Phaseolus sp.*	bean
**Phaseolus lunatus*	lima bean
**Phaseolus vulgaris*	common bean
Phytelephas macrocarpa	vegetable ivory
**Polymnia sonchifolia*	yacon
Prosopis chilensis	mesquite
**Psidium guajava*	guava
Sapindus sapaonaria	sapinda
Scipus tatora	tatora
**Solanum muricatum*	pepiño (fruit)
**Solanum quitoense*	lulo (fruit)
**Solanum topiro*	cocona
**Solanum tuberosum*	potato
Tillandsia gilliesii	Peru grass
Tillandsia latifolia	Peru grass
Tillandsia maculata	Peru grass
**Tropaeolum tuberosum*	aña (root)
**Ullucus tuberosus*	ulluco
**Zea mays*	corn

*Indicates domesticated plants according to Harlan (1971); others by Towle (1961).

these subareas are many regions—Ayacucho, Junin, etc.—each with its own numerous ecozones.

As is true of other major Centers, potentially domesticable plants occur in some of the various environmental zones. Because we do not know exactly which plants were native to which specific zone, table 2.1 lists the plants native to the entire Andean area alphabetically rather than by area or subarea. The list is impressive, even if our knowledge of the domestication of each plant is limited.

I believe that any true understanding of the development of village agriculture (or for that matter of civilization itself) in the Centers requires complete, well-documented sequences with domesticated plant remains from each of the major ecozones or subareas of the larger Center or culture area. In the Andean area, this means all seven subareas—the selva, montaña, north coast, south coast, west Andes, high Andes, and plateau-valley. Only by having adequate chronologies in each can we test our hypotheses about how and why village agriculture came into being there (see table 2.1).

Data for the Andean area are more complete than for any of the other major areas or Centers with which I shall test my hypotheses, but as we shall see shortly, even here they are woefully inadequate. At best, my trilinear hypothesis can be only partially tested at this time. There is always the chance that new explorations will modify these basic hypotheses or even be the basis for new ones. With these limitations in mind, let us examine the available data and attempt to apply them to my hypotheses.

RELEVANT ARCHAEOLOGICAL DATA (see table 2.2)

Let us start with the area that so far has yielded the least relevant data—the selva, or jungle, just below the eastern slopes of the Andes. My comments will be brief, for this area is part of the New World Tropics, which are discussed in more detail in chapter 10. No preceramic remains have been reported, although Donald Lathrap and his students have found a long ceramic sequence dating to perhaps as early as 2000 B.C., when village agriculture occurred elsewhere in the Andean area (Lathrap 1970).

Unfortunately, in the tropics preservation of plant remains is extremely unlikely. In fact, to get any inkling of the ancient subsistence system, one would have to use interdisciplinary techniques such as pollen and phytolith analysis, carbon–12-13 and nitrogen-15 isotopic studies of skeletons, soil studies, and flotation for plant remains. So far, few attempts have been made in this direction. I suspect, however, that even the earliest phases now known represent village agriculture and that the crucial stages necessary to discuss its development are still unknown. Finding these remains, which are probably preceramic, will not be easy in these jungle regions and will need much interdisciplinary help—remote sensing, study of ancient river courses, geomorphology, and so on—to provide relevant data.

The montaña—the area adjacent to the selva—has yielded similarly limited results. The only relevant materials come from the Kotosh site in the Huánuco Valley on the eastern slope of the central Andes (Izumi and Sono 1963). Here,

Table 2.2. Relevant archaeological sequences of the Andean area

WESTERN COASTAL LOWLANDS				
DESERT				
South Central region	Central region	North Central region	North region	Far North region
Hachas	Colinas			
		Sechin	Los Aldas	
	Chira			
Asia	Gaviota	Gavilanes 2	Huaca Prieta	
Otuma	Chuquitanta	Aspero	Alto Salavarry	Salinas
	Conchas		Padre Abán	
	Playa Hermosa			
	Encanto			
Cabeza Larga		Gavilanes 1		
	Corbina			
	Paloma			
	Canario			Talera
San Nicolas			Paijan	
	Luz			
	Arenal			

under the direction of the late Seiichi Izumi, a long sequence dating back to the late preceramic Mitos phase was uncovered. Unfortunately, this site represents the final stage—village agriculture—and the all-important earlier developmental stages have not yet been found.

My brief survey in the area with Rogger Ravines and Wayne Wiersum suggests that intensive survey of the terraces and rock-shelters along the Huánuco River may yield some preceramic sites (Ravines 1965). Ravines's finding of the preceramic Ambo site with Middle Archaic Canario points is an example. In the montaña, as in the selva, humidity is high, few plant remains are preserved, and special data-collecting techniques are required. Izumi's team attempted these techniques in their Kotosh expedition. They floated many Kotosh and Mito floors and hearths for carbon remains, some of which were identified by their botanists (Matsutani 1972). Obviously, more and better research must be done along this line.

It should be noted that many do not consider either the selva or the montaña part of the Andean interaction sphere. However, many of the plants basic to village agriculture have their nearest relatives in the tropics, and numerous tropical domesticated plants—lucuma (*Lucuma bifora*), coca (*Erythroxylon novograna-*

ANDES				EASTERN SLOPES	
DRY WESTERN SIERRA		PLATEAU VALLEY	PUNA	MONTANA	
Callejon de Hualas region	Upper Lurin region	Ayacucho region	Junin region	Huanuco region	Dates
		Wichqana		Kotosh	
			San Blas		
Toril		Andamarka		Wayra-jirca	
					2000 B.C.
		Cachi	Cuchimachay 1-3	Mito	
			Telarmachay 5		
Quishqui Punco?		Chihua	Pachamachay phase 5		3000 B.C.
			Pachamachay phase 4		4000 B.C.
		Piki	Pachamachay phase 3	Ambo	5000 B.C.
Guitarrero IIe	Quiché	Jaywa	Pachamachay phase 2		6000 B.C.
Guitarrero IId			Pachamachay phase 1		
Guitarrero IIc					
Guitarrero IIb					7000 B.C.
Guitarrero IIa	Tres Ventanas	Puente	Lauricocha I		
Guitarrero I					8000 B.C.

tense), manioc (*Manihot esculanta*), sweet potatoes, peanuts (*Arachis hypogaea*), and perhaps cotton—diffused from the tropics into coastal areas during the late preceramic. Common beans (*Phaseolus vulgaris*) may also have diffused out of this tropical subarea at an earlier date. Thus, to me, the selva and montaña are integral parts of the Andean area, and I suspect that plant remains relevant to the problem of the origins of agriculture will one day be uncovered there, perhaps even by the research Don Brockington and I have proposed downslope from Cochabamba, in eastern Bolivia.

The northern coastal region also has tropical vegetation and will be treated extensively in chapter 10. As already noted, the presence of tropical vegetation means that preceramic sites are difficult to find and that preservation of plant remains is poor. Good evidence for the final stage—pristine village agriculture with ceremonialism—exists at Real Alto (Lathrap, Marcos, and Zeidler 1977) and dates from about 3300 B.C. to 2700 B.C. In addition to ceramic evidence—the earliest type of Valdivia pottery—we have the results of interdisciplinary techniques that include phytolith analysis and flotation studies. These reveal that the site's human inhabitants had agriculture involving corn (*Zea mays*), coca (*Erythroxylon novogranatense*), cotton (*Gossypium barbadense*), achira (*Canna edulis*), and canavalia beans (*Canna valia ensiformis*) (identified by use of a scanning electron microscope), as well as lima (*Phaseolus lunatus*) and/or common beans (*Phaseolus vulgaris*) (Damp, Pearsall, and Kaplan 1981). As we shall see, this culture entity is similar to Kotosh and to cultures of my stage 3 from the drier southern coasts. Unfortunately, for the northern region, as for the selva and montaña, data on the preceding preceramic developmental stages are lacking, so little can be said about why village agriculture came into being.

There are, however, some preceramic remains located south of Valdivia in the Santa Elena peninsula, and Karen Stothert believes these were ancestral to Valdivia (Stothert 1985). Here, Lanning found a preceramic sequence that he named as follows: Exacto, dating at more than 10,000 years ago; Manantial, about 9000 B.C.; Carolina, about 7000 B.C.; and Vegas, at roughly 5000 B.C. Stothert has now dug site 80, the major Vegas component, which is full of burials and crude artifacts (Stothert 1985). Unfortunately, the site yielded only a few hints—such as a corn phytolith in Vegas (Pearsall 1988)—as to what the subsistence systems were. Stothert's work on the Santa Elena peninsula looks promising, although as yet the necessary "accumulated facts (preserved plant remains from secure archaeological contexts)" (Lathrap 1977), as well as sequences leading to village agriculture, are lacking, so I have no data with which to test my hypotheses.

Like the first three Andean subareas, the puna also lacks the necessary data for testing hypotheses (see table 2.2). The puna, however, is the most promising subarea, because many perceramic sequences have already been identified there in a series of stratified cave excavations. The major excavators of the area, Danielle Lavallée and John Rick (Lavallée 1977), both skilled excavators, are well aware of interdisciplinary techniques for obtaining information about the utilization of plants. In fact, Rick used some of these techniques (palynology, phytolith analysis, and flotation studies) to obtain information about ancient sequential subsistence systems (Rick 1980; Pearsall 1980, 1988). Although few relevant domesticated plant remains have been uncovered so far, I believe it is worth discussing the findings further. Not only are their finds closely related to the materials from Guitarrero Cave in the dry western sierras and the plant remains from the Ayacucho Valley (which, as we shall shortly see, yield much information useful in testing my hypotheses), but also because the altiplano cultural changes (and perhaps causes for changes) closely parallel those in these two adjacent regions.

The puna subareas have one of the longest histories of research on the preceramic for any part of the high Andes. It started with the excellent pioneering work of Augusto Cardich in Lauricocha Cave in 1958 (Cardich 1964). He found beautiful stratigraphy, of which zones L through R contained preceramic materials, which he divided into three periods. Since no one has had the audacity to set up cultural phases for this subarea (and I hope this book encourages someone to do so), I shall place each of these materials from the altiplano excavation into the periods I used for Ayacucho (MacNeish, Patterson, and Browman 1975). Lauricocha I seems equivalent to Ayacucho period 4, the Puente phase (9000 to 7100 B.C.), and Lauricocha II to period 5, the Jaywa phase (7100 to 5800 B.C.). While there was a gap to Lauricocha III, its points suggest it fell roughly into my period

7, the Chihua phase (4400 to 3100 B.C.). The cave also has a Lauricocha IV (with Kotosh and Chavín sherds) that corresponds to my period 9, the Andamarka phase, when definite village agriculture began almost everywhere.

Shortly after Cardich's pioneering efforts, Ramiro Matos took over the leadership for research in this subarea, digging some caves right around Junín and, more importantly, encouraging others to carry out his program. Early on, Peter Kaulicke dug Uchumachay, which had good stratigraphy and could also be related to Ayacucho dates: level or zone 7 matching my period 3 or Huanin phase (11,000–9000 B.C.), and level 4 matching Ayacucho or Chihua phase period 7 (4400–3100 B.C.), while Uchumachay levels 1 through 3 fit in the final preceramic period (3100–1750 B.C.) (Kaulicke n.d.). Panaulauca Cave, which had been dug previously, was retested later by Rick (Pearsall 1988). It had a similar early preceramic sequence, with level 7 fitting Ayacucho period 4, levels 4 through 6 matching Ayacucho period 5, and levels 1 through 3 belonging to Ayacucho period 6. Cuchimachay had stratigraphy that yielded later materials relating to my periods 6, 7, and 8, while Acomachay seemed to have materials only of the latest perceramic. Ramiro Matos also dug the San Blas site near Cerro de Pasco, which has in its earliest layers Waira-jicra sherds like those from the early village agriculture period (1750–900 B.C.) from nearby Kotosh. Telarmachay, being investigated slowly and carefully by Lavallée (1977), had by 1985 produced only late preceramic materials of periods 7 and 8 under Formative (village agriculture) remains.

Although a few preliminary reports have come out about these excavations in the Junín area (Wheeler, Pires-Ferreira, and Kaulicke 1976), the only one that has been adequately reported was Pachamachay, dug by John Rick (Rick 1980). In terms of stratigraphy, it added little new. Rick found five preceramic phases: his phase-1 levels related to my period 5, and I would redate his phase to 7000–6200 B.C.; his phase 2 (levels 26–30), redated to 6175–4800 B.C., roughly equals my period 6; his phase 3 (levels 21–25), roughly 4800–4400 B.C., fits my period 7; and his final preceramic phase, phase 5 (levels 12–17), dated by me at 3600–2800 B.C., would also be of my period 7 but last into the beginning of my period 8. Rick's phase 6 has pottery of the initial ceramic period, roughly from 1500 to 800 B.C. His description of lithic types correlates with the poorly reported ones from the other stratified sites as well as with our well-dated types from Ayacucho, putting the preceramic culture sequence on a firm footing. More significantly, Rick's descriptions and analyses reveal trends and information, albeit with insufficient domesticated plant remains, that are relevant to the testing of my hypotheses about how plants became domesticated. His sequence seems to document the shift from system A, Hunting-Collecting Bands, to system B, Destitute Foraging Bands, via route 1a, as well as the shift via route 1b from system B, to Foraging Bands with Incipient Agriculture (system B1) via route 2b to Village Agriculturists (again poorly defined for this subarea), culture system E.

On a general level, Rick's carefully analyzed data tend to confirm my primary developmental hypothesis from Pachamachay. In the shift from his phases 2 and 3 (6175–4400 B.C.) to 4 and 5 (4400–2800 B.C.), the increased number of storage pits and manos and metates (really mullers and milling stones) would seem to indicate increased foraging. Throughout his sequence from phases 2 to 5, Rick indicates that there was evidence of decreased residential mobility—including longer stays by macrobands which increase in size (Rick 1980, 318–24).

Further, Pearsall's studies show people in this period "utilizing seeds" (chenopods) "that are genetically unstable (susceptible to domestication)" like quinoa (*Chenopodium quinoa*). Her table 9.0 (Pearsall 1980) shows a dramatic shift in seed size between levels 18 and 12 (phases 4–5, my period 7) of Pachamachay. The earlier levels 18 through 32 (phases 1–3, my periods 5 and 6) suggest that these puna or altiplano people were selecting larger seeds (of quinoa) as longer stays increased demand for food, and that eventually (in phases 5 and 6, my periods 7–9) they began planting domesticates or cultivars. More recent analysis of Chenopodium seed sizes in nearby Panaulauca or Panalaqua Cave shows similar trends in the same periods or slightly later (Pearsall 1988). There is evidence, therefore, that there existed in the Junín area between 4400 and 2800 B.C. a positive-feedback situation involving longer stays by larger groups needing more food, which they obtained by seed selection of chenopods, resulting in longer stays by still larger groups, needing still more food, and so on (Pearsall 1980). Thus my hypothetical feedback system explaining how and why foraging bands developed incipient agriculture and/or domesticated plants seems to be confirmed by these Junín data.

In fact, data from the puna part of the altiplano are sufficient to test my hypotheses about why Foraging Bands with Incipient Agriculture (system B1) developed from Hunting-Collecting Bands (system A), although such sufficient conditions as development of broad-spectrum subsistence options (including seed storage and collecting), environmental changes reducing the availability of staples, and decreased residential mobility seem to have been occurring in the Junín region from 8000 to 3100 B.C.

Studies of the other feedback system leading to village agriculture, which should involve considerable information about population increases and settlement patterns as well as in-depth studies of seasonality, size, and length of occupation as causative factors, have not been undertaken. In fact, in terms of interpretations in this realm, Rick and I have some major disagreements. Certainly, this is a problem worthy of investigation and a future research program for the altiplano.

Testing of the later part of my primary developmental hypothesis in a Center—the reasons for the shift from Foraging Bands with Incipient Agriculture to Village Agriculturists—is hampered by a similar dearth of knowledge. I have assumed that the villages of the early Formative, such as San Blas and others in the highlands, had agriculture utilizing such highland plants as potatoes (*Solanum tuberosum*), añu (*Tropaeolum tuberosum*), quinoa (*Chenopodium quinoa*), lupines (*Lupinus mutabilis*), ullucu (*Ullucus tuberosus*), cañihua (*Chenopodium pallidicaule*), oca (*Oxalis tuberosa*), maca (*Lepidium meyenii*), and amaranth (*Amaranthus caudatus*). I have also assumed that the Formative levels in the caves of Telarmachay, Lauricocha, and Pachamachay had a similar subsistence system. As of now, however, there is almost no solid contextual archaeological evidence to reinforce those assumptions. Further, knowledge of the subsistence exchange and settlement patterns and of the populations of the transitional stage—my period 8, from 3100 to 1750 B.C.—is extremely limited, because those data from the altiplano sites either have not been found or have yet to be analyzed. Thus, testing of hypotheses about why village agriculture developed from Bands with Incipient Agriculture must await data on the basic sufficient conditions mentioned above before we can discuss positive-feedback cycles involving those conditions. At present, the data are not satisfactory for testing hypotheses, but at least they indicate the directions that future research should take.

Like the central altiplano, the dry western sierra subarea in the western highland flanks of the Andes has limited data. Only three regions have yielded materials relevant to the origins of village agriculture: (1) the upper Lurín, where two caves—Quiché and Tres Ventanas—have been dug (Engel 1972), (2) the Cordillera Huayhuash in the basin of the upper Huayura River to the north of the Lurín, where Limpio and Pucayau preceramic sites have been investigated (Kozlowski and Krzonowski 1977), and (3) the Callejón de Huaylas, where Tom Lynch dug Guitarrero Cave as part of a long-term program concerned with early agricultural developments (Lynch 1980). Guitarrero had some domesticated plant remains that were relevant to agricultural origins, and some of the artifacts from the various stratified caves gave evidence of preceramic sequences. Nevertheless, data from this area are not nearly as complete as those from the central altiplano nor can most of them be used to test my hypotheses. However, since this subarea did yield relevant early domesticated plant remains, let me discuss the data.

The Huayura River materials are basically Middle Archaic, dating from 6000 to 4000 B.C. We shall dispense with them because most do not come from excavated contexts, nor were there plant remains. Tres Ventanas Caves I and II as well as Quiché Cave were located at an elevation of about 4,000 meters at the headwaters of the Lurín and Chilca rivers, which drain into the Pacific just south of Lima, Peru.

Quiché Cave (Engel 1972) had some stratigraphy with a couple of layers on top containing ceramics dating to about the time of Christ. These overlay two strata, dated at about 8000 and 6000 B.C., which had projectile points and tools similar to Jaywa and Puente phases (periods 4 and 5) of Ayacucho, but there were no plant remains.

Tres Ventanas Cave I had many more preceramic levels with even more tools of these phases as well as some from perhaps the Piki phase, which dates to 6000 to 4000 B.C. In this cave there was some preservation of plant remains, most of which have not been adequately described. Some of the levels dating to about 7000 to 6000 B.C. had plant fragments identified as possible ullucu (*Ullucus tuberosus*), wild jícama, wild sweet potatoes, and very probably domesticated white potatoes (Hawkes 1988). Other than indicating that people at this early date had started collecting plants that could be domesticated, these findings are of little help in testing my hypotheses about origins of agriculture.

Somewhat the same may be said of the findings in the Callejón de Huaylas. Guitarrero Cave is located at about 2,580 meters above sea level, and about

52 kilometers north of Huaras. Tom Lynch found some good stratigraphy there, which he named—from early to late—complex I, IIa, IIb, IIc, IId, IIe, III, and IV (Lynch 1980). Plant and pollen preservation occurred in most of the stratigraphic units, but unfortunately complex III was badly mixed with complexes II and IV, and "complex IV is made up of redeposited strata" (Kautz 1980). Varying carbon-14 dates (from 12,560 to 9140 B.C. or 10,610 to 7190 B.C.) suggest that complex I may also be mixed. Complex II, however, had good stratigraphy of ash floors with preserved plant remains with some carbon-14 dates (the earliest of which are not consistent with those of complex I). On the basis of these, Lynch dates the earlier deposits one way (Lynch 1980, 32); while I, on the basis of his dates plus my Ayacucho dating of his point types in complexes I and II, date them (in radiocarbon time) as follows:

Complex IIe: 6000–5400 B.C.
IId: 6600–6000 B.C.
IIc: 6900–6600 B.C.
IIb: 7200–6900 B.C.
IIa: 7400–7200 B.C.
I: 7900–7400 B.C.

The earliest domesticated plant remains may occur in complex IIa (7400–7200 B.C.). They include an example of chile pepper (*Capsicum chinense*) and possibly a common bean (*Phaseolus vulgaris*), but Larry Kaplan (1980) had compunctions about their cultural context. Both plants must have been domesticated elsewhere, as neither is native to this ecozone, nor were there examples of them in deposits of complex I. The same may be said of the lucuma (*Lucuma bifora*) specimens (whose relatives come from the tropics) that C. Earle Smith identified in complex IIb (7200–6900 B.C.) as well as in IId (6600–6000 B.C.), IIe (6000–5400 B.C.), and the mixed complexes III and IV (C. E. Smith 1980a).

Smith also identified a piece of rind from complex IIc (6900–6600 B.C.) as *Cucurbita maxima* squash, while from what Kaplan considers mixed levels IIc or IIb come examples of types 2 and 3 lima beans (*Phaseolus lunatus*) as well as common beans (*Phaseolus vulgaris*, type 5). It is in complex IIe (6000–5400 B.C.), however, that Kaplan feels there is the best evidence for both lima and common beans. Specimens of pacae (*Inga endlicheri*), oca (*Oxalis tuberosa*), and ullucu (*Ullucus tuberosus*) also occurred in these complex II deposits, but Smith was not sure they were domesticated.

The specimens of corn (*Zea mays*), gourds, manioc, galactia, peanut (*Arachis hypogaea*), and, possibly, quinoa (*Chenopodium quinoa*) pollen from complex IV are from mixed deposits that can be neither dated nor definitively associated with any culture complex. (Lynch also dug Quishqui Puncu, a later preceramic site, but its deposits were secondary and without plant remains or reliable cultural contexts.) Thus, while the Callejón de Huaylas preceramic excavation indicates that domesticated plants appeared early, around 6000 B.C., little of the information is complete enough to be used to test hypotheses about origins of either domesticated plants or agriculture or villages.

The final highland region is in the plateau valley subarea as represented by the Ayacucho basin. While not having as early or as numerous plant remains, this area has much information relevant to the problem of the origins of agriculture. As previously mentioned, my Tehuacán team, supplemented by other scientists, undertook a well-planned project in this region from 1969 to 1975 and one of their main contributions was defining the ecozones of the Ayacucho basin (see fig. 4).

More than six hundred sites were found in reconnaissance, and 15 stratified sites with preceramic and/or earliest (initial period) village remains were uncovered (MacNeish 1981a). This research yielded a sequence of twenty-two culture phases (see figure 2), spanning a period of 25,000 years, in four or five of the six ecozones (see fig. 4). Six phases were identified for the period from 9000 to 900 B.C. These were Puente, Jaywa, Piki, Chihua, Cachi, and the initial ceramic phase of Andamarka-Wichqana. They show Hunting-Collecting Bands (system A) developed via route 1 into Incipient Agricultural Bands (system B1), then via route 2 to Agricultural Villagers (system E). Since such data are pertinent to hypothesis testing, I will briefly summarize the five relevant preceramic phases and thc initial ceramic phase at Ayacucho (MacNeish, Vierra, et al, 1983).

1. *Puente* (9000–7100 B.C.). Six radiocarbon determinations, ranging from about 9000 B.C. to 7100 B.C., suggest that the Puente cultural complex may have existed from about 9300 B.C. to 7700 B.C., sidereal time. This would place it after the end of the Andean Pleistocene and the extinction of Pleistocene animals. In our project, the Puente phase was represented by thirteen excavated components, representing sixteen occupations (see fig. 4 for ecological zones). Eleven of these excavated ones were in the humid woodlands, but one each occurred in the thorn forest riverine and thorn forest scrub. All but one were microband occupations, as were two surface survey components from the thorn forest scrub and one from the mantaro xerophytic. Only one possible macroband site occurred (MacNeish 1981b, fig. 8.2, 226–27).

Food estimates indicate that occupations were short, perhaps two to three weeks in length, except at the end of the phase, when they may have lasted as long as ten or twenty weeks. The longer occupations probably occurred in the summer and fall in the humid woodlands, where the carrying capacity for human populations was high because of a greater concentration of animals, birds, and various fruit trees. Although information on the settlement pattern is far from complete, there is a suggestion that during the winter the Puente peoples may have resided in the thorn forest riverine ecozone, perhaps moving three to five times during that season, for occupational floors had bones representing only enough meat to last a family a couple of weeks. In the spring the people may have migrated into the thorn forest scrub and sometimes into the humid woodlands, but their movements were still relatively frequent. In the summer they moved up out of the valley to the humid woodlands, where they moved only half as often, and where they remained throughout the fall, moving only once or twice in that season as well. So although data are limited, evidence exists that the Puente people were seasonal nomads, and we can, at least in part, reconstruct their seasonal movements.

Like their predecessors, these microband groups were also hunters, mainly of

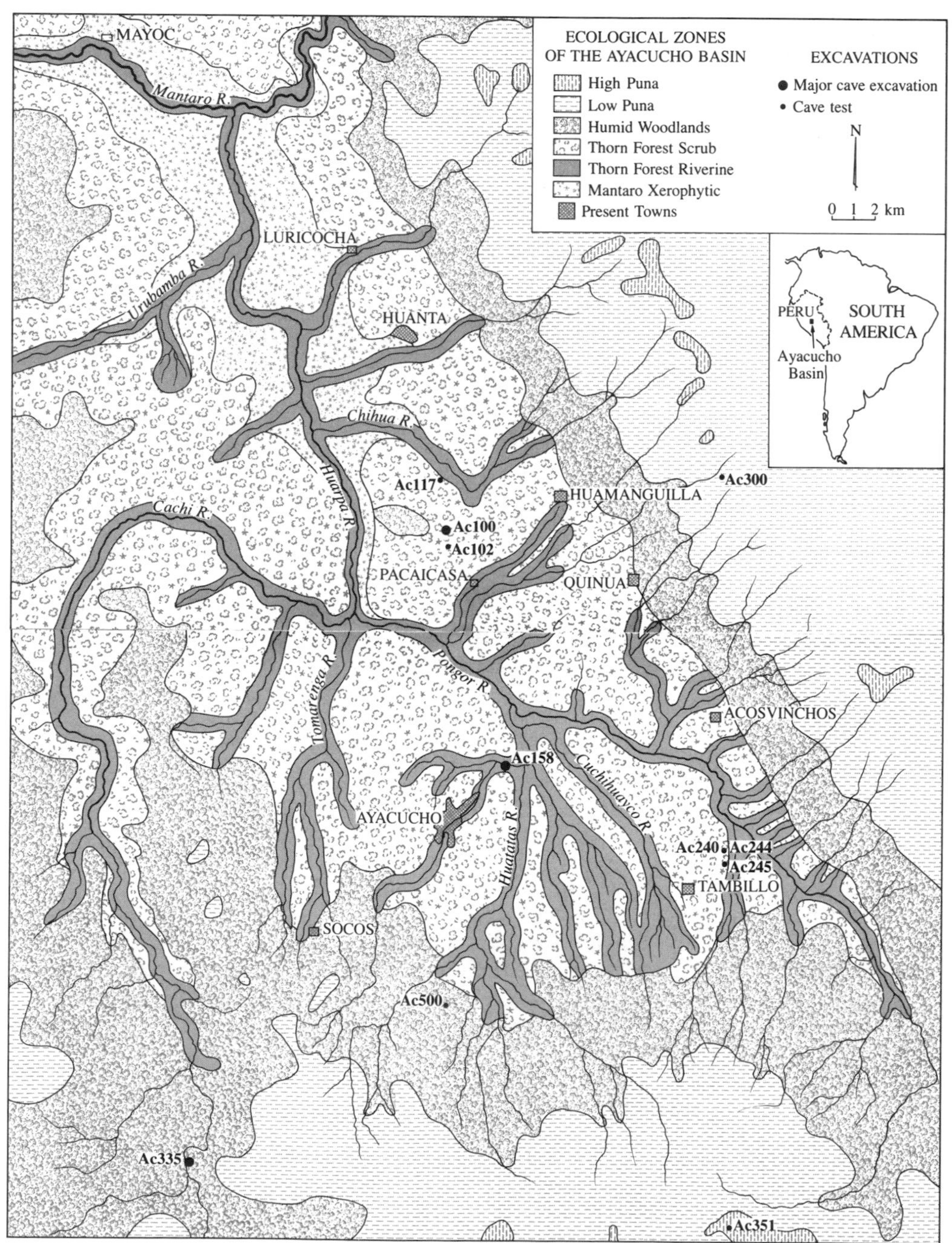

Figure 4. Excavated preceramic sites in the ecozones of the Ayacucho Valley

deer and a few camelids, because other forms of fauna, such as giant sloth, horse, and camel, were rapidly disappearing or had just disappeared. The two hunted species and the projectile points used both suggest at least two kinds of hunting. The first is indicated by large, unbarbed points with incipient stems and by the large pentagonal points that were perhaps used in ambushing herd animals such as camelids. The second kind of hunting is suggested by the small, stemmed points with barbs that served to keep the points in the animals; these were perhaps used to kill large, individual animals such as deer by some sort of stalking method, perhaps using atlatl (spearthrower) darts.

The bones of guinea pigs and other small mammals found in sites of low elevation suggest trapping, perhaps during the wet season. This seasonal data, in turn, suggests that microbands were beginning to schedule their subsistence activities. A few mullers, dating from near the end of the phase, may indicate seed collecting.

Analysis of artifacts and ecofacts suggests a number of technological activities: butchering, skin working, bone and pecked-stone toolmaking, string making and weaving, flint knapping, cutting, drilling, and sawing. A few flakes of Quispisisa obsidian from near Huancavelica indicate that long-distance trade occurred even at this time. Other activities may also have taken place, but they could not have consumed much time or energy, for these nomadic hunters lived in an environment of a diminishing biomass (in terms of large mammals). The Puente phase is thus an example of system A—Hunting-Collecting Bands—which is the base from which I believe developments toward agriculture started.

2. *Jaywa* (7100–5800 B.C.). The Jaywa phase, which followed the Puente, saw an attempt to solve the dilemma of increasing human populations and decreasing numbers of animals. Six radiocarbon determinations indicate Jaywa peoples may have lived from 7100 to 5800 B.C. (7700–6700 B.C., solar time). The phase is represented by materials from thirteen excavated components as well as from seven surface sites. Further, sites of these microband peoples existed in all microenvironmental zones, including the puna, which was not previously exploited (perhaps because of more glacial conditions), and the sites were occupied longer. Two macroband thorn forest encampments were also discovered. There is a suggestion of seasonally scheduled subsistence patterns, with deer and camelids being hunted in the dry season at higher elevations and guinea pigs and other small mammals being trapped in the wet season at lower elevations. Evidence of plant collection included an achiote seed (*Bixa Orellana*); berry stem, grass seed, and monicot and dicot plant fibers were found in three feces; only limited remains of hunted deer and camelids were present. In sites at lower elevations were pits, which suggest storage of surplus seeds collected and possible seasonal penning of guinea pigs.

Two of these lower-level sites are large enough to be considered possible macroband occupations. One of the excavated occupations had a series of posts across the mouth of the cave, making it an enclosed habitation area.

Although the subsistence-settlement pattern and many of the point types found are different from the Puente phase, the actual kind of technological activities was similar. Flint knapping was much the same, as was the much enlarged bone toolmaking industry. Skins were still worked, string making and weaving continued, and animals were butchered, albeit with different tools, while a new ground-stone tool industry came into being.

Obsidian from Quispisisa indicated trade, perhaps of the indirect procurement type. Two burned child burials suggest ceremonialism and perhaps infanticide. However, although the artifact types were different, the Jaywa phase was a continuation of the general Puente way of life.

3. *Piki* (5800–4400 B.C.). The Piki phase, dated by seven radiocarbon determinations as between 5800 and 4400 B.C. (about 6700–5000 B.C. solar time), is perhaps our best represented preceramic phase in terms of numbers of artifacts. In addition, the assemblage represents a break from the Puente-Jaywa development, which saw more shifts quantitatively than qualitatively. First of all, the Piki phase is represented by many more components than the previous horizons—forty components compared with twenty for Jaywa. While most (thirty-one) were microband occupations, there were nine macroband camps, compared to two for the Jaywa phase.

Unfortunately, our larger sample of twenty-four excavated components is heavily skewed toward wet-season occupations, so the seasonality of sites is based on less direct evidence than is the case for the Puente and Jaywa phases, which constitute the general model we apply to Piki. Wet-season occupations, possibly involving only one move per season, are found mainly in the thorn forest riverine and thorn forest scrub ecozones, with some of the latter being macroband occupations. Although evidence is speculative, my guess is that dry-season occupations occurred in the humid woodlands and puna, with about two moves per season. Therefore, while there may have been six moves per year, groups also stayed in one spot longer in the wet season, and the larger groups often coalesced into macrobands during the lusher season. There seem to be three clusters of sites, each concentrated around macroband encampments; this grouping may indicate the existence of band territoriality.

One other major difference from the earlier phases is in the sustenance at Piki. For the first time there is evidence, albeit meager, for the use of domesticated gourd, quinoa (*Chenopodium quinoa*), and, perhaps, squash (*Cucurbita andina*). Further, on some floors, as well as in many pits (hutches?), there were hundreds of bones of guinea pigs, suggesting penning of this tamed animal. I say tamed, because morphological studies of the bones reveal that these guinea pigs were probably not genetically different from modern wild ones. Thus, I believe that the animal at this time was tamed and taken out of its natural habitat, but not domesticated. Both of these activities using domesticated plants and tamed animals seem to have been scheduled for the wet-season occupation and occurred in the lower ecozones of the valley.

Numerous boulder mullers, single- and double-edged rocker manos, hoe weights or *majanas,* and spherical and ovid mulling stones, as well as fecal remains, indicate that considerable plant collection and preparation of plant food occurred during the wet season of the Piki peoples' annual migrations. While limited numbers of deer or camelid bones suggest hunting was not important, many new types of projectile points were being manufactured.

There are, then, hints not only at a seasonally scheduled subsistence system but even that the flint knapping, ground-stone toolmaking, and perhaps other such activities as string making, weaving, and textile making were scheduled to be undertaken mainly in the wet season in the large camps at low elevations. Bone toolmaking, both for fashioning woven fabrics as well as for skin tailoring and woodworking, may have occurred in all seasons in all microzones. However, hunting, both of the ambushing type (that is, specialized hunting of camelids) and the individual deer-stalking type, were mainly dry-season activities in camps at high elevations. Interconnected with these activities in various high, dry microband sites would have been butchering, fleshing, and skin scraping.

Further, although difficult to document archaeologically, extraregional exchanges and interactions seem not only to have increased but to have been of a different and perhaps more interstimulating nature. Besides obsidian being traded, perhaps the various regionally domesticated plants began to diffuse to other regions. Burial ceremonialism continued. It would seem that a new cultural system (system B1) with Incipient Agricultural Bands was evolving in the Ayacucho Basin.

4. *Chihua* (4400–3100 B.C.). Many of the significant changes initiated in the Piki phase reached their fruition in the following period, known as Chihua (about 5000 B.C.–4000 B.C. solar time). Seven radiocarbon determinations give dates that range from 4520 B.C. ± 125 years to 3300 B.C. ± 105 years. This was the millennium in Peru's history when many of the highland plants, domesticated in one highland or selva region, spread to other regions. Although few plant remains were excavated and only about a dozen feces have been analyzed, there seems to be some evidence that in addition to the gourd, squash, and quinoa used in the previous phase, the occupants of the Ayacucho basin had by this phase acquired common beans (*Phaseolus vulgaris*), achiote (*Bixa Orellana*), tree gourds (*Crescentia cujete*), lucuma (*Lucuma bifora*), and perhaps coca (*Erythroxylon novogranatense*), potatoes (*Solanum tuberosum*), and, at the end of the phase, primitive Ayacucho-type corn (Galinat, personal communication), which Alexander Grobman calls proto-Morocho-confite (Bonavia and Grobman 1988). Certainly there was horticulture, if not agriculture.

Guinea-pig bones were found in some profusion. Study of some of the skulls suggests that some of these are different from modern wild examples and similar to modern domesticated ones. This evidence indicates that the people were now penning and pastoralizing domesticated animals rather than taming them as in the previous (Piki) phase. Llama bones also suggest that some sort of control of those animals occurred, such as selective hunting, killing of male juveniles, and breeding of wild and/or tamed and/or semidomesticated camelids. Herding of fully domesticated llamas or alpacas, however, was still not practiced.

Although our number of Chihua components (eighteen excavated and twenty surface) is slightly smaller than those found in Piki, populations still seem to be increasing, and seasonal occupations may have lasted longer. At least six large macroband camps were found, and a number of subsistence options seem to have been available in the various microenvironments in different seasons of the year. In the wet season, populations could have gathered plants and/or engaged in limited hunting at lower elevation microzones. There are suggestions that some of these lowland occupations may have lasted into the dry season because of the food surplus that probably was stored in the various pits we uncovered.

The puna and humid woodlands again seem to have been regions mainly amenable to dry-season occupation, although a few chipped-stone hoes, which appear near the end of the phase, suggest that potato and/or root crop cultivation was supplementing the people's diet. Otherwise, their diet seems to have been based mainly on the products of their hunting deer and camelids. They even may have been herding wild camelids.

Accompanying the shifts in population, settlement pattern, and subsistence were major changes in the technological activities. Chihua points are less numerous than those of the Piki phase. In the Chihua phase many of the older types of points disappeared suddenly; only the small Cachi stemmed point seems to be a new addition. Chihua points are noticeably different from Piki in that the three most popular types are ovoid and unstemmed, and they are smaller and show less well controlled chipping. To some degree, these changes reflect shifts in hunting patterns. Perhaps ambushing and stalking were replaced by bow and arrow hunting. The point changes also reflect changes in flint-knapping techniques, which show less pressure-flake retouching and an increase in percussion flaking. Accompanying this shift is the disappearance of antler flakers or bone tools that could have been used for retouching, such as antler hammers for soft percussion.

Although the whole bone-tool industry seemed to be in decline during the Chihua phase, the techniques for making these kinds of tools continued. Needle-like bone net-making tools and weaving swords were still made, an indication that the weaving industry continued. In fact, a couple of pieces of twined netting hint that weaving was on the increase. However, all the awls, needles, and pins that seem to have been associated with the tailoring of skin were disappearing from the bone-tool assemblage.

Perhaps woven fabrics were replacing skins. Possible confirmation of such a hypothesis may be the decrease noted in fine haftable end scrapers for dressing skins. The remaining scrapers show less well controlled chipping techniques than were found on earlier types. These scrapers may have been used not on skins but rather for scraping plants, planing wood, butchering, and the like.

The chopping-tool complex continued relatively unchanged during the Chihua phase, although a new picklike type occurred. The polish on a number of the hoes, choppers, and scrapers hints that a log-working industry was nascent (perhaps because people needed to clear fields and build fences or houses?). All these changes in tools suggest that great changes were taking place in the way of life. Increased trade and cultural exchanges suggest that something new might have been happening in the realm of the Chihua social organization, but what it was, we cannot fathom. Nevertheless, these changes led to even more fundamental ones and resulted in the development of the Cachi phase, village agriculture (system E).

5. *Cachi* (3100–1750 B.C.). The Cachi phase has only three radiocarbon deter-

minations, and they fall in the range of 3100 B.C. to 1750 B.C. (solar time, about 3900 B.C.–2200 B.C.). Although the number of excavated components (sixteen) is only slightly less than for the Chihua phase, the number of surface sites has increased to about thirty-six, with perhaps nineteen macroband camps. Architectural remains also suggest seven or eight small hamlets. More important than these features, however, is the indication that a whole new scheduling system was brought about by different subsistence patterns. All the puna and humid woodlands sites seem to be seasonal camps, often with corrals. Preliminary analysis suggests that camelids, perhaps semidomesticated guanacos, were herded in the puna zone during the dry seasons, when some hunting and plant collecting were also taking place there. With the coming of the wet season, the people moved to camps in the low puna or humid woodlands to grow potatoes, and there was a corresponding de-emphasis on hunting and herding.

This seminomadic herding lifestyle was linked to a second major settlement subsistence pattern that took place at lower elevations, probably through kinship and ritual ties that served economic functions. Evidence for such an exchange system includes (1) the large number of camelid bones found at low-elevation sites where camelids could not live well; (2) the rocker manos (for grinding maize) found at high-elevation sites, where corn cannot be grown; and (3) the stone hoes, for root cropping, found at high-elevation sites but made from lower elevation sedimentary rocks. Additionally, obsidian projectile points, used at all elevations, seem on the basis of obsidian *debitage* to have been manufactured only in high elevations and then traded to the other regions. In other words, the highland and lowland subsistence systems were linked together by an exchange network, the beginning of a vertical economy.

As mentioned, this second part of this economy took place in the lower elevations, where all the macroband camps and hamlets appeared. Many of these sites were sedentary bases, or base camps, from which groups went out sporadically to plant, harvest, collect, hunt, and trap. Foodstuffs from Pikimachay Cave included (and some feces contained) corn (*Zea mays*), squash (*Cucurbita sp.*), beans (*Phaseolus sp.*), gourds, taro, and lucuma (*Lucuma bifora*), and, possibly, cotton, pepper (*Capsicum sp.*), quinoa (*Chenopodium quinoa*), and achira. Also found were bones of domesticated guinea pigs, camelids (possibly tamed and/or domesticated), deer, and small mammals, and also wild plant remains. Whether these lowland groups were practicing horticulture or agriculture is difficult to determine from the meager plant remains, but the sedentary hamlets and three sites with agricultural terraces suggest it was the latter.

Many of the technological activities of the Cachi phase were generally similar to those of Chihua, although the end products were different. Flint knapping was not done well, and scrapers were mainly large and crude. What little skin scraping occurred was probably accomplished with small blades or end scrapers that were hafted into handles. Edge-wear evidence suggests that some of these scraper planes were used either on plant remains or to pluck wool from camelids. In fact, a major advance seems to have been in the realm of weaving, as evidenced by bone weaving swords. There are indications that both wool and cotton were twine woven into fabrics, and nets and baskets probably were also made. A few bone tools were still made, but the industry was waning; perhaps the wooden-tool activity was also lessening. A number of distinctive side scrapers were manufactured, but whether they were used for butchering, hide working, woodworking, tailoring skins, or other activities has not been adequately determined by microscopic use-wear studies. Ground-stone tools were manufactured as well.

The building of terraces and circular houses outlined by larger rocks represented new architectural activities. Ceremonial patterns also seemed to be changing during the Cachi phase. Gone were infant, group, and bundle burials; instead people were interred with flexed bones and accompanied by burial goods and red ochre. One suspects from the ceramic phases that followed that ceremonial activities may have been moving in new directions during Cachi. From many standpoints the Cachi phase may be the first representative of village agriculture (system E), but if not, Andamarka-Wichqana, the first ceramic phase (or phases), certainly is.

6. *Andamarka-Wichqana* (1750–900 B.C.). Stratigraphically, the earliest phases of Village Agriculturist systems have two kinds of ceramics. The earlier is Andamarka (about 2213–1670 B.C., solar time), characterized by thick orange ware with occasional rectangular red paint decoration on large small-mouthed seed jar vessels. The later ceramics, called Wichqana (1670–1100 B.C., solar time), are predominantly seed jars, bowls, and bottles. They have stick-polished brown to almost black surfaces that occasionally have wide-line incised decoration. Although found in stratigraphic excavations at Wichqana and Chupas, the Andamarka ceramics were located in zones under those with Wichqana sherds. Many of the fourteen stratigraphic zones of Wichqana and Chupas had varying proportions of these two wares, as did most of the thirty-nine surface sites. In this brief description, then, I shall treat the two styles as a single period, although, of course, future analysis may distinguish differences between them.

The limited foodstuffs found at these sites suggest that the high elevation herding–potato growing and low elevation seed–agriculture subsistence patterns remained about the same as Cachi. The amounts of primitive Ayacucho-type corn grown at lower elevations may have increased slightly, as did the number of hamlets (to fourteen) at lower elevations. Some of these hamlets also had agricultural terraces watered from large storage tanks.

During the phases a number of architectural features were constructed with boulder-type masonry, and pyramids with the same kind of architecture were constructed at four sites. Associated with these four ceremonial centers, which were found at intermediate elevations, were clusters of hamlets and camps. It has been suggested that these clusters were tied to the ceremonial centers by kinship, ceremonial, and economic arrangements. Circular storage buildings and corrals, often containing camelid bones, may indicate that these sites at intermediate elevations were redistribution centers. They may have served as social links between low and high elevation communities and also united the lower elevation agricultural economies with those of the seminomadic high-elevation herd and root crop subsistence patterns.

THEORETICAL CONSIDERATIONS

We have now dealt in some detail with the Ayacucho Archaic sequence. It should be readily apparent that this development fits one part of my model (see tables 2.3 and 2.4). The Ayacucho phases show that Hunting-Collecting Bands, system A (here represented by Puente and/or its predecessor), evolved via route 1a to Destitute Foraging Bands, system B (represented by the Jaywa phase), to Incipient Agricultural Bands, via route 1b to system B1 (represented by the Piki and/or Chihua phases). These then evolved via route 2a to Agricultural Bands, system B2 (represented by the Chihua and/or Cachi phases), culminating in village agriculture, system E (as represented by Cachi and/or the Andamarka-Wichqana phases). Since chronological control over many of the variables (such as population pressures, environmental changes, subsistence changes, exchange systems) is good, we can determine which came first and, therefore, which were conditions or determinants (causes) of change in the whole system, leading to other systems with other kinds of variables (MacNeish 1977). In fact, Ayacucho provides just the kind of data needed to test my various hypotheses.

The Andean Primary Development

Looking at the ideal necessary and sufficient conditions needed to bring about the development of system A to system B via route 1, one can see that in the change from Puente (system A) to Piki (system B1) all the conditions not only occur but occur in the right sequence, thereby confirming my hypothesis (see table 2.4). Between about 9000 B.C. and 4400 B.C. the following conditions existed in the Ayacucho Valley:

NECESSARY CONDITIONS

1. A region with great ecological diversity, ranging from deserts to high puna tundra areas (see MacNeish 1981b), within mountainous zones that differ from each other (see fig. 4).

2. Domesticable plants appear to have been present: the puna has and probably had wild potatoes growing in it while the thorn forest and humid woodlands ecozones had quinoa, both of which could be domesticated (see fig. 4 and table 2.3).

3. The distance from the puna to the river bottom was sufficient (over thirty kilometers on steep slopes) to prevent its exploitation from any single base.

4. There were seasons in some of the ecozones—particularly the uplands—when harsh, rainless winters yielded few food staples.

5. Access to the desert lowlands to the west and the tropics to the east, as well as to other highland regions, was relatively easy, allowing the Ayacucho Valley to become a sphere of interaction.

SUFFICIENT CONDITIONS

On to the preceding set of necessary conditions were thrust sufficient conditions that triggered changes in the cultural system. These sufficient conditions seem to have been the following:

Table 2.3. Sequence of domesticated plants of the Andean highlands

Callejón de Huaylas	Ayacucho	Junín	Range of dates	Chile pepper	Lucuma	*Cucurbita sp.*	Achiote	Common bean	Lima bean
Guit. IV			ceramic mixed	x	x	x		x	x
Guit. III			ceramic mixed	x	x	x		x	x
	100	G	ceramic mixed	x	x	x	x	28	?
	100	A	1200 – 1600						
	100	B	1100 – 1500						
	244	A	100 – 1400						
	100	C	950 – 1350						
	100	D	900 – 1300		x				
	100	E	900 – 1300	f					
	100	I	AD 400 – 600					6	
	100	E4	AD 250 150 BC			x			
	100	II	AD 300 300 BC					4	
	100	ES	AD 50 550		x	x		1	
	117	A?	200 400 BC						
	244	C	300 500						
	335	B	150 650	f					
	100	ESA	400 600					1	
	100	E6	600 1000					1	
	244	D	950 1250					1	
	240	G	1000 1200						
	Ac100	F	1780 2020					1	
	Ac100	H	2100 2500			x			
	Ac100	VI	1850 2650	f				1	
	Pach	12–17	2800 3300						
	Ac117	D	3200 3600						
	Ac100	VII	3000 3700	f	x				
	Ac100	VIII	3100 4100	x				1	
	Ac100	X	4000 5000			x			
	Ac100	U	4400 5400						
	Ac100	V	4600 5600						
	Ac100	W	4800 – 5800			x			
IIc			5400 6000		x			x	x
	Ac102	VII	5435 5735			x			
IId-c			6000 6400		x			x	x
	Ac335	D	6105 – 6435	f			x		
IIc			6600 – 6900	?		x		?	?
IIb			6900 – 7200		x				
Guit. IIa			7200 – 7400 BC	x					
Tres Ventanas			7000 – 8000 B.B						

			?		?		?																													?	x	x	Galactia, oca, ullucu
					?																	x										x							Pacae
									x																														Gourd
										x	x	f	f		x	f?	f?									f?						5?							Quinoa
x?													f						f				f													x			Potato
													f																							x			Coca
												4?	4?	1		1	1?	1	1	1	3			1	2	3				1?	1	2	x	x	x	x	x	x	Corn
													?				x														x					x	?	?	*Cucurbita moschata*
													x														x									x			Taro
												?														x										x	x	x	Cotton
																		f					x			?													*Cucurbita ficifolia*
																																				x	x	x	Achira
																																					x	x	Peanut
																																					x	x	Manioc

Table 2.4. Necessary and sufficient conditions as a positive-feedback process in primary developments in the Ayacucho Valley of Peru

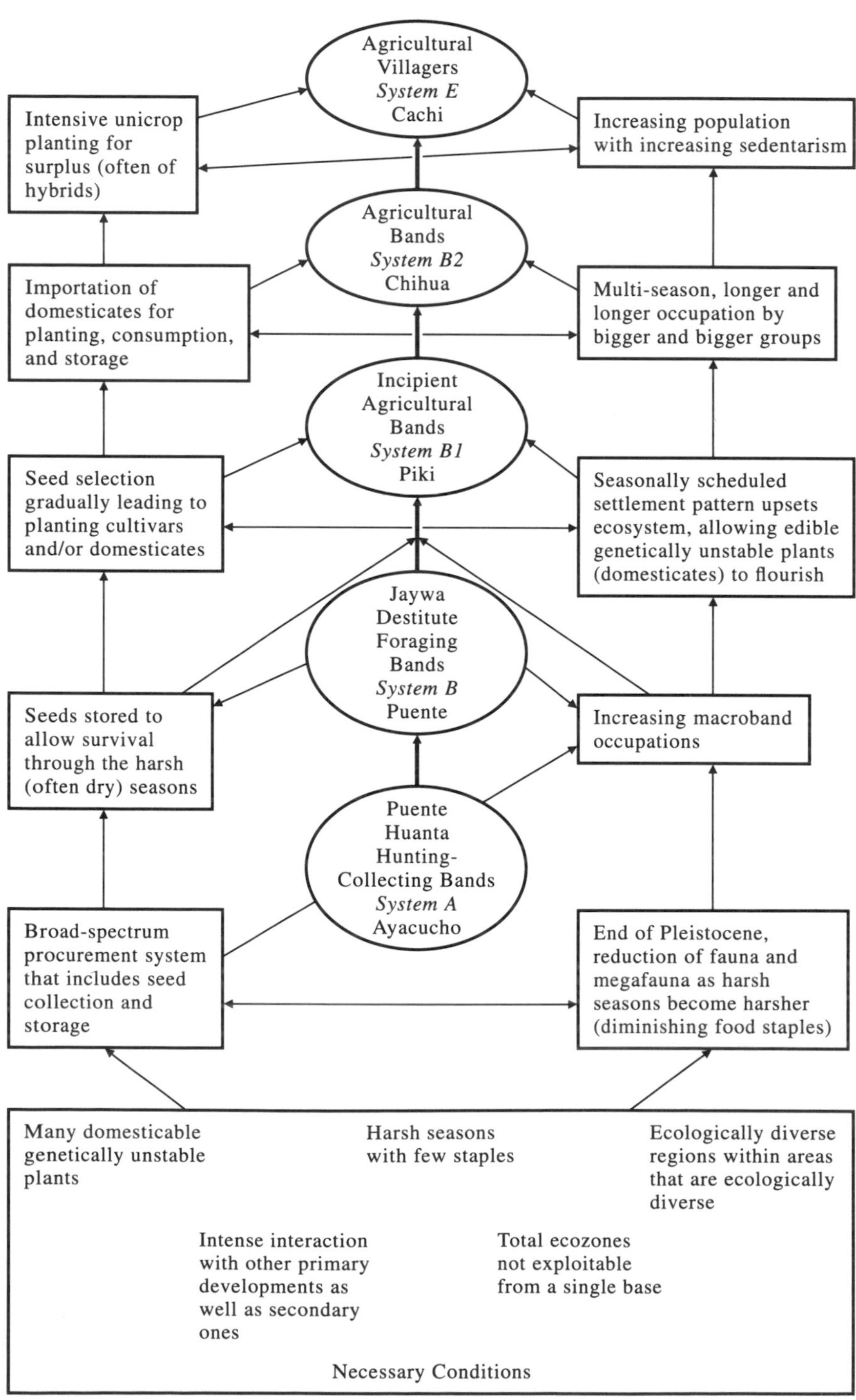

1. The environment at the end of the Pleistocene changed, causing large food animals, such as horses, deer, and camelids, as well as their grassland habitats, to either decrease or disappear.

2. Winter seasons became drier.

3. Jaywa and late Piki groups settled in the lusher areas for longer periods and in bigger groups, thereby decreasing residential mobility.

4. Seasonally scheduled activities put groups in the right ecozone at the right time, when foods reached fruition.

5. Refuse and used tools in seasonal camps indicate that Jaywa peoples developed a number of new subsistence options, including trapping, seed collection and storage, and a broad-spectrum procurement system.

6. They also may have taken domesticates from other areas, as they did in Piki times.

7. They domesticated local plants such as we see evidence for at Pachamachay (Pearsall, 1980), where quinoa seeds (*Chenopodium quinoa*) were selected for bigger size and probably planted, thereby making these people, by Piki times, Foraging Bands with Incipient Agriculture (system B1).

The above conditions are a startling confirmation of my primary developmental hypothesis. Although they fit the positive-feedback cycle of routes 1a and 1b, the seed or plant remains needed to confirm or deny the feedback cycle that involves genetic changes in plants have not been found. The closely related Pachamachay and Panaulauca (Pearsall 1980, 1989) data, however, did seem to show that this positive-feedback cycle occurred in the highlands. This evidence indicates that the subarea of the Andean highlands had one of the primary developments that takes place in Centers. Furthermore, the other part of the primary development—from Incipient Agricultural Bands (as represented by Piki) to Village Agriculturists (as represented by Cachi)—seems to have occurred in these same highlands roughly from 4400 B.C. to 1750 B.C.

At Ayacucho, the Piki and Chihua phases (systems B1 and B2) exhibited the following:

NECESSARY CONDITIONS

1. The domesticable plants—potatoes (*Solanum tuberosum*) and quinoa (*Chenopodium quinoa*)—were native to the area.

2. Piki and Chihua peoples had exchange access to other parts of the highlands (such as the Callejón de Huaylas, where beans had been domesticated) as well as to the selva, where lucuma (*Lucuma bifora*), achiote (*Bixa Orellana*), squash (*Cucurbita sp.*), and the like grew wild and/or were domesticated.

3. Food storage increased, as indicated by the presence of more storage pits. These developments seem to have led to further change, brought on by the conditions discussed below.

SUFFICIENT CONDITIONS

1. Longer stays in one spot (as evidenced by Chihua-Cachi seasonality) by more and larger macrobands in the Chihua phase.

2. More importation of domesticated plants (lucuma, achiote, corn, common beans, coca, and perhaps potatoes).
3. The selection of quinoa seeds for size (making quinoa a cultivar and/or domesticate).
4. Growing of more local plants (quinoa and squash) for storage and as a surplus.
5. Camelid herding.
6. Longer stays in one place by Cachi or Andamarka peoples.
7. Settlement in agricultural villages (system E).

My hypotheses about primary developments in the Andean Center have thus been confirmed by data from the poorly known preceramic periods in the highlands. Now, what about the secondary developments in the other part of the Center, namely, the Pacific Coast?

The Andean Secondary Development

As indicated previously, the basic cultural or human ecology of the highlands is fundamentally different from that of the coastal regions. Highland valleys like Ayacucho have layered ecozones—high puna, low puna, humid woodlands, thorn forest scrub, thorn forest riverine, and mantaro xerophytic (see fig. 4). Each zone contains not only different potential domesticates but different plants and animals in different seasons. Furthermore, the steepness of the slopes—roughly from 2,000 meters to more than 12,000 meters—prevents the different ecozones from being exploited on a daily basis from any single base or camp. The Callejón de Huaylas region of the dry western sierra subarea, for example, is said to have "three major complexes of vegetation" (Lynch 1980): (1) the lowest—"open xerophytic shrub and cactus with no major trees" (except along major streams, where there were gallery forests, which should be considered another ecozone); (2) "transitory grass-steppe with scattered (mostly rain-green or periodic) shrubs"; and (3) mountain grasslands grading into puna, "which again can be divided into high and low Puna ecozones" (Lynch 1980, 7).

The higher puna ecozones, according to Rick, have still other ecozones—high or rocky puna on steep slopes, low puna on rocky slopes, grassland punas on open plateaulike or rolling landscapes, and streamside puna with gallery shrubs and lakeshore communities with marsh and moors, both with distinctive congeries of seasonal fauna and flora; all types of puna have varieties of root crops susceptible to domestication (Rick 1980). Further, horizontal distances between ecozones prevent their easy exploitation from a single base.

In summary, the highland environs (1) have great ecological diversity; (2) have potentially domesticable plants (and animals) in many ecozones; (3) can rarely be completely exploited from a single base; (4) often have harsh seasons when certain ecozones lack food; and (5) include varying elevations, soils, and climates that cause many of the regions to have different sets or complexes of ecozones. There is therefore both regional *and* areal diversity of ecozones.

The desert coastal subarea, on the other hand, I characterize as a unit that may have regional diversity but usually has areal *uniformity*, not diversity (MacNeish, Patterson, and Browman 1975). In other words, the river regions—such as Chicama, Virú Valley, Casma, Chancay, Chillón, Lurín, Rímac, Asia, Ica, and Nazca—that cut from the Andes across the deserts to the Pacific Ocean have similar ecozones (Lumbreras 1974).

Although various authors have defined them slightly differently, these ecozones generally include (following Patterson 1973): (1) the upper rivers that form narrow canyons cutting into the western bases of the Andes; (2) the middle river regions with gallery forests often growing between the hills dividing the valleys; (3) the lower riverine environs with more dense gallery forests; (4) the lomas that are the foothills of the Andes on the coastal plain and that also separate (in part) one river from the next; (5) the river deltas, often rich in seasonal maritime resources, that braid and flow into the Pacific Ocean; (6) the points or peninsulas that stick out into the ocean and also have abundant seafoods; and (7) the bays located between the rocky points.

Michael Moseley further subdivided these areas and went into great detail on the food resources available in each one (Moseley, 1978). From his studies and others by Thomas Patterson (1973) and Sheila and Tom Pozorski (1979a), it is apparent that most of the coastal ecozones can be exploited from a single, strategically located base. Furthermore, usually one or more of these ecozones has large amounts of (seasonal or permanent) foodstuffs available within a circumscribed region. Few, if any, of these abundant food resources, however, are susceptible to domestication (see table 2.5).

Ideally, this coastal subarea represents one end of an ecozone dichotomy, while the highlands represent the opposite end. As might be expected, the Archaic developments in this subarea are equally different, although for quite different reasons.

Edward Lanning rather arbitrarily divided this generally uniform subarea into a number of subdivisions or regions, each with its own set of river drainage systems (Lanning 1967a). The archaeological programs in each of these regions, however, have tended to differ, and although all have late preceramic remains with village agriculture (and/or horticulture), many do not have sequences going back to nomadic collecting bands. Moreover, few full reports have been published on preceramic site excavations in the Peruvian coastal area (see table 2.5).

I discuss the Peruvian coastal regions below, progressing from those with the poorest remains to those with the best. In the north-central region a number of preceramic sites have been dug between the Huaura and Santa rivers, yet all seem to be of the final preceramic stage with village agriculture (system E). The huge site of Huaynuma had samples of white potatoes, achira, gourds, and sweet potatoes; Las Haldas had gourds, avocados, peanuts, and, maybe, corn; Culebras had corn, lima and common beans, cotton, pacae, and gourds; Gavilanes had achira, pacae, peanuts, amaranth, and gourds in period 1 (about 3000 B.C.) and chile peppers, avocados, peanuts, achira, pacae, amaranth, manioc, guava, pepper, cotton, beans, lucuma, warty squash, and gourds in period 2, about 2000 B.C. (Bonavia 1982). (More recently, Michael Malpass has found early preceramic

Table 2.5. Sequence of domesticated plants on the Peruvian coast

South and south central	Central and north central	North and far north	Range of dates B.C.	
Asia				1640
Casavilca			1750	1900
		Salinas	1750	1900
		Huaca Negra	1750	1900
		Huaca Prieta	1750	1900
	Gaviota		1750	1900
		Huaynuma	1750	2000
	Aspero		1750	2000
Pampa de S.D.			1800	2000
	Chuquitanta		1800	2000
		Alto Salaverry	1800	2100
Otuma			1810	2000
		Los Aldas	1800	2054
		Culebra	1800	2100
		Rio Seco	1900	2300
	Conchas		1900	2100
	Chira Villa		2000	2300
		Padre Abán	2000	2300
	Gavilanes II		2000	2400
	Playa Hermosa		2100	2400
	Pampa		2400	2800
	Gavilanes I		2500	3100
Cabeza Larga			2950	3190
	Encanto		2800	3500
	Chilca		2500	3700
	Corbina		3500	4200
		Talera	4000	5500
	Paloma		4000	5500
	Canario		4200	5500

sites, such as the Paijan complex at the Campanario and Mongocillo sites, which date about 5000 B.C., but these early hunting or foraging sites had no plant remains.)

Almost as incomplete in terms of sequential data is the south region, which lies south of the Pisco River, and the south-central area from the Mala River to the Pisco. The final village agriculture stages seem to be represented at the preceramic Asia, Otuma, and Casavilca sites, which have square houses and domesticated plant remains that include cotton, lucuma, chile pepper, guava, pacae,

Gourd	Squash	Lima bean	Common bean	Pepper	Guava	Cotton	Achira	Amaranth	Peanuts	Pacae	Sweet potato	*Cucurbita ficifolia*	Pepper (*Capsicum baccatum*)	Avocado	Manioc	*Cucurbita moschata*	Lucuma	Bean (*Phaseolus sp.*)	Corn	Jicama	White potato	*Cucurbita andina*	Chile pepper (*C. frutescens*)	Chile pepper (*C. chinense*)	Jack beans
x		x	x		x				x	x							x								
x	x									x															
x																			?						
x	x					x	x																		
x	x	x	x	x	x	x	x					x	x			x	x						x	x	x
	x	x	x	x	x		x		x	x	x	x					x					x			
x							x				x										x				
x	x		x	x	x		x		x	x				x	x				x						
															x										
x		x	x		x		x			x	x					x	x			x					
x	x	x		x	x	x				x				x			x								
x	x	x				x			x	x															
x									x					x					x	x					
?		?	?			x													x						
x	x		x			x													?						
x		x		x	x		x				x						x								
						x				x							x		x						
x	x					x																			
x	x	x		x	x	x	x	x	x	x				x	x	x	x	x							
x				x	x	x	x				x		x						?						
x	x				x	x				x	x	x													
x		?					x	x	x	x															
		x	x			x																			
x	x			x	x	?																			
x	x	x																							
?																									
x																									
?																									
x																									

and lima beans (Engel 1966). The Cabeza Larga site, seemingly without agriculture, is basically a burial site. It probably dates to an earlier stage, but the reports on it are confusing. The remains at San Nicolas and Playa Chiba may be of the same period, but both are basically unexcavated.

The final region with insufficient sequential data is the far north, which extends from north of the Lambayeque River to the Ecuador border. Final preceramic village communities seem to be unrepresented and are said to be preceded by the Honda and Siche site materials. Both these sites, however, lack plant remains and are poorly described. The inhabitants of these sites could have been earlier foragers but certainly not early hunters.

The north, from the Lambayeque to the Virú River, yields much better data, but its Hunting-Collecting Band remains, called Paijan, are controversial and there seems little continuity from them to the later preceramic of the region. The Pozorskis have recently described a late preceramic sequence from Padre Abán (roughly 2500 B.C.) to Alto Salavarry (about 2000 B.C.) to Gramalote of the initial period between 1800 and 1400 B.C. (S. Pozorski and T. Pozorski 1979a). All have architectural remains, with those from Padre Abán being relatively simple. All had slightly different plant complexes: at Padre Abán were found cotton, gourds, and several squash species, while Alto Salavarry had these plants plus common and lima beans, chile peppers, avocados, pacae, cansaboca, guava, and lucuma. The Gramalote site added corn and peanuts to the above.

Huaca Prieta and Huaca Negra, which Junius Bird and others dug early on (but reported inadequately), seem to be late preceramic village agriculture sites. They had about the same plants as other north coast sites, as well as achira and definite evidence of *Cucurbita maxima* (Whitaker and Bird 1949). There are, however, no reports of early preceramic remains that could tell us how this late stage evolved.

In fact, only the central coast region from the Chilca to the Chancay River provides an adequate sequence (see table 2.6). Much surveying has been done by Ed Lanning, Tom Patterson, Mike Moseley, and Frederic Engel, yet few of the many excavated sites have been reported sufficiently. Nevertheless, we do have a sequence from Hunting-Collecting Bands to Agricultural Villagers. Although recent research by Robert Benfer and others (Benfer 1983) may update the original sequences of Lanning (1963) and Patterson (1973), the latter seem to be holding up well enough to be used here.

Instead of considering the controversial remains from the Red Zone, Chivateros, Oquendo, and so on, I will begin with the Arenal remains. These seem to date to just after the end of the Pleistocene (8000–7000 B.C.) and the extinction of the megafauna (that the earlier, hunterlike Chivateros-Conchita may have hunted). Arenal sites seem to have been mainly in the lomas or coastal treeless hills and are characterized by distinctive projectile points (like those used by the Jaywa peoples of Ayacucho), suggesting that these people were mainly hunters. A few sites have shells, which suggests some, perhaps seasonal, maritime adaptations.

The Luz complex that followed, known chiefly from surface collections in the lomas near Ancón, has projectile points (some like Paijan and Jaywa), some maritime remains, and also milling stones, mullers, small mortars, and cobble pestles. These suggest that perhaps by this time (6000–5500 B.C.) the coastal inhabitants were Affluent Foragers (system C).

The first reliable Archaic remains from excavated contexts, first mainly by Engel and later carried forward by Benfer at Paloma (Benfer 1983) were dubbed the Canario complex by Lanning (Lanning 1963). Food remains from Paloma and other sites suggest that the people's subsistence was based mainly on wild plants collected from the lomas in the wet season and from along the flanks of the

Table 2.6. Necessary and sufficient conditions as a positive-feedback process in secondary developments in the central region of coastal Peru

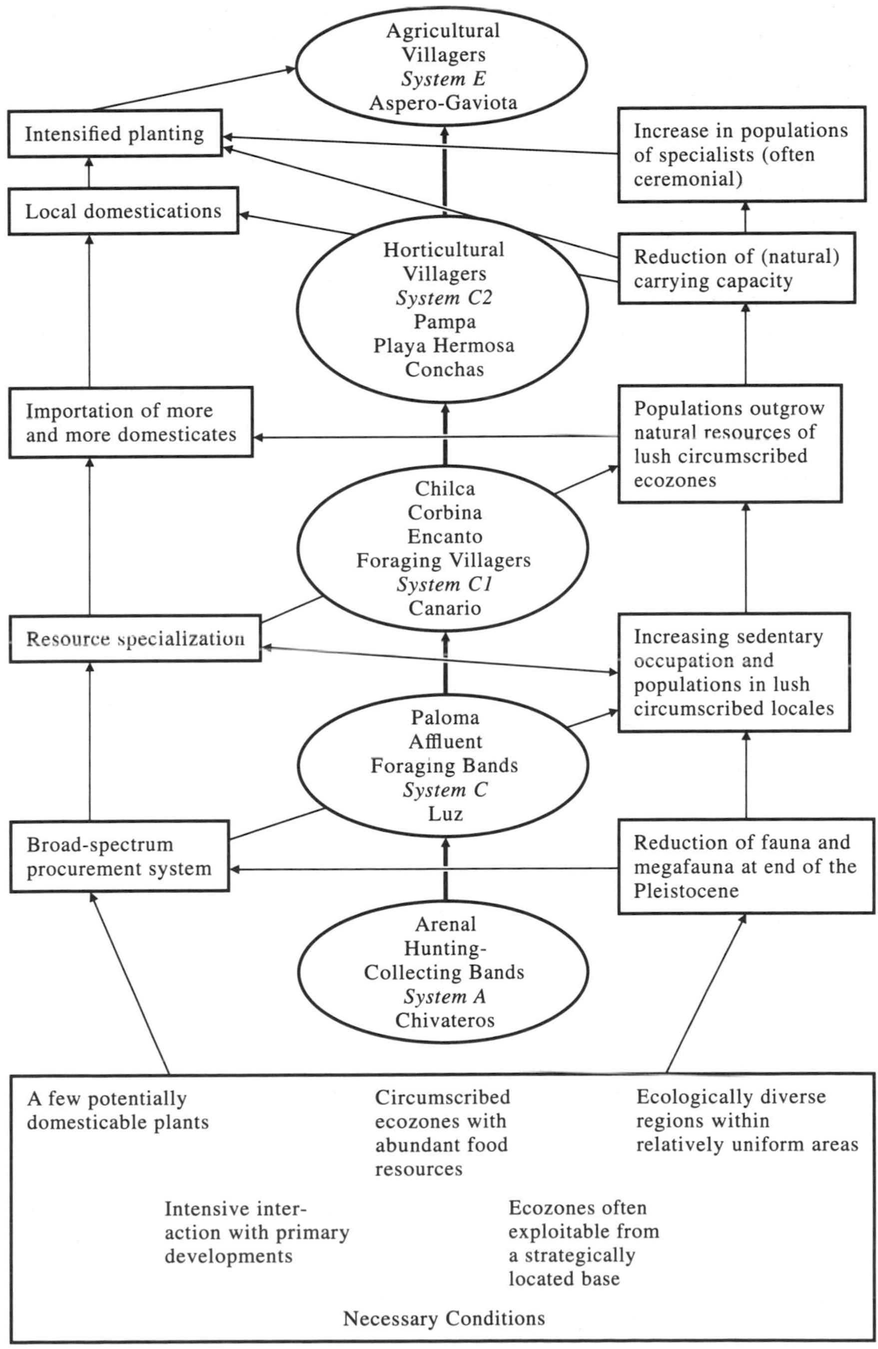

middle section of the rivers in the dry season. The only domesticates found were gourds, and these do not seem to have been used as foods. Bones and large leaf-shaped projectile points indicate hunting occurred throughout the year, and there is evidence of shell gathering and fishing on the coast.

Although more survey needs to be done, Patterson's Lurín data (Patterson 1973) suggest that the major settlements were hamlets (Paloma had shallow ovoid pit houses) or base camps located along middle sections of the rivers so that the adjacent ecozones could be exploited radially. The food remains from Paloma suggest that the inland base camps or hamlets were occupied throughout the year, but Patterson's survey data from Lurín suggest that there were forays to form temporary camps. In the Ancón region, wet-season camps and some inland and dry-season ones may have been located on the coast. All in all, the picture one gets from this meagerly reported data is that the Canario complex represents Foraging Villagers—my system C1—who by strategic location of sites had a fairly successful Affluent Forager economy that allowed them to be semisedentary if not sedentary.

The Canario complex that appears to have existed from radiocarbon dates 5500 to 4200 B.C. seems to have evolved into those called Corbina and Encanto (4200–3500 B.C.). Much of the data on these periods comes from surveys by Benfer, Engel, Lanning, Moseley, and Patterson, with Engel being responsible for most of the site excavations for this period. Except for the Ancón region, hamlets with house structures (Donnan 1964) far outnumber the camp sites. This evidence led Mark Cohen (in opposition to Patterson) to suggest there was a jump in population from perhaps 50 to "200 to 300 persons" (Cohen 1977b). As with Paloma, there is a suggestion that sites were located strategically to allow exploitation of the largest possible number of ecozones. For instance, the Chilca Monument 1 was at the junction of the delta and the lower stream area not only so that the Chilca River valley and lomas could be exploited but also so that it was less than a day's walk to the sea.

Superficial analyses of foodstuff remains suggest resource specialization in terms of intensive seed exploitation as well as more use of marine resources. At the same time there was a decline in hunting and lomas plant collecting—perhaps due to overexploitation of some of the inland ecozones (Cohen 1977b). It is perhaps significant that the first domesticate—the squash (*Cucurbita ficifolia*)—and the date of its introduction on the coast occurs long after (over 2000 years) domesticates were being used in Ayacucho and the Callejón de Huaylas.

The real jump in both population and use of domesticates—mainly imported from elsewhere—occurred in the period from 2800 to 1900 B.C. with the appearance of the Pampa (2800–2400 B.C.) complex. Throughout this period there is thought to have been a rapid increase in population (according to Mark Cohen's estimates, which are based on survey data (Cohen 1977b)—unfortunately not excavation data—by Lanning, Patterson, and Moseley in the Chillón and Ancón regions). The Pampa site in that region represents the first village on the coast. The presence of many domesticated plants—such as squash (*Cucurbita ficifolia* and *Cucurbita moschata*), pacae, achira, guava, galactia, jack beans, and cotton, which comprised but a small part of the diet—suggests that there were also inland hamlets. Most of these domesticates were probably imports, since only pacae and perhaps cotton have wild relatives on the coast.

Data from Playa Hermosa (2400–2100 B.C.) show a continuation of these trends, with the only new domesticate being chile peppers. Conchas (2100–1900 B.C.) culminated this transitional period—from my developmental system C2 to system E—a sort of horticultural rather than agricultural village stage. As population increased still further, lima beans (imported from the highlands) and lucuma and sapinda seeds (imported from the tropics) were added to the subsistence system that was still based mainly on wild seed collecting and on maritime resources. Also, the number of valley villages may have begun to outnumber the coastal ones.

The Conchas phase led to the Gaviota phase (1900–1750 B.C.), which really does represent village agriculture—system E—in the coastal subarea. The population of the area, according to Mark Cohen (Cohen 1977b), may have risen to six thousand people. Perhaps three to four thousand of these were in Chuquitanta, which was located inland along the Chillón River and could serve as the exchange center in the network linking coastal villages and valley hamlets. In addition to all the previously mentioned plants, imported jícama (and perhaps white and sweet potatoes and peanuts) now became major items of subsistence and indicate that true agriculture was practiced.

This scenario from Hunting-Collecting Bands to Village Agriculturists perhaps represents what happened in other coastal developments in Peru. Data for those areas, however, are even flimsier than the data for the central coast, where the available information leaves a lot to be desired. However, what data we do have clearly indicate that this secondary development on the coast was decidedly different from the primary development in the highlands. Data from good dated archaeological contexts (and adequate site reports) are insufficient for both areas, yet both are complete enough to test my basic hypotheses. Now let us examine my hypotheses about secondary developments of village agriculture with the limited coastal data now available (see table 2.6).

Coastal development from Hunting-Collecting Bands—system A—(as exemplified by Arenal) seems to have led to Foraging Villagers—system C1—by Canario times. Our exact understanding of the process is vague, but it may have been that at the end of the Pleistocene, roughly 8000 B.C., worsening environmental conditions on the coast reduced food staples, with the ground sloth, horse, and mastodon becoming extinct. This caused Arenal to develop, first, seasonal scheduling as well as a broad-spectrum procurement system and resource specialization—an Affluent Foraging Band culture (system C)—that led to increasing population in circumscribed lush zones by Canario times and to the exploiting of a series of ecozones from a base by specialized wild food procurement systems and a Village Forager system (system C1). The question now becomes, Does the shift from Canario village foraging (system C1) to village horticulture (system C2) to Chuquitanta's village agriculture (system E) occur due to the same conditions as in my model? In other words, can my model be tested with Peruvian coastal data? It would seem it can, for the necessary conditions seem much the same:

NECESSARY CONDITIONS

1. Canario people were sedentary foragers with specialized resource-procurement techniques.

2. They lived in an area that was ecologically uniform from one coastal river valley to the next, although each region had distinct ecozones.

3. They could use locations from which all ecozones could be exploited; Canario served as the midriver location and Encanto the upper deltaic portion.

4. Some ecozones, like the coastal rocky points and deltas, while circumscribed in area, had abundant fish and shellfish foods.

These conditions, triggered by the sufficient conditions discussed below, caused development to agricultural villages.

SUFFICIENT CONDITIONS

1. Populations increased due to sedentarism in Encanto, Pampa, and Playa Hermosa times.

2. Resources of the riverine and lomas ecozones diminished.

3. People imported domesticated plants from other parts of the Peruvian interaction sphere.

4. This increase in food supply led to further increases in population in the Concha period.

5. The people further outgrew their maritime village resources, which were limited because they were in circumscribed areas.

6. To get food, the people did more and more planting in river areas, and by Gaviota times had become Agricultural Villagers.

Thus data from both the coastal and the highland areas do, in fact, test my primary and secondary developmental models in the Peruvian Center of village agricultural evolution.

Obviously, more data are needed from the Andean area. To change my hypotheses into generalizations, tests of both Centers and non-Centers from independent areas are needed. Let us, then, move on to another Center that is also in the New World—Mesoamerica.

CHAPTER 3

Mesoamerica

Mesoamerica is the Center that interests me most, if for no other reason than that I began my research in Mexico and have continued it in that region for almost forty years. Over these years my techniques and methods have changed as I have tried to solve the various problems of the origins of village agriculture. Although we have come a long way, much remains to be done in Mesoamerica. Data are still so meager that I can test only one of my trilinear hypotheses. Before assessing the evidence, let me first relate what I think has happened in research there.

A BRIEF HISTORY OF RESEARCH ON THE PROBLEM

As I mentioned in the Preface, my first contact with the problem of the origins of agriculture was finding those corncobs in La Perra Cave in 1949. However, that find was preceded by the 1945–1946 season when I found the first preceramic sites and caves in coastal Tamaulipas (MacNeish 1947). Despite limited funds and rough terrain I was able to find about two hundred sites.

One of the reasons for my success was hard work. It helped too that this rattlesnake-infested desert had limited vegetational cover, so that sand dunes, arroyo banks, and cross-sections provided frequent exposures of sites. Some of our success was also due to the fact that, as Alex Krieger put it, "MacNeish was learning to think like an [preceramic] Indian and asking the right questions"—questions such as, "Where would I live one thousand to ten thousand years ago in this desert if I were an Indian dressed in a loincloth or naked and all I had for making a living was a stone-tipped atlatl dart, a few collecting bags and baskets, some crude stone tools, and my wits?"

After finding the sites, the next step was to analyze them to get a preliminary sequence to better define where and what I should do next to get a firm stratigraphic foundation.

So, in 1949, I went back to dig up the rock-shelters that I had found in Canyon Diablo in the Sierra de Tamaulipas in 1946. I knew that one of them had stratified remains, for I had tested it. In my five-foot square I found Diablo (and Abasolo) preceramic materials under Pueblito (Classic Period) sherds. To discover a preceramic sequence, I dug stratified rock-shelter sites by the excavation techniques I had learned in Arizona from Ralph Beals, George Brainerd, and Watson Smith; in Illinois under Roger Willis, Fay-Cooper ("Poppa") Cole, Moreau Maxwell, William Hagg, James Ford, and others; and in New York from Mortimer Howe and William Ritchie. Thanks to their teaching, I invented or developed new and better digging and recording techniques in Tamaulipas.

Although we lived, to quote Helmut de Terra, "lower than the level of the cave dwellers we were digging up," in the 1948–1949 season we had great success, for we excavated three beautifully stratified small caves and three open sites. This gave us a sequence of eight complexes, five of them preceramic—Diablo, Lerma, Nogales, La Perra, and Almagre (MacNeish 1950).

My dating was overly conservative, and I had colleagues who were skeptical of the sequence (Taylor 1960). At first Don Pablo Martinez del Rio was the only person in Mexico to show interest in the preceramic. After Alex Krieger, Glen Evans, J. Charles Kelley, and Thomas Campbell in Texas had seen the materials, they too became firm believers in my finds. Then Helmut de Terra, with the generous financial backing of the Viking Fund (later the Wenner Gren Foundation), under Paul Fejos and Don Pablo Martinez del Rio of Mexico's National Institute of Anthropology and History (INAH), further kindled interest in Early Man in Mexico. At last somebody besides me had recognized that the preceramic was important.

In the next ten years I began to widen my experience. First, I made contact with Paul Mangelsdorf at Harvard and began to learn about economic botany. By then my interest had turned to the origin of corn agriculture—where and when and how maize was domesticated. During our 1954–1955 excavations in the nearby Sierra Madre of Tamaulipas, however, it shifted to the origins of domestication of all kinds of foods—beans, squash, peppers, amaranth, and so on. To this problem I brought my growing experience in archaeological analysis, for I was moving from data collection to data description, particularly in the chronological realm. Jim Ford helped me learn to find chrono-types to show the cultural sequence, and he was ably assisted by James B. Griffin, Alex Krieger, Joseph Caldwell, and sometimes by Albert Spaulding.

About this time the then-new radiocarbon method of dating came into use, and James B. Griffin, Frederick Johnson, and even the inventor himself, Willard Libby, were my teachers.

The third Tamaulipas expedition, 1954–1955, represented a new approach. Not only was I looking for a preceramic sequence, but I also sought one with preserved plant remains that might show the evolution of corn domestication and, inferentially, a sequence of subsistence systems, in the words of Robert Braidwood, from food collection to food production. Pushing the Old World analogy to the limit, not only did we seek pyramids (caves with stratigraphy) but also ones containing a King Tut's tomb (with preserved plant remains and golden corn) within them.

Working with me in the Sierra de Tamaulipas were the geologists Glen Evans, Grayson Meade, and Helmut de Terra, as well as a soil expert, Loren de Witt, a coprolite expert, Eric Callen, and a physicist, Willard Libby, who had dated some of our La Perra levels. We had even begun to use the word "interdisciplinary" (MacNeish 1978).

Our 1954 archaeology, mainly in southwest Tamaulipas, progressed successfully. Our surveyed sites tallied over 350, and in fewer than sixty days we found three deeply stratified caves with unbelievable masses of preserved plant remains—not just one King Tut's tomb but the equivalent of three—the impossible dream come true.

We dug these caves with great care, by methods I called the La Perra technique—stripping living layers from a vertical face rather than using arbitrary levels and bagging the various materials in separate foil wrappings that were marked with location and materials. We also drew exact profiles and did *in situ* floor plots, square by square, along with square descriptions and photos. These techniques gave us great control over what we dug.

Besides getting a good preceramic sequence (see table 3.2), we began to realize that there were concentrations of types of artifacts for certain activities in identifiable areas. These data, combined with all the remains of subsistence (garbage, feces, bones, etc.) were a great help in reconstructing the ancient way of life, occupation by occupation.

While the digging went well, it was impossible to keep up with the contextual analysis. In large part this was because most of the food remains had to be sent out for analysis. It became clear to me that successfully attacking the problems of the origins of agriculture required well-organized and controlled interdisciplinary studies.

At the National Research Council conference held by the Committee on Archaeological Identification in Chicago, I found that most other archaeologists were having similar troubles. This particular conference gave me a chance to get to know Fred Johnson much better. I realized that he was the one I should emulate, since he had been successfully doing interdisciplinary archaeology since 1940, when he had worked on the Boylston Street Fishweir in downtown Boston, Massachusetts.

From 1955 to 1961, when the Tehuacán Project started, we obtained little information that helped us learn about the origins of agriculture. During this period, however, the preceramic was becoming better known as growing numbers of archaeologists became interested in this realm. In northwestern Mexico, William Massey had started exploration for preceramic remains in lower California, while Donald Lehmer, following up on Robert Lister's work at Swallow Cave in Chihuahua, was looking for Cochise materials. George Fay also reported on the Peralta (Cochise) complex in Sonora (Fay, 1950). Farther to the south, J. Charles Kelley found some Cochise-like surface sites in Durango. In 1959, Cynthia Irwin-Williams excavated two stratified caves—Tecolote in Hidalgo and San Nicolas in Querétaro—and uncovered a complete Archaic sequence, but unfortunately without plant remains. Even more unfortunate, the results were never published.

Still farther south, in the Valley of Mexico, Luis Aveleyra made a great discovery at Itzapàn. There, for the first time in Mexico, artifacts were found in direct association with mammoth bones (Aveleyra and Maldonado-Koerdell 1953). More important, a department of prehistory was formed under his direction at the National Institute of Anthropology and History (INAH), and it set about excavating several preceramic sites, among them Yuzanú, in Oaxaca, in 1955 (Lorenzo 1958), Chicoloapan, in the Valley of Mexico, in 1947 (de Terra 1949), and Comitán, in Chiapas, in 1959 (Lorenzo 1961).

Meanwhile I continued the great corn hunt. First, in 1958, I tried to find preceramic caves in the southern part of Mesoamerica, namely, in the caves of Copán and the Comeagua Valley of Honduras, then in the Zacapá Valley of Guatemala. In 1959, we briefly visited Oaxaca and the Río Balsas Valley of Guerrero. Finally, in 1959, I excavated Santa Marta rock-shelter near Ocozocautla, Chiapas (MacNeish and Peterson 1962). We found some fine preceramic tools in the 7000–4000 B.C. range as well as some pollen, but no preserved plant remains.

By 1960, I had narrowed the corn hunt to three highland valleys of Mexico, between Chiapas to the south (where I had worked in Santa Marta Cave) and the Valley of Mexico to the north, where Paul Sears showed that the earliest corn pollen was not much older than corn found in Tamaulipas. Then, in 1960, we found Coxcatlán Cave in the third valley, which was Tehuacán (fig. 5).

Excavation of this cave was the start of the Tehuacán Project in 1961. It was an unbelievable success. Not only did our surveys locate 454 sites between 1960 and 1963 (now supplemented by Edward Sisson and Robert D. Drennan to over 1,000 sites), but we dug or tested sixteen stratified sites that had early village or pre-village preceramic remains (see fig. 5 and table 3.2), and at least one site occurred in each ecozone.

In total we had over sixty preceramic components, representing perhaps one hundred occupations, and another forty preceramic surface sites; these spanned the period from roughly 20,000 to 2000 B.C. They contained some 50,000 lithic artifacts, more than 100,000 plant remains, over 10,000 bones, and some 250 human feces. As you can see from the accompanying figure (fig. 6), we had a great cultural sequence—at least nine preceramic phases—which yielded enough plant remains to reconstruct the subsistence systems as well as to learn about the cultural systems.

From many standpoints (although there are still disputes about the possible relationship of corn and teosinte), we had solved the problem of the origins of corn and were able to attack the how and the why of many other domesticated plants in highland Mesoamerica. By 1965, we had developed a series of hypotheses about how and why village agriculture happened in Mesoamerica and had begun attempting to test these hypotheses with comparative data from other regions and areas (MacNeish, Fowler et al. 1975).

While I was testing these hypotheses in Peru, work on the preceramic of Mesoamerica continued. Across the mountains from Tehuacán, Kent Flannery conducted a project in Oaxaca (Flannery, Marcus, and Kowalewski 1981). Although his preceramic plant remains were less abundant and his survey of this period less complete than mine at Tehuacán, his excavation of four stratified rock-shelters gave an equally long sequence (Flannery 1970, 1986). Flannery's project lasted for many years, roughly from 1964 to 1982, but only a few preceramic results

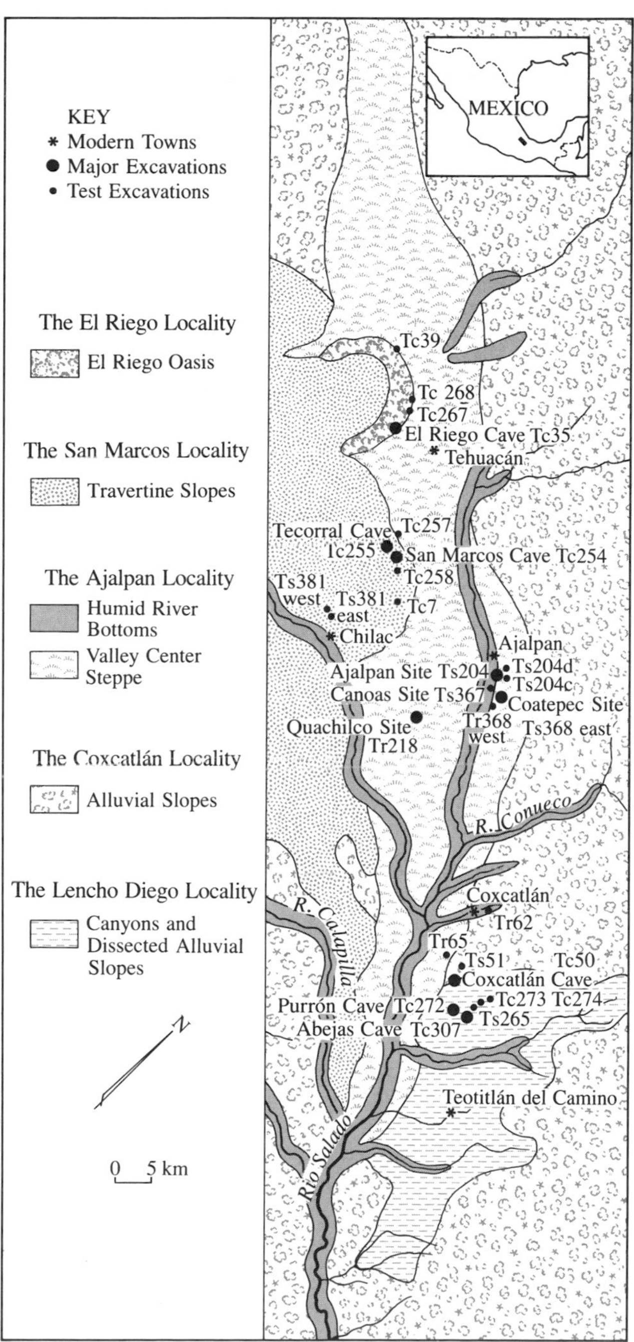

Figure 5. Tehuacán Valley, showing principal towns, major excavated sites, and tests in the ecozones

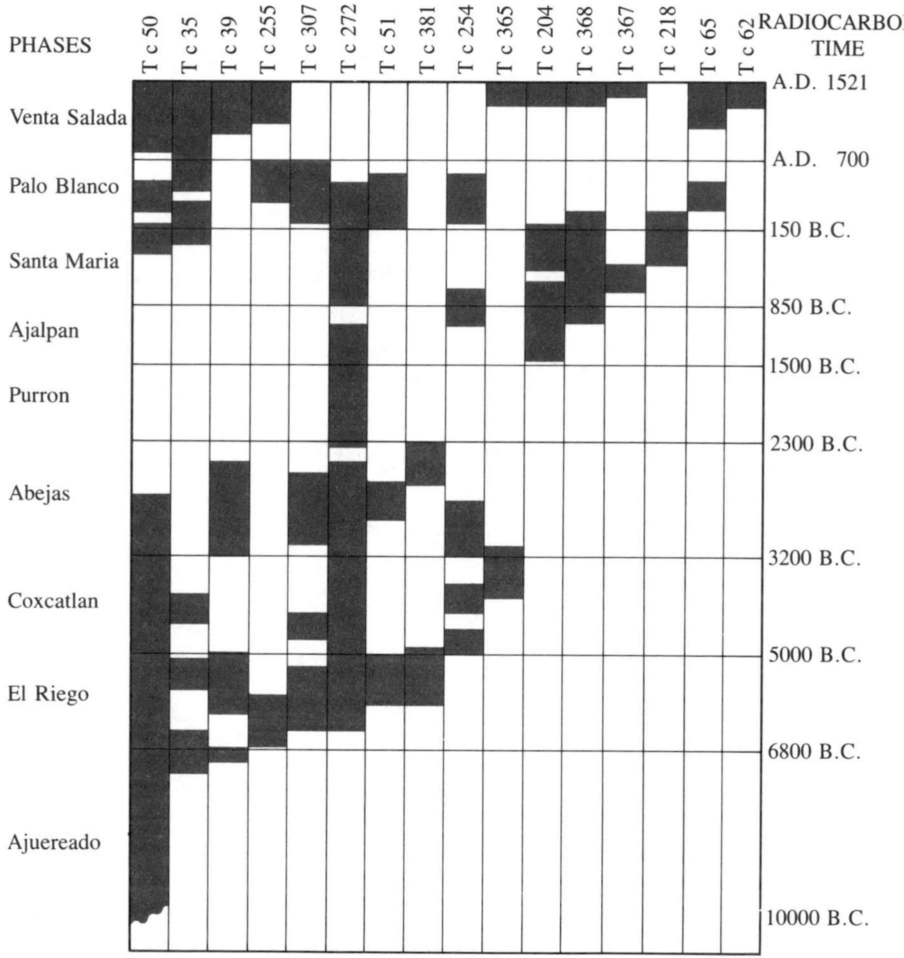

Figure 6. Complete cultural sequence of the Tehuacán Valley, with occupations of various stratified sites

have been published (Flannery 1986). In outline, his sequence of five preceramic phases based upon his excavation of stratified sites is shown in table 3.2.

Just north of Tehuacán, in the state of Puebla, excavations took place in the Valsequillo region outside the city of Puebla. These occurred during the final years of the Tehuacán Project, and the most important were the excavations under the direction of Cynthia Irwin-Williams, which concerned Early Man (over 20,000 years old) in the Valsequillo area (Irwin-Williams 1967a, 1967c). Unfortunately she was forced to end her project in 1964. Later the INAH Department of Prehistory moved into the territory and began testing caves to find preceramic remains. One cave, Texcal, did yield some evidence of an Archaic sequence (Garcia Moll 1977), but it lacked the preserved remains needed to bear on the problem of the origins of agriculture.

The other major program in this area, which started about the time the Tehuacán Project ended, in 1965, was Tlapacoya. It was under the direction of José Luis Lorenzo and Lorena Mirambell, of the INAH Department of Prehistory (Mirambell 1973). This project was interested mainly in Early Man from 10,000 to 40,000 years ago; it lasted nearly as long as the Oaxaca project with almost as few site reports. Christine Niederberger dug one of the subsites in this area, Zohapilco, which had some Archaic materials under early Formative sherds; and what is more, she published an excellent description of these materials. Although the number of artifacts and ecofacts she uncovered was small, she was able to delineate a Middle and Late Archaic sequence of Playa I, II, and Zohapilco (Niederberger 1976). Her research finally gave us some reliable Archaic remains from the Valley of Mexico. It was a great beginning for the study of the origins of agriculture in this key region in Mexico, but unfortunately the project has not been continued.

Starting slightly later was a renewed interest in the preceramic of the lowland areas of the Pacific and Gulf coasts. Early in 1947, Phillip Drucker had discovered preceramic shell middens in the Chantuto region of the Pacific coast of Chiapas (Drucker 1948). Later, José Luis Lorenzo (1955) had put in some soundings in the same area, getting the same limited types of materials. In the mid-1940s, I had discovered preceramic sites in coastal Tamaulipas and had even found a Nogales site on Laguna Chila in northern Veracruz in 1949 (MacNeish 1950). Charles Brush had tested another shell mound at Puerto Marquez in coastal Guerrero (Brush 1969), and in 1963, Ford and Medellin Zenil had collected possible preceramic materials from sand-dune sites north of Palma Sola along the Gulf Coast of Veracruz (MacNeish 1967).

In spite of these sporadic findings, no one had actually set up a program of research on the Archaic of the lowlands until Barbara Voorhies and Barbara Stark did so. Barbara Stark carried out her research in the Alvarado Bay area of central Veracruz, and although she got some fine information about marine adaptations, no preceramic remains were uncovered (Stark 1977). Between 1971 and 1974, however, Barbara Voorhies, in Drucker's Chantuto region of Chiapas, dug five deep shell heaps, dating from 2000 to 3000 B.C. She uncovered the first definitely preceramic artifacts from reliable contexts in the lowlands of Mesoamerica (Voorhies 1976).

In the early 1970s, Joseph Mountjoy excavated a shell mound with a few artifacts and seafood remains at the Matanchon shell midden at Ceboruco, Nayarit, which dated to 1960 B.C. (Mountjoy, Taylor, and Feldman 1972). In the late 1970s, Jeff Wilkerson dug up some preceramic (3000 B.C.) remains called the Palo Hueco complex at the Santa Luisa site in central Veracruz (Wilkerson 1975).

In the early 1980s, Wilkerson joined me (and many others) on the coast of Belize. Here, after doing a survey that netted about one hundred preceramic sites, we excavated nine stratified sites, which yielded a tentative sequence of six preceramic complexes: Lowe-ha, sometime before 7500 B.C.; Sand Hill, from about 7500 to 6000 B.C.; Orange Walk, from perhaps 6000 to 5000 B.C.; Belize, from perhaps 5000 to 4000 B.C.; Melinda, from perhaps 4000 to 3000 B.C.; and finally, Progreso, lasting up to ceramic times at about 2000 B.C. (MacNeish and Nelken-Terner 1983b). This cultural sequence gave us a start at understanding the preceramic development of the lowlands of Mesoamerica, but we still have a long way to go.

Perhaps more complete was the research carried out at the opposite end of Mesoamerica, in Nuevo León and Tamaulipas, under the direction of Jeremiah Epstein (Epstein, Hester, and Graves 1980). From roughly 1965 until well into the 1970s this project tested the San Isidro site, as well as three stratified caves in Nuevo León with a preceramic sequence from 9000 to 1000 B.C. While this research project provided great preceramic data, little of it had any bearing on the problems of the origins of agriculture because preserved plant remains were lacking (in fact, perhaps no domesticated plants were ever used prehistorically in this region).

Other archaeological studies of slightly more recent date have tended to focus on Early Man rather than on the origins of agriculture. Among them was the research carried out in nearby San Luis Potosí by the INAH Department of Prehistory, under the direction of José Luis Lorenzo (Lorenzo 1981) and their endeavors in the late 1970s at Santa Marta Cave and Los Grifos in highland Chiapas (García-Barcena 1981). Here one might expect some Archaic remains because of our earlier endeavors (MacNeish and Peterson 1962). In fact, in 1974, Angel García Cook and I had planned and were funded to redig the Santa Marta Archaic levels and float all of the charcoal-filled floors for plant remains that would be relevant to agricultural origins (García Cook 1973). The project, however, was taken over by the INAH Department of Prehistory, and its results have not yet been reported sufficiently for us to know if their finds are relevant. Now there also are plans for investigation of caves in Guerrero (where I was in 1966) and western Mexico to study the corn problem further, but so far little has been accomplished (McClung de Tapia 1989).

At the southern limits of Mesoamerica, Kenneth Brown found about 190 preceramic sites in the Quiché Valley of Guatemala (Brown 1980), and in 1983, I surveyed Ripley Bullen and William Plowden's La Esperanza site in highland Honduras and found 20 preceramic sites. These may belong to four complexes spanning a 20,000-year period (Bullen and Plowden 1963). However, real field research is yet to be done in this area.

With the background set by this brief review of the nearly fifty years of preceramic work in Mesoamerica, I will briefly describe the area itself before moving on to the relevant archaeological data.

THE ENVIRONMENT

In summarizing the environments of Mesoamerica, I will follow closely the divisions Robert West established (West 1964), modifying them only to better meet my particular problems. West divided Mesoamerica into three major areas: (1) the extratropical dry lands; (2) the extratropical highlands and extratropical appendages; and (3) the tropical lowlands. All of these were further divided into subareas (fig. 7).

The extratropical dry lands, which we did try to subdivide (see fig. 6), consist of four subdivisions: (1) the Mesa del Norte in the central plateau of northern Mexico (the states of Nuevo León, Chihuahua, and Coahuila); (2) the mountainous regions of Sonora and northern Sinaloa; and (3) Baja California. These three subdivisions do not lie mainly in Mesoamerica, nor did their inhabitants develop village agriculture or the Mesoamerican cultural complex. Because they are not considered part of the Mesoamerican Center, they are not discussed further here.

The fourth extratropical dry land subarea that seems relevant is the subdivision called Tamaulipas subhumid lowlands. Because most of our early agricultural finds came, ironically enough, from the southern highland portion, I really consider this zone a transitional one between the Mesoamerican highlands and lowlands.

Tamaulipas has distinct seasons. The major rains (some of them hurricanes) occur in the late summer and fall, while the winters are dry and quite cold, with occasional frosts. Although 400 to 1,500 millimeters of rain fall each year, the spring has little or no rainfall even in the mountains. When it does rain, the runoff is great, and the arroyos and eastward flowing rivers (San Fernando, Soto la Marina, Río Grande, and so on) change from a fordable trickle to raging torrents in a matter of minutes.

Like highland areas, the Tamaulipas extratropical dry lands subarea has great ecological diversity that varies with the season. While the ecozones of the region range from coastal deserts to tropical cloud or rain forests in the central Sierra Madre, generally speaking the area is desert with thorn scrub vegetation in which grasses, mesquite, acacia, organ cactus, yucca, and opuntia predominate. Deer, peccary, and jaguar are the predominate larger mammals, and there are many small mammals such as coatimundi, opossums, rabbits, and rodents, as well as many reptiles—iguana, lizards, turtles, and snakes (including some of the biggest rattlesnakes I have ever seen). Although rivers and lakes have limited amounts of fish, the coastal area seasonally has fish and shellfish. In the mountains and valleys there is great ecozone diversity that varies with the season, making the region more like highland areas than the lowlands.

West divided the extratropical highlands into six subdivisions, which are similar to the subdivisions of the Andes in their elevational and areal diversity, although the Andes have much higher life zones (see figure 7, first four and last two). Perhaps most comparable to the Andes are the Sierra Madre Occidental and

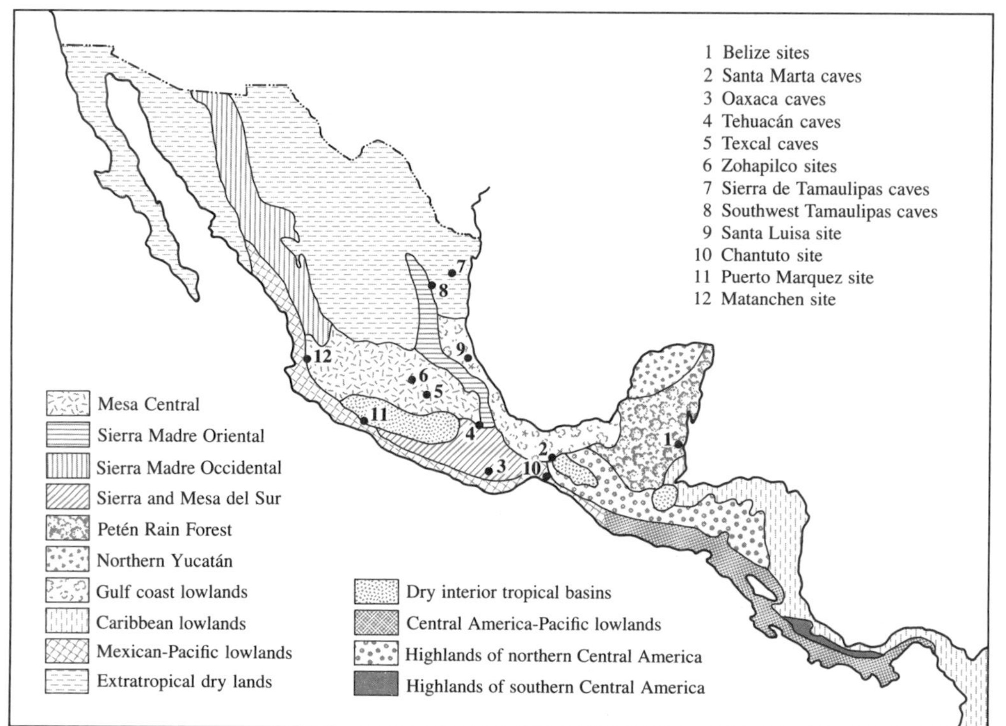

Figure 7. The environmental life zones and relevant sites of the Mesoamerican area

Sierra Madre Oriental subdivisions. Above 2,200 meters, pine and oak forests predominate, although in the southern division of the eastern range, the Sierra Madre Oriental, a few locations have tropical rain forest above the oak-pine zones. Generally speaking, rainfall is heavy, with the heaviest rains coming in the summer and fall. There is also rainfall in winter, when there are occasional frosts and, at the highest elevations, snow. The eastern range has high peaks with tundra vegetation and glaciation. The western range, the Sierra Madre Occidental, is slightly drier and warmer and often has ecozones with grassland and desert vegetation. Small and large game and seasonal food abound in these two mountainous subdivisions. Unfortunately, few archaeological studies have been undertaken on these two flanks of the Mesoamerican highlands.

West's third highland subdivision is the Mesa Central, which may be further divided into at least three major basin regions: (1) the Valley of Mexico, (2) the Valley of Puebla and Tlaxcala, and (3) the Valley of Toluca. These highland valleys are at slightly different elevations and have a host of rather different ecozones, little valleys, basins, and canyons within them. Here there is spatial diversity with regional diversity *par excellence*. Generally speaking, rainfall is moderate and falls in a warm summer-fall wet season, while the winter-spring season is cool and dry—but even this pattern varies. The eastern flanks of the basin (particularly Puebla) fall in the rain shadow of the Sierra Madre Oriental, causing them to be desertlike, although at higher elevations there is basically more rainfall that flows into the basins, often forming huge lakes. Soil, climate, and elevation differ within this subdivision, and these differences are reflected in a great diversity of temperate flora and fauna, not only from valley to valley but also from one part of any one valley to the next so each may have many different ecozones that may be relevant to the archaeological sequences.

West's fourth highland area, the Mesa del Sur, lies to the south of the Mesa Central, mainly in Oaxaca and adjacent states. Like the three valleys of the Mesa Central, it is also flanked by high mountain ranges, although neither the plateau nor the valleys of Oaxaca or Tehuacán is quite so high, and generally there is less rainfall in the wet and dry seasons and temperatures are warmer. These factors plus different soils and more limestone than volcanic rock cause this region's vegetation to be more xerophytic and its game less abundant than that of its northern neighbor. Like the Mesa Central, however, within any of the regions, basins, valleys, or canyons of the Mesa del Sur there is great diversity in ecozones, ranging from hot cactus desert in some valley bottoms to cloud forest regions about 3,000 meters high on the eastern flanks of the plateau.

The fifth major highland subdivision of Mesoamerica, which West calls the highlands of Central America, runs from Chiapas, Mexico, to Costa Rica. It does not seem to have as much diversity as the other subdivisions, and it is generally at lower elevations with more rainfall and has more tropical (lush) flora and abundant fauna. There is a definite dry season and an intense wet season. Oak-pine forests grow at higher elevations, while basins and valleys vary from lush grassland to dry desert. Except for MacNeish and Nelken's attempt to follow-up on William Plowden's and Ripley Bullen's work in the Esperanza Valley of Honduras, few preceramic remains have been found here.

West's final highland subdivision, called the highlands of Costa Rica and Panama, is really not part of the Mesoamerican culture area. Like Tamaulipas, it is a transitional zone; it is at lower elevations and has a longer wet season and often a tropical cloud forest vegetation. Instead of the mountain fauna of the other highland areas, it has such tropical animals as ocelot, margay, spider monkeys, anteaters, and many reptiles.

From the standpoint of necessary conditions for development of agriculture, the Mesoamerican extratropical highlands have the same ecological characteristics as the Andes. There is both areal and regional diversity. There is great seasonality, often with harsh seasons, and all or most of the ecozones of a valley are difficult to exploit from a single base. Instead, they must be exploited by living in them at different times of the year. Finally, many of the highland subdivisions have domesticable plants (see table 3.1) in one or more ecozones.

Distances between one highland valley and the next are short, so the valleys can easily interact not only with one another but also with the lowlands below them. As in the Andes, all these general ecological characteristics of highland Mesoamerica contrast with those of the lowlands.

West has divided the tropical lowlands of Mesoamerica into three major divisions: (1) the Caribbean and Gulf coastal lowlands, (2) the Pacific coastal lowlands, and (3) the dry interior tropical basins. He further subdivides the first and second regions into seven subdivisions and the third into two. I shall deal mainly with the major, threefold division and merely touch on the subdivisions, because preceramic data from the lowlands are minimal.

The Caribbean and Gulf coastal lowlands, on a lower mountain escarpment of hills (below 1,000 meters) in Belize and central Veracruz, are *tierra caliente* ("hot lands") with abundant rainfall and a short dry season. Vegetation is typically a dense rain forest, although Veracruz, central northern Belize, and the Mosquito Coast of Nicaragua and Honduras have patches of savanna-type grassland. The dry northern section of the Yucatán Peninsula, with its underlying limestone rocks, has scrub forest. Fauna is typically tropical and abundant, as are sea resources. Except for the Yucatán, large rivers dissect the relatively wide coastal plains. Connected with them are a series of ecozones that provide local or regional diversity, although one drainage looks much like the next, and the subarea is relatively uniform ecologically.

Much the same may be said for the narrow Pacific coastal lowlands. Coe and Flannery in their classic article (1964) point out that the Ocós region ecozones of the volcanic lowlands of Central America stretch from the sea to the mountains. The ecozones include sandy beaches, low scrub, estuary with red mangrove forests, gallery forest, riverbank "monte," salt playas, mixed forests, low "monte," tropical savanna, and upper estuary forests. Each has a rich complex of floral and faunal food resources. On a more general level, this subdivision is characterized by definite wet and dry seasons of about equal length, and except for the Panama region with its rain forest, it usually has semideciduous forests with scattered patches of savanna and abundant animal food resources.

The final subdivision is, like Tamaulipas, a transitional zone. Called the dry interior tropical basins, it includes such areas as the Balsas-Tepalcatepec River

Table 3.1. Some native cultivated and/or domesticated plants of Mesoamerica*

Scientific name	Popular name
**Achras zapota*	sapodilla
Acrocomia mexicana	coyol
**Agave atrovirens*	maguey
**Agave fourcroydes*	henequen
Agave latissima	maguey
Agave mapisaga	maguey
**Agave sisalana*	sisal
**Agave tequilana*	tequil maguey
**Amaranthus cruentus*	quelite
**Amaranthus leucocarpus*	amaranth
Annona cherimolia	cherimoya
**Annona diversifolia*	ilama
**Annona glabra*	anona
Annona muricata	soursop (guanabana)
**Annona purpurea*	anona
**Annona reticulata*	anona
**Annona squamosa*	sweetsop
**Biomarea edulis*	sansilla
**Brosimum alicastrum*	ramon
**Byrsonima crassifolia*	nance fruit
Calocarpum mammosam	mamey fruit
Calocarpum viride	sapote (fruit)
**Canavalia ensiformis*	jack bean
**Capsicum annuum*	chile pepper
**Capsicum frutescens*	chile pepper
**Carica papaya*	papaya
**Casimiroa edulis*	white sapote
**Casimiroa sapota*	matasano
Chamaedorea tepejilote	tepejilote palm (vegetable)
Chamaedorea wendlandiana	pacays
**Chenopodium nuttaliae*	apazote grain
Cnidosculus chayamansa	chays herb
**Crataegus pubescens*	tejocote (fruit)
**Crescentia cujete*	calabash, tree gourd
Crotalaria longirostrata	chipilín (herb)
**Cucurbita ficifolia*	squash
**Cucurbita mixta*	walnut squash
**Cucurbita moschata*	cashew squash
**Cucurbita pepo*	pumpkin
Cyrtocarpa procera	chupandilla
Dahlia excelsa	dahlia
Dahlia lehmannii	dahlia
Dahlia pinnata	dahlia
**Datura stramonium*	jimson or loco weed
**Diospyros ebenaster*	black sapote
**Gossypium hirsutum*	cotton
Hylocereus undatus	night-blooming cereus, pitahaya fruit
**Hyptis suaveolens*	chia grande seeds
**Indigofera suffruticosa*	indigo añil dye
Jatropha curcas	piñoncillo (medicine)
**Lagenaria siceraria*	bottle gourd
**Lophophora williamsii*	peyote

Table 3.1. (continued)

Scientific name	Popular name
**Lycopersicon esculentum*	tomato
**Manihot esculenta*	manioc
Manilkara zapotilla	sapodilla fruit, chicle
**Maranta arundinacea*	arrowroot
Nicotiana rustica	tobacco
Nopalea cochenillifera	cochineal cactus
**Opuntia ficus-indica*	prickly pear
**Opuntia megacantha*	prickly pear
**Opuntia streptacantha*	prickly pear
Pachycereus emarginatus	pitayo (fruit, protection)
**Pachyrrizus erosus*	jicama (vegetable)
**Panicum sonorum*	panic grass cereal
**Parmentiera edulis*	cuajilote (fruit)
**Persea americana*	avocado
**Persea schiedeana*	avocado
**Phaseolus acutifolius*	tepary bean
**Phaseolus coccineus*	scarlet runner bean
**Phaseolus lunatus*	lima bean
**Phaseolus vulgaris*	common bean
**Physalis ixocarpa*	husk tomato
Polianthes tuberosa	tuberose
Pouteria campechiana	yellow sapote (fruit)
Protium copal	copal (resin)
**Prunus serotina*	capulin cherry
**Psidium guajava*	guava
Psidium sartorianum	guayabilla, guava
**Salvia hispanica*	chia seeds
Sambucus mexicana	elderberry
**Sechium edule*	chayote (fruit, roots)
Sideroxylon tempiscue	cosahuico
**Spondias mombin*	hog plum
**Spondias purpurea*	jocote (fruit)
Tagetes erecta	marigold (medicine)
Tagetes patula	marigold (medicine)
Taxodium mucronatum	ahuehuete, cypress
Theobroma angustifolium	cacao
Theobroma bicolor	cacao
**Theobroma cacao*	cacao, chocolate
Tigridia pavonia	tiger flower
**Vanilla planifolia*	vanilla
Yucca elephantipes	yucca (vegetable)
**Zea mays*	corn

*Domesticated plants according to Harlan (1971); others, Mangelsdorf et al. (1964).

Basin in southwest Mexico, the Valley of Chiapas, and the middle Motagua Valley in eastern Guatemala. All are rain shadow areas having limited rainfall in very well marked seasons and thorny scrub-cactus vegetation with a limited desert faunal assemblage. This subdivision does not have quite the areal uniformity of the other lowland areas, nor does it have their many (or any) lush circumscribed ecozones that give abundant food resources for limited groups over limited periods. Nor do the dry interior tropical basins generally have strategic locations that allow groups to exploit a series of ecozones from a single base; nor do they have seasonality with few harsh seasons.

The characteristics of the two major lowland subdivisions of Mesoamerica resemble conditions in the Andean lowland areas and, like them, could well be necessary conditions for the shift from Hunting-Collecting Bands to Village Agriculturists. Unfortunately, present research in the lowlands of Mesoamerica makes it difficult to ascertain whether such necessary conditions pertain.

Next let us consider, subarea by subarea, how the Mesoamerican Archaic materials test my model as well as the primary and secondary developmental hypotheses in this Center.

RELEVANT ARCHAEOLOGICAL DATA

As mentioned earlier, for one reason or another many of the Mesoamerican subareas do not pertain to the problem. In fact, of the twenty-eight subdivisions of West's three major divisions, eighteen lack relevant Archaic sequences by which to test my models or hypotheses. Baja California (West's subareas 3 and 12f) has little Archaic sequential material, and agricultural villages seem never to have developed there. Sonora and northern Sinaloa, in northwest Mexico (West subdivision 2), have some Archaic Cochise materials but for the most part lack sequences from Hunting-Collecting Bands like Clovis (Haury 1962) to Village Agriculturists. Even when and where sequences are present, village agriculture (Hohokam et al.—see Cordell 1984) is of the southwestern, non-Center type rather than the Mesoamerican Center type, so other hypotheses pertain. Much the same may be said of the Mesa del Norte (West's division 1), particularly its western margin subdivision, which is mainly in Chihuahua and Durango. The other subdivision (1b) of the Mesa del Norte, the steppe lands of eastern Coahuila and Nuevo León, is slightly different. Thanks to the research of Walter Taylor (1966) and Jeremiah Epstein (Epstein et al. 1980), we have a long Archaic sequence there. Some of the Coahuila sites even have undescribed foodstuffs; none of them, however, leads to a Mesoamerican type of village agriculture.

Even within Mesoamerica proper there are a number of subdivisions with no relevant archaeological materials. These include Los Tuxtlas (West's 11c); the southern Veracruz-Tabasco rain forest area (11b); northern Yucatán (11e—although there may be Early Man remains in Loltún Cave); the Río Balsas area (13a); the Sierra Madre Occidental (subdivision 5); and the highlands of northern Central America, mainly in Honduras, El Salvador, and Guatemala (subdivision 9). This last region seems to have considerable potential for yielding relevant results. The Early Man site of Los Tapiales has been dug in Guatemala (Gruhn and Bryan 1977), and Kenneth Brown has found many preceramic sites in the Quiché Valley of the same country (Brown 1980). Following Bullen and Plowden, I have seen similar sites in the La Esperanza area of Honduras.

Five of the remaining subareas have sparse materials. Only one subdivision in the Petén-Yucatán rain forest area of northern Belize (West's 11a) has even the semblance of an Archaic sequence that might be used to test my hypotheses about secondary developments in Centers (see table 3.2).

Table 3.2. Relevant archaeological sequences in Mesoamerica

HIGHLANDS					LOWLANDS		
Extratropical dry lands subarea	Sierra Madre Oriental subarea	Mesa Central subarea		Mesa del Sur subarea	Dry interior tropical basin subarea	Petén rainforest subarea	
Sierra de Tamaulipas region	Sierra Madre de southwest Tamaulipas region	Valley of Mexico region	Tehuacán valley region	Oaxaca valley region	Chiapas region	Belize region	Dates
						Swazey	
	Mesa de Guaje Guerra	Ayotla Nevada	Ajalpan Purrón	Tierras Largas Espiridon			2000 B.C.
	Flacco	Zohapilco				Progreso	
Almagre La Perra				Martinez		Melinda	
			Abejas				3000 B.C.
	Ocampo	Playa 2					
				Blanca			4000 B.C.
Nogales			Coxcatlán			Belize	
		Playa 1					5000 B.C.
				Jicaras		Orange Walk	
					Santa Marta		
	Infiernillo		El Riego				6000 B.C.
						Sand Hill	
Lerma		Tlapacoya (2)		Naquitz			
							7000 B.C.
			Ajuereado		Los Grifos	Lowe-ha	

With these limitations in mind, let us consider the Pacific Coast. As mentioned before, Mountjoy dug the Matanchen shell midden in Nayarit—West's subdivision 12f, coastal lowlands of Nayarit-Sinaloa. Here, in layers dated 1760 B.C., were found four flakes, three worked cobbles, and seafood remains (Mountjoy, Taylor, and Feldman 1972). Equally limited were the Ostiones complex remains from the lower levels of a midden at Puerto Marquez, Guerrero, in the subdivision West calls the coastal lowlands of southwestern Mexico (12e). Here in layers dated as early as 2920 B.C. ±130 were found flakes, a piece of ground stone, a possible metate, a burned floor of perhaps some sort of house in association with shells of both the wet and dry season, suggesting the remains of a preceramic hamlet (Brush 1969).

In fact, our only relevant information comes from the investigations in the same subdivision made by Barbara Voorhies in the Chantuto region of Chiapas (Voorhies 1976). Following Drucker (1948) and Lorenzo (1955), she tested five shell mounds and obtained a series of carbon-14 determinations of the general period from 2000 to 3000 B.C. There are suggestions of house remains, and she concludes these were hamlets (perhaps seasonally inhabited). Voorhies interprets the crude manos and metates as indications of agriculture. The other artifacts were relatively simple and consist of worked and unworked crude obsidian blades and flakes brought in from the highlands in trade. If this interpretation is correct, then this Chantuto phase represents village agriculture—my system E. Because this system is the end of my model, the Chantuto materials date after the relevant development and are of little help in testing hypotheses about how and why agriculture came into being.

West's final Pacific lowland category, the dry interior tropical basin (subdivision 13), has two subdivisions—the Balsas-Tepalcatepec River Basin (13a), where preceramic remains have yet to be found, and the Valley of Chiapas (13b). The region around the Ocozocautla Basin of Chiapas, where Santa Marta and Los Grifos caves were excavated, can be included in this subdivision, but here again there are problems.

Santa Marta Cave, excavated in 1959 (MacNeish and Peterson 1962), has yielded a few early Archaic remains from five floors, two of which were dated 6780 B.C. ± 400 and 5370 B.C. ± 300. The INAH group that worked at the site next found earlier Archaic remains in Santa Marta (Santamaria 1981) in addition to Hunting-Collecting Band remains in Los Grifos (Santamaria and Garcia-Barcena 1984). As yet these finds have not been adequately reported nor has adequate analysis been published to determine ancient subsistence systems. Potentially we have a sequence from Hunting-Collecting Bands (system A) to, possibly, Foraging Bands (system B, C, or D) but little or no reported evidence of a transition to village agriculture, although zone D of Santa Marta Cave has been dated at 1330 B.C. ± 200 and has Early Formative sherds (Cotorra phase). Further, the ecology of the region has not been studied adequately. My brief foray into it, however, suggests that it does not fit the lowland pattern of regional diversity, nor does it have areal uniformity and/or many lush resource areas that can be exploited from a single base. In fact, there are suggestions that its ecological pattern is much like that of the highlands. Thus, while the Pacific coastal lowlands region has great potential, as yet we cannot use information from it to test my hypotheses.

The same sort of thing may be said of the Gulf Coast lowlands, which has only two regions with Archaic remains—West's deciduous forest area of northern Veracruz (11d) and the Petén-Yucatán (Belize) rain forest area (11a). Jeff Wilkerson has investigated the Santa Luisa site on the Tecolutla River in central Veracruz. His preliminary ecological studies suggest that this subarea has all the lowland characteristics of my model. Wilkerson's site yielded preceramic remains, which he calls Palo Hueco, in its earliest stratified deposits (Wilkerson 1975). In these early deposits, with dates of 2930 B.C. ± 100 and 2150 B.C. ± 800, there were abundant mollusks, seafood remains, cracked cobbles, flakes and core choppers, a drill, gravers, many El Paraiso and Querétaro crude obsidian blades and flakes as well as net sinkers, but no grinding stones. This evidence suggests a possible Village Forager (system C1) occupation at about 3000 B.C. Unfortunately, the crucial remains indicating how these alleged Village Foragers became Village Agriculturists are unknown, as is evidence of the earlier stages showing the evolution from Hunting-Collecting Bands to Foragers.

Secondary Development

Perhaps the lowland materials that are nearest to being relevant or usable for testing hypotheses are those from the northern coast of Belize. These materials, however, have dating problems, and the final necessary interdisciplinary studies (particularly of the subsistence and ecological aspects) never were funded and thus were not completed. About 90 to 100 sites of Wilkerson's and my survey of 260 could be classified in one of six tentative preceramic phases. We dug nine stratified sites with twenty-six stratified zones, many with more than activity areas. Although these provided a sound chronology (see table 3.2), unfortunately they were cross-dated rather than directly carbon-14 dated.

Let me summarize our Belize findings (MacNeish, Nelken-Turner 1983b), emphasizing the ecological subsistence and settlement patterns and the population data, the factors most relevant to testing secondary development of Centers.

1. *Lowe-ha* (9000–7500 B.C.). Six Lowe-ha phase components were excavated in the bottom levels of five sites; stratigraphically, therefore, this phase is the earliest in our sequence. Soil studies suggest that these strata were laid down at the end of the Pleistocene, some 10,000 years ago, while the end scrapers and points—Madden Lake fishtail, El Inga fluted, and Plainview—indicate that this complex existed roughly between 9000 and 7500 B.C.

Five of the six components were found on sandy ridges with pine and savanna vegetation. The sixth was on a ridge with sandy humic soil and a covering of deciduous and marshy seasonal forest vegetation. All of the six excavated components were relatively thin deposits existing in small areas. Further, the surface sites covered equally small areas. The limited numbers of artifacts as well as the thin occupation strata suggest that stays were brief; in other words, the Lowe-ha people were Hunting-Collecting Microbands. As yet, we cannot determine precisely what their seasonal scheduling was.

Because we have not yet studied the floated plant remains or investigated the phosphate in the soils, we can draw only extremely tentative conclusions about their subsistence system. Preliminary analysis of artifacts suggests that hunting was possibly more important than plant collecting. Reliable conclusions about subsistence and food preparation, however, must await intensive use-wear studies of the limited Lowe-ha remains. These were begun by George Odell and his team but have not been completed. To date, we have use-wear evidence for such activities as cutting, slicing, chopping, butchering, flint knapping, and scraping.

2. *Sand Hill* (7500–6000 B.C.). Sand Hill, our next phase, is much more adequately represented, having eight stratified components and eight surface components with many more artifacts and ecofacts, which were found in more than a dozen activity areas. We still do not have good chronometric dates for this phase. While the number of Sand Hill components is twice the number of Lowe-ha components, the former did not encompass a much larger area, suggesting that we are still dealing with a basic microband population. The Sand Hill population differs, however, in that the sites occur in a much larger number of ecozones. Artifacts in occupations at one site seem to pertain to the wet season, while others, on the coast, may belong to the dry season. Definitely determining that these people were exploiting different ecozones in different seasons, however, will depend on analysis of pollen and flotation samples.

Analysis is also needed to determine subsistence. Preliminary analysis suggests that plant collecting increased at the expense of hunting, and our coastal site may indicate some sort of marine adaptation. Evidence also exists of other activities including butchering, food preparation, and skin working. The flint-knapping technique is rather different, much of it emphasizing the making of macroblades. Some of these macroblades (as well as various snowshoe-shaped adze-like tools) indicate that woodworking, possibly including boat building, was now a major industry. There is also evidence of bone working.

3. *Orange Walk* (6000–5000 B.C.). Of all our tentative phases, Orange Walk is represented by the most artifacts, but it has fewer excavated components (five) from only four sites with about ten or eleven activity areas. Its stratigraphic position over Sand Hill at three sites and under Belize remains at site BAAR 31 firmly establishes its chronological position. Good chronometric dating does not exist; however, by cross dating La Mina, Trinidad, and San Nicolas points, scraper planes, and grinding-stone types, we can estimate that it existed in the general period from roughly 6000 to 5000 B.C.

Although the five excavated components are all of modest size, some of the many surface sites found during reconnaissance were much larger and perhaps were macroband encampments. There is certainly a suggestion of some sort of population increase. Both microband and macroband sites were located mainly inland on sand ridges with pine forest savanna vegetation. Some of these seem to suggest occupation during the wet season, while the relative lack of grinding stones suggests the coastal sites may have been dry-season occupations. One microband and four macroband sites were found on islands off the coast, while a single macroband occupied a coastal freshwater lagoon near the sea. Definite evidence of settlement patterns, however, await soil and pollen analyses.

Our artifacts and use-wear studies, most of which are incomplete, do give some hints about subsistence practices. Plant collecting again was on the rise, and grinding stones suggest seed collecting had become more important than hunting. Other activities, such as food preparation and the working of skin, bone, and wood continued much the same, although the types were slightly different. One new activity was the importation of obsidian (perhaps from the distant Guatemala highlands) and of tufa for grinding stone (perhaps imported from the nearby Maya Mountains).

4. *Belize* (5000–4000 B.C.). From many standpoints—particularly in terms of the number of artifacts and components—the Belize and later phases are not as well documented as the earlier three phases. Only three Belize components were excavated, two of which contained few artifacts. Absolute dating is still lacking, but cross-dating stone bowls with the Coxcatlán phase of Tehuacán allows us to estimate that the phase existed somewhere between 5000 and 4000 B.C.

Although the number of excavated components was small, those found in the survey were as numerous as those of Orange Walk. There is a suggestion of a changing settlement pattern and an increasing population. Apparently there were wet-season forays of microbands into the sand-ridge country away from their dry-season macroband base camps, which lay on sand ridges along rivers and on the coast. Artifacts suggest hunting was on the wane; and although coastal sites may point to some exploitation of marine resources, most of our positive evidence indicates that use or collection of vegetal materials not only was on the increase but perhaps was a dominant subsistence activity.

Other technological activities were much the same as Orange Walk, but butchering, skin working, and the making of macroblades were on the wane. The flint-knapping techniques changed slightly, and there was more importation of both obsidian and volcanic rocks for stone tools.

5. *Melinda* (4000–3000 B.C.). Although more extensive Melinda materials from the Betz Landing site were carbon-14 dated from 2700 to 1200 B.C., and the materials cross-date with preceramic remains at Arenal, Costa Rica, to 3000 to 3500 B.C., we still feel uncomfortable with our definitions of this phase (MacNeish, Wilkerson, Nelken-Turner 1980). There were few diagnostic Melinda tools, and our remains could still be aceramic rather than preceramic.

Assuming that this artifact complex is real, it would appear that the settlement pattern for Melinda is quite different from that of previous horizons. Half of the components were in maritime environs, while three were in riverine ecozones. Only two sites, on a sand ridge of Lowe's Farm, were far from the sea. Further, most of the Melinda sites are much larger than previous sites. Two of them were especially large and had rocks and wattle and daub, an indication that they had architecture and permanent settlements. During the Melinda phase, then, in addition to moving closer to the sea, people may have been becoming more sedentary. In fact, this phase could represent Village Foragers—system C1.

The Melinda subsistence system was apparently as different from previous phases as its settlement pattern was. Projectile points were rare, but deer bones and antlers indicate that hunting was still being undertaken. Much more numerous, however, were shells of both freshwater and saltwater species, suggesting that shell fishing had become a more important subsistence activity. Net sinkers were another sign of fishing. A number of petaloid hoes and picks showing ground polish indicate digging (possibly for roots).

Despite considerable evidence for plant collecting, more definite proof of subsistence patterns must await the identification of the floated plant remains, the pollen and phosphate studies, and the amino-acid studies of the bone and shell. So far, there is no evidence among the plant remains of any kind of plant domestication or agriculture. I would speculate that the Melinda people subsisted mainly on plant foods and maritime resources, perhaps on a seasonal basis; in other words, specialized foraging of system C1. Other activities changed slightly from trends that started in the Belize phase.

6. *Progreso* (3000–2000 B.C.). While we have some qualms about the Belize and Melinda phases, we have even more about Progreso. Stratigraphically it lies between preceramic Melinda and the ceramic periods, and many Progreso artifacts resemble those of Swazey, the earliest ceramic period (1100–800 B.C.) so far defined for Belize. For the moment, we estimate that Progreso existed in pre-Swazey late-preceramic times, from roughly 3000 to 2000 B.C.

Real data on Progreso subsistence must await further analysis. Projectile points suggest hunting, and large net sinkers and coastal sites indicate the use of maritime resources. Milling stones and mullers may indicate seed collecting, and scraper planes suggest plant scraping and plant collection, while large ellipsoidal hoes may be a sign of root digging.

Perhaps the most important artifacts (because they suggest a change in the economy) are the manos and metates. In Mesoamerica such tools are almost always used to grind corn; their presence is therefore indicative of corn agriculture. Supporting evidence comes from nearby Lamini, which is dated at 3000 to 2000 B.C.—roughly the time of Progreso. The Lamini finds contain corn pollen (Pendergast 1971, 1975, 1981a, 1981b).

The hint of corn agriculture in Progreso times at Lamini and at Pulltrouser Swamp, where the first corn dates to about 2500 B.C., is not in conflict with the limited settlement-pattern data available for Progreso locations similar to those of the Maya Formative. Furthermore, there are huge sites on points projecting into freshwater lagoons, and these have good agricultural soils. Most of these sites, however, still seem to have been mainly maritime. Three large ones and one small one were on islands in the sea; others were located on coves. Only three sites (at Ladyville) were still in the sand-ridge country with pine-forest savanna vegetation, but even these are only a kilometer or two from salt water. In this way the Progreso settlement pattern may have changed, and the population grown still larger.

A number of the technological aspects of Progreso seem rather different from those of previous periods, although the same general activities were undertaken. Unfortunately, our funds ran out before our use-wear studies were completed, so it is better not to attempt to describe Progreso technology at present.

Now the question becomes, Can this rather incomplete Belizean coastal preceramic data be used to test my model and hypotheses for secondary developments in Centers? Although the data are incomplete, they suggest that the Lowe-ha people had a hunting-collecting system—type A of my model—and slowly

Table 3.3. Necessary and sufficient conditions as a positive-feedback process for secondary development in coastal Belize

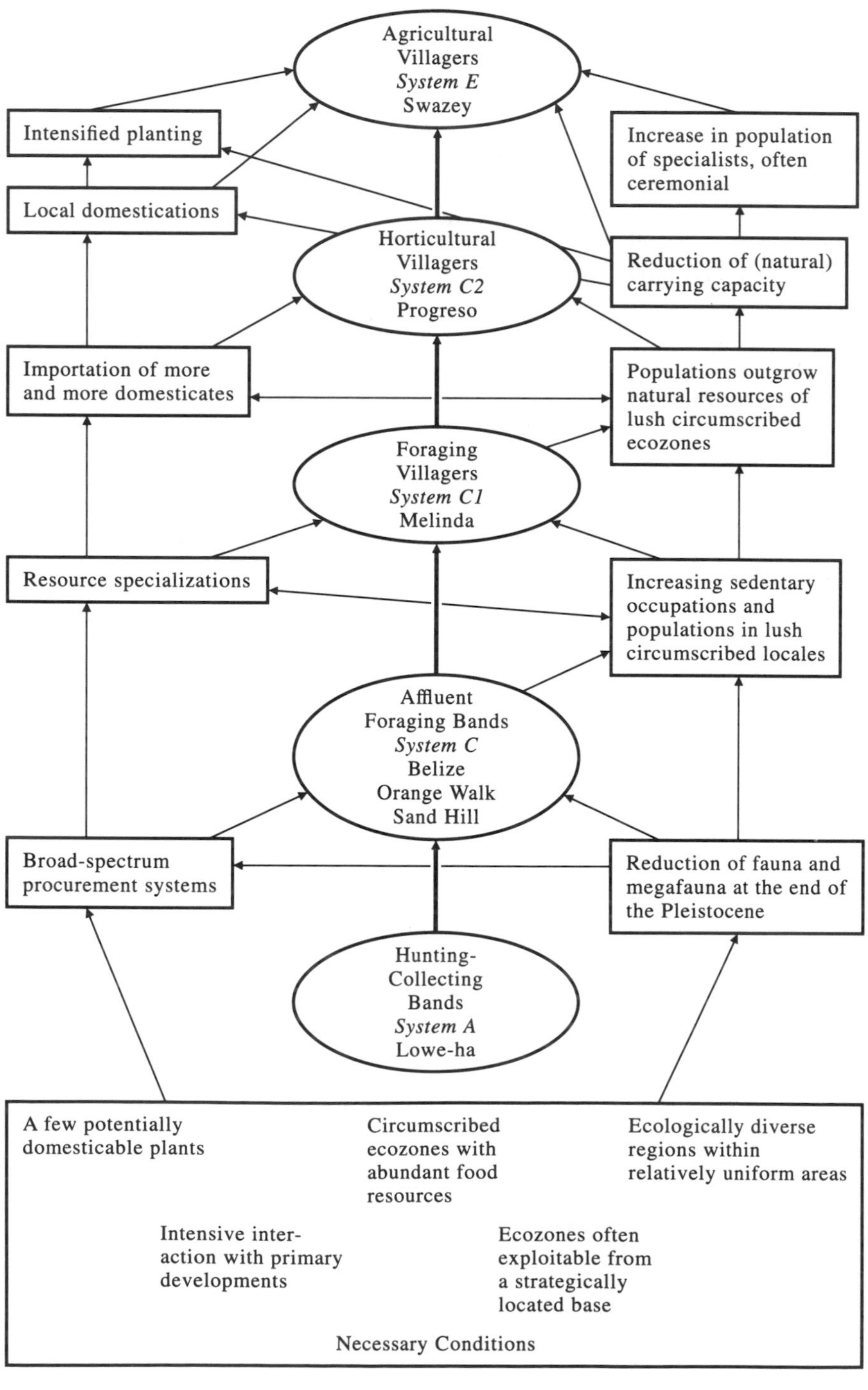

developed via the Sand Hill, Orange Walk, and Belize phases into Foraging Villagers (system C1), as exemplified by the Melinda phase. The data further suggest (although even less clearly) that the Foraging Villagers of Melinda evolved into the Progreso phase of Village Horticulturists (system C2), who became Village Agriculturists (system E) in the early ceramic Swazey phase. Thus, the Belizean data seem to confirm the secondary developmental model (table 3.3).

The necessary conditions for the change from Village Foragers to Village Agriculturists seem to pertain, even though the proposed interdisciplinary studies of the ecology of coastal Belize were never undertaken. Certainly, coastal Belize seems to have regional diversity with areal uniformity. From one river valley to the next the environment seems quite uniform, yet within each valley are marine and deltaic ecozones, including mangrove swamps along the rivers near the shore and surrounding salt marshes. Upriver are pine-forest savannas and narrow- or broad-leaf jungles, depending on the soils and topography. Often the whole set of ecozones, including certain relatively circumscribed marine zones that easily could provide multiple marine resources, can be exploited from one riverine base. Furthermore, evidence of obsidian trade indicates that this Belizean area had interaction with highland Guatemala or Mexico, where plants were domesticated early. In this way the Melinda people may have been foragers with resource specialization in marine life as evidenced by net sinkers and the site locations.

Now comes the question, Are the sufficient conditions of my model sustained by data involved in the shift from Melinda (Foraging Villagers, system C1) to Progreso (Horticultural Villagers, system C2)? While there is limited evidence of an increase in population from Melinda to Progreso, our research (MacNeish, Wilkerson, and Nelken-Turner) does not adequately demonstrate that the inhabitants had decreased the carrying capacity or outgrown the resources of any of their lush, circumscribed ecozones, which might have caused them to practice agriculture (as evidenced in the corn-pollen profiles at 2500 B.C.) during the time of Progreso.

All in all, the evidence concerning Progreso and Melinda is not sufficiently rigorous to test my secondary developmental hypotheses, although the general developmental model does seem to pertain. My hypotheses are not, of course, therefore invalidated, but at present they cannot be tested adequately.

Primary Development

In contrast to the lowland data, the data for the highlands of Mesoamerica do seem to confirm my primary developmental hypothesis. In fact, the highland data provide a better test and a more adequate validation of this model than do the data from the Andean area.

The first data (and, I might add, the least convincing) came from a region that West classifies as part of the extratropical dry lands—the Tamaulipas subhumid lowlands (subdivision 4). Ironically, my major investigations were done not in the Tamaulipas lowlands but in the highlands—the Sierra de Tamaulipas. According to my limited ecological studies, this region has all the necessary conditions of my highland model. Tamaulipas has coastal, steppe, and mountain zones, which give it obvious areal ecological diversity. In fact, the Sierra de

Tamaulipas region, particularly the Canyon Diablo locale, where I did most of my excavation, has even more diversity. Its mountaintops have oak-pine forests; the high slopes, deciduous forest; and the lower parts of the valley, thorn forest or steppe (depending upon soil and topography). In the valley bottom are either gallery forests or oasislike vegetation around spring waterways.

Because of their ranges of elevations, not all of these zones can be exploited from a single base. The great seasonality of the region, which includes a harsh dry season, means that people must move from one locale to another to get sufficient food. Among the native foods are wild squash, wild pumpkins, and scarlet runner beans, all of which are potentially domesticable. In addition to these characteristics, the region has had continuing interaction with the Huasteca in northern Veracruz and the Sierra Madre, both of which had domesticates early. This part of Tamaulipas thus has all the necessary conditions for primary development in the highlands. What is more, even the badly out-of-date geological studies by Glen Evans, Helmut de Terra, and Harold DeWitt still indicate that at the end of the Pleistocene this environment underwent a change (MacNeish 1958, pp. 195–99). Along with the reduced availability of such food staples as horse, buffalo, and mammoth, rainfall diminished, making the harsh season harsher. Both are sufficient conditions for the shift from Hunting-Collecting Bands (system A) to Destitute Foraging Bands that used domesticated and/or cultivated plants (system B).

Let us examine the whole Sierra de Tamaulipas sequence to see if the other sufficient conditions pertain (MacNeish 1958).

1. *Diablo Complex* (20,000 B.C. ± 10,000 years). The Diablo complex is represented by two laterally worked flakes, a split-pebble end scraper, and eight crude bifaces in gravels of the upper terrace, which suggests a late Pleistocene date, since horse bones also occur in the deposits. Because materials are limited and occupational contexts poor, little can be said about the way of life of these peoples.

2. *Lerma Phase* (8000–6800 B.C.). Five excavated sites and one small surface component of the Lerma phase were dug. Layer 5 of Diablo Cave was dated at 7320 B.C. ± 500. Artifacts were limited. Bones included those of mazama deer and beaver (no longer in Tamaulipas) and white-tailed deer and a few small mammals. These were in association with Lerma points, suggesting that hunting was a major activity; however, some plant collecting still may have occurred. Many butchering tools and snub-nosed and stemmed end scrapers for working animal skins further indicated the use of animals for food. Other artifacts suggest butchering, skin working, and woodworking, and there are gravers that may indicate tailoring. Because neither the tools nor the debitage have been subjected to use-wear studies, reconstruction of technological activities is not very accurate.

Bones were not numerous, occupational layers were thin, and areas of refuse were small. These data suggest that occupations were brief and by small groups. It is probable that the Lerma people were basically Hunters who lived in nomadic microbands—my system A.

3. *Nogales Phase* (5000–3000 B.C.). There were five excavated components and thirteen surface sites of the Nogales phase, and five of the surface sites were

large enough to suggest that they were occupied by macrobands. Although there is no direct evidence of seasonality, the seasonal data from the following La Perra complex suggest that the Nogales phase had some sort of seasonally scheduled micro-macroband settlement pattern. The increased number of components also suggests a population increase.

Projectile points found in conjunction with bones of white-tailed deer and jaguar indicate that hunting still occurred. Coatimundi bones indicate possible animal collection or trapping. The presence of mortars, mullers, milling stones, and pebble manos suggests that seeds were collected and ground; large plano-convex end scrapers, disc scrapers, and gouges may have been used on other plants collected.

Other technological activities—skin, bone, and shell working—occurred, as did flint knapping and the making of ground-stone tools. The presence of shells suggests trade with the Gulf Coast. These artifact types give ample evidence that Nogales evolved into La Perra.

4. *La Perra* (3000–2000 B.C.). The five excavated components of La Perra phase all look like microband seasonal occupations; the three surface sites could be macroband occupations, but our sample is small and inadequate, and one cannot tell if population grew, shrank, or stayed the same since Nogales times.

Among the five components excavated, one had preserved plant remains, perishable tools, and feces. The preserved foodstuffs and feces allowed us to reconstruct La Perra food consumption fairly accurately. In terms of bulk, we estimated 76 percent for wild plant foods, 15 percent for meat, and 9 percent for domesticates—Nal-Tel corn and pumpkins. Putting these into food values—calories and proteins—meat would yield perhaps 20 percent, domesticates 25 percent, and wild plants 55 percent or less.

These foodstuffs plus certain tools (and ethnographic analogy) suggest what the La Perra subsistence activities were. White-tailed deer, tropical red deer, peccary, and jaguar bones found in conjunction with atlatl dart tips indicate hunting was still practiced. Canis, rabbit, rodent, and coatimundi bones found with slip loops of agave suggest that these animals were trapped, while turtles and snakes may have been collected.

Squash seeds (*Cucurbita pepo*) and small corncobs (primitive Nal-Tel) indicate that the La Perra peoples did some gardening. However, the small number of domesticated plants in proportion to the large amounts of wild foods shows that gardening was an unimportant activity.

During the course of the La Perra occupation, a child died and was buried in a pit in the back of La Perra cave. Another small pit with a petate (mat) in it could have been used for storage.

Among the normal technological activities undertaken were flint knapping, food preparation, skin scraping, and wood and bone working. In addition, the people made agave string, loop-twilled mats, nets, and coiled baskets. Conch shells give evidence of trade with the coast; the corn and squash also could have been originally imported.

5. *Almagre* (2200–1500 B.C.). Nine excavated and seven surface components of the Almagre phase were found. Seasonal cave and camp occupations were

most common, but two sites seem to have been villages with wattle-and-daub houses. Only one temporary camp and one cave occupation were found. The dominant community pattern has been classified as semisedentary macrobands. An increase over the previous horizon in agricultural activities may have contributed to the size and sedentarism of groupings, although in leaner months (evidenced by the two temporary occupations), microbands may have continued to separate off to hunt and forage. At this time the villages or base camps (so designated by their wattle-and-daub remains) probably had not yet achieved true permanence, for it seems likely that agriculture was still too limited to rely on, and the depletion of local resources by intensive food collecting would have necessitated moving from time to time.

Comparison with La Perra (its ancestor) and with Flacco (its relative in southwest Tamaulipas) has enabled us to estimate the subsistence pattern of Almagre. It seems likely that 70 percent of the food, by bulk, came from collecting. Twenty percent was from agriculture and 10 percent from animals.

Burials seem to have been treated in the same manner as those of La Perra. Two flexed burials with a few chipped artifacts were found in pits in the back of the caves.

The other industries of the Almagre people generally seem to have been similar to those of La Perra and Nogales, although the artifacts were different. Obsidian again indicates trade, as do the domesticates.

6. *Laguna* (500 B.C.–0). There may have been a long temporal gap after Almagre. The Laguna phase that followed it is, in number of components and artifacts (mainly ceramics), the best represented in the Sierra de Tamaulipas. Eight excavated and fourteen surface components are from this phase. All these sites, except the La Perra Cave occupation, are large ruins with round house platforms and pyramids. Certainly village life was a fact. Furthermore, preserved foodstuffs indicate that these people were full-time agriculturists. Technological activities, including pottery making, were well advanced. There were a few links to the early preceramic. Not only had village agriculture diffused to Tamaulipas from the southwest in Mesoamerica, but in the millennia before this phase there must have been a gradual development of such a way of life, and understanding of this development would be crucial to our understanding of the emergence of village agriculture.

The Sierra de Tamaulipas does seem to have some of the conditions for the development from Hunting-Collecting Bands to Incipient Agricultural Bands, thereby tending to confirm my primary hypothesis. However, certain crucial data or conditions were not found, among them seed storage pits, evidence of genetic changes of potential domesticates, and population increases. The shift from Incipient Agricultural Bands to village agriculture is therefore not well demonstrated here. Like the Belizean data, data from the Sierra de Tamaulipas are not sufficiently rigorous to test my developmental hypothesis adequately. They do suggest, however, that the model about the primary development of incipient agriculture is perhaps valid. In no way do the data invalidate my hypotheses; they do suggest the need to collect the information required for an adequate test of my primary models.

Somewhat the same may be said of the next subarea under consideration—the Oaxaca valley region of the Mesa del Sur. Here Kent Flannery and his group made a great start in Central Oaxaca, but there are bad gaps in the sequence, and the results of all their rock-shelter excavations some twenty years ago need to be published in full.

The ecological studies of Kent Flannery, James Schoenwetter, C. Earle Smith, and others (Flannery 1986) indicate that the Oaxaca region of the Mesa del Sur has all the environmental characteristics and necessary conditions of my primary development hypothesis. As figure 7 shows, the Mesa del Sur has the following necessary conditions:

1. Great regional ecological diversity and areal differences from the ecozones of Tehuacán (to be discussed shortly).
2. Potentially domesticable plants, such as wild corn (in Schoenwetter's early pollen profiles [Flannery, 1986]), pumpkins, and scarlet runner beans.
3. Slopes sufficiently steep and valleys wide enough that multiple resources are inaccessible or nonexploitable from a single base.
4. A harsh dry season when few food resources are available.
5. Relatively easy access to areas (such as Tehuacán) with domesticable and/or early domesticated plants.

Moreover, Schoenwetter's pollen studies (Schoenwetter 1974; Schoenwetter and Smith 1986; Flannery 1986), Robert Whallen's ecological analysis (Whallen 1973), and Robert G. Reynolds's simulation model (Flannery 1986) show that in the early part of the sequence (in Naquitz times), the environment was changing. This change reduced the availability of food staples and made the harsh (i.e., dry) season harsher (drier) as well as the wet season shorter and less predictable—good sufficient conditions for the shift from Hunting-Collecting Bands (system A) to Incipient Agricultural Bands (system B1). Now let us examine the preceramic sequence of Mesa del Sur to see if the other sufficient conditions pertain.

1. Possible *Ajuereado* (10,000–8000 B.C.). Surface collections of Lerma points in Cueva Blanca zone F and the same Ajuereado-type of extinct Pleistocene fauna—giant tortoise, fox, and jackrabbit suggest we have the presence of Hunting-Collecting Bands—system A—in the Oaxaca Valley. During this time, as well as in the following early Naquitz phase, the Pleistocene was coming to a close, the above animals and others were disappearing, and the valleys were getting drier. Although more investigation is needed, even these limited facts show that the valley fits my model.

2. *Naquitz* (8000–6000 B.C.). Zone E of Cueva Blanca and zones C, D, and E of Guilá Naquitz Cave are components of the Naquitz phase. Unlike Flannery, I would place zone B of Guilá Naquitz in the Jicaras phase, because of dates and the presence of Pedernales points and scraper types. All the Naquitz sites are microband occupations. The Cueva Blanca site seems to have been occupied in the winter; those of Guilá Naquitz, in the fall. This suggests a seasonally scheduled micro (macro?) band settlement pattern.

Bob Whallen (1973) has speculated about the seasonally scheduled subsistence pattern, but the archaeological evidence for it is suggestive rather than positive, for we lack well-documented, excavated sites in the ecozones that he sees as being exploited. Certainly the association of deer bones with points suggests hunting, and zone E of Cueva Blanca could well be a winter hunting encampment. Cottontail and turtle bones suggest collection, and the broken leg bones of the rabbits hint that they were trapped.

The refuse from Guilá Naquitz, however, suggests that plant collecting was a major activity (at least in the fall). Found there were maguey quids, mesquite beans, wild runner beans, opuntia, wild onion, *nance, susi* nuts, pine nuts, and acorns; some of the last-named occurred in two pits in zone D of Guilá Naquitz. Grinding stones and scraper planes, for the preparation of such food, were present. Also present were pollen grains of maize (probably wild) in the earliest zone C levels of Martinez Cave (Schoenwetter and Smith 1986) and other grains that Flannery (not Schoenwetter) called teosinte (Flannery 1986). The presence of a few pumpkin seeds indicates that some plant domestication had begun, but we do not yet have evidence from these levels of how this process may have occurred. Other technological activities such as flint knapping, butchering, et al., are indicated by the few illustrated artifacts.

3. *Jicaras* (6000–5000 B.C.). Zones B1, B2, and B3 of Guilá Naquitz, a fall microband encampment, and zones A and B of Gheo-Shih, a summer macroband encampment, suggest that the macro-microband settlement pattern was firmly established by this time, but whether there were population increases we cannot know without more adequate reconnaissance.

Although there is more evidence of storage pits, the subsistence pattern seems to have been about the same. The food resources from zone B1 of Guilá Naquitz, however, do not represent an adequate year-round sample. Again, deer bones in association with Pedernales, La Mina, Trinidad, and San Nicolas points suggest hunting; the same types of wild plant remains indicate plant collecting, and pumpkin seeds suggest incipient agriculture.

Other technological activities occurred, but we must await adequate description of them. Again, in addition to written description, more "good old cave digging" of the remains of this phase needs to be undertaken in Oaxaca.

4. *Blanca* (5000–3500 B.C.). The only components of this poorly documented phase are zones C and D of Cueva Blanca. Because of the carbon-14 dates, Flannery has suggested that there is a gap between Jicaras and Blanca, but I feel the Blanca phase should be 5000–3500 B.C. Its point types—Tilapa, La Mina, Trinidad, and San Nicolas, of zone D—would, in Tehuacán, fit right into the period of 5000 to 4500 B.C. (where Flannery speculates there is a gap), and the Coxcatlán points of zone C would fit in the period 4500 to 3500 B.C. Still, the only settlement pattern data indicate that the Blanca people had winter microbands.

The points and associated deer bones suggest hunting, and the grinding stones hint at plant collecting and/or agriculture. The lack of preserved plant remains for this period is particularly unfortunate, because there were Foraging Bands with Incipient Agriculture in neighboring Tehuacán during this period.

5. *Martinez* (3500–2000 B.C.). This phase is represented by zone B of Martinez Cave, which was some sort of seasonal microband occupation or occupations, and perhaps by the Yuzanú open-macroband or hamlet site from near

Yanhuitlán, Oaxaca, and dated at 2000 B.C. ± 100. Again, there is a suggestion of hunting, collecting, and agriculture, but no concrete data have been reported.

6. *Esperadian Complex* (1800–1300 B.C.). In the Esperadian complex we have definite evidence of village agriculture with pottery, my system E. Thus, in Oaxaca we have the framework of the sequence of my primary development model from Hunting-Collecting Bands to Foragers with Incipient Agriculture to Village Agriculturists. This further confirms the hypothetical sequential model; however, while many of the necessary conditions of the primary developmental hypothesis pertain, most of the sufficient conditions are absent or inadequately documented.

The best early Archaic data at present come from the final two subareas—the Mesa Central and the Sierra Madre Oriental. Here is where most of our preserved domesticated preceramic plant remains were discovered (see Table 3.4).

Although we did not do interdisciplinary studies of the Ocampo region of the Sierra Madre in 1954, Paul Martin and his ecological group did a transect through the region from the high mountains of Gomez Farias to the east (Martin 1958). According to their classification, as one descended westward from the cloud forest of the highest peaks, one found a series of ecozones—cloud forest mountaintops, oak-pine deciduous forest at the high elevations, thorn forest slopes, and well-watered canyon bottoms with lush gallery forest or harsh oasislike patches. Thus the Sierra Madre of southwestern Tamaulipas has all the environmental prerequisites for a highland area in my model. Furthermore, all the necessary conditions, as described for the Oaxaca region, also pertain. The question becomes, Do all the sufficient conditions occur, thereby allowing me to test my primary development hypothesis? The following description of the sequence (see Table 3.2), shows that they do.

1. *Lerma* (8000–7000 B.C.). Although no Lerma phase remains were excavated in the Sierra Madre, a single surface component, the east end of the open site Tmc278, had a Lerma point in it. Lerma components also occur to the east, in the Sierra de Tamaulipas, as well as to the southwest, at El Cedral in San Luis Potosí. Thus we do have some remains of the Lerma Hunting-Collecting Bands (system A) in the Sierra Madre Oriental, and I hope future investigations will clarify this manifestation.

2. *Infiernillo* (7000–5000 B.C.). Thus far, the Infiernillo phase is known only from excavated components: occupation 1 of Tmc274, Ojo de Agua Cave; occupation 1 in layer 1 of Tmc247, Romero's Cave; and occupations 1, 2, and 3 of Tmc248, Valenzuela's Cave. All except occupation 2 of Valenzuela's Cave were short-term microband occupations. The numerous hearths of the latter suggest that it was a short-term macroband occupation. Most components seem to have been summer occupations in the caves.

Faunal remains occurred at all components; plant remains at four from Romero's Cave and Valenzuela's Cave; and feces in occupations 2 and 3 of Valenzuela's Cave. Therefore we can estimate food consumption as well as subsistence practices. Garbage remains suggest that foods from plants and animals made up about equal portions of the diet, although one feces suggested that two-thirds of the food was from meat and the rest from plants. The deer, skunk, and bison bones and associated Infiernillo Diamond and contracting stem-shaped Almagre points suggest hunting. Agave, opuntia, a few small setaria seeds, seeds and rinds of *Cucurbita foetidissima,* wild runner beans, and six small pumpkin seeds give evidence of the wild plants eaten and collected in Fuegian nets and baskets as well as in loop-twine nets.

In later levels were storage pits full of wild plant remains. One pumpkin seed in the latest Infiernillo level is larger and may be from cultivated and/or domesticated pumpkin, while two gourd rinds and pepper seeds were definitely from domesticated and/or cultivated plants—perhaps 1 percent of the diet if calculated in terms of bulk and 5 percent if calculated on the basis of calories and grams of protein.

Artifacts include scraper planes, flake butchering tools, flake choppers, netted and woven mats, fire tongs, digging sticks, and wooden atlatl fragments. A few shells suggest trade with the coast.

3. *Ocampo* (4000–2300 B.C.). This phase is represented by three occupational layers from Tmc247 and the same number from Tmc248, as well as by three surface sites (Tmc268, 275, and 285), and two other cave sites (Tmc247 and Tmc276). Most (seven) were short-term microband occupations, but two excavated components (occupation 4 of Tmc247 and occupation 5 of Tmc248) and one open site (Tmc285) were large enough and had enough hearths to be macroband seasonal camps. Most cave occupations were short-term seasonal habitations used in the spring or summer, but the latest one of the phase, zone L (occupation 4 of Tmc247), may have lasted from spring until well into the fall.

Throughout the phase, the proportion of meat in the diet seems to have decreased from perhaps 40 percent to 20 percent, but the hunting of deer, skunk, and coatimundi with Tortugas, Nogales, and Abasolo atlatl points continued. The Ocampo people used Fuegian and full-coiled nets and twilled and interlocking loop-coiled baskets to collect such plants as acorns, opuntia, agave, wild squash, runner beans, setaria, and wild pumpkin, which gave them more than 60 percent of their subsistence. Numerous pits provided food storage.

Throughout this phase, however, the use of domesticates increased from perhaps 5 percent to more than 25 percent. Some of these were stored. Gourds and chile peppers continued to be used, and fragments of corn (in feces) and two varieties of common beans appeared.

Callen's analysis of feces revealed two interesting trends. Throughout the phase the size of the setaria seeds increased until they became very large. Cucurbita seeds at the beginning of the phase were evenly divided between small and large; in later levels all were big. Both trends may be interpreted as indicating seed selection, as well as the planting of domesticates (squash) and cultivars (setaria).

Artifacts that indicate technological activities include gouges, antler wedges, bifacial knives, mortars, grinding stones, and shell beads. The Ocampo phase people were Incipient Agricultural Bands—System B1.

4. *Flacco* (2300–1800 B.C.). The Flacco phase was a relatively short one, represented by six floors of Tmc247, a single floor with plant preservation (occupation 6 of Tmc248), and two surface macroband sites. One of the floors—number 3 of Tmc247—had four hearths and many pits and may have been a macroband

encampment. Length of occupation is difficult to determine, but it was still seasonal.

Many trends of Ocampo continued into this phase. Feces suggest that meat constituted 10 to 20 percent of the diet. The hunting of deer, skunk, coatimundi, jaguar, and the like with Flacco and Gary-stemmed points seems to have been decreasing.

The collecting of agave, opuntia, acorns, setaria, aloe, tripsacum grass, and other plants continued at about the same level (50 percent), and the collectors used twilled and Fuegian baskets, simple coiled bags, and carrying loops. Some plants were prepared by grinding on milling stones with mullers and by pounding in mortars. Food from domesticates continued to increase (from perhaps 20 to 40 percent) with incipient agriculture, but a very subtle change was about to take place.

5. *Guerra* (1800–1400 B.C.). There were four excavated components of the Guerra phase, and all four—zone occupations 5 through 8—came from Romero's Cave (Tmc247). Two were microbands; the others macrobands. All represented lengthy occupations lasting from two to four seasons. Further, two of the open sites were macroband sites. One had some wattle and daub, suggesting that it was a permanently occupied hamlet. Although the evidence is not overwhelming and more investigation is needed, there seems to have been a trend toward longer occupations by larger groups of people.

This trend seems to have been the result of increased agricultural produce (often found in the storage pits in our excavations), which, according to Callen's feces analysis, constituted 25 to 50 percent of the Guerra people's diet. In addition to such older domesticates as gourds, pumpkins, common beans, chile, sunflower, amaranth, and corn, domesticated plants included annual Río Balsas teosinte, cotton, and warty squash. Corn increasingly dominated the diet, however, and about half of this corn was the hybrid of corn and teosinte called *Tripsacoid chapalote*. Interestingly, the large setaria seeds had disappeared. Callen believes that people no longer cultivated it because corn was more productive.

Large amounts of food (40 to 50 percent) still came from such wild plants as agave, opuntia, setaria, aloe, and runner beans, but these foods were on the wane. Also diminishing was meat from deer, now hunted with Matamoros, Catan, and Palmillas corner-notched atlatl points. Collectors used split-stitch interlocking loop and simple-stitch baskets and simple nets. Well-decorated twilled mats, cotton loom weaving, and other techniques came into use. The Guerra people were Agricultural Bands (system B2) on the threshold of village agriculture.

6. *Mesa de Guaje* (1400–1100 B.C.). Although the sample from the Mesa de Guaje phase is not large, there were microband and macroband occupations from the excavation of Tmc247, a large macroband and/or hamlet site (Tmc347), and a definite hamlet with stone house platforms (Tmr233, which I tested in 1955). These leave no doubt that these pottery-making people lived mainly in hamlets or villages. In addition, the preserved foodstuffs indicate that almost 50 percent of their diet came from agricultural plants. Village agriculture—system E—had definitely arrived in the Sierra Madre.

This sequence from Tamaulipas confirms my model well and supplies much data for testing to validate my hypotheses. However, before doing that, let us consider the key sequence in the Mesa Central—that from Tehuacán—which further tests my primary development hypothesis.

The Mesa Central subarea has yielded the most Archaic materials, and they come from a variety of sites in Hidalgo, Querétaro, the Valley of Mexico, and Texcal in northern Puebla, as well as from Tehuacán. All the sites have similarly diverse highland environments. The data from Hidalgo and Querétaro, however, have never been reported, and in Texcal Cave material was from arbitrary levels (Garcia Moll 1977), so none of that data can be utilized to reconstruct the sequence of ancient subsistence systems.

The information from the Valley of Mexico, although more thoroughly reported by Christine Niederberger (Niederberger 1976), is not much better. The earliest relevant phases, Playa 1 (5000–4500 B.C.) and Playa 2 (4200–3800 B.C.), are represented by few artifacts and almost no ecofacts, although pollen suggests amaranth may have been grown in both periods, and corn (and/or teosinte) in the later one. Further, Niederberger has interpreted these occupations as sedentary, which I have shown they were not (MacNeish 1981c). The Zohapilco phase (3500–2300 B.C.) may have had year-round occupations, but it is poorly documented.

Fortunately, all the materials from the Valley of Mexico seem related to those from Tehuacán, and there we have abundant data, so that both the environmental and the cultural sequences can be reconstructed with some accuracy. Since both have been described in considerable detail elsewhere (Byers 1967a), I shall only outline them here.

Seven general ecozones exist in the Tehuacán survey area, the highest being the eastern mountainous oak-pine ecozone. It has lush vegetation and abundant animal life, because it caught rain and moisture from the Gulf Coast. This mountainous zone puts the next ecozone to the east into a rain shadow, forming the alluvial slope thorn-forest, which covers not only the eastern slopes of the Sierra Madre but also the higher elevation and slopes of the hills on the west side of Tehuacán Valley. Rainfall in this zone drops dramatically from over 2,000 millimeters to 400 to 800 millimeters annually, and there are great seasonal variations in temperature, with frosts possible in the winter and a limited rainy season from June through September. Vegetation is mainly thorny trees, scrub, and grasses, which yields mainly wet-season foods and provides forage for herbivorous game.

The southeastern portion of the valley encompasses the narrow canyon and dissected alluvial slopes ecozones. Their flora and fauna are like those of the previous zones, but the annual temperature generally is warmer (without frosts), and during the spring and summer the canyons fill with water—often runoff from the mountains to the east.

These two ecozones contrast with another at roughly the same elevation in the northwest sixth of the valley. Called the travertine slopes desert ecozone, it is characterized by slightly lesser amounts of rainfall in a brief, rainy summer season. Its underlying travertine rocks have a limited soil cover that results in a Sonoran desert-like vegetation. Without irrigation, plant and animal foods are available only during the brief wet season.

The travertine slopes desert ecozone, however, surrounds the El Riego oasis, a small set of cliffs with springs, located at the north end of the valley. Here the vegetation is green and lush throughout the year, and animals—particularly in the dry season—are fairly abundant. The region is crescent-shaped and limited in size, being only a few kilometers long and less than a kilometer wide. This narrow strip of lush vegetation, which provides a haven for animals in the dry season, is similar to the humid river bottoms ecozone that flanks the Río Salado, which runs north and south through the Tehuacán Valley. Like the surrounding valley steppe ecozone, the humid river bottoms has slightly warmer temperatures but even less rainfall—less than 500 millimeters—which falls in a few brief, torrential showers during the summer. Because of the permanent water and rich alluvial soils, however, the humid river bottoms is a lush zone. Surrounding it is a barren grassland with a few scrubby trees; this zone yields food (grass seeds) mainly in the spring and wet seasons.

Obviously, the Tehuacán environment had great ecological diversity. It contained many potential domesticates, and its multiple resources could not be exploited from a single base. Its dry season was very harsh; in fact it became harsher after the Pleistocene, when horses, antelopes, mammoths, jackrabbits, and giant turtles disappeared, thereby reducing food staples. On the other hand, Tehuacán was centrally located; it had easy access to Oaxaca, the Valley of Mexico, and both coasts. Therefore it had all the necessary conditions of my primary development model. As we shall see in the following description of the phases, it also had all the sufficient conditions, or triggering causes.

1. *Ajuereado* ($12{,}000^{+}$–7600 B.C.). We discovered more than twenty Ajuereado-phase occupations, but only eleven came from excavation. Two were in the El Riego Oasis, eight stratified sites were from Coxcatlán Cave (four of which were in the then Pleistocene grasslands), and the four upper sites were in the thorn forest scrub, as was the one from the test in cave Tc391. All were microband occupations of short duration occurring in various seasons, but no real scheduled system could be determined. These were unscheduled seasonally nomadic microbands.

Subsistence activities varied slightly throughout this phase, because the early part of it occurred during the Pleistocene, when later extinct animals—horses, giant jackrabbits, antelopes, giant wolves, and giant turtles—still existed. At this time, meat was a major part of the diet (in terms of calories, over 70 percent), and it was garnered by rabbit drives, ambushing herds with lances, and stalking animals with atlatl darts. Later, as many animals became extinct, most of these practices ceased, except the stalking of deer with darts. The trapping and collecting of small animals increased, along with wild fruit picking (chupandilla), leaf (agave) and pod (opuntia) picking, and seed (setaria and other grass) collecting.

In addition, in some of the latest occupations we uncovered pits in which seeds that had been ground up with mortars and mullers were stored. This surplus probably helped carry the people through the lean seasons, which at the end of the Pleistocene had become even leaner because of the scarcity of animals.

Other activities included flint knapping and blade making, skin working with haftable end scrapers, woodworking, bone working (with burins), and agave string making. Obsidian from Orizaba indicated that trade occurred even at this time. In terms of my model, the early Ajuereado-phase people are examples of Hunting-Collecting Bands—system A.

2. *El Riego* (7600–5000 B.C.). Of the forty-one El Riego–phase components, thirty came from excavation and included occupations in all seven ecozones (in certain seasons). Ten were macroband camps; twenty-one were brief microband encampments; and ten were indeterminate. The El Riego components contain evidence of seasonal scheduling—that is, the preferential occupation of certain ecozones in certain seasons so that a seasonal macro-microband settlement pattern existed.

A study of the found materials seems to indicate seasonally scheduled subsistence options for certain environs—a major shift from the Ajuereado system. The stalking of animals with darts was a major subsistence activity, supplemented by ambushing with lances in the winter. Hunting techniques seem little changed from Ajuereado times, except that a wider variety of projectile point types was used.

The other two winter subsistence options, trapping and leaf cutting, both minor, continued in a lesser way throughout the year. These practices were little changed from Ajuereado times, although nets may have been used in addition to spring snares, and leaf-carrying equipment may have included carrying loops, nets, and baskets. Seed collecting, as important as hunting, began in the spring and continued into the summer. Fruit picking was a minor subsistence option in the spring and summer but became the equal of hunting in the fall. Some seed storage took place.

Although rabbit drives seem to have disappeared in the middle Ajuereado with the extinction of the Pleistocene jackrabbit, other subsistence options were about the same in El Riego as in Ajuereado; however, they occurred in different proportions and were neatly scheduled.

Only two new subsistence options occurred, both late in El Riego: seed planting, which had grown imperceptibly out of seed collecting and was used as a major option only in the summer; and fruit-pit planting, the harvesting of which occurred as a minor option in the fall.

One might speculate that the process of annual migration, which entailed a shift from seed collecting to seed planting, began as El Riego people returned seasonally to some seed or fruit area each year. This process would have led to some clearing, enrichment, and improvement of the habitat of the seeds and fruits. Selection by humans (out of the random genetic variation of the plant community) of larger seeds for food, together with the changes accompanying growth in the disturbed habitats of camp areas, would have led to changes—some possibly genetic—in the seed and fruit population. Eventually the process may have led to the use of the domesticates (genetically changed food plants) and, finally, to some planting of individual seeds (of amaranth, chile, *mixta* squash) or pits (avocado) in special plots or gardens (horticulture).

Another practice that would have developed is storage, which made life more secure in hard times or bad years and allowed larger groups to come together and stay longer at particular spots. In fact, the concept of storage might well be considered a necessary condition for the advances that took place in later systems.

In spite of these new subsistence options, El Riego subsistence technology and tools were little changed from Ajuereado times except for the tools used in food preparation. Butchering and barbecuing continued as before, but new chopper types and pebble hammers were used to crush bones for gruel. Food, particularly leaves, was roasted in special large pits. Some seeds were ground round and round in milling stones by mullers; others were ground back and forth on metates with manos. Seeds and nuts were pounded into a more palatable form in three kinds of mortars by conical, long, and cylindrical pestles. These activities usually took place during the spring and/or summer.

A number of other new activities were undertaken in spring and summer. One was making ground-stone tools by gouging, pecking, and grinding used chipped-flint tools. Another was making Marginella shells into beads by piercing them with chipped-stone drills or gravers. Also, paint (possibly red) was ground in small hemispherical dishes. Handmade string or yarn, often of chewed agave fibers, now was sometimes made of soft bast fiber and often woven into two-ply cord and four-ply rope. The rope was sometimes knotted, and some knotted nets, as well as knotless nets, were made into bags, while others were woven into twined baskets, blankets, bags, or kilts. Coiled baskets with interlocking or non-interlocking stitches were also manufactured.

Other activities similar to those of Ajuereado continued throughout the year, but some techniques and tools were slightly different. Bone tools now included three kinds of awls, as well as needles, hammers, and flakers. These were not only whittled and sliced but scraped, sawed, and drilled. Woodworking was much the same, but some sticks were ground to points, and flint gouges came into use.

Skins were fleshed by a new, wide assortment of bifaces, blades, and side scrapers, and they were scraped by new, small, haftable scraper types, as well as by larger scraper planes. Flint knappers continued to use some of the older techniques with the same tools; abrader hammers and antler hammers were used for new percussion techniques on prepared platforms, and retouching techniques were more widely used.

Group burials suggesting infanticide indicate that ceremonial activities were important. The presence of obsidian and shells is evidence of trade. In terms of my model, the El Riego phase was approaching the Destitute Foraging Band type—system B.

3. *Cocaxtlán* (5000–3400 B.C.). There were twelve Cocaxtlán-phase components from excavations and the same number from surface collections. The smaller number of occupations than in the El Riego phase may seem to indicate a diminution of population, but in fact, it does not. Twelve of the occupations were by macrobands, all for two or more seasons (on the travertine slopes, one spring-summer

and one fall-winter occupation; in the valley steppe, a single spring-summer occupation; in the humid river bottoms, two fall-winter and four spring-summer encampments; on the alluvial slopes, two spring-winter camps; in a canyon on the dissected slopes, a spring-summer open site).

The twelve microband occupations were much the same as those of El Riego, but one on the alluvial slopes lasted from spring until fall; one on the travertine slopes was spring and summer; and the other three were summer only. Two other single-season microband occupations were discovered in the oasis (in the dry seasons), and there were camps in the Lecha Diego Canyon (one summer, two spring, and two winter camps).

The Coxcatlán subsistence system was much like that of El Riego in terms of seasonal scheduling. Winter was still the time of hunting, mainly by stalking with darts but also by ambushing with lances. A new hunting technique came into being, however; it entailed the use of thin points with serrated edges and sharp barbs. Using this method, a hunter could wound an animal and then trail it until it bled to death. Leaf cutting and trapping also occurred in the winter, and all three activities continued during the rest of the year in a diminished way. Seed collection was still important in the spring and summer. Fruit and pod picking started in the spring and became dominant by the fall.

One major difference from the El Riego phase was the planting of cultivars or domesticates in the spring or summer. This practice yielded storable surplus foods for at least the summer. Unlike the previous phases, this one used a wide variety of plants, including corn, common beans, chupandilla, white and black sapotes, walnut and cashew squash, gourds, avocados, chile peppers, and amaranth. Avocado, chupandilla, and corn show gradual increases in size as well as genetic change throughout this phase, changes that definitely were the result of human manipulation.

There is considerable evidence that corn was first domesticated in or near Tehuacán and that it definitely did not evolve out of teosinte (Wilkes 1988). Although teosinte is the nearest wild relative of corn and has the same chromosome number, it has certain genes that are dominant and that *cannot* become the recessive ones of corn. The evidence that corn came from Tehuacán (or from nearby Oaxaca) is that corn, and Oaxaca's corn pollen (Schoenwetter and Smith 1986, p. 228, table 15.25)—not teosinte—has been dated earlier here than anywhere else (MacNeish 1985).

During the Cocaxtlán phase the inhabitants had more storable foods, which allowed bigger groups to spend more than one season in a place. They also began pounding their food in new kinds of stone bowls and grinding it back and forth with a number of different kinds of manos on better made metates.

The butchering and cooking of animals were done much as in the previous phases, but the tradition of making bone gruel by pounding was going out of style. Many of the year-round activities, such as woodworking and bone working, were the same as those of El Riego, although more kinds of products were manufactured. Flint knapping was only slightly changed.

Cremations indicate that ceremonialism was still practiced. Large amounts of obsidian, volcanic tufa, flint, and shell were signs that trade had increased greatly. The host of new plants—probably domesticated elsewhere—that this group began to use also is evidence of increased trade. In terms of my model, this phase is a prime example of foraging with Incipient Agricultural Bands (system B1).

4. *Abejas* (3400–2300 B.C.). Nineteen of the thirty Abejas phase components were from excavation. While there were still eleven microband encampments and sixteen macroband camps, seemingly all were occupied for longer periods of the year. In the humid river bottoms were found three large sites with architectural (slab) features, suggesting hamlets occupied not only year-round but possibly for a number of successive years.

Because food supplies were better in the Abejas phase, I believe that most of the macrobands (sixteen) could remain in one place for two or more seasons and that the three hamlets with pit houses could have been occupied year-round. I further believe that eleven microband camps (six of which were for a single season) represented groups who worked out from the larger base camps. Thus we have classified their settlement pattern as being the central-based band type.

Most of the information on Abejas subsistence (which comes from the four multiseason camps on the alluvial slopes and from the one on the travertine slopes) indicates that the seasonal scheduling of this subsystem was much like that of Coxcatlán. Trapping and leaf cutting were minor activities throughout the year, hunting dominated in the winter, seed collection was most important in the spring and summer, and fruit picking and hydro-horticulture (the planting of fruit trees on watered areas) were more important in the fall.

The planting of seeds of the same plants as during the Cocaxtlán phase, plus tepary beans, was only slightly more popular in the spring, summer, and perhaps fall than it was in Coxcatlán times. There was, however, an increasing emphasis on planting new, more productive types of hybrid corn. Some large concentrations of corn in certain activity areas in association with digging sticks, plus the occurrence of many of the larger sites near fertile flats, suggest that crops were beginning to be sown in fields (barranca agriculture). Cache and storage pits suggest that some seasonal surpluses were grown so that sites could be occupied longer, thereby allowing greater security.

In spite of a slightly larger population, a new settlement pattern, and a subsistence pattern that produced some surplus, the food preparation activities changed little. Butchering techniques, cooking, storage, and pounding and grinding round and round were about the same. The back-and-forth grinding method was still used, but large metates and manos requiring two hands appeared. For the first time there was some evidence that food was boiled in stone bowls.

Many of the previous year-round technological activities continued: bone working, fleshing, scraping, and tailoring. Although our sample is poor, there is a suggestion that less energy was expended on these activities, perhaps because of a greater use of woven fabrics. Woodworking remained about the same, using whittling, sawing, and grinding techniques on small sticks. Polished and rubbed pebbles seem to be new tools used for woodworking, and one carefully carved fruit pit was found.

One technological difference is that logs (perhaps for house construction) were chopped beaver-fashion with some kind of adze, perhaps the flake and discoidal choppers, hafted or unhafted. The other year-round activity, flint knapping, continued older traditions and also started a new one: the making of fine blades (often obsidian) by pressure flaking neatly made conical cores with a prepared striking platform.

Of the other technological activities that recurred from spring through fall, some—such as string making, knotting, and making twined mats—continued in much the same way, as did the making of ground-stone objects and the grinding of paint, although sometimes the latter was done on flat slab palettes. Disk-shaped stone beads were made in a new manner: stone cylinders were ground into shape, drilled from opposite ends, and then sawed into disks.

Perhaps the most important technological innovation of the Abejas phase was the construction of pit houses. Pits were dug—perhaps with wooden slab shovels—some five or six meters long, three meters wide, and one meter deep, and poles were cut for a tentlike roof frame. The poles were then tied or laced together and finally covered with a brush roof.

The presence of foreign objects suggests trade, of both the direct and indirect procurement types. Cremations in a few winter sites suggest that some sort of ceremonial system was still important.

We now seem to have Agricultural Bands (system B2), a transitional stage between Incipient Agricultural Bands (system B1) and village agriculture (system E).

5. *Ajalpan* (1500–900 B.C.). The Ajalpan phase had sixteen occupations (only two from the surface)—fourteen hamlet sites with wattle-and-daub houses and both a macroband and a microband encampment. Abundant foodstuffs indicate that these people were full-time agriculturists. Other remains show that they made pottery, figurines, and fine blades, built houses, and had a growing ceremonial complex. With the Ajalpan phase, Village Agriculturists (system E) arrived in Tehuacán.

The Tehuacán sequence and the previous one from the Sierra Madre of Tamaulipas both validate my hypotheses about the causes of primary development of village agriculture, thereby confirming the hypothetical model of development from system A (Hunting-Collecting Bands) through system B (Destitute Foragers) to system E (village agriculture). Now let us reexamine my hypothesis in detail by testing it with the best data available from highland Mesoamerica.

THEORETICAL CONSIDERATIONS

In my model I am looking for regions with great ecological diversity and resources that cannot be exploited from a single base. Also important are a harsh dry season, the presence of potentially domesticable plants, and access to areas with different resources.

Both the Tehuacán Valley of southern Puebla and the Canyon Infiernillo of southwest Tamaulipas have great regional diversity and are slightly different areally. To review, Tehuacán has an oasis (El Riego), humid river bottoms, valley steppes, travertine and alluvial slopes, canyons, and oak-pine forest ecozones. These zones contrast with those of the Infiernillo region, which ranges from a fertile spring and riverine gallery forest zone, through a steppe terrace zone and slopes covered by thorn and deciduous forest, to oak-pine forest and/or cloud forest ecozones on the mountain summits.

In Tehuacán some of the ecozones—for example, the oasis and dissected alluvial slopes—are too far apart to be exploited from a single base camp. In addition, differences in elevation make foot travel impossible between the various extremes—for instance, humid river bottoms to oak-pine forest—in one eight-hour day. This condition is even more characteristic of the Canyon Infiernillo, where the distance from canyon bottom to mountain peak is greater and the terrain much more rugged than at Tehuacán.

Both regions have a harsh dry season when little or no wild food is available in one or more of the ecozones. The Tamaulipas region also has an extremely wet season at the higher elevations and in the cloud forest; as a result, living and traveling in the region are difficult, and food is scarce. To survive in this area, therefore—at least since the Pleistocene—one had to move seasonally from one zone to the other.

In both Tehuacán and Canyon Infiernillo one or more of the ecozones has yet another important feature: potentially domesticable or cultivable plants (table 3.4). In Tehuacán, according to Paul Mangelsdorf (Mangelsdorf, MacNeish, and Galinat 1967a), wild corn grew in at least one of the canyons of the travertine slopes, while the thorn forests of the alluvial slopes had wild chupandilla, cosahuico, and coyal fruit trees. Tom Whitaker and Hugh Cutler found wild squash (*Cucurbita mixta*) there (Cutler and Whitaker 1967), and C. Earle Smith, Jr., reported wild avocado in the oak-pine forest mountain summits (C. E. Smith 1967). Tamaulipas had a different complex, with wild runner beans and setaria on the terrace steppe (Kaplan and MacNeish 1960) and wild pumpkin (*Cucurbita texana*)—the probable wild ancestor of *Cucurbita pepo*—in its thorn forests (Whitaker, Cutler, and MacNeish 1957).

Finally, both Tehuacán and Canyon Infiernillo have (and had) easy access to other subareas, where still other plants probably were domesticated. Tehuacán has access both to Oaxaca, where pumpkin (*Cucurbita pepo*) and runner beans were domesticated early, and to the Valley of Mexico, where pollen studies suggest that amaranth was cultivated by about 5000 B.C.

Further, Tehuacán's location gave it access to Veracruz and to the Sierra Madre Occidental and the Pacific Coast via the Río Hondo and the Río Balsas. The influx of many new domesticates not native to the Tehuacán area—common, runner, and tepary beans, squash (*Cucurbita moschata* and *C. pepo*), and sapotes (C. E. Smith 1967)—during Coxcatlán times suggests that diffusions of domesticates from other areas did indeed take place.

The data for Tamaulipas do not support my hypothesis quite as well, because in the deserts to the north and west few if any plants were domesticated. This subarea did, however, have easy access to lowland Veracruz, where various fruits may have been domesticated, as well as to the Mesa Central (of which Tehuacán

Table 3.4. Sequence of cultivated and/or domesticated plants in highland Mesoamerica

Tehuacán		Tamaulipas	Oaxaca	Valley of Mexico	Range of dates B.C.		pumpkin (*C. pepo*)	corn	gourd	chile	amaranth	avocado	mixta squash	moschata squash	black sapote	white sapote	common beans	chupandilla	teosinte	cotton	canavalia beans	tepary beans	sunflower	cosahuico	coyol	jitomate	runner beans	peanuts	sieva lima beans	guava
		247			1000 B.C.-	1521 A.D.	x	x	x	x	x	x	x	x	x	x	x	x	x	x	x	x	x	x	x	x	x	x	x	x
Ajalpan	Tc254 – C				825	1225	x	x												x						x				
"	Tc272 – J				1250	1500		x		x	x									x				x?	x?					
Purrón	Tc272 – K1				1500	1700		x																						
		Guerra 247			1200	1800	x	x	x	x	x			x			x			x										
"	Tc272 – K				1800	2300		x																						
		Flacco 248			1800	2500	?	x	x	x	x								?				x							
		La Perra 174			2215	2775	x	x	x								?													
Abejas	Tc50 – VIII				2870	3200		x	x	x	x	x	x	x	x	x	x	x			x	x		?	?					
"	Tc50 – IX				3071	3295		x	x	x	x			x	x	x		x												
"	Tc254 – D				3050	3550		x																	?					
"	Tc50 – X				3100	3300		x	x	x	x				x			x		x										
		Ocampo 248			2800	4000	x		x	x						x														
Coxcatlán	Tc254 – E				3950	4350		x				x																		
				Playa 2	3500	4200		P			P								x											
"	Tc50 – XI				4025	4217		x	x	x		x			x	x	x	x												
"	Tc50 – XII				4570	4700		x	x	x	x		x	x																
"	Tc50 – XIII				4700	5250		x	x	x																				
				Playa 1	4200	5000					P?																			
"	Tc254 – F				4800	5200		x																						
El Riego	Tc50 – XIV				4800	5300				x	?	x	x																	
"	Tc50 – XV				4900	5400				x	?	x								?										
"	Tc50 – XVI				5430	5800				x	?																			
(Tmc 248)		Infiernillo			5000	7000	x		x	x																				
			Naquitz C		6670	7450	x	P	x																					
			Naquitz D		7840	8750	x																							

x = macro fossil
p = pollen

is part) and to the Río Balsas and the Pacific Ocean via the Río San Luis Potosí. As in Tehuacán, various nonlocal domesticates—corn, common beans, squashes, gourds, and sunflowers—indicate that such routes were used.

Both Tehuacán and Tamaulipas therefore had all the hypothetical necessary conditions or prerequisites to allow Hunting-Collecting Bands to become Village Agriculturists. But did the sufficient conditions also exist in each region? (table 3.5).

In these harsh areas the hunting-collecting stage—the Lerma phase in Tamaulipas and Ajuereado in Tehuacán—seems to have ended at the close of the Pleistocene. In Tehuacán, evidence seems to indicate that the end of the Pleistocene involved a slight decrease in moisture, causing long dry seasons and fewer winter frosts, both of which brought about a reduction in the grassland areas (or the invasion of those areas by more xerophytic vegetation) and the disappearance of the herds of horses, antelopes, jackrabbits, and other animals that had been food staples in early Ajuereado times.

The data from Tamaulipas are not so complete but do seem to indicate that the end of the Pleistocene brought decreased moisture and led to the disappearance of beaver, horse, and perhaps bison. The exact proof of diminishing food staples is suggested, but not proven, in this region.

There is no doubt, however, that in El Riego and Infiernillo phases both regions experienced an increase in logistic mobility. People in Tehuacán started collecting the fruit of avocado and chupandilla and began to store food. Moreover, from late Ajuereado times well into Coxcatlán, there was an increase in seed collection, not only of wild grass, setaria, mesquite, and cactus seeds but also of such potential domesticates as chile, corn, amaranth, and squash (*Cucurbita mixta*).

Going hand in hand with the rise of the broader spectrum subsistence system was the development of a seasonally scheduled subsistence pattern. Changes in the size and structure of avocado and chupandilla pits and of corncobs and squash seeds (*Cucurbita mixta*) indicate that the genetic structure of these plants changed gradually from wild to domesticated. There was a change as well from plants being collected to being planted.

In Tamaulipas, indications of the above changes are clear in the period from Infiernillo to Ocampo. Seasonality, increased mobility, and seed collecting and storage started in Infiernillo. There was a shift from wild pumpkins with small seeds to domesticated pumpkins with large seeds and a shift from black seed (wild) chile to red seed (domesticated) chile. Furthermore, setaria grains increased in size. These changes certainly seem to indicate seed selection, and the presence of domesticates in Ocampo times suggests planting.

Evidence of macrobands in both areas suggests population increases, and we may deduce that such increases further stimulated these foragers and provided another sufficient condition for continued development of incipient agriculture.

Thus, Tamaulipas and Tehuacán both meet all of the hypothetical conditions for the evolution from Hunting-Collecting Bands (system A) to Incipient Agricultural Bands (system B1). In fact, highland Mesoamerica provides a better val-

Table 3.5. Necessary and sufficient conditions as a positive-feedback process for primary development in Tehuacán

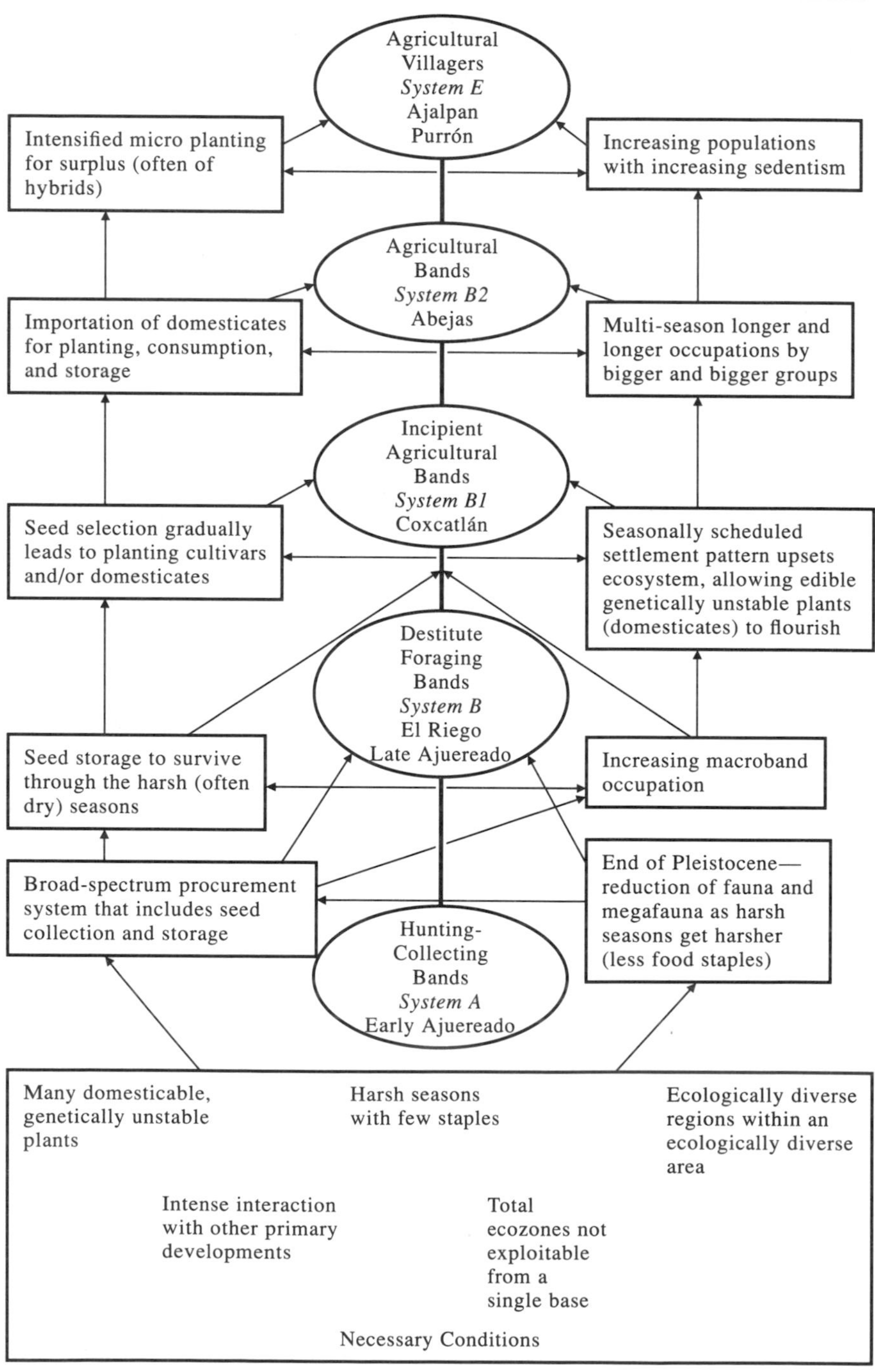

idation of the first stage of the primary development model than does the Andes region.

The second stage of primary development—from Foragers (system B) to Incipient Agricultural Bands (system B1) to Village Agriculturists (system E) via route 2 occurred in both Tamaulipas and Tehuacán. During the Ocampo phase of Tamaulipas and the Cocaxtlán phase of Tehuacán the people were foragers with a seasonally scheduled system. They used increasingly domesticated plants for storage, not only to sustain them during the harsh season but also to allow them gradually to decrease their residential mobility as the macroband populations slowly increased—that is, to make spring-summer stays longer and longer. Decreased residential mobility led to a positive-feedback cycle in both Ocampo and Cocaxtlán. More and more food from the planting of domesticates resulted in longer and longer seasonal stays, which further increased population. This created a need for more food from planting, and so on.

In both phases, increased agriculture, which often used hybrids and specialized in corn, caused incipient agriculture to become subsistence agriculture that in time supported a population in one location year-round, that is, village agriculture. Let us summarize the scenario for Tamaulipas.

In Ocampo and Flacco times (from 4000 to 1800 B.C.), people in Tamaulipas added to the domesticates they already planted (i.e., chile, gourds, and pumpkins). First came common beans, then corn, then squash and tepary beans. By the end of Flacco times, people were able to live as macrobands for two or more seasons. This longer period of residence with an increased population probably required more corn production, and the people experimented to obtain more productive hybrids. By Guerra times (1800–1400 B.C.), their storage pits were full of just such hybrids rather than a variety of wild and domesticated plant remains, as in the previous two phases. Increased planting raised consumption of planted food from 20 or 30 percent by volume to almost 40 or 50 percent. This allowed people of the Mesa de Guaje phase (1400–1100 B.C.) to store enough food that, during the harsh season, they could stay in the same location, so that their base camp now became a hamlet and/or village—with an increasing population that required more food production. The Tamaulipas data in this final preceramic phase confirm my primary hypothesis and seem somewhat sounder than the data from Tehuacán.

In Coxcatlán and Abejas times (5000–2300 B.C.) at Tehuacán, more and more domesticates were planted—corn, beans, moschata squash, black and white sapotes, chupandilla, and jack beans. With these came a shift from an even number of macroband and microband encampments for one or two seasons to a pattern of fifteen microband encampments for two seasons or more.

Our excavations brought to light evidence of nine macrobands staying for two to four seasons and three staying year-round. We also dug up thirteen hamlets or village occupations with evidence of but few (five) seasonal forays as macrobands (one) or microbands (four) away from their homes. The evidence is limited for the use of increased hybrids and storage pits. The same sufficient conditions of my hypothesis, however, seem to have caused the Coxcatlán Foraging Bands with Incipient Agriculture to become Village Agriculturists by at least Ajalpan times.

Tehuacán and Tamaulipas give evidence that supports my hypothesis about primary development, that is, why Hunting-Collecting Bands became Village Agriculturists (via routes 1 and 2) in two highland regions of Mexico. There are hints that future research in the other subareas of the Mesoamerican Center—Oaxaca, the Valley of Mexico, the lowland Sierra de Tamaulipas, and the lowland basins of the Río Balsas of Guerrero and the Río Grijalva of Chiapas—will further confirm these findings.

Unfortunately, data from lowland Mesoamerica are not yet adequate to test my hypothesis about secondary development in Centers. The one sequence we do have, from Belize, although lacking sufficient subsistence information, suggests that more research very well might validate the hypothesis. Whether other lowland subarea sequences will do the same remains to be seen.

CHAPTER 4

The Near East

From the standpoint of modern civilization, the Near East Center is significant, because it is where the plants that make our civilization possible were first domesticated. It is also one of the areas where the world's great archaeology has been accomplished. Yet we still do not have all the answers about how and why village agriculture occurred in this culture area. In fact, we do not even have enough data from some subareas adequately to test my primary developmental hypothesis. Our knowledge of other subareas such as the Levant, however, is much fuller. There we have excellent archaeological sequences of the Epipaleolithic and/or Mesolithic and Neolithic with both artifacts and some relevant plant remains. These data allow adequate testing of my secondary developmental hypothesis.

The data on the Near East are more or less opposite of those for Mesoamerica, where the highland sequence is well documented and the lowland inadequate. They also differ from the Andes, where both the highland and the lowland subareas have data adequate for testing hypotheses.

A distinctive aspect of the Near East is that the usual development leading to village agriculture happened much earlier, roughly from 10,000 to 6000 B.C., than it did in the Andean and Mesoamerican Centers (6000 to 2000 B.C.). Besides this temporal difference, there is a spatial one. There were secondary developments in the Aegean, Egypt, and the Indus Valley that were intimately connected with developments in the Near Eastern Center itself. Thus, the sphere of interaction of the Near East extends over a much wider area than that of either the Mesoamerican or the Andean Center. In addition, the Near East has been the focus of interest in archaeology and the problem of the origins of agriculture for much longer than has the New World. In fact, the history of archaeology relevant to this problem in the Near East is so long and complex that we can touch on only some of its more salient features.

A BRIEF HISTORY OF RESEARCH ON THE PROBLEM

In chapter 1, I mentioned the various theories that have developed over the past century about the origins of agriculture in the Near East, from Alphonse DeCandolle in 1884 to Charles Redman in 1984. Here I shall touch upon only the technical-botanical developments and relevant archaeological discoveries in this same period. First let us consider the botanical developments.

As early as 1826, C. Kunth attempted to identify plant remains from tombs in Egyptian pyramids (Kunth 1828). By the latter half of that century the famous German botanist L. Wittmack was identifying plants from excavations not only in Anatolia but also in Italy, Germany, Egypt, and Peru (Wittmack 1880). It was during this period that DeCandolle speculated about the origins of cultivated plants (DeCandolle 1884). Later, human geographers such as Eduard Hahn did the same (Hahn 1896).

At the beginning of the twentieth century, excavations in the Near East increased, and further identifications were made by many botanists. E. Schiemann, on the basis of her identifications, wrote an article entitled "New Results on the History of Cultivated Cereals" (Schiemann 1951). Also starting work in this period (about 1938) and continuing until the present was Hans Helbaek, who started with Swiss lake dwellers (Helbaek 1948) and moved on to Danish materials (Helbaek 1954). Later he began to identify Near Eastern materials, first for Robert Braidwood (Helbaek 1960) and later for James Mellaart in Anatolia (Helbaek 1964), for Diana Kirkbride in Palestine (Helbaek 1966), and for Frank Hole and Kent Flannery in southwest Iran (Helbaek 1969).

Since the 1950s, interest in the problem of the origins of agriculture has increased enormously. Many programs have attacked this problem with the cooperation of a growing body of botanists and geographers—in Syria, Willem Van Zeist and S. Bottema (1966), at Tell Ramad, and Willem Van Zeist and W. A. Casparie (1968) at Tell Mureybit; Maria Hopf (1969) at Jericho and other sites in the Levant; and Charles Vita-Finzi and Evie S. Higgs (Vita-Finzi and Higgs 1970) and Jane Renfrew (Renfrew 1973) in the same area.

Going hand in hand with these increased interdisciplinary studies has been a series of conferences concerned with early agriculture: the British Academy Research Project in the 1950s and 1960s, the International Work Group for Paleoethnobotany in 1968, the Institute of Archaeology Conferences in 1968 and 1970, the Ninth International Congress of Anthropological and Ethnological Science in Chicago in 1976 (Reed 1977), the Southampton Conference of 1986 (David Harris and Hillman 1988), and many others.

Accompanying this increased interest in the origins of agriculture are archaeological studies of the Mesolithic or Epipaleolithic and early Neolithic of the Near East (see table 4.2). Although the Paleolithic and Neolithic were recognized in the Near East in the nineteenth century, a real breakthrough occurred in the 1920s when Flinders Petrie and Arthur Keith recognized the Natufian at Wadi Natuf in the Levant. This discovery was followed in the late 1920s and 1930s by the pioneering efforts of Dorothy Garrod at Mount Carmel. She located Natufian remains in their stratigraphic context and described them in her work on Zarzi Cave in the Zagros Mountains (Garrod 1930). Between 1930 and 1940, John Garstang discovered preceramic Neolithic remains in the lower levels of the historic Jericho site (Kenyon 1957).

Probably the next big program—really the first to attack directly the problem of village agriculture—was that of Robert Braidwood, which began in the 1940s and carried on well into the 1980s (Braidwood and Howe 1960). This interdisciplinary effort in the "Hilly Flanks" not only made great discoveries and stimulated others to attack this problem but also developed a host of students who have become leaders in the field—Linda Braidwood, Bruce Howe, Robert McCormick Adams, the late Joseph Caldwell, Halet Gambel, Charles Reed, Herbert Wright, Jack Harlan, Patty Jo Watson, Kent Flannery, Frank Hole, Charles Redman, and others.

The other big advances started just after World War II with the work of Kathleen Kenyon at Jericho (Kenyon 1957), Ralph and Rose Solecki at Shanidar (R. S. Solecki 1961), Carleton Coon at Hotu Cave (Coon 1956), and the French under Jean Perrot in the Levant (Perrot 1960a). The 1960s brought others into the field—Philip Smith and Cuyler Young in Iran (Smith and Young 1972), Diana Kirkbride (Kirkbride 1966), Fekri Hassan, Ofer Bar-Yosef, and others in the Levant (Perrot 1968), and Maurits Van Loon, James Skinner, and Willem Van Zeist in Syria (1970). In the next decade James Phillips (Phillips 1970), Anthony Marks (Marks 1971), and others moved into the Levant, while Dexter Perkins and Patricia Daley (Perkins and Daley 1968) and Braidwood's crew investigated Turkey (Braidwood et al. 1974). There James Mellaart and James Pritchard (Mellaart 1975) made their famous discoveries at Çatal Hüyük and Hacilar. Jean Perrot, Frank Hole, Kent Flannery, Phillip Smith, and others continued their fine work in Iran until political troubles brought almost everything there to a standstill.

Yet, despite the unrest that plagues the Near East, archaeology goes on in the Levant, Arabia, Turkey, and elsewhere, and one day we should have better data for our hypotheses. Before we discuss the relevant archaeology of the area, however, let us consider the region itself.

THE ENVIRONMENT

Various authors use roughly the same attributes—climate (temperature, rainfall, and seasonality), topography, and hydrology—to classify the major geographical subdivisions of the Near East, but each classification is slightly different. Charles Redman has eight divisions (Redman 1978), Karl Butzer and H. B. S. Cooke have five (Butzer and Cooke 1964), Herbert Wright has four (H. Wright 1968), and W. B. Fisher (Fisher 1968) and George Cressey (Cressey 1960) have still others. All, however, note a highland-lowland dichotomy.

Using the same criteria, I have divided the Near East into eight subareas, four in the lowlands and four in the highlands (see fig. 8). My classification emphasizes the kinds of ecozones that characterize the regions of each major subdivision. Unfortunately, the quality of the ecozone studies of each of these major subareas varies considerably, as does the quantity of archaeology accomplished. Thus, my classification has limitations. Nevertheless, it does provide a basis from which to attempt to solve the problem of the origins of village agriculture.

I have divided my eight Near Eastern environmental life zones or subareas into three general groups based upon the amount of archaeological data each has yielded that is relevant to the problem of agricultural origins. Three subareas (one lowland and two highland) have produced the fewest materials. These are

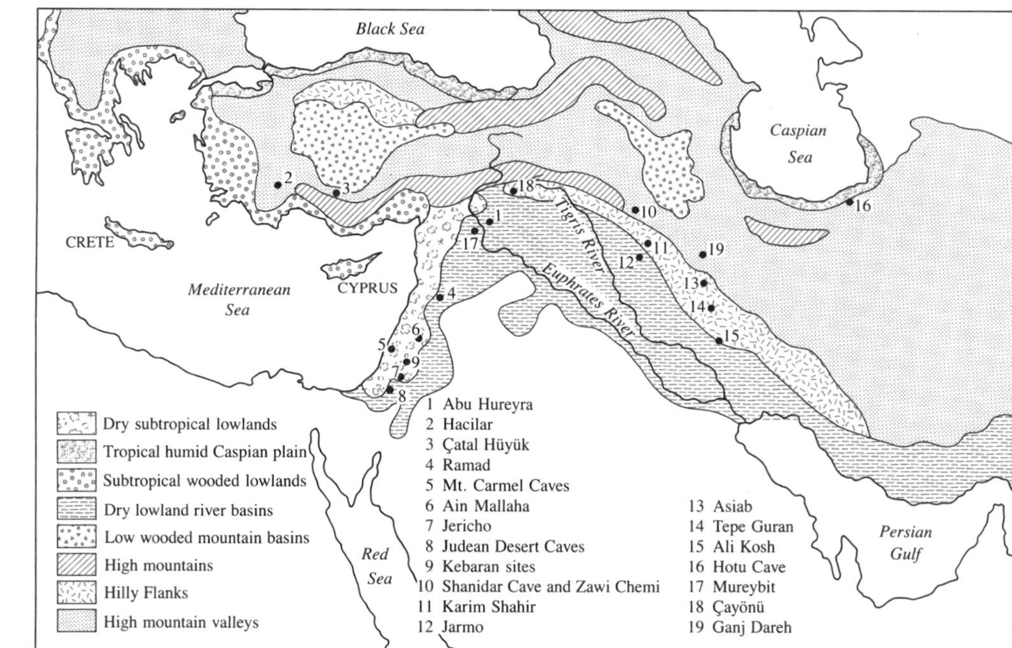

Figure 8. The environmental life zones or subareas and relevant sites of the Near East

the tropical humid lush Caspian Plain along the southern shores of the Caspian Sea in Iran; the very high mountain valleys that are comprised of the Taurus, Zagros, Pontic, and Elburz mountains; and the low wooded mountain basins or plateaus that are in the Iranian and Anatolian plateau regions.

The Caspian tropical humid plain is a long (300 kilometers) and narrow (30 kilometers) strip north of the Elburz Mountains in Iran. Except for foothill cliffs and a few raised beaches with narrow northward-flowing streams cutting across them, it is relatively flat and has little variation in topography. Rain falls throughout the year, and because of the steep mountains to the south, the plain is deluged with winter rains—over 2,500 millimeters per year. Frosts are rare, and summers are hot and humid. These conditions create lush vegetation—thick groves of oak, pistachio, and olive trees—that is Mediterranean in type. There are a few records of domesticable grains in the area, and Daniel Zohary does note barley and wild einkorn from the shores of the Caspian in the U.S.S.R. just north of this area (in Harlan and Zohary 1966).

Carleton Coon's excavation of Belt Cave indicates that game, including gazelle, oxen, sheep, and goats, was abundant, along with seals from the sea and other marine (shell) foods (Coon 1956). The wild vegetation would have provided abundant nuts, berries, olives, and the like. While this environmental subdivision contains some different ecozones that roughly parallel the coast or the streams, the region is relatively uniform. Most resources of the coastal plain could have been exploited from a single base located at a central point. Ecologically, the Caspian plain region, as an extremely lush zone, ideally fits my secondary development model. Unfortunately, only limited archaeological investigation has been undertaken there.

The relative dearth of archaeology in the nearby high mountain valley subarea (only the Behistunsi site has been excavated) is more understandable, because transportation is problematic and the living is uncomfortable much of the year. Generally speaking, the mountains run east-west and rise above 1,500 meters; they are steep and cut by large valleys. One chain is made up of the Pontic Hills south of the Black Sea in Turkey and the Elburz Mountains along the Caspian Sea. To the south of these mountains are the Taurus Mountains of Turkey and Syria, which connect with the Zagros Mountains in northern Iraq and southern Iran.

Rainfall varies slightly, with the northern chain, the Elburz, getting more than the southern, which has deserts in its southern reaches. There is, however, great seasonal variation, ranging from frosty winters to dry seasons. Vegetation is related to elevation and rainfall. At lower elevations there are evergreen oak and pines, while deciduous oak, cedar, maple, pine, juniper, and fir grow at higher elevations. The wetter, northern range is dotted with coniferous forests, while the peaks, often capped with snow year-round in the Taurus, approach a scrub-Alpine vegatation.

While plant foods are limited and seasonal, some steppe areas in the Zagros do have wild wheat and barley. Generally speaking, however, this is the homeland of sheep and goats, and animal foods are more abundant than plant foods. The high mountain subarea is thus at the opposite end of the ecological dichotomy from the tropical humid Caspian subarea. The high mountains have domesticable plants, great ecological diversity that, because of elevation, cannot be exploited from a single base, and great seasonality, with harsh seasons that often force movement from one ecozone to another in the search for food. Butzer and Cooke show no major excavations in the zone (Butzer and Cooke 1964, p. 140), but this is mainly true (Çatal Hüyük excepted) of our two mountain basin areas that I consider to be a single subarea.

The high mountain subarea surrounds two major plateaus or low wooded mountain basins—the Anatolian to the west and the Iranian one at Dasht-i-Kavir to the east. Both have low rainfall, because of the surrounding mountains, and a steppe and/or grassland and/or xerophytic vegetation. Changes in elevation, however, cause shifts in the relative amounts of grassland to deciduous forest, contributing to great ecological diversity. There is great seasonal variation in rainfall, often with a harsh winter season as well. In spite of this austerity, wild grasses and game are fairly abundant seasonally. The high basin zone fits well into the harsh ecological zone of my primary development model. This ecological subarea is closely related to the three zones that follow, all of which have had some important archaeology undertaken in them.

The first of the low wooded mountain basin subareas is the Anatolian Plateau-Basin region. Here Hacilar, Çatal Hüyük, Suberde, and Asikli Hüyük were excavated (Mellaart 1967, 1970, 1975). Surrounded by mountains, this region is semiarid and has a noticeable seasonal change with a dire dry season. The plateau itself is hilly with a series of canyons leading into its various basins. Much of it is covered with deciduous mixed forest interspersed with large areas of grasslands. Wild peas, einkorn, lentils, and possibly rye (but not barley) are found in this subarea. Game, including wild oxen, was probably once abundant. Certainly the plateau is a subarea of great ecological diversity with potentially domesticable plants, and its ecozones are relatively far apart. Further, seasonality would necessitate movement from one ecozone to another to ensure survival. This subarea, then, would seem to fit my primary development model.

Resembling the Anatolian Plateau region of the low wooded mountain basin subarea is the region just north of the Zagros Mountains in the high mountain valley subarea of Iran. This basin region to the east is more rugged than that of Anatolia; rivers have cut steep canyons through the mountains, and the numerous depressions often contain lakes. Elevation ranges from 800 or 900 meters to 2,500 meters. The climate is semiarid, and there is marked seasonality in the higher elevations, which are cool and wet in the winter, encouraging people to migrate to the lowlands in that season.

Vegetation varies considerably across the plateau; the valley floor is grass covered, while the higher elevations have deciduous woodlands with oak and pistachio. The many wild relatives of domesticates include einkorn, emmer, barley, peas, and lentils. The high mountain valley subarea also has abundant game, including sheep, deer, wild goats, and wild pig. It clearly is a subarea of great ecological diversity with ecozones that cannot be totally exploited from a single base. Furthermore, it has great seasonality and many potential domesticates. Significant agricultural sequential data have come from this subarea—

from Shanidar Cave (R. S. Solecki 1961), Zawi Chemi (R. L. Solecki 1964), and Ganj Dareh (H. F. Miller 1988). Pollen cores taken at nearby Lake Zeribar (Van (Van Zeist and Wright 1963) show that just before evidence of agriculture appeared in Shanidar, there was a major shift in vegetation (invasion of the forests) toward greater diversity. These archaeological and palynological materials seem intimately connected with our final similar highland area, which Braidwood termed the "Hilly Flanks" (Braidwood 1951).

Located on the south slopes of the Zagros Mountains, between 500 and 1,500 meters above sea level, the Hilly Flanks region overlooks the alluvial subarea of the Tigris and Euphrates drainages to the south. Topography is rugged; many steep canyons cut through the flanks of the mountains. I envision this zone as running from southern Iran with Tepe Guran in the southeast to Çayönü in southeastern Turkey.

Among the many sites dug in the Kurdish Hills region of this subarea are Jarmo, Karim Shahir, and Palegawra Cave. Because these sites are on the flanks of the mountains, rainfall is relatively heavy, 500 to 1,000 millimeters per year, and mixed deciduous woodlands are denser and more subtropical, while steppe grasslands are less bountiful than in the Highland subareas. Many of the wild ancestors of domesticates found in the adjacent area also appear here, and game is only slightly less abundant than in the high mountains.

The "highland" subareas discussed above have much in common, yet contrast with the tropical humid Caspian coastal plain. The highlands also contrast with the three final subareas, which I have grouped together because abundant archaeology has been completed in them: Dry subtropical lowlands of the Levant, the subtropical wooded lowlands of the northeastern Mediterranean, and the Dry Lowland River Basins subarea of the Tigris-Euphrates of the Mesopotamian alluvial plain.

The dry lowland river basins subarea is the least well known as far as the crucial Epipaleolithic time period is concerned, although the subarea abounds in later "tells" that have been excavated. Fortunately, the principal Early Neolithic site—Ali Kosh—was well excavated, and the ecology of the region intensively studied (Hole, Flannery, and Neely 1969).

Basically, this Khuzistan area is dominated by the drainages of the Tigris and Euphrates rivers south of the Piedmont area of the Zagros Mountains and north of the Arabian Desert. It is a relatively dry area with marked seasonal changes: a dry and intensely hot summer with strong winds contrasts with a wet and cool winter. South of the broad, flat river valley, the levees, marshland, and open plains with sandy soil give way to desert, while to the north, Flannery and others have defined a series of "microenvironments" or ecozones that roughly parallel the river and climb up the hillside to a height of about 300 meters. Each of these long parallel zones has its own abundance of distinctive plants and animals. The different zones, however, are not far apart and often could all be exploited from a strategically located base. Among the wide variety of wild plants are emmer, einkorn, barley, peas, and others that could be domesticated, but they represent a small minority of the food resources in this lush subarea.

Although the parallel ecozones create regional diversity, the same five or six zones recur as one moves up and down the river. Thus there is considerable subareal uniformity, even though there is regional diversity; that is, each region has the same kind and number of ecozones. This Mesopotamian Riverine region fits well at one end of the dichotomy in my model and contrasts with the previously described highlands.

The dry lowland river basin subarea (see fig. 8) is often connected with the coastal Levant region of the dry subtropical lowland subarea, for in terms of important necessary conditions for the origins of agriculture the Levant has significant similarities, although there are marked ecological differences. From west to east, we find the Mediterranean coast ecozone with its sea resources (and a sinking shoreline that has been burying sites since the Pleistocene); next, a coastal plain ecozone that varies in width from south (wide) to north (narrower). This plain receives considerable rainfall seasonally (500–1,000 millimeters), and it has warm humid summers and cool dry winters. Its soils are alluvial and fertile, and probably at one time it had subtropical woodland vegetation with an accompanying complex of forest animals.

To the east of the north-south plain and paralleling it is the foothills ecozone, generally about 500 meters high, although parts of the northern Lebanon Mountains have peaks reaching about 3,000 meters. The hills are mostly rounded and covered with grassy vegetation surrounded by large areas of woodlands. Soils are fertile and food resources abundant, including many edible plants such as einkorn, emmer, barley, and vetch.

Farther eastward are the ecozones of the narrow interior depression. They receive less rainfall because of the hilly barriers to the west, and much of this area is therefore desert with limited grassy vegetation, although there are oases and lakes (like the Sea of Galilee) that support aquatic plants and dense stands of willows.

East of this depression rises the semiarid uplands ecozone that leads to the Arabian Desert and forms a transition zone. These uplands have limited rainfall, are hot and dry in the summer, cold and wet in the winter, and have relatively limited seasonality and steppe vegetation with limited fauna. Like the riverine or dry lowland river basin subarea, then, the Levant subarea has six or seven parallel ecozones and shows regional (east-west) diversity but areal (north-south) uniformity. Most of the ecozones, as well as the catchment basin areas that Higgs and Vita-Finzi described (1972), are not far apart, and many of them can be exploited at different times of the year. There is a long history of successful foraging in this subarea until agriculture finally enters the scene.

The final subarea is the subtropical wooded lowlands, on the northeastern Mediterranean Coast, which encompass the coast north to Turkey. Although the islands of Crete, Cyprus, and the Aegean, as well as the south coast of Greece, are intimately connected with this region, they are discussed elsewhere since they lack the dichotomous highland ecozones and are basically in a non-Center. This Anatolean Mediterranean region has narrow strips of fairly well watered coastal plain (often backed by steep mountains) with relatively heavy rainfall, fertile soils, and either Mediterranean vegetation or subtropical woodland flora with limited grasslands. In pre-agricultural times there was abundant game as well as marine resources and varied vegetal foods, including some susceptible to domestication—olives, grapes, pulses, barley, einkorn, and a few others. Few ecological studies have been made of this subarea, but there are suggestions that we are once again dealing with a subarea that has lush food resources and regional diversity as well as subarea uniformity. Moreover, this narrow coastal strip can be exploited easily from a single base.

In all the subareas of the Near East Center, the basic dichotomy that occurred in the previously discussed Centers—the Andes and Mesoamerica—also exists. As in all Centers, the two major divisions of the Near East share two basic characteristics: domesticable plants (see table 4.1) and subareas with easy access to one another, making them potential spheres of interaction. Yet the differences between the two divisions far outweigh the similarities. The highland subareas have numerous ecozones that are not exploitable from a single base. Moreover, in each subarea, changing elevations, climate, and so on have created different kinds of ecozones. In contrast, although the lowland subareas contain diverse ecozones, from subarea to subarea or region to region within each subarea the ecozones are relatively similar. Further, because of the relative narrowness of each zone or lack of rugged terrain, these ecozones may be exploited from a strategically located base.

Furthermore, the highland and lowland areas themselves are qualitatively different—the lowland subareas have one or more ecozones with abundant food staples and usually lack a harsh season that requires one to move to obtain food. The highland subareas, on the other hand, generally have the opposite attributes and so are harder places in which to live.

These attributes are among the necessary conditions for the primary and secondary developments of village agriculture. The differences in attributes of the two divisions were significant and limited the directions that development might take rather than determining how it happened. The basic deterministic factors were the sufficient conditions, which commenced when a changing environment at the end of the Pleistocene generally reduced accessibility of food staples. The results placed more subsistence tensions on the highland groups than on the lowland ones and helped push some toward village agriculture. Let us therefore examine the basic archaeological material from these environmental divisions.

Table 4.1. Some native cultivated and/or domesticated plants of the Near East

Scientific name	Popular name
**Agropyron sp.*	wheat grass
Avena abyssinica	Ethiopian oats
Avena brevis	short oats
Avena byzantina	red oats
Avena nubibrevis	naked oats
**Avena sativa*	common white oats
**Avena strigosa*	sand oats
Avena weistii	desert oats
**Cannabis sativa*	hemp
Cappareis spinosa	caper

Table 4.1. (*continued*)

Scientific name	Popular name
Carum carvi	caraway
Carthamus tinctorius	safflower
Celtis australis	hackberry
**Cicer arietinum*	chickpea
**Coriandrum sativum*	coriander
**Cucumis sativum*	cucumber
**Cuminum cyminum*	cumin
**Ficus carica*	fig
Hordeum distichion	naked two-row barley
**Hordeum vulgare*	naked six-row barley
**Juglans regia*	English walnut
**Lathyrus sativus*	grass pea
**Lens esculenta*	lentil
**Linum usitatissimum*	flax
**Lupinus albus*	lupine
**Medicago sativa*	alfalfa
**Olea europea*	olive
Onobrychis viciifolia	sainfoin
**Papaver somniferum*	opium poppy
**Phoenix dactylifera*	date palm
Pistacia atlantica	pistachio
**Pistacia vera*	pistachio
**Pisum sativum*	garden pea
Prosopis stephaniana	prosopis
**Prunus amygdalus*	almond
**Prunus armeniaca*	apricot
Prunus dulcis	almond
**Punica granatum*	pomegranate
**Pyrus communis*	pear
Quercus sp.	acorn
Raphanus sativus	radish
**Secale cereale*	rye
**Trigonella foenum-graecum*	fenugreek
**Triticum aestivum*	bread wheat
Triticum carthlicum	Persian wheat
Triticum compactum	club wheat
**Triticum dicoccum*	emmer
Triticum durum	hard wheat
Triticum macha	macha wheat
**Triticum monococcum*	einkorn
Triticum sphaerococcum	short wheat
**Triticum timopheevi*	timopheevi wheat
Triticum turanicum	kharsan wheat
**Triticum turgidum*	tetraploid wheat
Triticum vavilovi	vavilov wheat
Triticum vulgare	common wheat
**Vicia sp.*	vetch
**Vicia ervilia*	butter vetch
**Vicia faba*	broad or horse bean
**Vitis vinifera*	grape

*Domesticated plants according to Harlan (1971); others, J. Renfrew (1973).

RELEVANT ARCHAEOLOGICAL DATA (table 4.2)

As in previous chapters, our discussion of the relevant archaeological data starts with the regions or subareas with the least information and ends with the Levant region, which has the most data. Two subareas—the high mountains and the low wooded mountain basins of eastern Iraq-Iran—have yielded no relevant archaeological materials, while the subtropical wooded lowlands of the northeastern Mediterranean and the tropical humid Caspian coastal plain have limited relevant data. Somewhat better are the data from the high mountain valley of Anatolia, the dry lowland river basins, the Hilly Flanks, and the high mountain valley subareas of the Iranian Plateau region of the Zagros. Only the dry subtropical lowland of the Levant has excellent data relevant to the development of village agriculture (see table 4.2).

Let us begin with a brief description of the materials from the lush southernmost Caspian Coast region of the tropical humid Caspian plain subarea. Many years ago Carleton Coon worked in Iran on the shores of the Caspian Sea. At Belt Cave, on the Caspian, beneath Neolithic remains dated at 6135 B.C., he found Upper and Lower Mesolithic remains that dated from 8180 to 6054 B.C. Neither had preserved plant remains (Coon 1956). This area obviously needs more investigation and perhaps even a reanalysis of the original materials, now at the University of Pennsylvania Museum.

The Anatolian northeastern Mediterranean coastal region of the subtropical wooded lowland subarea has produced even fewer materials, although the adjacent Aegean has produced a fine sequence from Paleolithic to Neolithic at Franchthi Cave. Thus far, only a few Proto-Neolithic remains have been found along the coast of Anatolia, mainly at Belbasi and Beldibi on the south coast. Microliths have been reported for both of these sites, but little else has been written up that is relevant to our problem of the origins of agriculture.

A third region—the Anatolian Plateau-Basin of the high mountain valley subarea—has yielded some relevant materials from the general time period from 7000 to 6000 B.C. The sites include Hacilar (Mellaart 1970), Asikli Hüyük (Todd 1966), and Suberde (Bordaz 1966), as well as Çatal Hüyük, Can Hasan, and Grikshaciyen (Mellaart 1967). The first of these, Hacilar, is in southwestern Turkey, and in its lowest levels Mellaart (1970) uncovered the remains of a small village from the period between 7000 and 6500 B.C. Its people lived in rectangular houses, some of which had plastered floors with red painted designs associated with ceremonialism. In the later Neolithic (6500–5000 B.C.), when pottery arrived, this village grew and its ceremonialism became more intense, but even in the earliest period there was a stable agricultural economy using emmer, lentils, naked six-row barley, einkorn, hulled two-row and six-row barley, and bread wheat (see table 4.5). Obviously, village agriculture—system E—is represented here.

Asikli Hüyük, in central Anatolia (Turkey) was similar, giving evidence of a preceramic mud-brick village with agriculture of emmer, peas, einkorn, bread wheat, and naked six-row barley (see table 4.5). Fragments of red burnished plaster again suggest speical ceremonial rooms even in 7000 B.C., while later levels with pottery reflect a more ornate development of this complex.

Table 4.2. Relevant archaeological sequences in the Near East

	HIGHLANDS			LOWLANDS			
High Mtn. Valley subarea	Hilly Flanks subarea	Low Wooded Mountain Basin subarea	Subtropical Wooded Lowlands subarea	Dry Lowland River Basin subarea		Dry Subtropical Lowland subarea	
Eastern Zagros Mt. region	Kurdish Hills region	Anatolian Mountain Basin region	Anatolian Mediterranean Coast region	Khuzistan Tigris River region	Upper Euphrates River region	Levant region	Dates
Shimshara				Sabz			
	Hassuna			Sawwan	Amouq		
		Çatal Hüyük				Munhata	
				Jaffir			
	Jarmo				Bouqras		6000 B.C.
		Hacilar					
				Ali Kosh			
					Ramad		
Tepe Guran						Jericho PPNB	7000 B.C.
				Bus Mordeh			
	Çayönü						
							8000 B.C.
Ganj Dareh	Karim Shahir		Beldibi		Mureybit	Jericho PPNA	
					Abu Hureyra	Ain Mallaha (late Natufian)	9000 B.C.
Zawi Chemi							
Shanidar B1			Belbasi			El Wad (early Natufian)	
							10,000 B.C.
	Zarzi						
Shanidar B2							
						Këbäran Geometric	11,000 B.C.
Palegawra						Kebaran	12,000 B.C.

The third preceramic mud-brick village is Suberde, which lies just east of Hacilar. There is evidence there of domesticated animals, but we can only guess that there were domesticated plants as well. All three sites represent the village agriculture stage—system E—but we lack evidence of earlier stages that might help us understand how or why agriculture developed.

Explanatory conditions for the dry lowland river basin subarea of Asia Minor are equally hard to come by. We have no Hunting-Collecting Band remains at all, and our earliest remains, from Ali Kosh—Bus Mordeh, dated at 7200–6800 B.C.—are already villages with agriculture of such plants as einkorn, emmer, and naked two-row barley. The later horizons—Ali Kosh, Jaffir, and Sabz—imported even more domesticated plants (see table 4.6) for their agriculture (Hole, Flannery, and Neely 1969).

At the western end of this dry lowland river basin subarea, in the upper Euphrates region at Mureybit, in Syria, there are equally unsatisfactory data (Van Loon, Skinner, and Van Zeist 1970). At both Mureybit and Tell Abu Hureyra there is evidence of pre-agricultural villages with large round houses dating from roughly 8500 to 7000 B.C. The little evidence we have indicates that these people were successful Foraging Villagers (system C1) who did not use domesticated plants, although they did use wild einkorn, emmer, and lentils, which were potential domesticates.

This fits my type C1 system. In the upper levels of Mureybit, the village gets bigger, the houses become rectangular, and horticulture—system C2—occurs. Also, in the area are the large village agriculture sites of Ramad, 6500 and 6000 B.C. (Contenson 1971), and Amouq, 5800 to 5000 B.C. Both villages (see table 4.5) contain such imported domesticates as emmer, einkorn, hulled two-row barley, lentils, bread wheat, and oats (J. Renfrew 1973). Certainly these represent village agriculture (system E), but information is presently insufficient to tell us why these Foraging Villagers became Agricultural Villagers. Yet this incomplete sequence seems to fit the more recent end of my secondary developmental model. To really test my hypothesis about why agriculture developed, we must look at the Levant. Before doing so, however, let us consider the Hilly Flanks.

The western Hilly Flanks, whose inhabitants were perhaps interacting with those of the nearby lowlands on the Euphrates River in Syria, had one site, Çayönü, located in present-day southeast Turkey. Although Çayönü seems relevant to our problems, it too was mainly of the time period of village agriculture—8500–7000 B.C.—and basically after the time period of the major development of agriculture (Gambel and Braidwood 1980).

Çayönü is a valley site at 830 meters above sea level, located in a steppe-forest combination of pistachio and oak trees with many potential domesticates, such as wild wheat and barley. Fauna included wild aurochs, pigs, sheep, and goats. Excavation revealed a sequence of five phases, all of which had evidence of the domesticates einkorn and emmer wheat, peas, lentils, and vetch. Only the final four phases had latticelike structures as well as agriculture, indicating system E—village agriculture—had arrived. The earliest and most poorly defined phase was characterized by "basal pits," and it "may be that the community at that time was not yet a well-established village" (Redman 1978, 155), and perhaps represented an Agricultural Band—system B2.

Over time the population grew and agricultural technology (microblades, bone and ground-stone tools) developed. Pollen was found in the late phase. All these factors are characteristics of village agriculture. As yet we do not understand the process involved in the shift from bands to villages, nor do we have sites or phases that represent the earlier Mesolithic or Epipaleolithic periods when plant domestication occurred; however, the region still has great archaeological potential.

Since this subarea, like the previous ones, lacks sufficient sequential dates and domesticated plant remains, let us turn to the easterly Hilly Flanks or Kurdish Hills and the related Zagros Mountain regions, where there is fine sequential information, although plant remains are insufficient.

I shall treat the abovementioned two regions as if they were one for two reasons. First, the data on the early period are relatively sparse for each; second, comparisons of microlithics suggest that we are dealing with but a single cultural area at this time period. Underlying our relevant sequence are Paleolithic remains. Mousterian remains occurred at various places; in Shanidar Cave they were found under Baradostian remains that date between 33,000 and 26,700 B.C. (R. S. Solecki 1961). Bones, blade points, and other blade-flake butchering tools suggest that these people were hunters. Pollen studies (Leroi-Gourhan 1969) suggest major climatic fluctuations at the end of the Pleistocene that probably affected the megafauna and diminished the available food supply (table 4.3). Settlement pattern data are not convincing, but the occupations in Shanidar as well as those in nearby Warwasi Cave in Iran seem to have been by nomadic microbands. Perhaps these people represent Hunting-Collecting Bands—system A—during the Late Pleistocene, but we cannot be sure because more evidence is needed. Whether these Paleolithic groups developed into the later Zarzi groups is unknown because there is a 10,000-year gap in our sequence.

The next stage, Zarzi, dates roughly from 14,000 to 9500 B.C. and can be divided into an earlier and a later phase with significant developmental differences. The earlier stage (14,000 to 11,000 B.C.) is represented in the lower levels of Palegawra Cave, which date between 12,450 and 11,400 B.C. in radiocarbon time (Turnbull and Reed 1974). (The materials from the lower levels of nearby Pa Sangar Cave may also date from this time period.) Bones of gazelle, red deer, wild cattle, wild goats, onagers, wild sheep, pigs, foxes, wolves, lynx, and a dog-like creature, together with microblade points, show that these people hunted. The variety of animals represented by the bones further suggests that they exploited a number of different ecozones and perhaps had a seasonally scheduled subsistence system. Snails were collected also, and these as well as the celts, querns, and lack of fireplaces suggest a seasonal spring occupation. Querns, or grinding stones, and possible storage pits suggest seed collecting and/or agriculture, further strengthening my belief that this was a seasonal spring occupation. Further, a celt occurred—the kind that could have been used to clear fields, perhaps for wheat (cerealia), causing the weed plantago to grow, as evidenced by plantago pollen in the nearby contemporary pollen profile. The combination of querns and celts with microblades (possibly sickles) is significant, for it occurs in later sites with agriculture (Pullar 1977).

Other artifacts are mainly microliths of varying shapes—trianglar, lunate, trapezoidal, and rectangular. As far as typology is concerned, these peoples may have been nomadic collecting bands, but the other evidence indicates that by this time they had gone one step further and become Destitute Foragers—system B. Also, they may well have stored some seeds, as well as collecting and grinding them.

Pollen from the Lake Zeribar core (see fig. 9) further suggests that at this time the steppe environment of the highlands was shifting to woodlands with more oak and pistachio, with a possible concomitant diminution of such herd animals as onagers and gazelle (Van Zeist and Wright 1963). I believe that all of these features or conditions—the possible development of seasonal scheduling, a broader spectrum subsistence that included seed collecting and storage, and a changing environment that led to harsher seasons and a reduction in the availability of food staples (herd animals)—combined to bring about a significant new development.

This new stage is characterized by the Shanidar B2 zones, located stratigraphically under zone B1 of Shanidar (R. S. Solecki 1961, 1971). (Another component, at Pa Sangar, is above Palegawra-like lower Pa Sangar remains. The remains in Zarzi Cave [Garrod 1957] have similar microliths and may also be of this time period.) This phase might be considered late Zarzi; radiocarbon determinations suggest it lasted from roughly 10,000 to 9500 B.C. Large mammal bones associated with triangular microblades indicate hunting continued, while grinding stones (not found in Shanidar, since it probably represents a winter season) and storage pits suggest plant and seed collecting. Unfortunately, the pollen from Shanidar B2 was limited (see table 4.3), although that from Lake Zeribar suggests subsistence changes (Van Zeist 1967). Here at Lake Zeribar in the contemporaneous lake levels *Cerealia* pollen began to increase, suggesting the selection of domesticable (or domesticated?) plants, while increases in *Plantago*—a weedy plant that grows in cleared fields—suggest planting (see figure 7). The number of subsistence options seems to have increased, and a broad-spectrum subsistence system probably occurred (Pullar 1977). Our data, however, are too limited to tell if these people lived in macrobands or whether population was increasing.

Late Zarzian flint technology is represented predominantly by various kinds of microliths (like those of Palegawra). Obsidian continues to turn up, indicating trade and relatively easy access to other subareas where domestications may have been taking place. In terms of typology, these people were definitely Foragers, but there are not enough hard data to determine whether they had incipient agriculture or domesticates, though there is a suggestion that they were experimenting with plants. Perhaps this horizon is that of Foraging Bands with Incipient Agriculture—system B1—a transitional stage between foragers and farmers.

The following horizon, Karim Shahir, dates roughly from 9500 to 6500 B.C. It has an early and a late phase: Zawi Chemi, 9500 to 8000 B.C. (R. L. Solecki 1964) and Ganj Dareh, 8000 to 6500 B.C. (P. Smith 1975). The best examples of the early phase are the Zawi Chemi summer base-camp sites and the nearby mi-

Table 4.3. Pollen grains of *Gramineae* and *Cerealia* from Shanidar and Zawi Chemi, suggesting early cultivation and/or domestication (after A. Leroi Gourhan 1969)

Site	Level	Gramineae grain diam. $<40\mu$ ann. diam. $<8\mu$	"Cerealia type" grain diam. $<40\mu$ ann. diam. $>8\mu$	"Cerealia type" $>40\mu$ $>8\mu$	Total nos. of pollens	Dates
	Recent (A)					
Shanidar	0·60 m.	203	73	36	364	
						8500 B.C.
	Proto-Neolithic (B1)					
Shanidar	0·50 m.	212	25	20	377	
	0·70 m.	128	15	25	726	
	1·00 m.	215	11	19	525	
	Bottom	36	2	1	151	
	Proto-Neolithic					
Zawi Chemi	0·90 m.	110	1 +	5	697	
	0·95 m.	12	3 +	2	419	
	1·00 m.	61	3 +	4	556	
	polished celts 1·20 m.	116	2 +	2	422	
	(C4) 1·60 m.	6			86	

	domestic sheep? 1·70 m.	38				377	
	1·90 m.	84	4	+	5	501	
	1·95 m.	29	3			705	
	2·05 m.	34	4	+	1	632	
	(C4) 2·15 m.	28	2	+	4	241	
	2·20 m.	51	1	+	1	602	
	(C4) 2·21 m.	13				376	
	2·35 m.	83				365	
	2·45 m.	22				112	
							9500 B.C.
	Mesolithic (*B2*)						
Shanidar	2·85 m.	22				70	
							11000 B.C.
							26000 B.C.
	Baradostian						
Shanidar	3·00 m.	20				61	
	3·40 m.	7				24	
	3·75 m.	4				25	
	4·00 m.	54	1			361	
							33000 B.C.
							44000 B.C.
	Mousterian						
Shanidar	4·25 m.	7				36	
	4·35 m.	3				13	
	6·30 m.	3				10	
	6·40 m.	4				28	
	7·30 m.	10				66	
	7·50 m.	20				71	
	7·57 m.	15			1	61	
	Neanderth. IV	22				min. 1816	
	8·60 m.	12				188	
	9·65 m.	18			1	107	

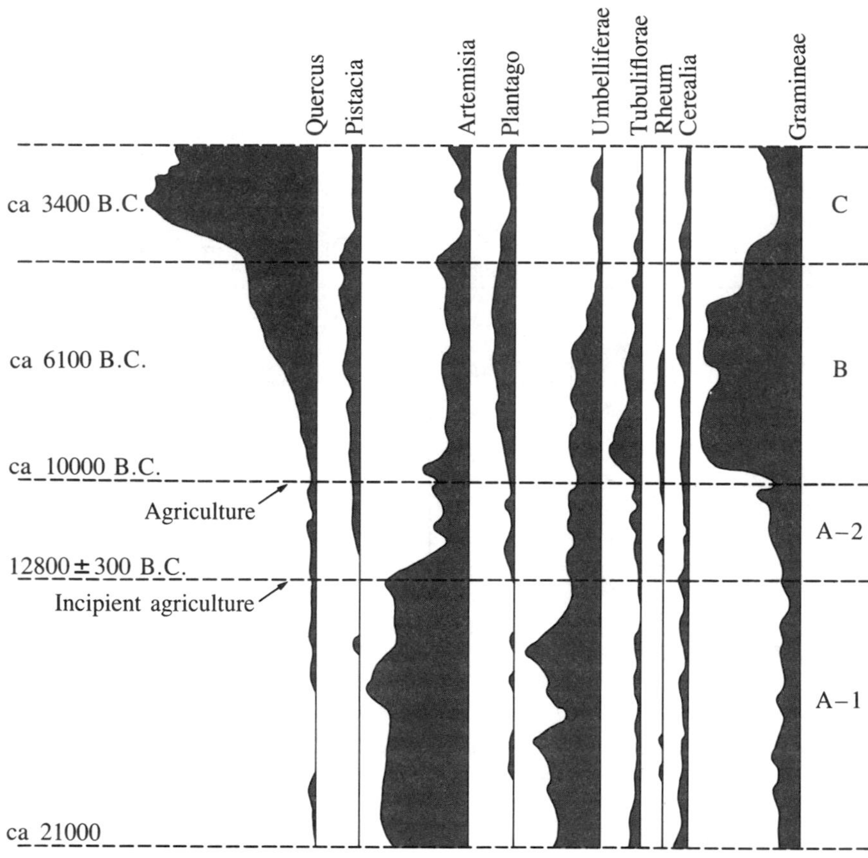

Figure 9. Pollen diagram of the Lake Zeribar core, indicating early agriculture in the highlands of the Near East (after Van Zeist and Wright, 1963)

croband winter cave occupations of zone B1 of Shanidar Cave. Moreover, the nearby sites at Karim Shahir and M'lefaat may also be of this time period. Obviously more components of this type need to be found and excavated. Although the data are inadequate, there is a suggestion of increased numbers of seasonal macrobands, minor population increases, and possibly longer occupations. Equally provocative is the evidence on the subsistence pattern, which definitely indicates a broad-spectrum system, perhaps with agriculture. The presence of bones suggests a continuation of hunting. There are also small mammal bones that may indicate trapping or collecting, while some sheep bones suggest the penning of tamed animals, if not herding of domesticated ones.

Mortars, pestles, grinding stones, and storage pits suggest the collection or cultivation of a number of kinds of plants. Although the crucial domesticated plant remains have not been found, there is good pollen evidence from both Zawi Chemi and zone B1 of Shanidar (Leroi-Gourhan 1969) indicating that *cerealia* plants were being consciously selected for increased size (see table 4.3). Whether this shift indicates domestication (with genetic change) or cultivation (size selection from among a wild species), we do not know, but probably the people were practicing agriculture. Further confirmation of this hypothesis comes from the contemporaneous Lake Zeribar pollen profiles, where increases in *Plantago* pollen indicate field clearing and use, while dramatic increases in both proportion and size of *Cerealia* and *Gramineae* indicate planting and probably incipient agriculture (Pullar 1977).

As far as tools are concerned, microliths continue to dominate, but celts in Zawi Chemi (R. L. Solecki 1964) give further evidence of tools for clearing fields (agriculture) or gardens (horticulture). Obsidian indicates trade was still increasing, but whether it included the spread of domesticates along easily accessible routes remains unknown. Although the evidence is far from complete, I believe this horizon should be classified as system B2, Agricultural Bands, rather than system B1, Foraging Bands with Incipient Agriculture.

The later Karim Shahir phase, the Ganj Dareh period, dates roughly from 8000 to 6500 B.C. and seems to be transitional between systems B2 and E. Components for this period are more numerous: in Iran, a number of levels at Ganj Dareh, and in Iraq the Asiab site, as well as the Tepe Guran in Iran and perhaps the Sarab sites. Although pottery comes somewhat earlier, the shift in residence as seen from zones E through C at Ganj Dareh (P. Smith 1975) shows not only a tendency toward a more sedentary way of life but strongly suggests an increasing population. The possible population growth seems to have been accompanied by an increasing use of domesticates, namely emmer, peas, lentils, and vetch (see table 4.6), and in the lowest zone of Ganj Dareh there is definite evidence of domesticated barley (Miller 1988). By the end of the period there were also domesticated pigs, goats, and sheep. Increasing numbers of storage pits (for domesticates?) suggest crops grown for surpluses. However, there is little evidence for *in situ* domestication in terms of changing seed size, and all possible domesticates may have been imported. Throughout the levels there is evidence of increasing trade in copper, shells, and obsidian, an indication that food may also have been imported. Microliths still predominated in tool assemblages, but the number of ground-stone tools increased throughout the phase. By the end of the period at Ganj Dareh we almost have village agriculture, but it did not really begin until Neolithic times.

The Neolithic, with and without pottery, is represented by the Jarmo phase (see table 4.2), which dates from 6500 to 5000 B.C. The ceramic levels of zones B through D of Ganj Dareh, Sarab, and Tepe Guran are also representative of this stage. These were true villages with pottery and a full complex of agriculture with crops of emmer, einkorn, peas, vetch, lentils, and perhaps others, as well as domesticated animals. Village agriculture—system E—has arrived, and my model, which shows a shift from Foraging Bands with Incipient Agriculture to Village Agriculturists, seems confirmed, even though the data are limited.

The only really adequate data come from a region that has been investigated more intensively than any other part of the Near East—Syria, Israel, and Jordan (Redman 1978), or the Levant. Israel has the longer sequence, beginning with a series of pebble tools—choppers, picks, hand axes—that were found in stratified levels of the Ubeidiya site in the Jordan Valley. These tools might be from half a million to more than a million years old. Next came the Tayacian industry of crude flakes, taken from the lower levels of el Tabun Cave near Mount Carmel, followed by Acheulean hand-axe assemblages, which occurred at Dsir Jacob site and in the next-to-lowest levels of el Tabun. Recently a Jabrudian industry of thick flake scrapers has been defined in the levels of Art Jelinek's redigging of el Tabun Cave, and this is overlaid by Amadian flake-blade industries (Jelinek et al. 1973). All these early levels are thought to be of middle Paleolithic times, roughly from 300,000 to 50,000 years ago.

While finds of these earliest periods are relatively rare, we have more numerous examples of the Levalloisian-Mousterian period, including small, fine hand axes, triangular points, and Levalloisian discoid bifacial cores. From es-Skhūl Cave near Mount Carmel and Qafzeh Caves near Nazareth, Israel, we even have Neanderthal-like and modern skeletons (Garrod and Clark 1965). Furthermore, there is evidence of the extinction of Pleistocene fauna (rhino, horse, etc.), of which we have only hints in the related Hilly Flanks.

The transition to the Upper Paleolithic is not well understood, but blade cultures in Israel, like Aurignacian and Gravettian of Europe, follow the earlier phase in places such as the stratified El Wad site (Garrod 1957) and are followed by the Ahmarian phase between 20,000 and 30,000 years ago (Phillips 1986). This phase could be considered the Hunting-Collecting Band—system A.

Somewhat better defined is the Kebaran stone-tool industry, which began about 20,000 years ago (Bar-Yosef 1975), with microlithic blades, bladelets, fine traingular points, and mortars. The development from Hunting-Collecting Bands began in what I have called the Kebaran period, roughly from 20,000 to 9500 B.C., during the closing phases of the Pleistocene, which does not seem to have had quite so telling an effect on the food supply here as it did in the highlands (Bar-Yosef 1975). This Kebaran period, like Zarzi, represents a shift to my model's second stage—Affluent Foraging Bands—system C. As I did with the Zarzi phase, I have divided Kebaran into two general periods—Kebaran and

Kebaran Geometric—although the difference between the two is not as great as it is in the highlands.

Kebaran seems to have existed in the general time frame from 20,000 B.C. to about 12,000 B.C. It is represented by many components: zone C of Kebara Cave; zones Ca through Cf of Hayonim Cave; zone A2 of Sefunim Cave; layers 8 and 9 of Nahal Oren; the top levels of El-Khiam; level 7 of Jabrud III rock-shelter; and Hefsibah and Fazael III sites. More sites are being found all the time (Marks 1971). The sites are concentrated mainly on the coastal plain and are generally small hunting microband cave camps—but whether the people were seasonally nomadic is difficult to determine. Evidence for hunting abounds (goat bones), but ground stone and pits suggest plant collecting as well, and seashells are also found. The material culture is characterized by microliths, mainly bladelets and small narow points. This is our system A—Hunting-Collecting Bands.

The Kebaran Geometric phase lasted from about 12,000 B.C. until about 9500 B.C. (Bar-Yosef 1975; Bar-Yosef, Kislev 1988). Microliths still abound but take on a wider variety of geometric forms. The sites seem larger, and more are located outside caves than in them. Hunting still predominated and shells are found, but there are signs of seasonal scheduling and more evidence of plant collecting, particularly at the Nahal Oren site (Stekelis and Yizraely 1963). Perhaps the people had begun to develop a slightly broader spectrum of subsistence options and had become Affluent Foraging Bands—system C—rather than Hunting-Collecting Bands.

This tendency reached its culmination in the following Natufian phase, 9500 to 7000 B.C., which has good evidence of broad-spectrum subsistence even in the early stage—here called El Wad—that dates roughly from 9500 to 8500 B.C. (Garrod 1957). Evidence for a broader spectrum subsistence system (besides evidence of hunting with microlith points) includes mortars, pestles, querns, pounders, stone vessels, and storage pits, suggesting a wide variety of plant-collecting techniques. Sites have become larger and the people are definitely seasonally nomadic, a further development of system C. In general there is a movement away from the coastal plain up to the hills, which may then have been more heavily forested.

By 8500 to 8000 B.C. this phase gradually developed into late Natufian, called Mallaha (Perrot 1960). There is more evidence of specialized hunting at various sites. At Ain Mallaha and Nahal Oren, gazelle bones predominate, while goat bones are numerous at Beidha. More types of grinding stone appear, and there is evidence of specialized plant collecting as well. In other words, there is considerable resource specialization and the possibility of domesticated emmer at Nahal Oren (Miller 1988). Accompanying this subsistence change is a gradual settling down in villages of round pit houses or in base camps near springs, which, of course, have circumscribed resources. A population increase goes along with this sedentarism. We now have reached the stage of Foraging Villagers—system C1.

In the general period from 8000 to 7000 B.C. these Natufian populations grew rather rapidly into the pre-pottery Neolithic A type (PPNA), with large villages that began to adopt agriculture or horticulture from other regions outside. During this period the great towns and villages of Jericho and Beidha developed. The first plants they adopted were einkorn, emmer, and hulled two-row barley, but only a small part of their diet came from agriculture. In fact, these were really incipient agricultural villages or horticultural villages (possibly system C2).

As time went on, both Beidha and Jericho, as well as other pre-pottery Neolithic B sites (PPNB), expanded their populations considerably until they were living in towns. To meet this growth, they acquired more plants: emmer, naked two-row barley, oats, peas, lentils, and vetch, giving them a true agricultural economy (see table 4.5). By about 7000 B.C. the people of the Levant had become true Agricultural Villagers—my system E.

As the foregoing survey indicates, there is considerable data indicating how village agriculture developed in the Near East. The question remains, Why did it happen? To consider this problem, let us try testing my primary and secondary developmental models with the Near East data.

THEORETICAL CONSIDERATIONS

The developments for the Near East seem to be similar to the dichotomous developments in other Centers (table 4.4). As we have just seen for the Levant and perhaps for the Riverine regions, there is a five-stage development. It begins with the Hunting-Collecting Bands—system A—of the Upper Paleolithic and evolves into the Affluent Foraging Bands—system C—of Kebaran, about 16,000 B.C. Artifacts as well as food remains indicate that the Kebaran people used a series of subsistence techniques: hunting, trapping, plant collecting, fishing, and marine exploitation in different seasons in different environments. This foraging way of life became more affluent and efficient until early Natufian times—El Wad phase—about 9000 B.C. In the following millennia a sedentary village way of life gradually evolved, in the Mallaha phase of the late Natufian period, as well as at Mureybit. The subsistence base, however, was still that of very affluent foraging—my system C1—Foraging Villagers. Gradually, in pre-pottery Neolithic A—PPNA, 8000 to 7000 B.C.—there is evidence at Jericho, Beidha, Nahal Oren, Mureybit, and Abu Hureyra that people began to supplement their foraging subsistence system with a few domesticated plants—einkorn, emmer, vetch, and two-row barley (Miller 1988). These people could be classified as Horticultural Villagers—my system C2. The Horticultural Villagers increased their use of domesticated plants rather quickly, both by importation of more domesticates (naked two-row barley and hulled brown barley) as well as by domesticating some local plants, such as peas, lentils, vetch, and horse beans. The increased use of more domesticates as well as more intensive planting soon led these people to become agriculturalists. Thus, by pre-pottery Neolithic B (PPNB) and Neolithic times, about 6,000 to 7,000 years ago, a village agricultural way of life—system E—had evolved. The Levant thus presents an instance of villages evolving before agriculture, as was the situation in coastal Peru and perhaps in Belize.

This situation contrasts with my reconstruction of development in the Hilly Flanks of the south slopes of the Zagros Mountains. The first step there—from Hunting-Collecting Bands (system A, represented by Baradostian) to Destitute Foraging Bands (system B, represented by early Zarzi or Palegawra)—seems similar to the development in the Levant. The type of forager occupation in the Hilly Flanks, however, seems rather different from that in Kebaran, although less survey and excavation have been done in the Hilly Flanks. Early Zarzian sites seem to be smaller and less numerous than Kebaran sites. The Zarzian sites lack the permanent pit-like hut structures present in early Kebaran, although they have many pits, showing that storage was necessary, and they have many more grinding stones. The animal bones excavated, however, were less abundant and from fewer species than at Kebaran. This evidence suggests a much less affluent subsistence in early Zarzian than in late Kebaran or early Natufian. The present evidence suggests that the highland people were Destitute Foragers—my system B.

The next stage that I perceive in the Hilly Flanks is equally in need of more data, and some may question my interpretation of the pollen evidence as indicating that late Zarzian and Shanidar B2 had incipient agriculture. If my interpretation is correct, however, then in the Hilly Flanks there were Foraging Bands with Incipient Agriculture—system B1—that were contemporaneous (11,000 to 10,000 B.C.) with the Affluent Foraging Bands of the Kebaran Geometric and/or early Natufian (El Wad) phase in the Levant. There would be great differences between the two systems and certainly their artifact complexes—more grinding stones in the Hilly Flanks, larger and more numerous sites in the Levant—tend to confirm those dissimilarities.

The next stage in the Hilly Flanks, lasting roughly from 9500 to 7000 B.C., is represented by early and late Karim Shahir sites (Shanidar B1, Zawi Chemi, Ganj Dareh), the earliest phases at Çayönü, and others. There is more evidence of agriculture, not only in pollen data but also in material remains of domesticated einkorn, emmer, peas, lentils, and vetch, as well as of domesticated sheep. Yet, while there are remains of pit houses and round structures at these sites, there is no evidence that any of them was occupied year-round for a number of consecutive years. Thus, these dwellings were some sort of base camps rather than villages or hamlets, and the sites of this stage were therefore those of Agricultural Bands—my system B2. This situation was different from both the contemporaneous Levant stages—Mallaha (late Natufian) and pre-pottery Neolithic A—as well as from the Riverine subsistence at Mureybit and Abu Hureyra (Moore 1988), where house structures were found that had been occupied year-round for a number of consecutive years. In the Levant we had a Foraging Villager stage—system C1—evolving into a Horticultural Villager stage—C2. We have no evidence for a similar development in the Hilly Flanks. Between 7000 and 5500 B.C., however, both areas evolved into village agriculture—system E.

These dichotomous routes for the development from Hunter-Collectors to Village Agriculturists in the Near East Center seem to follow an identical sequence to those described for the Andean and Mesoamerican Centers, thereby confirming my models. The question now becomes, Do the same conditions or causes of change pertain to the Near East that occur in the other Centers, and can we test my primary and secondary developmental hypotheses with these data from southwestern Asia? The answer is a weak "yes," for while the data are good for confirming the secondary development, the information is not so convincing for the primary development. Let us first discuss the latter.

Table 4.4. The two evolutionary routes to Agricultural Villagers in the Near East

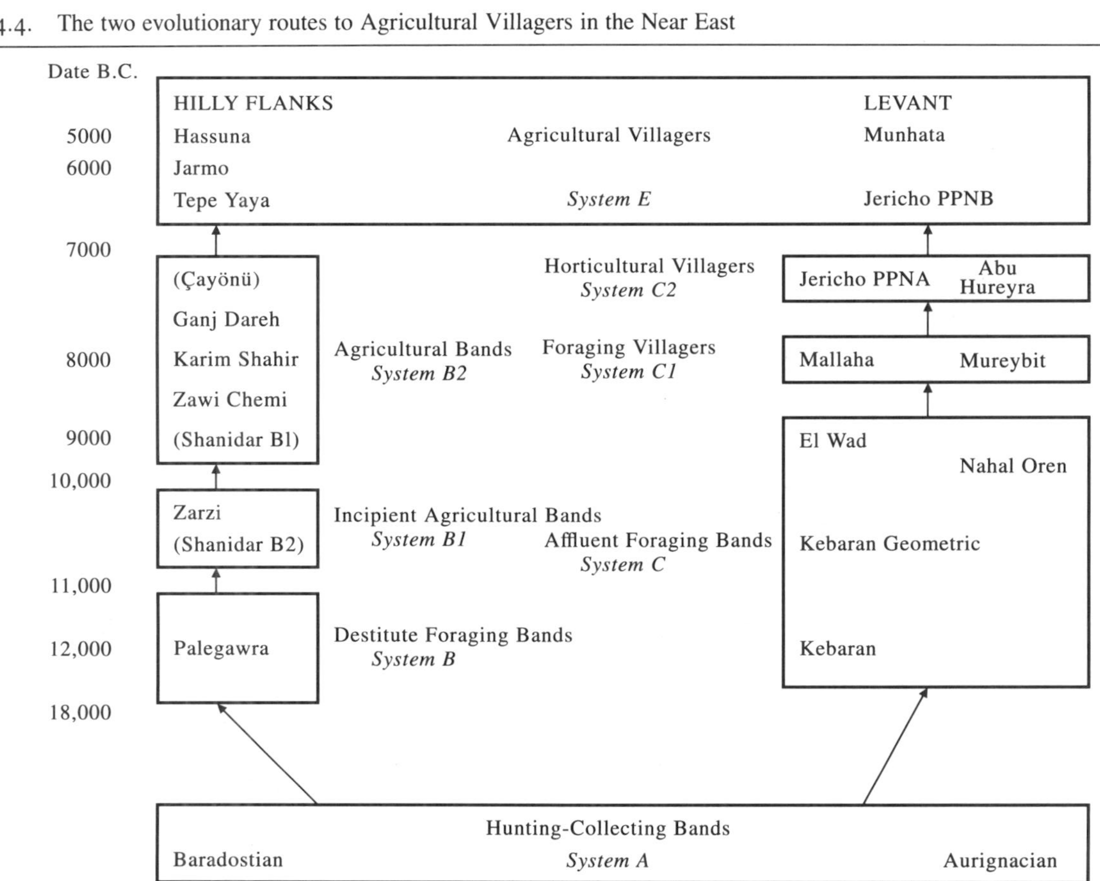

Table 4.5. Sequence of domesticated plants in the lowlands of the Near East

Levant Region	Upper Euphrates region	Khuzistan region Tigris River	Range of dates B.C.		Einkorn	Emmer	Hulled 2-row barley	Naked 2-row barley	Oats	Peas	Lentils	Vetch	Hulled 6-row barley	Bread wheat	Millet	Naked 6-row barley	Flax	Grapes	Olives	Dates
		Urbaid 4	3500	4000																x
Cadish			4000	5000														x	x	
Mersin			5000	5500												x				
		Sabz	5000	5500			x				x	x	x	x			x			
		Sawwan	5600	5800	x	x	x						x	x		x	x			
	Amouq		5000	5800		x	x	x	x											
		Jaffir	5600	6000		x	x	x	x			x								
	Bouqras		5900	6400	x	x				x	x		x	x	x	x				
	Ramad		6000	6500	x	x	x				x			x						
		Ali Kosh	6000	6800		x	x		x				x							
Beidha PPNB			6500	7000		x		x	x			x								
Jericho PPNB			6500	7200	x	x	x		x	x	x	?								
		Bus Mordeh	6800	7200	x	x		x												
Jericho PPNA			7250	8500	x		x													
Nahal Oren			7000	8500		x														
	Mureybit		7000	8500								?	x				x			
	Abu Hureyra		7250	8500							x	x								

Table 4.6. Sequence of domesticated plants and significant pollen in the highlands of the Near East

Anatolia Mountain Basin region	Zagros Mt. region	Kardish Hills region	Range of dates B.C.		Artemisa pollen	Plantago pollen	Cerealia pollen	Emmer	Peas	Lentils	Vetch	Naked 6-row barley	Einkorn	Hulled 2-row barley	Bread wheat	Hulled 6-row barley
Can Hasan			5000	5500				x	x							x
		Hassuna	5500	6000									x	x		
		Matarrah	5000	5500				x						x		
Hacilar ceramic			5000	5800				x		x		x	x	x	x	x
Çatal Hüyük			5600	5800				x	x			x	x		x	
	Tepe Guran		5500	6200								x		x		
		Jarmo	6000	6500				x	x	x	x		x			
Hacilar aceramic			6000	7000				x	x	x	x	x				
	Ganj Dareh		6500	8000								x				
		Çayönü	6500	8000				x	x	x	x					
	Shanidar B1		8500	9500			P									
	Zawi Chemi		8500	9500			P									
	Shanidar B2		9500	11000			P									
	Zeribar-Palegawra		11000	13000	P	P	?									

Although some of the sufficient conditions, or triggering causes, of my primary development model are not well documented, the necessary conditions of the primary model do seem to pertain to the Hilly Flanks of the Near East Center. Wild ancestors of many domesticable plants—emmer, einkorn, barley, peas, lentils, and others (J. Renfrew 1973)—existed in this period (H. Wright 1976) in the Hilly Flanks. Furthermore, there is evidence that intense interaction was possible between this subarea and others in the Center. For example, obsidian spread throughout the Center in Mesolithic or Epipaleolithic times (Renfrew, Dixon, and Cann 1966), and Mediterranean shells appeared in the Hilly Flanks (R. S. Solecki 1961) and copper in Çayönü (Redman 1978). The spread of many domesticated plants not native to the region might be taken as further evidence that the Hilly Flanks subarea was part of a sphere of interaction.

Although no ecological studies have been directly focused on this subarea, it does seem to have a cold winter season with freezing weather and/or snow when few food staples, particularly plants, would be available. So perhaps the necessary condition of a harsh season does pertain.

Also, most of the sites in the Hilly Flanks are in steep canyons surrounded by mountains, making it impossible to exploit all ecozones from a single base. For instance, Shanidar Cave is in the Zab River valley at about 700 meters above sea level. The valley bottom is grassland sustaining gazelles, onagers, and wild pigs, while the area around the cave, at 822 meters, was scrub oak forest with wild legumes, almonds, and aurochs. The surrounding Bardost Dagh Mountains, which rise to 1900 meters, are covered with oak-pistachio forest and abound with almonds, wild wheat, and barley as well as with wild sheep, goats, and red deer. Even a base camp located at 1300 meters could not be used to exploit the resources of both the river valley and the mountains in a single day's foray. Therefore, the necessary condition of the total ecozone not being exploitable from a base seems to pertain.

One further necessary condition—a variety of ecozones—also seems to pertain in the Hilly Flanks. While comparative ecological studies of the regions surrounding the various excavated caves have not been undertaken, the area around Çayönü seems to have at least a Riverine zone and a desert steppe below 700 meters plus oak forests above 800 meters. Palegawra Cave, at 900 meters, and Ganj Dareh, at 1400 meters, are both in high valleys in the oak-pistachio mountains with grasslands above and below them. These are indications of ecological diversity, with the areas at lower elevations containing a different set of ecozones from those at higher elevations.

Thus the Hilly Flanks subarea seems to have all the necessary conditions of my primary development model (table 4.7). The next question is, Do the sequences of the Hilly Flanks have the sufficient conditions of the primary model, that is, are the sequences of the causes of change in this Center the same as they were in highland Mesoamerica and the Peruvian Andes?

Obviously, the shift from Baradostian, at 25,000 B.C., with system A, Hunting-Collecting Bands, to the Destitute Foraging Bands (system B) of Palegawra, in about 12,000 B.C., fits my model. The sufficient conditions for this change in my primary development hypothesis do seem to have occurred in the Hilly Flanks. Pollen evidence from Lake Zeribar (H. Wright 1977) indicates that the environment underwent a change between Baradostian and Palegawra that reduced the availability of food staples and made the harsh season harsher (Pullar 1977).

In Shanidar B2 or even in Palegawra times there were hints of seasonal scheduling and the development of broad-spectrum subsistence options including (1) seed storage and (2) seed collection. Further, between Shanidar B2 (or Zarzi) and Karim Shahir we see possible evidence of a decrease in residential mobility and an increase in logistic mobility. Whether there were increasing numbers of macrobands has not been well documented for this period.

In Zawi Chemi times, as evidenced by the increase in *cerealia* pollen size (see table 4.3), seed collecting might have affected the ecosystem and the genetics of some of the seeds collected. This might have led to more planting of domesticates, thus making Zawi Chemi people Incipient Agricultural Bands, system B1. This confirmation of the first part of my primary development hypothesis is perhaps not as definite as it is for Tehuacán, but it is as good as or better than it is for Ayacucho.

Now let us examine the model for development from Incipient Agricultural Bands (system B1) to Agricultural Bands (system B2) to Agricultural Villagers (system E) as it applies to the highland areas of the Near East Center.

Both Çayönü and Ganj Dareh give ample evidence of food storage for a harsh season. The sequential phases of Çayönü are a prime example of a decrease in residential mobility that seems to require domestication or importation of more

Table 4.7. Necessary and sufficient conditions as a positive-feedback process for primary development in the Hilly Flanks of the Near East

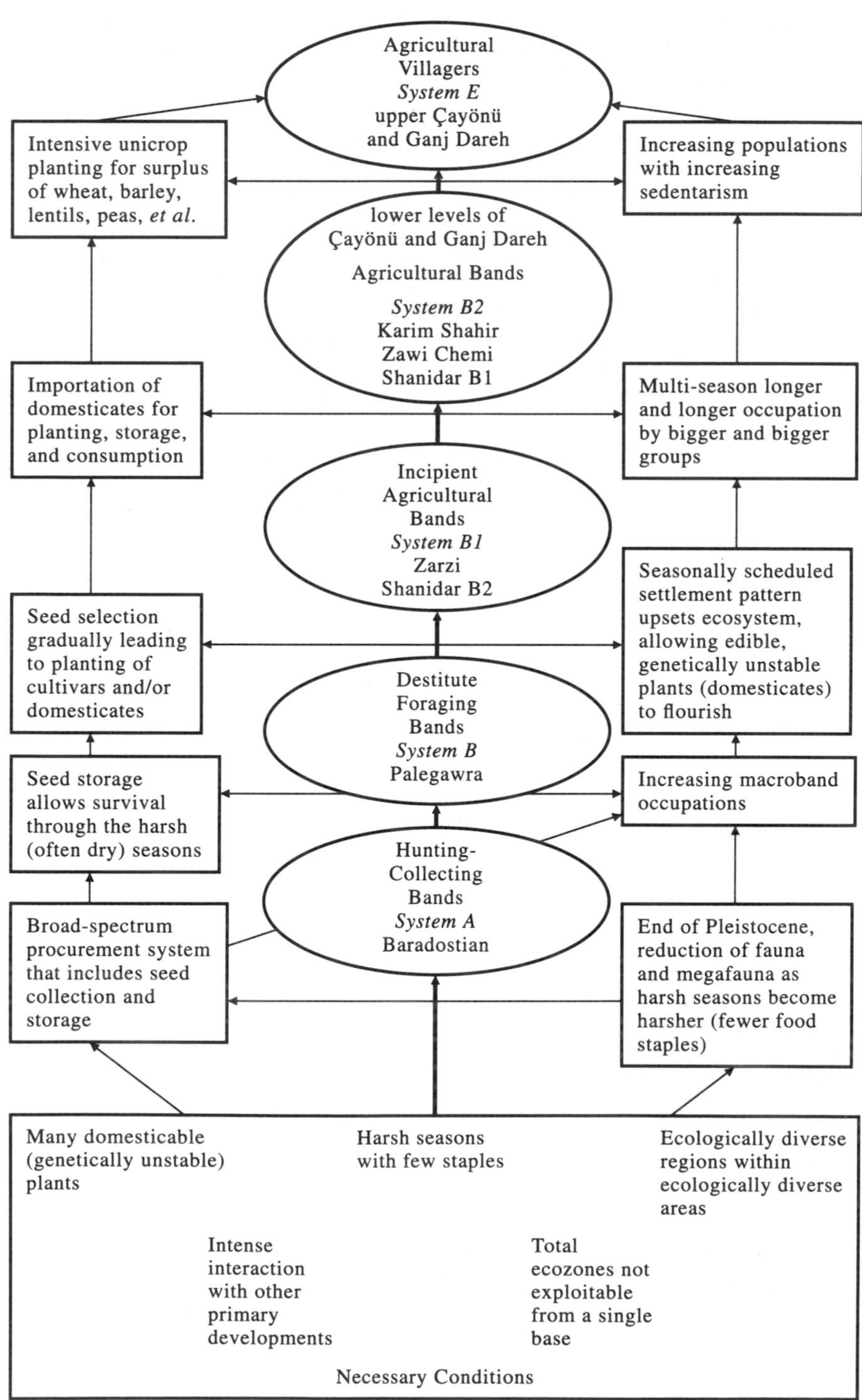

plants, such as emmer, einkorn, peas, and lentils. Data from the Çayönü upper-level phases give evidence that more food was grown for storage, resulting in increased populations living longer in one spot and leading to the agricultural village system of Jarmo. Whether use of hybrids or crop specialization increased is unknown, but certainly Redman suggests increased agriculture in a feedback system with longer residency and increasing population (Redman 1978). Once again there seem to be hints of confirmation for my primary development model, but the evidence is not as good as it is for either Tehuacán or Ayacucho. My primary model is not adequately tested by the data from the Hilly Flanks of the Near East.

My secondary development hypothesis, however, is confirmed by data from the lowland region, the Levant (see table 4.8). Some of the necessary conditions for the lowlands, like those for the highlands, show the presence of a number of potential domesticates: wheat, barley, peas, horse beans, vetch, lentils, and others. The presence in the Hilly Flanks of obsidian from the highlands and marine shells from the coast indicates that the lowlands were part of an interaction sphere that included the area of primary development. Unlike the highlands, however, the lowlands had abundant foods throughout the year. These resources often existed in relatively lush locales such as oases, lakeshores, and coastal plains, which were circumscribed by regions with considerably fewer resources, for example, deserts or mountains.

Further, these lusher bases often served as centers from which a number of other environmental zones could be exploited, as Higgs and Vita-Finzi have so aptly shown (1972). According to their study, while one "catchment basin" may contain a number of ecozones—seasonal marshes, sand dunes, arable land, and grasslands surrounding a spring or oasis—the next oasis will have much the same ecozones. The lowlands then may have regional ecological diversity like the Hilly Flanks, but unlike the highlands, the lowland subarea has areal uniformity. The necessary conditions of my secondary development model are thus very much in evidence in the Levant. What about the sufficient conditions, or triggering causes?

At the end of the Pleistocene the early Aurignacian and Ahmarian hunters of specialized game were faced with variable dry-wet-dry cycles as well as with rising sea levels that resulted in fluctuations in the food supply (Phillips 1986). These scarcities caused them to take on a broad-spectrum procurement system that allowed them to become noticeably more sedentary with increasing populations that required still further resource specialization. This positive-feedback situation lasted throughout Kebaran and early Natufian (El Wad) times and eventually caused people to develop from Affluent Foragers (system C). In Mallaha times, at about 9000 B.C., they reached the status of Village Foragers living in sedentary hamlets, system C1.

Apparently this sedentariness resulted in population increase, which meant that people outgrew their circumscribed oases or lush village areas and had to improve their food production still further. One of the ways this problem was solved by pre-pottery Neolithic A times (8000 B.C.) was by the use of domesticates such as emmer, einkorn, lentils, and others. These were probably imported

Table 4.8. Necessary and sufficient conditions as a positive-feedback process for secondary development in the Levant of the Near East

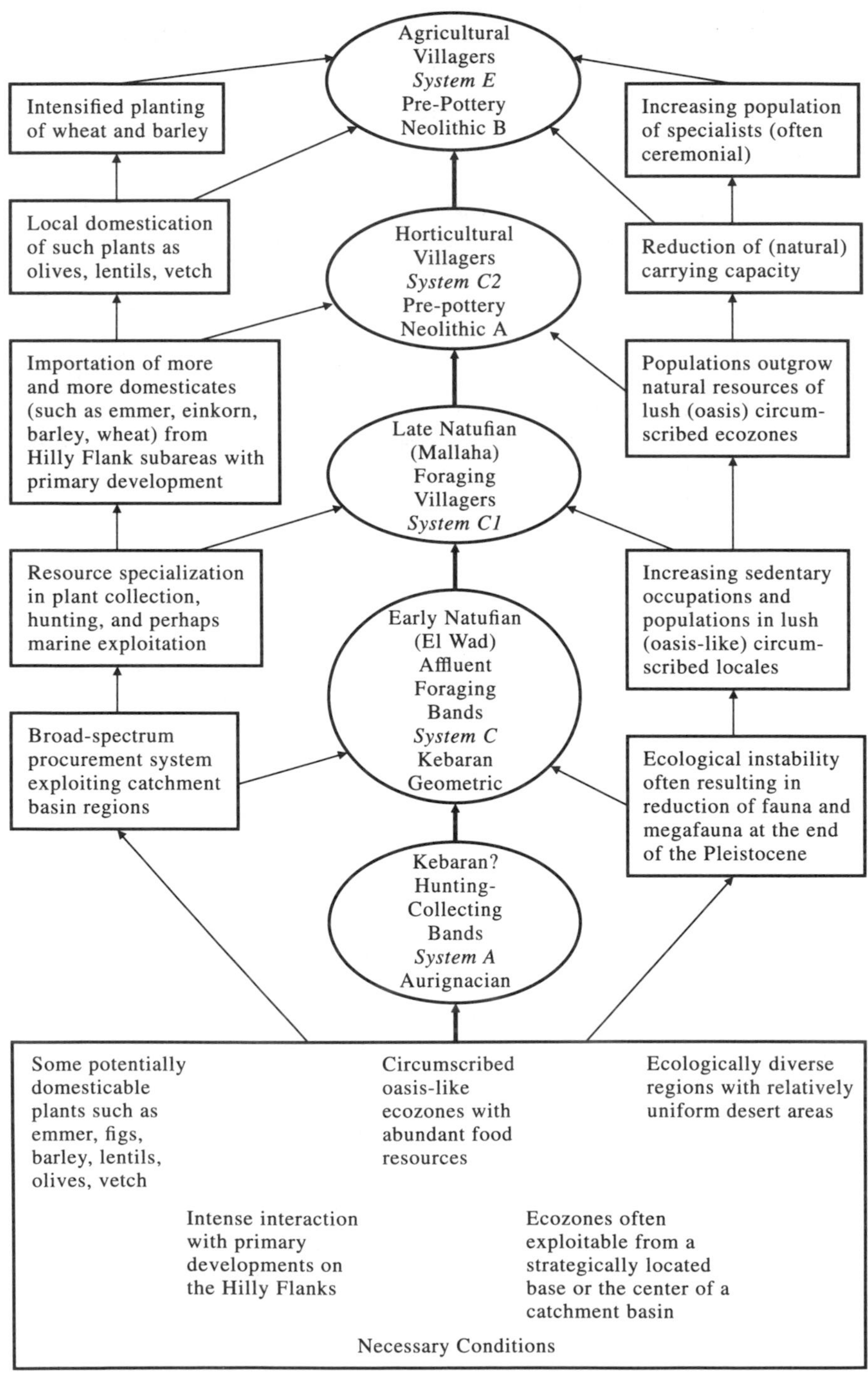

from the highlands, with which the lowland people had long been interacting for obsidian trade and the like. Thus, population growth, diminished carrying capacity of circumscribed hamlet locales, and importation of domesticates—all working as a positive-feedback situation—caused Village Foragers, system C1, to evolve into system C2, Horticultural Villages or hamlets. The food supply improved and increased as the people cultivated imported plants in addition to those domesticated locally (vetch, horse beans, oats, and hulled six-row barley). These developments in turn resulted in further population growth and lowered the carrying capacity of the surrounding areas so that more farming was needed. This positive-feedback situation resulted in people rapidly becoming Agricultural Villagers—system E—by Pre-Pottery Neolithic B times, about 7000 B.C. (see Table 4.8).

The Levant, then, is an excellent confirmation of my secondary developmental hypothesis. As we have already noted, the data from coastal Peru also confirm this same hypothetical development; the data from Mesoamerica were inadequate to either confirm or deny it. Future chapters will provide other examples of secondary developments that confirm this hypothesis, and most of these developments will be in non-Centers. The only other Center to be considered is the Far East (chapter 5), and here again, the basic evidence is lacking. However, the Far East data will, for the first time, introduce evidence to test the hypothesis of a tertiary development.

CHAPTER 5

The Far East

The Far East Center stands out in sharp contrast to the three Centers previously described in that it is totally lacking any materials pertinent to the crucial transition from Hunting-Collecting Bands—system A—to Village Agriculturists—system E. Further, the ecological studies needed to determine some of the Necessary Conditions for either the primary or secondary developments have not been attempted, and archaeological research on the problems of the origins of agriculture has been meager. Also, peculiarly enough, the only complete sequential data we have leading to village agriculture (this time rice) is not from the Center itself but from Japan, a peripheral area of the Far East Center.

The data from Japan, however, seem to test not the primary and secondary developmental hypotheses but a third, the so-called tertiary developmental hypothesis, which follows the route from Hunting-Collecting Bands to Efficient Foraging Bands, to Semisedentary Bands with Domesticates, to Horticultural Villagers to Agricultural Villagers. In testing this hypothesis, the materials from the Far East Center have considerable value, but much more research needs to be undertaken. Our knowledge of the Center is limited, and although I have been trying to work on the beginnings of agriculture there for a generation, I have never been to the area to obtain the needed firsthand knowledge.

THE ENVIRONMENT

I define the Far East Center as including all of present-day China, Korea, Taiwan, and Japan (see K. C. Chang 1971). Unlike some others, I exclude Southeast Asia and the Canton Hinterlands and also exclude Tibet and the Sinkiang-Mongolia region since the latter two have yielded no relevant archaeological remains or sequences (Treistman 1972). Following this reasoning, however, I would also have to eliminate the Khingon Mountains in Manchuria and Korea in the north and most of inland southwest China, that is, the Southwest Uplands, South Yangtze Hills, Central Mountain Belt, and the Szechwan Basin, all of which have potential for yielding relevant remains. My real reason for excluding Tibet and Sinkiang-Mongolia is that they do not have one of the main prerequisites for a Center—

potential domesticates or stable agricultural crops (K. C. Chang 1970). Although Southeast Asia does have some important native plants—for example, rice, yams, taro, gourds, and water chestnuts—I exclude it because it has little or no relevant archaeology pertaining to the crucial period of transition from Hunting-Collecting Bands to Agricultural Villagers.

In terms of relevant subareas or environmental life zones for discussion, we are severely limited; basically we are considering: (1) a northern inland zone (the loessland); (2) north(east) China, a coastal zone (the Yellow Plain); (3) a middle coast, roughly in the delta of the Yangtze River (Yangtze Plain); (4) a southern zone—namely, the southeast coast and Taiwan; and (5) Japan (see fig. 10). Before we consider the Far East mainland, let us look at Japan, the only subarea with relevant archaeological remains. Japan has a complex set of ecological zones. Of key importance in terms of relevant archaeological remains are the coastal zones, both on the Pacific (east) coast and on the west coast facing China. These coastal ecozones are unlike the coastal ones in China or, for that matter, in the Near East, Mesoamerica, and the Andes. Rather than being zones with marine resources in circumscribed ecozones, Japan's coastal zones are almost limitless.

The interior of Japan has a number of ecological zones—high mountain zones above the timberline, a coniferous forest zone, a deciduous forest zone, and, to the south, a subtropical laurel forest ecozone (Akazawa 1981). Unfortunately, we have few relevant archaeological materials from these inland zones; equally unfortunate, microenvironmental studies of the kind done in the Near East and Mesoamerica have not been attempted for the Japanese coast. With this sketch of Japan in mind, let us consider the environment of the heartland of the Far Eastern Center.

As with other Centers, I envision the heartland subareas as basically dichotomous, consisting of highland and lowland areas. The northern inland zone is relatively harsh, equivalent to the highlands of the Near East, Mesoamerica, and the Andes. The relatively lush, lowland coastal zone consists of the northern, middle, and southern subareas mentioned above (see fig. 10, Cressey 1951).

The northern inland zone called the loessland is in the northernmost area northeast of the Yellow River, or Huang Ho, but in its southern portions the provinces of Honan, Shansi, and Shensi lie both north and south of that river. This area is covered by fertile loess; while some of it is now grassland, much of it was once forested. The area is now under intensive agriculture, making it difficult to reconstruct the original ecozones. Some pollen profiles have been done, and these, plus the rugged topography and limited seasonal and annual rainfall, hint that here, too, we are discussing regional diversity with great seasonality, including a long harsh winter season and a short hot, wet summer. This region is the homeland of broomcorn and foxtail millet, soybeans, artichoke, garlic, leeks, and many other plants. Varieties of these are more prevalent in one part of the subarea than in another. There are many small drainages as well as big valleys of rivers like the Fen, Wei, and Wu-ting, which make regions with possible domesticates easily accessible to one another.

Table 5.1. Some native cultivated and/or domesticated plants of the Far East

Scientific name	Popular name
**Abutilon avicennae*	Chinese jute
Aleurites cordata	tung oil
**Aleurites fordii*	tung oil
**Allium bakeri*	scallion
Allium fistulosum	Welsh onion
**Allium ramosum*	leek
Allium sativum	garlic
Angelica kiusiana	leaf vegetable
**Aralia cordata*	udo
**Aralia quinquefolia*	ginseng
**Arctium major*	burdock
Arundinavia sp.	bamboo
**Boehmeria niveae*	ramie (fiber)
**Benincasa hispida*	winter melon
Brasenia schreberi	water shield
Brassica alboglabra	Chinese kale
**Brassica campestris*	rapeseed
**Brassica cernua*	leafy vegetable
**Brassica chinensis*	Chinese cabbage
**Brassica chinensis* var. *oleifera*	oil cabbage
Brassica japonica	water mustard
**Brassica juncea*	mustard seed oil
Brassica pekinensis	celery cabbage
**Brassica rapa*	turnip
Camellia oleifera	tea oil
**Camellia sinensis*	Chinese alum
Canarium pimela	Chinese olive
**Carya sp.*	Chinese hickory
**Cannabis sativa*	hemp
**Castanea henryi*	Chinese chestnut
**Chaenomeles sp.*	Chinese quince
Chrysanthemum coronarium	chrysanthemum (vegetable)
Citrus aurantium	sour orange
Citrus reticulata	Mandarin orange
Citrus sinensis	sweet orange
**Cinnamomum cassia*	cassia (spice)
Clausanea lansium	wampi (fruit)
**Corylus sp.*	Chinese hazelnut
Crataegus pinnatifida	Chinese hawthorn
**Cucumis conomon*	pickling melon
**Cucumis sativus*	cucumber
Dioscorea batatas	Chinese yam
**Discorea esculenta*	Chinese yam
**Echinochioa frumentacea*	Japanese millet
**Eleocharis tuberosa*	sedge tuber
**Eriobotyra japonica*	loquat (fruit)
Euphoria longana	longan (fruit)
**Fagopyrum esculentum*	buckwheat
**Fagopyrum tataricum*	tatar buckwheat
**Fortunella japonica*	kumquat

Table 5.1. (*continued*)

Scientific name	Popular name
**Ginko biloba*	ginkgo
**Glycine max*	soybean
Hemerocallis fulva	daylily (vegetable)
Ipomoea aquatica	water spinach
**Juglans regia*	walnut
Lactuca denticulata	Chinese vegetable leaf
**Lagenaria siceraria*	bottle gourd
**Lilium tigrinum*	tiger lily
**Litchi chinensis*	litchi
Malas prunifolia	Chinese prune
**Malva verticillata*	mallow
Morus alba	mulberry
Nasturtium indicum	nasturtium
**Nelumbium speciosum*	lotus
**Oenanthe stolonifera*	water celery
**Oryza sativa*	rice
Panicum miliaceum	broomcorn millet
Phaseolus angularis	adzuki bean
**Phyllostachys sp.*	food bamboo
Polygonum hydropiper	knotweed (vegetable)
**Prunus armeniaca*	apricot
Prunus mume	Japanese apricot
**Prunus persica*	peach
Prunus pseudocerasus	Chinese cherry
Prunus salicina	Chinese plum
Pueraria thunbergiana	kudzu vine
**Pyrus pyrifolia*	sand pear
**Raphanus sativus*	Chinese radish
**Rhus vernicifera*	lac tree (for varnish)
**Rheum palmatum*	medicinal rhubarb
Sagittaria sinensis	arrowhead (root)
**Sagittaria sagittifolia*	elephant ear
**Sapium sebiferum*	Chinaberry
**Setaria italica*	foxtail millet
**Stachys sieboldi*	Chinese artichoke
Stizolobium hassjoo	velvet bean
**Strobilanthes flaccidifolius*	indigo dye plant
Thea sinensis	tea
**Trapa natans*	water chestnut
Viola verucunda	violet (vegetable)
**Vigna angularis*	adzuki bean
**Wasabia japonica*	horseradish
Xanthium strumarium	cocklebur (vegetable)
**Zanthoxylum bungei*	Chinese pepper
**Zingiber officinale*	ginger
**Zizania latifolia*	Manchurian water rice
**Zizyphus sativa*	Chinese jujube (fruit)

*Domesticated in Harlan (1971).

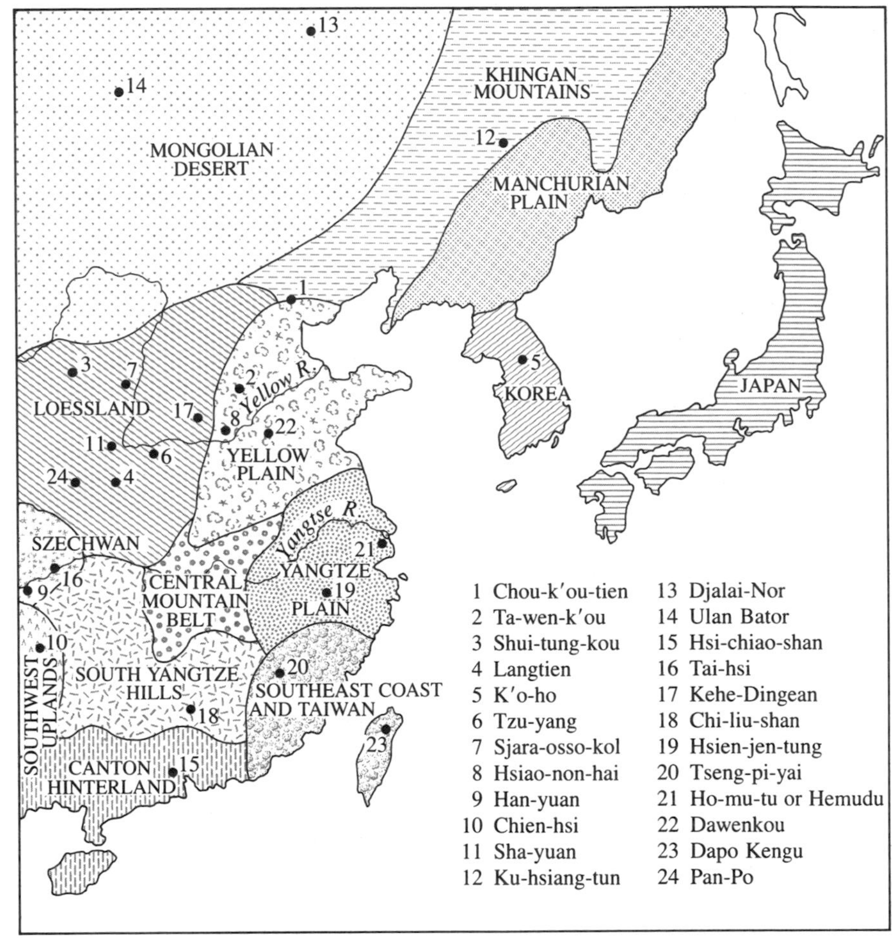

Figure 10. The environmental life zones and relevant sites of the Far East

This subarea has many of the characteristics of the harsh highland zones of the other Centers. However, it is difficult to determine just how comparable it is, for no basic ecological studies have been reported. K. C. Chang has long asserted that this area is where the first agriculture occurred (1971). Based upon the similarity of the poorly studied ecological factors (necessary conditions) to those in the other Centers, I find myself in basic agreement. Unfortunately, the archaeological data needed to test this hypothesis have not been uncovered (Zhimin 1988).

The same may be said of the three coastal subareas. Superficially they resemble the lush lowland areas of the Near East, coastal Peru, and coastal Mesoamerica, and they seem to have similar necessary conditions: lush, circumscribed ecozones with food staples that often can be exploited from a central base, more limited seasonality, and accessibility to the inland area as well as to other areas with domesticates. Whether there is ecozone diversity within each region and ecological uniformity in various subareas cannot be determined at this time, but each of the three coastal subareas is slightly different from the others.

The northern coastal subarea, called the Yellow Plain (see fig. 10), runs from Shantung province in the south to Liaoning in the north and includes Hopeh province on the coastal plain east of the loess hills. This area has limited annual rainfall (500–1,000 millimeters), which falls mainly in the winter; there are occasional winter frosts. Inland vegetation is deciduous forest, with aquatic food resources in the coastal and riverine areas. Chou-K'ou-tien and the Dawenkou sites are within this zone, but our archaeological knowledge of the Mesolithic period there is limited.

The middle coastal area, called the Yangtze Plain (see fig. 10), is becoming better known, at least for the early Neolithic time period. This coastal area runs south from Kiangsu to Chekiang province east of the uplands. The dominant feature of the area is the deltaic ecozone of the mighty Yangtze River. To the south, rainfall approaches 1,500 millimeters or more a year, and while some falls in every month, much of the rain falls in the winter. Frosts are rare; seasonality is not great; and vegetation is subtropical. There are many streams, marshes, and lakes full of resources, and seafoods are even more abundant. Richard Pearson (1977) has reported on the recent great archaeological advances in this subarea, and there is hope that the Mesolithic archaeological barrier soon will be breached. One of these days, some Paleolithic remains may also be uncovered.

The final coastal area (or perhaps two areas) consists of the southeast coast with Taiwan and the Canton hinterland (see fig. 10); ecologically this area extends from Fukien province well into North Vietnam and nearby islands, which are really outside our discussion. Much of what we know about the relevant archaeology of this area or areas comes from research on Taiwan. This is an area of abundant rainfall—1,750–2,500 millimeters annually—with winter being a wet season and fall a typhoon and/or monsoon season. The coastal plain is relatively narrow except in the far south, and there are a series of uplands before one reaches the hill country and/or plateaus of Hupeh, Hunan, and Kweichow provinces to the north and west. Vegetation is tropical, and marine resources are abundant.

In these Far Eastern coastal subareas there are hints of environmental conditions similar to the necessary conditions in our other Centers, but the resemblances are not well documented and are vague and unsatisfactory. Surprisingly, studies of changes in the environment at the end of the Pleistocene in the Far East are relatively good (see, e.g., Aigner 1972), for in the other Centers they are often poor. One of the major changes in the Far East at the end of the Pleistocene was the regression of sea level. At 10,000 B.C. the sea level was 40 meters lower than present levels, and 20,000 years ago it may have been 140 to 200 meters lower. At that time many islands (including part of Japan) were connected to the Asian mainland.

Temperature was also probably depressed. It is estimated that at the height of the final Tali glaciation, 20,000 years ago, temperatures were 8°C less in northern China than at present (Aigner 1972). Although it was not a great deal colder then, precipitation seems to have been heavier than it was during the Pleistocene, resulting in harsher winters with heavier snowfall and more glaciation in the north and in the mountains to the west as well as in Japan. The end of the Pleistocene, with these major changes in temperature and rainfall, brought about drastic changes in vegetation—at least inland, in the north, and on Japan—with a concomitant reduction in fauna. In the north, at Sjara-osso-kol, woolly rhinoceros, mammoth, and other animals disappeared fairly rapidly, while the Tzu-yang fauna of South China included many animals (including mammoth) that became extinct at this time. Thus one of the triggering mechanisms, or sufficient conditions, to start people on the route from Hunter-Collectors to Foragers (with or without agriculture) existed in the Far East in conjunction with a number of basic necessary conditions. Now let us examine the meager archaeological evidence of what seems to have happened.

RELEVANT ARCHAEOLOGICAL DATA

At the outset I should note that there are major problems with Far Eastern archaeology, not the least of which is whether the European-derived terms of Lower, Middle, and Upper Paleolithic as well as Mesolithic are applicable to the Far East. In Europe the basic divisions for Lower and Middle Paleolithic were based on hand axe types and Clactonian flake-type sequences, while the Upper Paleolithic was based upon blade industries. No such sequences of these tool types exist in East Asia, which is characterized mainly by sequences of chopping-chopper tools, with some flake and microblade tools possibly occurring in the Late Pleistocene. In fact, the only way of comparing the European and Asian Paleolithic is in terms of time: Lower, Middle, and Late Pleistocene and Recent. Fortunately, much of the Asian Paleolithic occurred well before the beginnings of agriculture, so we can sidestep the problem of terminology in this discussion.

The Asian Paleolithic, however, does eventually seem to evolve into a so-called Mesolithic that is pertinent to problems of origins of agriculture, so let us say a word or two about it (K. C. Chang 1971). The basic Lower Paleolithic was best represented at the famous Chou-K'ou-tien site, where the skulls and bones of *Sinanthropus* were discovered and excavated between 1924 and 1937. Localities 1, 13, and 15 there produced most of the stone tools, which consisted mainly of (pebble) choppers, end- and side-retouched flakes, pointed flakes, possible hammers, and awls. Since these early excavations, a few similar tools of roughly the same age—Middle Pleistocene—have been discovered at eleven or more sites near K'o-ho on the southwest Shansi plain, at the site of Shui-mo-kou and Huikssing Kou in Honan, at sites in Szechwan and Kwengsi in south China, and at Langtien in Shensi, where a Peking Man–like skull was also found. Some of the pointed pebble choppers of this last site are similar to those of the Middle Paleolithic remains that have been reported for the Ordos of Inner Mongolia and the Fen River valley in Shansi. From many standpoints, the Ordos area has the best representation of Middle Paleolithic remains.

While the Upper Paleolithic industries of the Late Pleistocene continued with a few chopper and flake tool types like those of earlier periods, the succeeding Ordos culture had basically unifacial industries, with flakes and blades struck from prepared discoidals, tortoise, and prismatic cores, with or without faceted striking platforms. Some of the flakes are triangular and retouched like Mousterian ones to the west, and the blades and burins bear resemblances to Aurignacian types found in Europe.

Perhaps contemporary with this industry was one found in Hsiao-non-hai near An-yang in northern Honan. Similar remains also occurred in a cave at Shantung. While some of the flake and blade tools at these sites were struck from cores similar to those of the Ordos culture, these sites differ in having fewer burins and more microlithic tools. The latter have led some to suggest that this stage might be a developmental one leading to microblade–tongue-core industries of the Mesolithic, but adequate proof is still lacking.

Generally resembling these relatively late northern industries are three contemporaneous ones from western China. Perhaps most similar is the Han-yuan assemblage from Szechwan, which has microblades and tongue-shaped cores as well as Ordosian flake tools. Here again dating is so inadequate that this industry may yet be found to belong to the Mesolithic rather than the Paleolithic. Still farther south and farther from the northern Upper Paleolithic were the industries that occurred at Chien-hsi in Kweichow province, where a series of large irregular retouched flakes was uncovered, and at Yi-liang in Yunnan, where large retouched flakes as well as blades and choppers occurred. It seems that both industries were developing in different directions from those of northern China, and perhaps the split between the so-called Mesolithic and Neolithic began at this time.

Now we must face the problem of the so-called Mesolithic, for in the Far East, as well as in the Near East and Europe, it is the period when village agriculture began to develop, roughly at the end of the Pleistocene or in the early Recent period. Again, a temporal division is one basis of comparison, for there are few similarities in the East to the artifact assemblage of the West. In the Far East there really are no microlith types like those found in Southwest Asia or in Europe; instead, in the northern Far East during this period there were microblades and tongue-shaped cores, while in the south there were pebble-tool industries. Perhaps the simplest way out of this dilemma is to define the Far Eastern Mesolithic vaguely (and perhaps unsatisfactorily) as the general transitional stage between the Paleolithic, with its chopping-chopper tools, and the Neolithic, with its village agriculture and ceramics. This definition would put the Mesolithic stage in the Far East somewhere between 15,000 and 6000 B.C.—perhaps earlier than that stage in the New World but not as early as, or only slightly later than, the similar stage in the Near East.

The Hunting-Collecting Band base (system A) out of which village agriculture developed would seem to fall in the Early Mesolithic (see table 5.2). There seems to be a definite dichotomy at this time between what Jason Smith called Nanamt—Northeast Asian–Northwest American Microblade Traditions (J. Smith 1971)—and the Hoabinhian (and/or Hoabinhian-like) pebble-tool tradition (Gorman 1969).

It is difficult to obtain adequate distributional data on Far Eastern sites belonging to the two traditions, but the microblade–tongue-core sites generally seem to be north of the Yangtze River. The southwesternmost site is Fu-lin in central Szechwan; a few sites are in Shensi and Honan provinces; and Hopeh and Shansi provinces have the Hutoulian sites (Pei 1977). Moving northwest into Manchuria and the lower Amur River region, we find such sites as Kuhsiang-tun and Djalai-Nor (see fig. 10). Connected or closely related are the Outer Mongolian sites near Ulan-Bator. These were found long ago by Nels Nelson and others in the Gobi Desert (Nelson 1937). Other related sites may be the Sha-yuan assemblages at Sjara-osso-kol, Shih-wu and Hsiao-non-hai in the Ordos area (K. C. Chang 1971), and the scores of sites that belong to what Y. A. Mochanov has called the Dyuktai tradition of the Trans-Baikal area and northeastern Siberia (Mochanov 1978). Many of the latest preceramic sites found in Korea (Pearson 1977), Manchuria, Japan, and the Kamchatka Peninsula are also characterized by tongue-shaped cores and microblades and are obviously related (Ikawa-Smith 1982).

It is mainly in these regions that sites have been adequately excavated and dated (Sha-yuan in Shensi province excepted), and these sites date from about 20,000 B.C. to 12,000 B.C., when the Japanese started making pottery. If these dates are applicable to the related sites in China, and I think they are, then there is a long gap between the Mesolithic and the following Pan-P'o or Yang-shao Neolithic village sites with agriculture and well-made pottery (K. C. Chang 1971). The complete lack of microblades in the following Yang-shao sites suggests that a gap of 6,000 to 7,000 years does in fact exist in the archaeological record. This may have been when the process of plant domestication actually began. We shall return to this matter shortly when we discuss the evidence for the Neolithic in inland northern China. Before we do, let us consider the other Mesolithic tradition in the Far East.

Like the microblade-tongue core (or Nanamt) tradition, the Hoabinhian pebble-tool tradition, characterized by choppers and adzes and/or hoes, was first defined outside the area of nuclear China—this time in Southeast Asia (Gorman 1969). The Hoabinhian tradition unfortunately is poorly dated even there, and its occurrences in China are not extensively documented. Some of the late preceramic sites in Taiwan, however, do seem to have this sort of complex (K. C. Chang 1967, 1969), while to the east, on the mainland near Hong Kong, is the Hsi-chiao-shan site of Swangtung as well as others to the north. Also, according to the catalogs and collections in the American Museum of Natural History in New York City, some of the unreported lower levels of the caves and sites that

Table 5.2. Relevant archaeological sequences in the Far East

Japan subarea	North China Interior or Loessland subarea	Yellow Plain or North China Coastal subarea	Yangtze Plain or Central China Coastal subarea	Southeast China Coastal subarea	Dates
Yayoi					
Final Joman	Shang	Shang	Shang	Shang	1000 B.C.
Late Jomon	Lung-shan	Lung-shan	Lungshanoid Shang dong Hu-shu		2000 B.C.
			Liang-chu		
	Miao-ti-kou				3000 B.C.
Middle Jomon	Yang-shao	Dawenkou	Late Ching-lien-kang	Dapon Kengu	4000 B.C.
Early Jomon	Peiilang-Cishan		Early Ching-lien-kang		
	Laoquantai		Ho-mu-tu (or Hemudu)		5000 B.C.
					6000 B.C.
Initial Jomon			Hsien-jen-tung		
					7000 B.C.
	Kehe-Dingean				8000 B.C.
			Tseng-pi-yai		9000 B.C.
					10,000 B.C.
		Upper Chou-K'ou-tien	---------------	Hoabinhian-like	11,000 B.C.
Microblade complexes					

Nels Nelson tested along the Yangtze River might belong to this tradition because of their hand hoes and choppers.

Other sites of the Hoabinhian tradition are found at Tai-hsi in the Red Basin of Szechwan (K. C. Chang 1971). Farther north in Shansi is the Kehe-Dingean tradition, which has large flakes and picks, hand axes, choppers, and stone balls, as well as small choppers, flakes and adzes from the upper levels of Chou-K'ou-tien, near Peking. To the south, the chopper tradition seems to occur in caves near Chi-liu-shan in Kwangsi province and in rock-shelters at Hei-ching-ling in Yunnan province and on the Pearl River delta of Kwangtung. Of particular significance in the central Yangtze region are the cave sites of Hsien-jen-tung in Kiangsi province, Wong-yuan-wei in Kwantung, and Tseng-pi-yai (Pearson 1979) in Kwangsi. The lower layers at these sites contained Hoabinhian-like pebble choppers, chisels, and/or adzes as well as polished stone disks, stone and bone points, harpoons and/or bared points, and, most important, cord-marked pottery.

There seems to be a continuous development from the earliest preceramic Hoabinhian sites to the ceramic horizon. Somewhat confirming this is the carbon-14 date of 6875 B.C. (± 200), which comes from a lower pottery layer of Hsien-jen-tung. An even earlier date, 9360 B.C. (± 180), comes from shells found in the lower levels of Tseng-pi-yai (Pearson 1979). Nai Xia (1977) has doubts about these dates, however, and we know little about the subsistence patterns followed by the peoples of those early levels. Although this is when the action on domestication and incipient agriculture should have occurred, it could be the Village Forager stage—system C1. Further, some of the stone (hoes) and bone tools as well as the cord-marked pottery give hints of continuity into the early Neolithic Ho-mu-tu (or Hemudu) culture (5000 B.C.) of the central coast in Chekiang province (Pearson 1981; K. C. Chang 1979, see table 5.2).

We are therefore left in the untenable position of trying to understand the process leading to Neolithic village agriculture by considering the Neolithic itself (table 5.2). This approach is not very fruitful even when our Neolithic data are relatively abundant, as they are for the Yangtze River delta, thanks to intensive investigations of burial sites and the Ho-mu-tu (Hemudu) stratified village sites (Pearson 1979, 1981). The Ho-mu-tu village itself seems to have consisted of a series of structures erected on piles around small lakes. One of the lowest layers yielded radiocarbon dates of about 5000 B.C. What is more, these dates were associated with remains of domesticated rice (*Oryza sativa*) and water caltrop, jujube, gourds, water chestnut, acorns, and animal remains of domesticated dog and water buffalo (K. C. Chang 1973, table 5.3). Also present were scapula hoes, stone adzes and chisels, wooden daggers and mallets, bone chisels, whistles, cylindrical arrowheads, needles and awls, and grey sandy paste pottery formed into distinctive vessels with plain and/or cord-marked surfaces. This site represents a possible example of early village agriculture—system E—but it also could be a horticultural village, system C2. In addition to these remains there were others (mainly burials) from the three regions of the Yangtze delta—Chekiang, Kiangsu, and Shantung. These were grouped together as the Ching-lien-kang (or Gingliongang) culture, which was thought of as having early

(5000–4200 B.C.), middle (4200–3500 B.C.), and late (3500–3000 B.C.) phases (Pearson 1981).

North of the delta, this culture may have developed into Dawenkou (3000–2000 B.C.). Later, Lung-shan people from the north invaded between 2000 and 1500 B.C., bringing with them millets. South of the Yangtze delta, a Liang-chu culture developed between 2500 and 2000 B.C., while the later Shangdong Hu-shu culture was contemporary with Lung-shan. People at the Chien-shan-yang site of Liang-chu times were also Village Agriculturists, reputedly using rice, chestnuts, sesame, broad beans, peaches, and melons.

Our major problem concerns how the Ho-mu-tu or Ching-lien-kang acquired its first agriculture. Did it develop from a forager base in Hoabinhian times and lead to forager villages in Hsien-jen-tung times, about 9000 B.C.? Was rice agriculture imported by these Foraging Villagers in the Tseng-pi-yai period? Answers to these problems are needed to test various hypotheses or devise new ones about the development of village agriculture in the Far East Center. Right now those data are not available, and no amount of speculation will give us the answers (Rindos 1980).

There are hints that domesticated rice was not imported into this region from the south, for we do have archaeological remains from this time period on Taiwan and in the Hong Kong region. Roughly contemporary with the early Chinese Ching-lien-kang culture was the Dapo Kengu culture (5000–4000 B.C.) on Taiwan. Like the northern manifestation, it had cord-marked pottery and polished adzes. This culture also had bark beaters, slate points, and pebble tools, among other traits (K. C. Chang 1969). There is evidence that these people were successful Foragers or "progressive fishermen" (K. C. Chang 1971) and that they cultivated such root crops as yams and taro, but there is no evidence of any kind of rice cultivation.

We also have some archaeological remains from roughly 4000 B.C. on the coast north of the Yangtze. The least described culture of this period is the Dawenkou, or Ta-wen-k'ou, culture, known mainly for its cemeteries (see table 5.2). Although its pottery bears some resemblance to Ching-lien-kang pottery, it is mainly plain and encompasses a variety of new forms—jars, tripods, and pedestal bowls. Some of the Dawenkou tools suggest agriculture, but the crop was probably millet, not rice. As we shall see, the domestication of millet occured earlier, in the interior to the north and west of this coastal region. The reason for assuming that millet agriculture existed is that the Lung-shan culture, which followed Dawenkou in this area, had evidence of this crop.

This brings us to the so-called nuclear area in the loess highland or loessland ecozone at the big bend of the Yellow River. The earliest traces of Village Agriculturists—system E—are found in three clusters of sites: Cishan in Hopeh province, Peiilang in central Honan province, and Laoquantai in Shansi province and northwestern Honan (K. C. Chang 1971). All have distinctive pottery characterized by red paint, cord marking, incising, and rocker stamping. Tools include grinding stones, sickles, and saddle-shaped querns. The houses of the villages seem to have been semisubterranean, with storage pits and associated burials containing grave goods. Carbonized remains of foxtail millet and the bones of

Table 5.3. Sequence of cultivated and/or domesticated plants in the Far East

North China interior	sesame	peas	cotton	soybeans	red beans	barley	wheat	mulberry	hemp	leaf mustard	broomcorn millet	foxtail millet	rice	gourd	water chestnut	jujube	acorns	melon	peach	sour date	Range of dates B.C.	Central China coast	Japan
Historic	x	x	x	x	x	x	x	x	x	x	x	x	x	x	x	x	x	x	x	x	500 Historic		
													x	x							300 1000		F. Jomon
Shang						x	x	x	x	x	x	x	x					x			500 1500		
					x	x	?						x	x							1500 2000		L. Jomon
													x						x		1000 1800	Piao-maling	
Lung-shan							x	x	x	?	?	x	?										
	x												x			x	x	x			1800 2300	Tawenkien	
		x			?														x		2000 3500		M. Jomon
	x												x			x	x	x	x	x	3000 5000	Chien-shan-	
					x									x	x						3500 5000	yang	E. Jomon
Yang-shao								x	x	x	x	x									3000 5000		
													x	?	?	?	?				4000 6000	Ching-lien-kang	
													x	x	x	x	x				5000 6000	Ho-mo-tu (Hemudu)	
Peiilang											?	x									5000 6000		

domesticated pigs and dogs were uncovered in association with remains of wild plants and bones of fish and wild animals. These remains all date in the general range from 6000 to 4000 B.C.

The big question is: What was the subsistence system and culture of the predecessors of this complex? We do not yet have a clue. Ping-ti Ho has suggested that the ancestor may have been the microblade–tongue-core tradition, for which there is a long temporal gap before the Neolithic villages appeared (Ho 1977). K. C. Chang's suggestion (1971) that this Early Neolithic came out of the Kehe-Dingean tradition of Shansi province, which might have been a plant-collecting culture, is poorly documented and undated.

The data from northern China thus leave us little better off in terms of testing my hypothesis about the origins of village agriculture than did the information from southern and coastal China.

We are somewhat better off in terms of knowing the agricultural developments that followed the first Village Agriculturalists in this nuclear area of the loess highlands along the Yellow River (K. C. Chang 1970). About 4000 B.C. the Yang-shao culture seems to have developed out of the Laoquantai culture. Yang-shao dominated the northern area until about 3000 B.C. Material remains include a wide variety of red-painted storage jars, hoes, spades, adzes, celts, bone harpoons, and fishhooks. Houses were large semisubterranean wattle-and-daub dwellings that were often clustered in villages protected by a moat and/or palisade. Agriculture was probably of the slash-and-burn type, with crops of foxtail millet, leaf mustard, Chinese cabbage, hemp, and perhaps broomcorn millet (table 5.3). Domesticated dogs, pigs, and cattle were also raised.

Following the Miao-ti-kou culture between 3500 and 2000 B.C. was the widespread Lung-shan culture with the first evidence of silk weaving, and possibly wheat and rice agriculture in the north. Obviously, extensive investigation into the origins of agriculture in northern China is still needed. The suggestion that in Kehe-Dingean times there was a development that was like the primary development in other Centers is speculative at best.

Our best evidence of early agriculture comes not from the Center itself but from one of the peripheries: Japan. Here our knowledge of the beginning of agriculture and the continuity from Paleolithic Hunting-Collecting Bands to Village Agriculturists is quite completely documented (Crawford 1989). The Paleolithic—perhaps because of geological conditions—does not seem to begin quite as early in Japan as on the Asian mainland. However, work on the middle Pleistocene site at Gongenyama I in Kanto Plain north of Tokyo, the Sozudai assemblage north of there, and horizons of Fukui Cave in Kyushu do reveal that in Japan too there were heavy chopper and irregular flake industries in the Paleolithic (Ikawa-Smith 1978). As in northern China, this technology seems to be followed by blade and microblade tools from tongue-core industries, perhaps starting as early as 30,000 years ago and lasting up until about 10,000 or 11,000 years ago. Basically these people seem to have been Hunting-Collecting Bands or Efficient Foragers who were semi- or seasonally nomadic.

Near the end of this period some people began to exploit the coastal marine resources and became more and more sedentary. In about 9000 to 10,000 B.C.

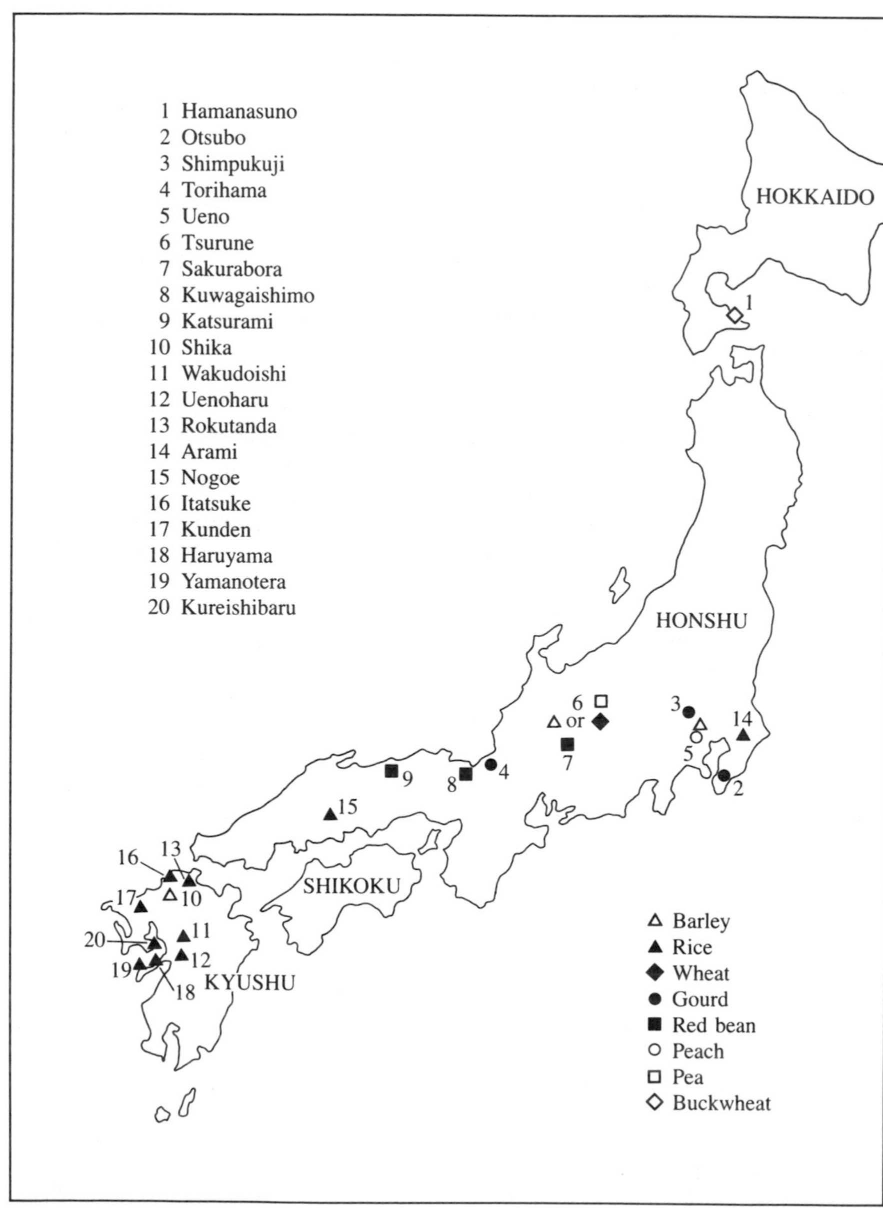

Figure 11. Jomon sites—from Early (top) to Final (bottom) periods—yielding cultivated and/or domesticated plants

this activity led to the Initial Jomon culture of Efficient Foraging Bands *par excellence* (Akazawa 1981) who had village life and pottery. By Early Jomon—5000 to 3500 B.C.—these Efficient Foragers (system D) had developed both a highly specialized maritime adaptation—mainly on the east coast—and a broad-spectrum foraging system in western regions, which often were inland. From a few of these southwestern sites fragments of buckwheat, red beans, gourds, acorns, and water chestnuts have been reported (see fig. 11). Remains of the last three plants named occurred slightly earlier in the deltoid Yangtze sites just across the South China Sea on the mainland (Kotani 1971). It should be emphasized that even at Torihama, which had the largest quantity of food remains, these domesticates represented a minor amount of the total subsistence diet and reflected highly incipient agricultural or horticultural practices in an economy that was mainly intensive broad-spectrum foraging.

In the following period—Middle Jomon, 3500 to 2000 B.C.—agriculture was in full swing on the mainland. In Japan, a few more domesticates appeared, perhaps including barley and/or wheat, peaches, and pears. Domesticates still represented only a small portion of the sustenance diet and appeared only in the western sites.

Late Jomon, 2000 to 1000 B.C., is little different in its subsistence pattern: gourds, possibly barley, and red beans were reported, mainly on the west coast, although a few appeared in the central western interior. In this period the first grains of rice occurred, possibly cultivated by dry-rice farming. Rice appeared in Japan far later than it had occurred on the mainland, where the domesticates quickly became part of an intensive agriculture complex instead of being part of minimal horticulture.

This pattern also seems to hold during the Final Jomon period, 1000 to 300 B.C., when domesticated rice and gourds reached the east coast. It has been suggested that the practice of wet-rice agriculture began in Japan during this period, but this theory is rather difficult to prove from archaeological data (see fig. 11). In fact, rice and agriculture do not seem to become the basic subsistence system until Yayoi times (mid-100–300 A.D.)—Japan's Iron Age, when village agriculture evolved. This practice occurred a millennium or two after civilization had arrived in China.

THEORETICAL CONSIDERATIONS

Obviously village agriculture did not develop in Japan as it did in the other Centers or even in China itself. There is no evidence of early local domestication or the use of domesticates before village life—my primary development—nor did early village foragers take on domesticates in a pattern that quickly led to agriculture, as in my secondary development model. The process in Japan seems distinct. Japanese data may therefore be used to test my hypothesis about the tertiary developments of village agriculture, a pattern that has not been encountered in the other Centers.

The ideal tertiary developments go from Hunting-Collecting Bands (system A)—the Japanese Paleolithic or Mesolithic with microbands and tongue cores—to Efficient Foraging Bands (system D)—the Initial Jomon times. The foraging stage in turn develops into Semisedentary Bands with Domesticates (system D1) in Early and Middle Jomon times. In Late and Final Jomon times the Foragers evolve into Horticultural Villagers (system C2). They do not become Village Agriculturists (system E) until Yayoi times. Thus we see that only reluctantly did these efficient maritime foragers become farmers. This tertiary evolution, which the Japanese development fits extremely well, was very slow; it took roughly 5,000 or 6,000 years for people to develop from Hunter-Collectors to Village Agriculturists.

Japan also seems to have had most of the necessary conditions for those tertiary developments. Certainly, each Japanese coastal section had a number of ecozones—rocky points, bogs, deltas, brush areas, coastal plain areas, and the like—but each part of the coast was much the same as another part. Thus, the coastal areas were ecologically diverse regionally but had overall uniformity. The coastal ecozones were lush and extensive in terms of marine resources as well as nearby land resources. In these lush, uncircumscribed areas, foods were readily available in all seasons. In addition, Japan was adjacent to and had casual exchanges with China, which had early cultivars, domesticates, and horticulture and/or agriculture in its secondary developments on the southern coast.

Although little acute analysis of the Jomon-Yayoi sequence has been undertaken to determine the sufficient conditions bringing about village agriculture, my hypothetical analyses of tertiary development do seem to pertain in Japan. Throughout Jomon times there was intensification of wild-food (marine) procurement systems that led to marine-resource and plant-collecting specialization, thereby allowing sedentary life that increased noticeably in the Middle and Late Jomon times. There seems to have been some sort of overexploitation of marine resources as well as a rising sea level, which caused flooding in the rich estuaries and upset the carrying capacity of the coast in Late Jomon and early Yayoi times, when much of the food gathering moved inland (Pearson 1977). The semifeudal Iron Age social system of the Yayoi may have further upset the chiefdomlike redistribution of the preceding Jomon periods. To support the Yayoi culture system, the Japanese may have been forced at last to shift their long-standing horticulture or use of domesticates to a wet-rice agriculture.

Although more analysis is needed, Early Neolithic evolution in Japan does seem to fit my tertiary development model, and the necessary and sufficient conditions do seem to pertain, thus confirming my hypothesis (table 5.4). In fact, routes 7, 5, and 4—the hypothetical positive-feedback cycle for tertiary developments—seems to explain the development of village agriculture in Japan. That cycle might be restated as follows: In Mesolithic Japan, bands developed during Initial Jomon into relatively efficient foraging systems within the relatively easily exploitable and noncircumscribed coastal environments. Then, although Early to Final Jomon had exchanges with China, where domesticates and/or horticulture and/or agriculture had developed, the Jomon people still continued to improve their efficient foraging system, which led to Semisedentary Bands with Domesticates (Crawford, Hurley, and Yoshizaki 1976). Only as this system began to overexploit its local ecozones and/or that environment's lush aspects diminished did that environment slowly change for the worse around 1000 B.C. In Final

Table 5.4. Necessary and sufficient conditions as a positive-feedback process for tertiary development in Japan

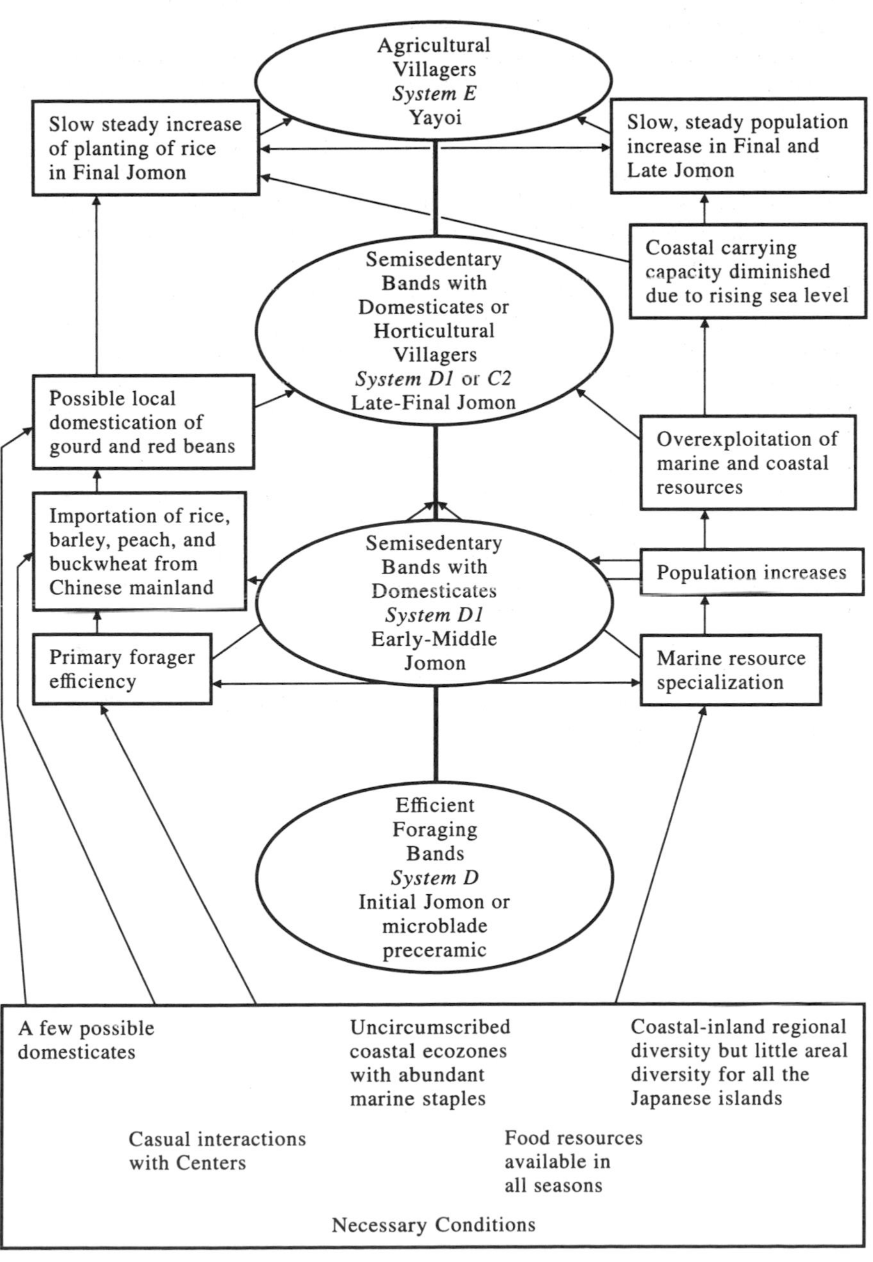

Jomon, the Japanese began to acquire first dry-rice and then wet-rice horticulture as a minor addition to the successful maritime foraging subsistence system that was the basis for their village life. Slowly, as their population gradually increased and the carrying capacity slowly diminished or the exchange or redistribution system broke down or changed (often due to social conditions in Yayoi times), they took on more and more horticulture until at last they became Village Agriculturists.

One test has thus been made of my tertiary development model, and the model seems to stand up. Other tests will be attempted in many of the secondary areas—the non-Centers of the U.S. Southwest, western Europe, the eastern United States, tropical South and Central America, Southeastern Asia, and Africa.

Unfortunately, the primary and secondary developmental models of this Far Eastern Center cannot be tested at this time, although there are hints that the interior loess highlands and China's coastal zones developed in different ways. Our knowledge of the northern loess development is so incomplete that we cannot tell whether it had a primary development; obviously no hypotheses can be tested with such data. Information for the south coast is almost as limited, but there are hints that the Affluent Foraging Bands of Hoabinhian times (system C) might have developed into Foraging Villagers—(system C1)—with coastal shell mounds. In Ho-mu-tu times they evolved to Horticultural Villagers (system C2) and then early in Ching-lien-kang times they became Village Agriculturists (system E). If this hypothesis holds true, then coastal China had a secondary development. Certainly its subareas had many of the necessary conditions for this development, but the sequential archaeological data are still too incomplete for us to know if such a development indeed took place, let alone determine what sufficient conditions or triggering causes pertained.

To test our secondary and tertiary developmental hypotheses, let us therefore turn to the non-Centers, starting with the temperate zones.

PART 2

The Temperate Non-Centers

We have now dealt with the four Centers, the areas of the world where agriculture began. In the previous chapters we noted that there were other areas where agriculture developed, although at a slightly later date; we also commented that these later beginnings owed much to the development in the Centers. For this reason we have called these other areas non-Centers. In dealing with their later development of village agriculture and the rather different causes that brought it about, we have divided the non-Centers into two general categories: (1) the temperate areas, such as the American Southwest, Europe, and the eastern United States, and (2) the tropical areas—Southeast Asia, tropical America, and Africa. Since we understand the latter group far less well than the former, we have relegated the discussion of the tropics to part 3. Of the temperate areas to be discussed in part 2, the one most intimately connected with a Center is the American Southwest, so we shall discuss it first.

CHAPTER 6

The American Southwest

Archaeological research in the southwestern United States has a long history of innovative investigations. The arid climate of the Southwest and other facts have meant that crucial plant remains are often preserved in archaeological contexts (Lipe 1978). Further, the area's inhabitants had long, intimate relationships and obvious interactions with one of our major primary Centers of plant domestication and village agriculture—Mesoamerica. The Southwest therefore should have yielded much relevant data that could be used to test my three hypotheses on the development of village agriculture, or better yet, it should have provided archaeological information for new or better hypotheses about this important problem. As the following discussion shall indicate, however, the area has not lived up to its potential (Ford 1981). Even more discouraging, the present archaeological dictums of Stephen Plog (1980), Michael Schiffer (1978), and others indicate that unless the direction of research changes radically, the situation will not improve appreciably. Since the Southwest does have great potential for testing my hypothesis, rather than speculating about the future, let us consider what past research has produced.

A BRIEF HISTORY OF RESEARCH ON THE PROBLEM

Interest in the archaeology of the southwestern United States started with some of the region's first explorers—John Wesley Powell, the Wetherills, Kit Carson, and so on—and culminated in the late 1920s with the first Pecos Conference, which provided a tentative sequence from Basket Maker II to Pueblo V and the historic Pueblo tribes of the northern part of the area (Kidder 1927). This synthesis was preceded by two decades of pioneering archaeological research, including Nels Nelson's stratigraphic digging in Pueblo villages in 1912 (Nelson 1916) and Samuel Guernsey's and A. V. ("Doc") Kidder's work in caves in the Tsegi region (Guernsey and Kidder 1921), which led to a long multidisciplinary program at Pecos sponsored by Andover's Peabody Foundation (Kidder 1924). Other significant research was done by Alfred Kroeber (1916), Leslie Spier (1917), Henry Mera (1938), and others. Unfortunately, the recovery of plant remains was accomplished without sufficient precision. Further, although solid data were lacking, a general hypothetical sequence was assumed for the development of agriculture: the people of the hypothetical Basket Maker I were collectors; Basket Maker II and III took on domesticated plants from Mexico and evolved finally to the village agriculture of the Pueblos (Kidder 1924).

This first period of research laid the foundations for the second phase, which ran roughly from 1930 to 1948. During this period many institutions did research in the Southwest, among them Harvard's Peabody Museum, the Gila Pueblo Foundation, the Smithsonian Institution, and the Field Museum of Natural History. During this time, sequences were worked out not only for the northern Anasazi but also for the Hohokam in southern Arizona and the Mogollon just east of it (McGregor 1941; Cordell 1984). The study of ethnobotany began under Volney Jones (1935), Edward Castetter and Willis Bell (1942), Alfred Whiting (1944), and others. George Carter provided a hypothesis about the multiple origins of agriculture in the area (Carter 1945), and although it was a neat case of the right theory for all the wrong reasons, at least theory had reared its tasseled head. Also in the theoretical and methodological realm was Walter Taylor's monograph, *A Study of Archaeology* (1948).

This Southwesterner's harsh criticism of the techniques, methods, and purpose of prominent southwestern archaeologists might be considered a fitting beginning for the next period, roughly from 1948 to 1964. During this period the number of archaeological projects increased, and they were increasingly sponsored by southwestern institutions rather than by "outsiders or Easterners." Critically important to this era were the advent of the system of radiocarbon determination, which could be used to date crucial Archaic period remains (back to which dendrochronology did not extend), and the excavation of Bat Cave. In the latter, the uncovering of tiny primitive corncobs—although misdated at 3600 B.C.—created a sensation and stimulated much new interest in the origins of agriculture (Mangelsdorf and Smith 1949). Also during the 1950s Paul Martin, from the Field Museum of Natural History, undertook his extensive investigation of the Mogollon of central Arizona and New Mexico (Martin et al. 1952). Like Bat Cave, which had the botanists C. Earle Smith and Paul Mangelsdorf analyzing ecofacts (Mangelsdorf, Dick, and Cámara-Hernández 1967), Martin had the Mogollon plant remains identified by Lawrence Kaplan (1963), Thomas Whitaker, Hugh Cutler, and others. Emil Haury and the University of Arizona carried on an even more ambitious program that had evolved out of the pioneering efforts of Gila Pueblo on the Hohokam (Haury 1962). Haury also dug Ventana Cave (Haury 1975), which supplemented the Archaic Cochise remains that E. D. ("Ted") Sayles had earlier investigated for Gila Pueblo (Sayles and Antevs 1941). Rivaling those programs was the research of Charles Di Peso of the Amerind Foundation (Di Peso 1956).

The beginning of the most recent research period—from about 1965 to the present—was marked by the arrival of Lewis Binford (Binford 1962) at New Mexico and the paradigms of the New Archaeology that replaced the cultural ecology of Julian Steward (Steward 1937). Also appearing in this period was a new element that has extensively affected the field of archaeological research. This endeavor goes under the title of Cultural Resource Management (CRM), or "contract archaeology." Since the 1970s a great deal of archaeology has been accomplished by dint of the funds earmarked for this type of endeavor, but the results (Plog, Plog, and Wait 1978; Goodyear 1975) with regard to the origins of village agriculture have been most disappointing and unsatisfactory.

Meanwhile, problem-oriented research has continued. William Longacre has applied the methods of New Archaeology to Grasshopper Ruin in the Mogollon subarea (Longacre and Reid 1974), while James Judge (1979) and Alan Simmons (1986) have investigated Chaco Canyon; Douglas Schwartz and Richard Lang have worked along the Grand Canyon (Schwartz and Lang 1973); the Huckells have worked in Arizona on Cochise remains (Huckell and Huckell 1988); Cynthia Irwin-Williams and Phillip Shelley have investigated the Rio Cuervo sites (Irwin-Williams and Shelley 1980), and others have worked on Black Mesa (Gummerman and Euler 1976) as well as in southern Utah (Matson 1988). From the standpoint of problems of origins of agriculture, perhaps the most important advances have been Cynthia Irwin-Williams's attempts to synthesize the Archaic materials of the Southwest (Irwin-Williams 1979) and the reevaluation of the Cochise remains by Sayles and Wamsley (Sayles 1983), as well as the reconsideration of the corn problem by Bruce and Lisa Huckell (1988), Paul and Suzanne Fish (1986), Michael Berry (1985), and Paul Minnis (1985; 1989). Also, the reevaluation and even reexcavation of the early botanical remains of Bat Cave and elsewhere have been undertaken by Richard Ford (Ford 1981), Vorsila Bohrer (1970, 1972), and others (Berry 1985; Minnis 1985). Finally, in 1985, Steadman Upham, David Carmichael (1986), and I began a program in the Organ Mountain region of southern New Mexico, investigating rock-shelters to learn about Archaic subsistence systems (Upham et al. 1987; MacNeish 1988; MacNeish and Beckett 1987). All this research allows us to hope that one day archaeology in the Southwest will realize its potential in the realm of problems of the origins, development, and spread of agriculture.

Before we discuss the data provided by this research, let us look at the natural features of the area.

THE ENVIRONMENT

Following the lead of many others (e.g., Lipe 1978), I divide the Southwest into four main subareas: the Rocky Mountains, the Colorado Plateau, the Mogollon Highlands, and the basin-range subarea (see fig. 12). The last may be subdivided into at least five parts: the lower Colorado River drainage, the Gila River drainage, the upper Rio Grande drainage, the middle Rio Grande or Chihuahua Desert, and the Great Basin itself, which is really outside the southwestern culture area.

The most extensive and perhaps the most typical environmental zone is the Colorado Plateau, which covers a large portion of northern Arizona, southeastern Utah, southwestern Colorado, and northwestern New Mexico. It is largely an upland composed of more or less horizontally bedded sandstone, shale, and limestone with a few volcanic extrusions. The relatively horizontal strata have been extensively eroded by the Colorado River and its tributaries, forming a multitude of buttes, mesas, canyons, and occasionally broad valleys.

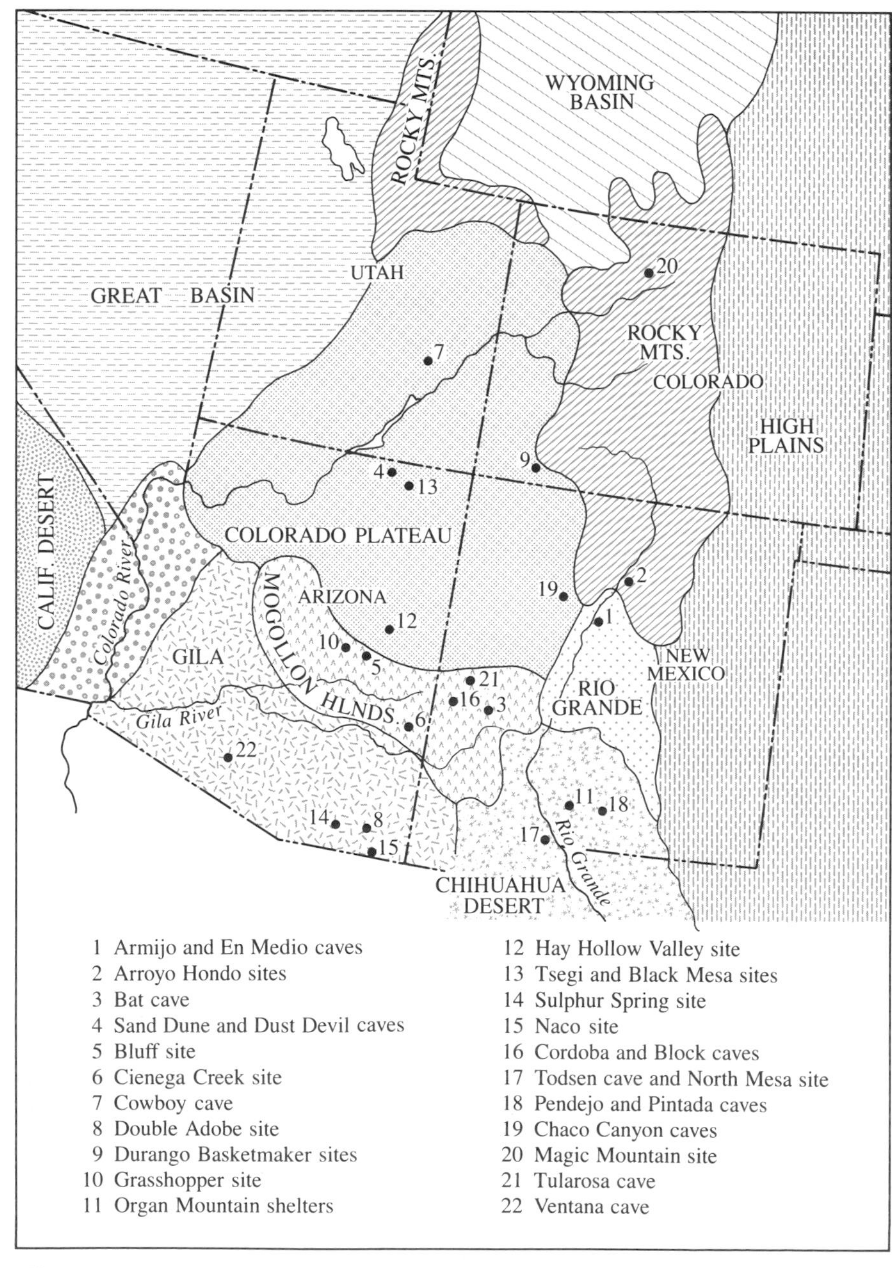

Figure 12. The environmental life zones and relevant sites of the American Southwest

In places bare rock is exposed, and soils are generally thin, but a few of the valleys have alluvial soils, while others have sand dunes.

The Colorado Plateau is a transition zone between two precipitation zones. The northwestern edge of the region receives its moisture from the Pacific in broad frontal storms that occur mainly in the winter and spring with only an occasional late summer storm. The southeastern part, on the other hand, which receives its rainfall from the Gulf of Mexico, has rains mostly in the late summer with a few in the spring and even fewer in the winter. The summer rains are often violent thunderstorms, which flood the dry arroyos with raging torrents in a matter of minutes and further erode the already much eroded terrain. Annual rainfall ranges from 300 to 1,016 millimeters per year, thus making the region a desert. Summers are hot and bright, and there are roughly 100 to 160 frost-free days per year.

These factors plus the range of elevation determine the type of vegetation. The higher peaks, which are often volcanic, have only alpine scrub or tundra vegetation, while the slopes below them are covered with spruce and fir. At a still lower elevation the vegetation is mainly ponderosa pines. Much of the Colorado Plateau lies between 1,500 and 2,100 meters above sea level and has piñon-juniper forests, with large patches of thorny scrub and grasses. At still lower elevations sage scrub is more common, while there is much saltbrush, blackbrush, and greenwood in areas of alkaline soil. In well-watered locales, oasislike patches with cottonwoods, willow, and other lusher plants occur. Unlike the other subareas being discussed, the Colorado Plateau and the Southwest in general had only a few plants susceptible to domestication.

Most of the archaeological exploration in the Southwest has been undertaken in the Colorado Plateau and the adjacent and closely connected upper Rio Grande subareas; in fact, our only complete Archaic sequence comes from its southeastern part, in the Rio Cuervo region just east of Albuquerque (Irwin-Williams 1979). Fortunately, ecological studies allow us to correlate these Archaic phases with different kinds of exploitations of the fourteen ecozones in the upper Rio Grande. Since these ecozones have the necessary conditions—the prerequisite causes—for cultural change, we will briefly outline them here.

Although the upper Rio Grande is a desert with limited resources, it has two lush circumscribed ecozones with food available year-round. These ecozones, which are similar to the oasis zone of the Levant, are the springs and ephemeral ponds in the canyon heads. They not only attract the animals of the region, including seasonal reptiles and waterfowl, but also have a wide variety of plants, including rushes, sedges, willows, composites, and such seasonal seed-bearing plants as grasses, amaranths, chenopods, Indian wheat, and hog potatoes. Surrounding and connected to these zones are the valley bottoms, with their gallery forest of juniper and willow, and the canyon bottoms with scrub oak, saltbush, and bearberry and squawberry bushes.

Surrounding these ecozones are the canyon rims and slopes with opuntia, yucca, other forms of cactus, grasses, and a few piñon and juniper trees, while the isolated mesas have mainly tall grasses and scattered juniper. The low mesalands and the sand-dune ridges have mainly short grass and cactus. To the east of the lusher zones, but not farther than a day's walk, are the short grasslands—the ceja slopes with occasional juniper and cactus, and the ceja uplands with occasional cholla and opuntia cactus. These ecozones contrast with the steeper, rising ecozones to the west: the mesa prieta slopes; the mesa prieta itself, which has some juniper and cactus; the mountain slopes with scattered forests of oak, pine, and juniper; and the volcanic peaks, which have similar vegetation as well as some ponderosa pine.

While these ecozones are far different from those of the Levant or coastal Peru, where secondary developments occurred, they do have the same or similar necessary conditions for that developmental model. The Rio Cuervo region of the upper Rio Grande (and perhaps many similar regions on the Colorado Plateau) has a few lush circumscribed ecozones with abundant food resources—the springs, ponds, and some parts of the canyon and valley bottoms—and all its ecozones can be exploited from a strategically located base. This pattern occurs again and again throughout the upper Rio Grande as well as the Colorado Plateau so that ecologically diverse regions are found within a relatively uniform subarea. Also, although only a few potentially domesticable plants are native to the region, interaction, albeit often indirect, occurred with Mesoamerica, where primary developments had earlier led to the use of such domesticates as corn, beans, and squash. And, as we shall see in my description of the Archaic period in the region, the various sufficient conditions for a secondary development also occurred. The archaeology of this region is therefore of key importance for testing my secondary development hypothesis with data from a non-Center.

Just south of the Colorado Plateau and the upper Rio Grande, in central-east Arizona and central west-southwest New Mexico, are the Mogollon Highlands. Much less archaeological work has been accomplished here, and most of it did not start until the 1940s and 1950s. In the Mogollon the plateau beds have been uplifted to form a series of rugged mountains with steep valleys and basins between them. The highlands are generally higher (above 2,100 meters) than the plateau to the north and therefore receive slightly more rain from the Gulf of Mexico. Because the highlands are farther south, temperatures are much the same in spite of the higher elevations, so vegetation is similar to that found at the higher elevations on the plateau: pine and Douglas fir in the upper elevations, pine-oak-juniper zones below them, and then mahogany–oak scrub with xerophytic patches occurring in rain-shadowed valleys, while others have lush grasslands.

The Mogollon Highlands subarea contrasts with the Rocky Mountains, which separate the Southwest from the high plains of eastern New Mexico and Colorado. Here the least amount of archaeology has been undertaken. The Rocky Mountains are part of the completely folded mountain chain that runs north and south through much of western North America. Elevations are high and slopes are covered with coniferous forest, while valleys are steep and narrow and often covered with grasslands. Rainfall from the Gulf of Mexico is heavier than in the previously mentioned zones, but the range of temperatures, although cooler, is much the same, as is the seasonality.

The fourth major environmental zone of the Southwest is called the basin-range subarea. It is larger and less well defined than the other subareas and can be best discussed in terms of its major parts. The highest and coolest of these parts is the Great Basin in western Utah and Nevada. The major west ridge of the Sierra Nevada, which rises over 4,000 meters, flanks the Great Basin on the west and keeps it in a rain shadow that blocks precipitation from the Pacific Ocean. Lying at about 1,350 meters, the basin receives limited rains from the west, mainly in late winter and early spring. To the east and west are mountains topped with pine, piñon, and juniper, while the basin itself is mainly a scrubland with grasses, saltbush, greenwood, and blackbrush. A major feature of its eastern portion is the Great Salt Lake, and drainage is into the lake, not out of the basin. Strictly speaking, this region is not part of the southwestern culture area, although it had some relationships with the Southwest throughout the Archaic period (Jennings 1978).

Just barely in the Southwest culture area is the second subdivision of the basin-and-range subarea, the lower Colorado River drainage. Its Archaic cultures were part of and interacted with those of the rest of the Southwest (Haury 1975). Although the Sierra Nevada to its west also placed the lower Colorado in a rain shadow, its elevations are much lower than those in the Great Basin, reaching sea level at the Gulf of California. The lower Colorado drainage is dominated by the relatively broad valley of the Colorado River and the canyons that enter it. Its soils are poor and sandy, and its vegetation is xerophytic with a monotonous creosote bush and bursage cover. Some grasses and a gallery forest, mainly of mesquite, occur along its waterways. Only limited archaeological investigation of the Archaic has been undertaken in this region.

Intimately connected with the lower Colorado to the east is the subdivision of the Gila River drainage. This drainage has many broad valleys and runs from the east into the lower Colorado. Although it too lies in the rain shadow of the Sierra Nevada, its eastern portions are at much higher elevations and support a rich flora of small shrubs. In addition to acacia and mesquite, many varieties of cacti occur—the giant saguaro, the organpipe cactus, bone cactus, cholla, and prickly pear. Like the lower Colorado, the Gila drainage is warmer than the Great Basin and the Colorado Plateau and rarely has frosts except at higher elevations. Thanks to the efforts of the Gila Pueblo Foundation, the Amerind Foundation, and the University of Arizona, considerable archaeology has been accomplished here.

The upper end of the Rio Grande Valley, a region similar to the Colorado Plateau, has been described above. It is relatively wide, but severe erosion in its tributaries has cut the region into steep canyons and many mesas and buttes, making it somewhat like the Colorado Plateau. As in that subarea, a few forests of spruce, piñon, and juniper occur at the higher elevations, and the valleys have poor soils and are mainly grasslands with shrubs. Along the river and its drainage appear gallery forests of cottonwoods and willows.

Similar to this subdivision is the final subdivision, the eastern Chihuahua Desert or middle Rio Grande drainage, which lies over the Continental Divide to the east and runs into the drainage of the upper Rio Grande. It too is in a rain shadow; its rains come in summer from the Gulf of Mexico. The region of the

Chihuahua Desert that concerns us is known as the Jornada region. Here a new program of research concerned with the subsistence development of the Archaic was begun in 1985 under the auspices of the Andover Foundation for Archaeological Research. Earlier, ecological studies by Melvin Satherwhite, Judy Ehlen, Thomas O'Laughlin, and others had been undertaken near Fort Bliss in this region (Satherwhite and Ehlen, 1980; O'Laughlin 1980), while in 1986 Sally Anderson and Patricia Crawford did botanical and ecozone studies there for the Andover Foundation. On the basis of a vegetational classification, the Jornada region might be considered part of the Chihuahua Desert domain. However, it has a number of distinctive ecozones, which have the necessary conditions for the tertiary developments that seem to have occurred in the Southwest from 5000 B.C. to A.D. 800. Unlike other parts of the Southwest, the Jornada has one ecozone that is distinct—the Rio Grande riverine ecozone. This zone is lush and uncircumscribed, with abundant food resources that are available throughout the year. While there is a harsh season (a cool, dry winter season), rush, sedges, willows, fish, and refuge game are available in some profusion even then. In the warmer, wetter parts of the year food can be obtained from cottonwoods, wolfberry, yucca, opuntia, soapberry, amaranth, grasses, and migratory fowl. Flanking this region are distinctive xerophytic lower and upper bajada zones with mesquite, creosote, sumac, and desert plants. These riverine and bajada ecozones allowed an efficient foraging system to develop. Surrounding this drainage is a series of basins and playas, each a little different, creating a whole series of ecozones with limited foods that are available only in certain seasons. These zones include the west mesa thorn forest and the similar upper and lower alluvial forest zones. These are interspersed with sand-dune grasslands and gallery thorn forests along the washes, which have water in them only in the wet seasons. Surrounding these zones is a desert mesa ecozone, which has few food resources, and forested dissected hills and mountains bearing oak, pine, piñon, and juniper. In a few regions these hills flank higher mountains that are forested with ponderosa pine, Douglas fir, tamarack, poplar, oak, and pine and populated with much game (mountain lions, sheep, and antelope).

Thus, while there is regional diversity, or diversity within each ecozone, each basin is a little different from the next. As in my primary development model, there is areal diversity as well as regional diversity, but the lush riverine parts make this subregion unique. Furthermore, few plants are susceptible to domestication (see table 6.1), and the contacts with Mesoamerica were casual and interaction often indirect. All these factors were necessary conditions for tertiary development.

The Southwest therefore has subareas with the necessary conditions for both secondary and tertiary developments. Although some subareas even approach the conditions for primary developments, they lack one important necessary condition, that is, the presence of many plants susceptible for domestication. (The lack of this prerequisite is even more pronounced in the desert area of California, the Great Basin, and Texas, which surround the Southwest.) This fact, plus some other necessary conditions that allowed the inhabitants to take up subsistence options other than agriculture, resulted in developments that did not lead to village

Table 6.1. Some native cultivated and/or domesticated plants of the American Southwest*

Scientific name	Popular name
Agave pariyi	agave cactus
Amaranthus sp.	amaranth
Amaranthus hypochondriacus	pigweed
Chenopodium sp.	chenopod
Cleome serrulata	Rocky Mt. bee weed
Cucurbita pepo ovifera	squash
Datura meteloides	Jimsonweed
Helianthus annuus	sunflower
Hordeum pusillum	desert barley
Nicotiania attenuata	Indian tobacco
Phaseolus acutifolias	tepary bean
Phaseolus coccineus	scarlet runner bean
Proboscidea parviflora	devil's claw

*After Harlan (1971) and R. Ford (1985).

agriculture. Thus, the necessary conditions determined by the unique desert environment led to specific kinds of prehistoric cultural evolutions in the U.S. Southwest.

Having seen what the Southwest is like ecologically, let us look at the relevant archaeology, starting with those subareas that have yielded the least information (see table 6.2).

RELEVANT ARCHAEOLOGICAL DATA

The eastern Rocky Mountain subarea has produced the least relevant materials and no plant remains, although two sites with Archaic remains have been excavated—a site near Sawuache, Colorado (Huscher 1941), and Apex D-E materials from central Colorado (Irwin-Williams and Irwin 1966). Perhaps the Lo Dais Ka site and Magic Mountain should also be included in this group, for both have Archaic levels (Irwin-Williams 1967b). Thus far most of these sites have produced mainly lithic materials—stemmed–indented-base and corner-notched points, choppers, and scrapers, as well as various kinds of grinding stones—but Lo Dais Ka did have a corncob dating to 1500 B.C. (Berry 1985). Ceremonialism is suggested by the practices of burying the dead under stone cairns (Apex C, Colorado) and by evidence of cremations (Irwin-Williams and Irwin 1966). Unfortunately, although the tools suggest that these people were foragers, we have little concrete evidence of their subsistence practices and even fewer studies of their settlement patterns.

Somewhat the same may be said for the Great Basin and lower Colorado River regions in the western part of the basin-range subarea, but for different reasons. In the Great Basin, largely thanks to the work of Jesse Jennings and his colleagues, we have a long preceramic sequence of the "Desert Culture" Archaic peoples, but there is little or no evidence of horticulture, and even in historic times (Fremont Culture excepted) agriculture did not occur (Jennings 1957).

Table 6.2. Relevant archaeological sequences in the American Southwest

CALIFORNIA DESERT SUBAREA	WESTERN BASIN RANGE SUBAREA			MOGOLLON HIGHLANDS SUBAREA		EASTERN BASIN-RANGE SUBAREA
	Colorado River part	Gila part				Chihuahua part
Mohave region	Grand Canyon region	Papagueria region	Gila River drainage region	Reserve region	Augustin Plain region	Jornada region
						Mesilla
		Hohokam		Mogollon		
						(North Mesa)
			(Hay Hollow)			
Silver Lake	Figurine Complex		San Pedro	(Cordoba)		(Pendejo)
Amargosa 3		San Pedro (midden layers)		San Pedro		Hueco
				(Block)	(Bat Cave)	(Todsen)
			(Matty Canyon)	(Tularosa)		(Tornillo)
						(Pintada)
Pinto			(Cienegas Creek)			Fresnal
Amargosa 2					(AKE)	(Todsen)
		Chiricahua-Amargosa 2	Chiricahua	Chiricahua		(North Mesa)
Amargosa 1		Armagosa 1	(Double Adobe)	(Wet Leggett) ?	(Bat Cave Sands ?)	Keystone (Keystone)
San Dieguito 3						
						Gardner Springs (North Mesa)
San Dieguito 2		Red Layer	Sulphur Spring			Angostura ?
						Folsom? Clovis? (North Mesa)
San Dieguito 1		Ventana	Clovis			Pre-Clovis (Pendejo)

Much the same may be said of the lower Colorado region of Arizona and adjacent California and Nevada.

Thanks to the efforts of Emil Haury in Ventana Cave in southern Arizona, we have a long sequence from Early Man through the Archaic on which to build in the Papagueria region of the adjacent part of the basin-range subarea (Haury 1975). Above the Early Man layers were ones with San Dieguito 2 remains, dat-

EASTERN BASIN-RANGE SUBAREA		COLORADO PLATEAU SUBAREA		ROCKY MOUNTAIN SUBAREA	
Upper Rio Grande part					
Rio Cuervo region	Chaco region	Tsegi Black Mesa region	San Juan River region	E. Colorado foothills region	Dates
Trujillo	Basket Maker III				A.D. 500
	(Sheep Creek)	Lolomai	Durango Caves Cedar Mesa		0
				Lo Dais Ka	
En Medio	En Medio-like				500 B.C.
			Grand Gulch		
		White Dog			1000 B.C.
Armijo	Armijo-like (LA 18091)	Black Mesa late Archaic			2000 B.C.
San Jose	(LA 18103) San Jose-like LA 17337-	Black Mesa San Jose	Burial 2 Sand Dune Cave	Apex complex	3000 B.C.
				Magic	4000 B.C.
		Black Mesa	Desha	Mountain complex	
		Desha-like			
Bajada		Black Mesa Bajada-like			5000 B.C.
Jay					6000 B.C.
Cody					
					7000 B.C.
Folsom					8000 B.C.
Clovis					9000 B.C
				Clovis	

ing roughly from 7000 to 5000 B.C.; these in turn were overlaid by deposits of Archaic/Amargosa I (5000–3000 B.C.) and Pinto Basin-Amargosa II (3000–500 B.C.), equivalent to Chiricahua and San Pedro Cochise, those most relevant to our problems. Outside the Southwest, the Stahl site, near Little Lake in southern California in the Mohave Desert region, has produced not only Pinto Basin stemmed–indented-base points, cobble choppers, scraping planes, and grinding stones, but also evidence of seven or eight pit houses, indicating a fairly sedentary lifestyle based on foraging, not horticulture. In the same region the somewhat later Rose Spring site of Amargosa III (500 B.C.–A.D. 500), which was contemporaneous with village agriculture sites in the nearby Southwest, showed a continuation of this settlement pattern and way of life without a hint of plant domestication.

Fortunately, the other part of this basin-range subarea, the Gila River drainage of southwestern Arizona, does have the relevant Cochise remains, and these extend into the Mogollon Highlands subarea. I shall therefore discuss the Archaic sequences of these two subareas.

The earliest Cochise remains, called Sulphur Springs, were originally thought to be associated with extinct animals such as mammoth, camel, horse, and dire wolf, making the people Hunting-Collecting Bands, but later excavation proved this belief untrue (Waters 1986). Numerous new dates place the Sulphur Springs culture between 7300 and 6000 B.C., which puts it well before our concern (Sayles and Antevs 1941). The Sulphur Springs complex occurs in the Gila River subarea but has not been found in the Mogollon Highlands. There seems to be a considerable gap until the second stage of Cochise, called Chiricahua (3500–1000 B.C.). Various scientists (e.g., Whalen 1971) have suggested that the Cazador complex may either fill (part of) that gap or, because of its leaf-shaped points, be the hunting aspect of the Sulphur Springs complex, which has no points and is dominated by choppers, scraper planes, and grinding stones. Some scientists (including MacNeish 1988 and Irwin-Williams 1979) disagree with this interpretation. Either way, we are talking about culture system C—the Affluent Foraging Bands who later developed into the more scheduled Foraging Villagers like Chiricahua—system C1—with or without gap.

The Chiricahua phase is known from many sites in New Mexico, such as Wet Leggett (Martin, Rinaldo, and Antevs 1949), Bat Cave (Dick 1965), Noquino (Irwin-Williams and Beckett 1973), Cienega Creek (Haury and Sayles 1947), and other sites in the San Pedro Valley (a branch of the Gila River). In Arizona perhaps some of the upper levels in Ventana Cave (Haury 1975) belong to this phase; probably the Peralta complex in Sonora (Fay 1950) and perhaps even some of the preceramic sites in Sonora, Sinaloa, and Durango (Byers 1967b) are also part of the phase. Whether Chiricahua of the Gila is different from that of the Mogollon Highlands remains to be seen.

Tools include Chiricahua side-notched points, diamond Pelona and contracting stem Augustin types, as well as a crude chopper, scraping planes, and a well-developed ground-stone complex that includes shallow-basined milling slabs and cobble manos or mullers (Irwin-Williams 1979). Data for reconstructing the Chiricahua subsistence system are not numerous and often are not in reliable contexts, so published reports about subsistence are confusing (Berry 1985; Minnis 1985). The people of the AKE site (Beckett 1980) in the Mogollon Highlands may have been Efficient Foragers (perhaps with domesticates)—system D—who used a wide variety of wild plants, while the people of the Gila may have been Foraging Villagers (system C1). Either way, we have little adequate or acceptable archaeological or palynological evidence that maize or squash (*Cucurbita pepo*) was utilized as part of the subsistence system (see table 6.3) in spite of persistent rumors to the contrary (Berry 1985).

Norman Whalen's study indicates that the Chiricahua subsistence system seasonally exploited two environmental habitats, the river valley and the hill slopes (Whalen 1971). Macroband wet-season base camps (such as Wet Leggett) were mainly in the valleys, while small sites were established mainly on the slopes at other seasons. Whalen postulates that the basic population unit was about twenty-five people (Whalen 1971). Evidence from Cienega Creek and other sites in the Gila drainage suggests that, near the end of this phase, shallow pit houses were being constructed (Huckell and Huckell 1988). Although the evidence is far from clear in latest Chiricahua or earliest San Pedro times in this Gila region, small foraging villages or hamlets without horticulture were possibly in existence—my system C1.

The San Pedro phase that evolved out of Chiricahua at about 1400 B.C. is a crucial one in our understanding of the development of the village agricultural way of life, but as yet our knowledge of it is most inadequate (Whalen 1973). Not only do San Pedro sites occur in southern Arizona and New Mexico, but surface collections indicate that these sites are also numerous in Sonora and perhaps Durango, Mexico (Lister 1958). San Pedro sites thus seem more numerous and larger than those of Chiricahua, but Tularosa and Cordoba caves (Martin et al. 1952) and Bat Cave (Dick 1965) were occupations that tell us little about settlement patterns or populations. Plotting the extent of the Hay Hollow, Sendoval, and Cienega Creek open pit-house village sites has not adequately demonstrated population size.

The domesticated plant remains suggest a major shift in the subsistence system during the span of the San Pedro phase, at least in the Mogollon subarea (see table 6.3). It would appear from the data from Bat Cave (Mangelsdorf, Dick, and Cámara-Hernández 1967) that primitive Early Chapalote corn, gourds, and pumpkins (*Cucurbita pepo*) were part of the horticulture of these groups, who were basically foragers early in the phase (perhaps 1200–300 B.C.). By Tularosa Cave times (300 B.C.–A.D. 150), beans and sunflowers had been added to the subsistence complex (Kaplan 1963), which continued on into at least the ceramic Pinelawn and Georgetown phases characterized by my system C2—horticultural villages or forager villages with horticulture. Real village agriculture—system E—did not occur in the Mogollon subarea until considerably after San Francisco times, at about A.D. 800, but of course it continued on into historic times. Thus we could possibly have a slow tertiary development, but we need more data on the crucial early part of the sequence.

The picture for San Pedro in the Gila or southern region of Arizona, where it appears to have developed into Hohokam, is even less clear (Lipe 1978; Fish et al. 1986). Many large pit-house sites—such as Milagro, Los Ojitos, and Donaldson occur—with San Pedro side- and corner-notched projectile points, scraper planes, core choppers, and mullers and milling stones of San Pedro type (Huckell and Huckell 1988). Further, tooth wear at the Los Ojitos site and the carbon

Table 6.3. Sequence of cultivated and/or domesticated plants in the American Southwest

Chihuahua subarea	Upper Rio Grande	Colorado Plateau	Gila	Mogollon Rim
		Historic		
El Paso				
			Sedentary	
		Pueblo 2-3		
			Gila Butte	
			Snaketown	
		Basketmaker II-Pueblo I		
				Georgetown
	Trujillo			
				Pinelawn
			Vahki	
		Lolomai		
Late Hueco				
	Late En Medio			
				Late San Pedro (Tularosa block)
		White Dog		
		Sheep Camp		
	En Medio			
				Early San Pedro (Bat-Cordoba Cave)
Early Hueco				
		Chaco Armijo		
			Early San Pedro (Cienegas I)	
Late Fresnal (Kneepad, Roller Skate)				
	Armijo			
Middle Fresnal (Tornillo zone D, Fresnal zone C2-F)				
	Late San Jose (Chaco sites LA 17337-18103)			
Early Fresnal (Keystone)				
Maybe early Fresnal (Swallows Cave levels 72-89)				
Keystone (zone J of Todsen Cave)				

13-12 ratios of 12.8 for a San Pedro site near Deming suggest that these people had become real agriculturists (Marino and MacNeish 1991). Some of these traits carry on into the Vahki phase (A.D. 50–150) of Hohokam, which had pit-house villages, corn, beans, squash, irrigation agriculture, figurines, distinctive pottery, and many other characteristics of the village agriculture way of life—my system E. The Hohokam way of life evolved into that of the Pima and Papago of modern times and a number of other cultigens were added: cotton at about A.D. 500, tepary beans at about A.D. 1000, and, later, sieva, runner, and ca-

Range of Dates					Pumpkin (*C. Pepo*)	Chapalote (pollen)	Proto-Maiz de Ocho	Gourds	Amaranth	Maiz de Ocho	Common Beans	Sunflower	Pima-Papago Corn	Pueblo Corn	Cotton	Tepary Beans	Sieva Beans	Runner Beans	Jack Beans	Walnut Squash	Warty Squash	Devil's Claw	Agave	Chile Pepper	Tobacco
Present		A.D.	1540		x			x	x	x	x	x	x	x	x	x	x	x	x	x	x	x	x	x	x
A.D.	1400	A.D.	1100		x			x	x	?	x			x	?										
A.D.	1350	A.D.	900		x			x			x		x		x	x	x	x							
A.D.	1300	A.D.	900		x			x		?	x	x		x	x		x	x	x	x	x				
A.D.	800	A.D.	600		x			x			x		x		x	x									
A.D.	600	A.D.	400		x			x			x		x		x	x									
A.D.	800	A.D.	500		x			x		x	x	x		x	?										
A.D.	650	A.D.	500		x						x	x													
A.D.	700	A.D.	500		x									x											
A.D.	500	A.D.	150		x			x			x	x			?										
A.D.	150	A.D.	50		x			x		?	x		x												
A.D.	500	A.D.	250	B.C.	x	?		x		x	x			x											
A.D.	250		400	B.C.	x	x	x	x		x	x		x												
A.D.	500		300	B.C.		?		x		x															
A.D.	300		300	B.C.	x	?	?	x		x	x	x													
	100	B.C.	750	B.C.	x	x	?			x	x			?											
A.D.	220		770	B.C.	x	x	x			x															
A.D.	300		800	B.C.		P																			
	390	B.C.	1060	B.C.	x	x	x	?	x	x	x	?													
	400	B.C.	900	B.C.		x	x	x	x	x	x														
	725	B.C.	1080	B.C.	x	x			?	x															
	900	B.C.	1400	B.C.		?	?		x	x															
	900	B.C.	1200	B.C.	x	x	x	x	?	x	x														
	800	B.C.	1800	B.C.		P																			
	1225	B.C.	1510	B.C.	x	x	x	x																	
	1610	B.C.	2035	B.C.		P																			
	1590	B.C.	2350	B.C.	x	P																			
		Circa	2500	B.C.		x																			
	2600	B.C.	2000	B.C.	?																				

navalia beans, mixta and moschata squash, amaranth, devil's claw, agave, chile, tobacco, and perhaps others (R. Ford 1985).

In the Gila River drainage, then, we do seem to have a secondary development: from the Hunting-Collecting Bands (system A) of Sulphur Springs to Affluent Foraging Bands (system C) in early Chiricahua times, and then to Horticultural Villagers (system C2) by the time of the late Chiricahua–early San Pedro phase. By the late San Pedro phase, there were Village Agriculturists (system E), who rapidly became Hohokam Village Agriculturists (Berry 1985).

However, we lack sufficient reliable contexts and have a long gap in our sequential archaeological data and no in-depth analysis to tell us why this development happened; at present we cannot really test my ideal model of secondary developments (Berry 1985; Minnis 1989).

Some of the necessary conditions do seem to pertain in the Gila, for by San Pedro times we do appear to have (1) sedentary agriculturists; (2) ecozones with areal uniformity but regional diversity; (3) many circumscribed locations at springs or riverbank areas; (4) a region that could be exploited from a single base; and (5) interaction during the Archaic period of peoples in the U.S. Southwest with Mesoamerica, where primary plant domestication had already occurred.

For the next part of the basin-range subarea, the Rio Cuervo region of the upper Rio Grande, the situation is much better, largely because of the research by Cynthia Irwin-Williams on what she calls the Oshara tradition (Irwin-Williams 1973) as well as her concerns with the Picosa culture horizon (Irwin-Williams 1967b).

In the Rio Cuervo region the sequence begins with Early Man remains—Clovis (10,000–9000 B.C.), Folsom (9000–8000 B.C.), and Cody (7000–6000 B.C.), with their respective distinctive projectile point styles. Following these, Cynthia Irwin-Williams (Irwin-Williams 1973) discerns an Oshara tradition composed of a number of sequential phases. The earliest, Jay phase (5500–4800 B.C.), had bifacial knives, distinctive scrapers, and contracting-stem shouldered projectile points like those of Mohave Lake. This phase developed into the Bajada phase (4800–3200 B.C.) with indented-base stemmed points, scraper planes, and grinding stones. Both phases represented early Hunters and Collectors—my system A—with little evidence of seasonal scheduling.

The first evidence of well-defined foraging comes with the San Jose phase (3000–1800 B.C.). Tools included stemmed points with concave bases (the bodies of many of them serrated), scraper planes, mullers, and milling stones. The number of sites increased, and some of the seasonal ones were large macroband base camps with numerous (five to fifteen) hearths; also, postholes indicated some sort of simple temporary structures. Population was definitely increasing, putting more pressure (at least seasonally) on the carrying capacities of the fragile and circumscribed environments around the ponds or springs at canyon heads where the major base camps or villages were located (system C1).

This trend continued into the Armijo phase (1800–800 B.C.), which had large macroband fall-winter camps near springs and hints of pit-house villages. Several objects suggest that the macroband base camps had magico-religious ceremonial functions. Population increases indicate that the people were outgrowing the natural resources of the circumscribed locales around springs, and one might guess that the carrying capacities for foraging were being sorely tested. Certain sufficient conditions were at work that resulted in maize agriculture. Pollen profiles at various sites in the Chaco Canyon region (Simmons 1986) as well as Rio Cuervo (Irwin-Williams 1973) suggest the presence of maize agriculture well before 1500 B.C., that is, early in the Armijo phase. Irwin-Williams (1973) sees "further aggregation possibly initiated by the presence of a small surplus produced by limited agriculture" (i.e., horticulture), so that late in the phase there would have been more pressure to adopt horticultural practices and acquire more domesticates—my system C2—Horticultural Villagers.

Next comes the En Medio phase (800 B.C.–A.D. 400), which Cynthia Irwin-Williams has divided into two subphases—early (roughly 800 B.C.–0), called the White Dog phase in the Tsegi (Smiley 1985), and late (0–A.D. 400), with the latter roughly equivalent to the traditional Basket Maker II phase of the Four Corners region of the Southwest (Kidder 1924), now called Lolomai and Cedar Mesa (Matson 1988). Population seems to have continually increased during En Medio. In the late substage appear shallow slab-lined pit-house villages of the sort Earl Morris described for southwestern Colorado (Morris and Burgh 1954). In addition to population pressures, the occurrence of many slab-lined storage pits suggests still further pressures on the carrying capacity of the circumscribed base camps or village areas. Again the response seems to have been more horticulture—first corn and squash (*Curcurbita pepo*) and later beans, sunflowers, and perhaps gourds, leading to agriculture.

By the end of Basket Maker II (late En Medio) times (A.D. 250–400) and in Basket Maker III we had village agriculture—system E. Pit-house villages were even larger, and some had ceremonial (kiva) functions. This characteristic suggests that the pressure for food surpluses may have been spiritual as well as materialistic. Analyses of refuse, dental wear, and carbon 13-12 isotopic studies suggest a rapid shift from horticulture to agriculture, with village subsistence agriculture continuing into Trujillo and Pueblo I, Irwin-Williams's Sky Village and Loma Alta phases (A.D. 600–850). In late Pueblo times (A.D. 1100–1300) there are more domesticates, and perhaps the first true villages with intensive agriculture. Tepary, runner, and sieva beans, mixta and moschata squash, cotton, and amaranth were added to the system (see table 6.3).

Recent investigations in three or four regions of the Colorado Plateau subarea tend to confirm this scenario and hint that Irwin's Oshara tradition with its sequence of phases pertains to these regions as well, albeit with modifications. This subarea is where "scientific" archaeology in the New World began. Guernsey and Kidder (1921), who defined Basket Maker II as the first people with corn, squash, and beans, representing the end of the Archaic, hypothesized that there was an earlier stage—Basket Maker I or Archaic peoples without domesticates or agriculture. For the next forty years—until the 1960s—the basic sequence in this area did not change, and the elusive Basket Maker I remained undefined. Although sites were found that seemed pre-Basket Maker II, they lacked good stratigraphic context or good dates, and definition of the phase eluded us.

Probably the first real breakthrough came in 1959–1962 with the find of the Desha complex in the excavation of caves along the Colorado River north of Navajo Mountain in the Kayenta region. Both Sand Dune Cave and Dust Devil Cave had artifacts, including sandals and baskets, that defined a new complex called Desha, which dated between 5500 and 3500 B.C. This complex lay under Basket Maker II remains, and burial 2 in Sand Dune Cave had San Jose artifacts (3500–1800 B.C.) between the Desha and Basket Maker remains (Lindsey et al. 1968).

In the late 1960s the Peabody Coal Company in the Kayenta district began the

Black Mesa project. On the basis of the analysis of the Basket Maker sites in the region, Smiley divided it into two phases—White Dog (800–100 B.C.), without pit houses, Maiz de Ocho, and few ground stone lithics, and Lolomai (100 B.C.–A.D. 500), with pit houses, many ground stone metates, much evidence of full-time corn, squash, and bean agriculture, as well as some sort of ceremonialism (Smiley 1985). The White Dog and Lolomai phases seemed very close to Irwin's early and late En Medio subphases, so the connection to the upper Rio Grande subarea seemed very secure. In 1983 (Perry and Christensen 1987) an analysis of the lithics from the 166 excavated sites was undertaken and resulted in the recognition of "early" and "late" Archaic sites preceding Lolomai and White Dog Basket Maker II phases. When I studied these materials further, I divided the early Archaic phases into Bajada-like followed by Desha-like ones, and separated the more numerous late ones into San Jose-like followed by Armijo-like ones. This sequence confirmed Irwin's but suggested there was a possible gap between a misdated Bajada and San Jose, a gap that was filled by Desha in the Kayenta region.

Further confirming the late part of the sequence, as well as giving us more information on early plants, was the investigation Simmons and others (1986) carried out in the late 1970s and early 1980s. Under Basket Maker II materials they found corn pollen with early San Jose sites and squash and Chapalote corn with Armijo-like remains. These investigations improved the picture for the Colorado Plateau and began to define an Archaic sequence of phases.

To the south, in the Chihuahua Desert subarea in south-central New Mexico and the state of Chihuahua, Mexico, research in the 1980s has produced some of our best evidence about the Archaic in the Southwest, evidence that has direct bearing on testing and modifying the tertiary developmental hypothesis. In the late 1940s Donald Lehmer dug La Cueva, a large rock-shelter on the west slope of the Organ Mountains, a few miles east of Las Cruces, New Mexico. He defined a late preceramic culture he called "Hueco," which he thought was separate from Cochise to the west, although he speculated that it was probably contemporaneous with the San Pedro phase of that tradition (Lehmer 1948). In the early 1950s Robert Lister excavated Swallow Cave in Cave Valley in Chihuahua, Mexico, some 300 kilometers south of the Jornada region (Lister 1958). Lister found 96 inches of stratified deposits, the lower half of which were preceramic. On the basis of the sequence of corn types he found—Chapalote followed by early Tripsacoid types (which we now would classify as Proto-Maiz de Ocho, Maiz de Ocho, and Pima-Papago)—he assumed those deposits were related to Cochise (Mangelsdorf and Lister 1956). My 1989 examination of a surface collection from the shelter, however, suggests the occupations were by Fresnal and Hueco phase peoples (to be described shortly).

The debate as to whether the Archaic remains of the Chihuahua Desert subarea were Cochise or a separate entity continued into the 1960s and the early 1970s. On the basis of survey sites around Las Cruces, Patrick Beckett, a student of Cynthia Irwin, concluded that they were Cochise (Beckett 1973). Under the auspices of Cynthia Irwin, then-students Mark Wimberly and Peter Eidenbach (1972) dug stratified Fresnal shelter near Alamogordo, New Mexico, in the late

1960s and early 1970s. Irwin concluded the Archaic remains there were a third tradition, separate from Oshara and Cochise. She first called this tradition Eastern; later, after consultation with me, she decided it was better called the Chihuahua tradition (Irwin 1979).

Further evidence on the Archaic came from Thomas O'Laughlin, who retested La Cueva rock-shelter, and the Archaic stratified pit-house site of Keystone on the outskirts of El Paso, Texas (O'Laughlin 1980). Other data came from David Carmichael (1986), who did a survey on Ft. Bliss, and Steadman Upham, who excavated rock-shelters in the south end of the Organ Mountains, east of Las Cruces. Of the more than 600 sites Carmichael found in the Tularosa Basin, 176 could be classified into our four Archaic phases, thereby providing crucial information about population, settlement patterns, seasonality, and scheduling of those ancient peoples.

Starting in 1983, Upham excavated a series of small rock-shelters with preserved plant materials and a few Archaic remains. In 1985 I joined his team and dug stratified Tornillo rock-shelter, which yielded a few late Archaic artifacts and many plant remains, including corncobs (MacNeish 1985). In conjunction with cobs from nearby shelters dug by Upham, these cobs gave a fine developmental sequence of corn races (Upham et al. 1987).

To obtain the necessary Archaic stratigraphy, in 1986 and 1987 we dug stratified Todsen Cave, which gave us the Archaic sequence of Gardner Springs, Keystone, Fresnal, and Hueco phases. Because we lacked sufficient information on the early Archaic as well as enough C14 dates, we excavated nearby North Mesa site in 1988 and 1989. In 1990 we moved into Pintada rock-shelter and Pendejo Cave, which yielded a long sequence of perishable remains (including those of domesticated plants) in the first season of excavation. Even though our studies are incomplete, the Chihuahua Archaic materials we do have are of importance for testing and/or modifying our tertiary hypothesis. Let me therefore briefly outline our relevant sequences for the Jornada region of the Chihuahua Desert.

GARDNER SPRINGS PHASE—6000 ± 500–4300 ± 300 B.C.

This phase is the most poorly understood of the Chihuahua tradition phases. It has only about 21 components represented by slightly less than 200 artifacts and about 1,500 pieces of debitage from excavation. Any questions about its origins must be, at most, speculative, particularly since its predecessors in the region—pre-Clovis, Clovis, Folsom, and possibly Angostura—are even less well known.

In terms of population and settlement patterns, the speculation that the Gardner Springs people were "Desert Culture" foragers fits very well with the limited site survey and seasonal data that we have. Most of the sites were on the desert floor and/or around the playas therein, but two were in mountains with oak/pine vegetation (fig. 12). The one at Fresnal could be a fall occupation, while the occupations on the alluvial slopes at Peña Blanca in the Organ Mountains could be summer occupations. The occupations in the upper bajada at North Mesa and Todsen seem to be summer and spring occupations, respectively. Except for a couple of sites on the desert floor, all are small, either microband or task-force occupations; even the larger ones could have been multiple microband occupations rather than macroband occupations. From our limited data, it would seem macrobands wintered on the desert floor and/or playa areas, but with the coming of spring, some may have moved out to at least the lower alluvial slopes and lower bajada ecozones and then moved on to most Jornada-region ecozones in the summer, going in the fall to high elevations (to collect acorns and piñon nuts), only to return to desert floor-playa zones as the season got colder. Task-force visitations seem to occur mainly in the summer months for specialized collecting, making stone-filled roasting pits, or flint knapping; the lack of storage pits seems to indicate the people were often on the move. The settlement pattern was of a simple scheduled seasonal round of only a few very small groups.

In terms of subsistence these groups were foragers who were hunting, collecting, or trapping animals, and collecting plants. This estimate is very speculative because we have no preserved plant remains, no coprolites, and no skeletons to analyze for C12-13 or N15. The few animal bones collected from Todsen suggest the people did more hunting of deer and antelope than collecting or trapping small animals such as jackrabbits. Our Abasalo, Jay-like, and Bajada (atlatl dart) points tend to confirm such a hypothesis. Pestles, mortars, mullers, and milling stones suggest the people collected seeds, perhaps on a seasonal basis, and broke or coarsely ground them. Numerous choppers and butchering tools suggest animal meat was more important than plant foods, and the few boulder-filled roasting pits could have been used to cook or roast both.

Use-wear analysis done on a few of the snub-nosed end scrapers suggests that the people scraped skins for clothing and perhaps footwear. Data for the contemporaneous occupations of nearby Hermit's Cave (Ferdon 1946) hint that these Archaic people may already have had (square-toed) sandals and done twine weaving with agave string or yarn to make clothing and blankets. Comparative data suggest twined baskets, nets, bags, and other textiles were used, but as of now we have found none.

Another industry indicated by our use-wear studies and the presence of denticulates suggests the people did some woodworking; Hermit's Cave had a grooved club or shaft straighter in its low level. Many of the choppers also could have been used for this activity. Undoubtedly the atlatl (dart) shafts and various handles were made of wood.

Some of these same tools as well as prismatic blades seem to have been used in bone working. In the refuse of Todsen Cave we found some polished and cut bone and/or antler. As with the textiles and wood tools, however, we have as yet found few bone tools in our excavations.

In fact, not only is our knowledge of the subsistence and technology limited, but hints about the people's religion and social organization, even from ethnographic analogy, are almost nonexistent. On the basis of our limited data, it is difficult to decide whether these tools were made and used by destitute foragers (system B) or efficient foragers (system D).

KEYSTONE PHASE—4300 ± 300–2500 ± 200 B.C.

Keystone is better dated (8 dates compared with 3 for Gardner Springs), has more survey sites (18 as compared to 13), and has more excavated components

(14 as against 8). It also has more artifacts (over 350 as against about 200) and more debitage (about 2,000 fragments as against fewer than 1,500 for Gardner Springs). Further, the Keystone site along the Rio Grande just north of El Paso has a possible pit house. Nevertheless, this phase remains poorly defined.

Most of the Keystone components seem to be brief seasonal microband encampments with a few task force sites at which about the same kind of activities were performed as at Gardner Springs. During Keystone times, however, five larger sites appear. Although they might be mainly multiple microband occupations, some may be macroband encampments. Pit house 2 of site 33 at Keystone tends to confirm this opinion; it may belong to a winter occupation, perhaps by three or four microbands living in similar houses and forming a macroband or base camp type of settlement. In spite of this change, a study of the seasonality of sites suggests a calendar-round pattern similar to Gardner Springs, with people wintering on desert floors or at pit-house base camps along the Rio Grande and then spreading out to other micro-environments in the other seasons of the year, to return to their base camps with the return of winter.

The settlement pattern seems to have changed little from Gardner Springs times, and the same seems to be true of the subsistence system as well, though our evidence is poor for we lack perishable plant remains, feces, and an adequate sample of skeletons to analyze isotopically for C13-12 or N14-15. One clavicle from the backfill of Chavez Cave may be of this horizon, however, and analyses suggest a diet containing considerable plant foods and limited animal remains, the reverse of what we postulated for Gardner Springs. The artifact complex—particularly the ground stone complex—tends to confirm such a hypothesis. In addition to more pebble-mullers and anvil-milling stones (known earlier), unifacial milling stones, well-made discoidal mullers, and narrow slab unifacial metates appear. Their presence suggests not only that plant grinding was increasing, but that new types of seeds were being ground to make finer flour. Further, the occurrence of scraper planes and the use-wear on unifaces suggest that leafy plants (opuntia, agave, and lechuguilla leaves) were being scraped to make them palatable. Supporting this belief are more and larger pits full of fire-cracked boulders or slabs, which suggest more roasting of plant foods.

As mentioned before, however, these pits could have been used to roast meat. Although our sample of bones from Todsen Cave, zone K, is not large, it does show an increase in small mammals (jackrabbits), and a decrease in large ones (deer and antelope). Slip knots of agave string hint that snares and spring traps caught these smaller animals, while the larger ones were probably hunted down with atlatl-propelled darts tipped with Bat Cave, Lerma, Pelona, Almagre-Gypsum, Todsen, and Amargosa-Pinto types of projectile points. The Keystone phase thus shows a subtle shift toward a more efficient desert foraging subsistence system as well as a possible exploitation of more desert plants from more ecozones.

Although the styles and functions of the types differ, the basic technology—flint knapping, wood working, and bone working—is much the same according to our use-wear studies. Skin or hide work, however, seems to have diminished noticeably, while the ground and pecked stone industry has increased. Both use-wear studies and collections from looters of local caves provide hints that the textile industry also is different from what we know of the previous horizon. Square-toed sandals were probably in use, and coiled baskets—often with interlocking or non-interlocking stitches—may have increased as twined ones decreased. Mats also may have been woven, as well as twined and coiled bags. More definite information from well-controlled stratigraphic excavation is badly needed.

Evidence on social organizations, although woefully inadequate, hints that some sort of exogamous band type may have been in existence by Keystone times. A bone bead hints that changes were occurring in the artistic and ceremonial realm. A further sign that such changes may have begun in this period is indicated by the following Fresnal phase, when these new features are better documented and more apparent.

FRESNAL PHASE—2500 ± 200 B.C.–900 ± 150 B.C.

In terms of classification, the Keystone phase seems to be Efficient Foraging Bands—my system D. The Fresnal phase appears to represent a time of major change in the Archaic, but whether this seems true because our sample is much larger—66 components with over 400 artifacts and 4,000 bits of debitage as well as burials and sites with preservation—or because fundamental changes occurred, has been a subject of some debate (Wills 1988). I suggest it is the latter, for at this time major shifts were occurring in other parts of the Southwest, both in San Jose in the Oshara tradition and in Chiricahua in the Cochise tradition.

The most obvious changes are in the settlement patterns and related population increases. Not only do we have more survey sites (43), but we have about 20 excavated components that are noticeably different from Keystone. Over 20 of the sites are task-force sites; an equal number are occupations by much larger macrobands, pit-house base camps, or some multiple-use microband sites. These larger sites appear in the riverine environment of the Rio Grande or the desert floor-playa and may represent both winter and summer (perhaps occasionally both) and seasonal forays. From these sites, mainly task-force encampments or microband occupations radiate out at all seasons and serve a variety of purposes. The Fresnal phase, during which the population may have tripled, thus marks a shift from the old calendar-round system to a base camp radial system.

Closely connected with these changes seem to be shifts in the subsistence system, although the Fresnal people still have basically a seasonally scheduled foraging subsistence system. Evidence of this pattern is provided by a few feces, preserved plant remains from Fresnal, Tornillo, and some Organ Mountain sites, bones and artifacts reflecting subsistence practices, and also two, or possibly three, skeletons from the same sites that have been analyzed isotopically for proportions of C12-13 and Nitrogen 15. All these remains suggest that the people scheduled their collecting of a wide variety of plants by making forays into different ecozones in different seasons. Even the incipient agriculture of corn and squash was a sort of wet season plant supplement to their collecting rather than a main item of their diet. Both collecting and planting, however, were sufficiently successful that people began to store their surpluses in pits for the lean winter seasons ahead. In addition, they made and used bags, baskets, carry loops, and nets to bring the plant foods home, where they were prepared in a variety of ways. Some seeds were ground to fine flour on metates; others were coarse ground on mortars or milling stone or were cracked in mortars. Some seeds and leaves (and meat) were roasted in a wider variety of roasting pits—boulder-filled ones, slab-lined ones, burnt rock middens, and the like. The Fresnal people were learning to exploit more and more of the plant kingdom in a variety of ways in a number of ecozones. They even selected the Chapalote corn they received from Mexico, so that by the end of the period (1225 B.C.) they had developed Proto-Maiz de Ocho, which was drought resistant and adapted to their unreliable cycles of rainfall.

They of course continued to have some meat in their diet; the majority of it seems to have come from small mammals (mainly jackrabbits) that they collected, killed in drives, brought down with rabbit sticks, or caught in spring traps or net snares. A few deer and antelope bones attest to hunting by a wide variety of dart-type spearpoint types—Todsen, Pelona, Augustin, Chiricahua, La Cueva, San Jose, Armijo, Fresnal, Nogales, Maljamar, and others. In fact, the number and variety of styles of points seem way out of proportion to the amount of big game animal meat in their diet.

Flint knapping, woodworking, and bone working industries stayed about the same, but as during Keystone times, the ground stone industry increased and skin working decreased. Also, Fresnal sites show evidence of a flourishing textile industry. A wide variety of kinds of string of agave, lechuguilla, and other substances were made and a wide variety of knots—overhand, slip, square, sheep bend, and others—were used. The Fresnal people also made square-toed and fishtail two-warp sandals, twined cloth and nets, and wove several kinds of baskets, mainly coiled, with a number of kinds of coiling elements of bundles and sticks that are stitched together by locking or non-interlocking elements or fibers.

Ceremonial or artistic objects include tubular bone beads (plain, notched, or painted), beads of olivella shell imported from the Pacific Coast, and wooden objects, sometimes painted. Also, there are hints of pictographs featuring circles, zigzag lines, snakes, stick "dancing" figures, and other animals.

In fact, we are beginning to get hints of the people's social organization, shamanistic rituals, religious organizations, and cosmology. Thus, we have a fuller, albeit preliminary, picture of the Fresnal way of life, which seems to be of Semisedentary Bands with Domesticates (system D1).

HUECO PHASE—900 ± 150 B.C.–A.D. 200 ± 100

We have only about 31 excavated components, but we do have 97 surface sites, over 800 artifacts and about 9,000 ecofacts, as well as many (25) chrometric dates. Thus, our Hueco sample is the best of any of our Archaic phases, even though it still leaves much to be desired.

The continuity of Fresnal trends is very apparent in the realm of population and settlement patterns. Most of our 25–30 excavated components represent seasonal microband or task force occupations, but Carmichael netted 85 Hueco sites in his survey. A major increase occurs in sites (16 to 43) that are obvious task-force occupations, but it is difficult to differentiate archaeologically between family microband occupations exploiting a single seasonal resource for a short period and occupation by one or a few individuals who do a special activity for a few hours or days.

The Hueco phase also has an increase in macroband base-camp sites, often bigger than in Fresnal, probably with more pit houses, and occupations lasting for longer periods. We cannot determine these occupations until adequate excavations of some of the large Hueco open sites are done. Unlike the previous periods, however, the base-camp sites, from which seasonal forays were made, now begin to occur in most of our ecozones. This trend seems to continue into the Mesilla phase, in which the base camps gradually develop into year-round pit-house hamlets, until by the end of the phase, they are sedentary hamlets with full-time agriculture (system E). The Hueco and Mesilla phases represent the middle step in this development—Horticultural Villagers—system C2.

The Hueco subsistence system is also a middle step. Evidence of it now comes from some feces (many of which from Fresnal have not been adequately analyzed), sites with preserved refuse, as well as bones, and at least three skeletons whose bones have been isotopically analyzed. These studies suggest further increases in the seasonally scheduled collecting of more edible plants in more ecozones, as well as increased seasonal planting of more domesticates—Chapalote, Proto-Maiz de Ocho, Maiz de Ocho, and Pima-Papago corn, pumpkins, common beans, and perhaps amaranth—as a supplement to plant collecting. During the phase the number and variety of storage facilities for these plants increased, suggesting longer occupations in one spot. More roasting pits of various kinds occur along with more kinds of grinding stones (including heavy manos for grinding corn) and more carrying loops, nets, bags, and baskets for collecting the increasing varieties of plants.

Analysis suggests meat and related subsistence activities decreased during this time. Deer, antelope, and large animal bones decrease, even though there are a large number of dart points of various types—Armijo, Hueco, San Pedro large, San Pedro small, Hatch, En Medio, and Padre Gordo. The last three of these are small enough to be arrowpoints, but real proof of such is lacking. Bones of small mammals (mainly jackrabbits) are more numerous than those of big animals, which were probably collected, hunted, and trapped. A new element is fish and turtle bones, suggesting an exploitation of the Rio Grande, a trend that continues to increase to El Paso times and is reflected in increasing N15 values.

Some of the technology connected with Hueco subsistence activities has changed, both in terms of function and style, but in the main the flint knapping, woodworking, and bone-working activities are much the same. Hide work decreases, while ground stone activity increases. The textile industry continues to increase and we now have: four warp scuffer-toed, two warp scuffer-toed, fish-tailed, fuller length heeled, and twilled scuffer-toed types of sandals. In addition to interlocking stitch baskets, we have a few split stitch ones.

Ceremonial objects such as notched beads (for a sash), shell beads, painted sticks, and pictographs increase; we have hints of some sort of exogamous band social organization; and pictographs hint of shamanistic ceremonial leaders. An adolescent burial with 276 notched bone beads and other burials without grave goods hint at social differentiations. Also, the kinds and variety of pictographs may be increasing; more insects, serpents, birds, and animals with possible ceremonial or supernatural connections occur. Again we have hints of more complex band-type social organizations, but much study is needed for fuller understanding.

In terms of classification, these people, as well as those of the following Mesilla phase (A.D. 250–400), seem to be Horticultural Villagers (system C2) and the ancestors of the people of the Doña Ana (A.D. 900–1100) and El Paso phases (A.D. 1100–1300), who were Agricultural Villagers (system E).

THEORETICAL CONSIDERATIONS

All in all, our research into the Chihuahua Archaic indicates a slow evolution through the Archaic and into the early ceramic (pit-house) Mesilla period, with real agriculture and village life not being attained until the time of Doña Ana and/or El Paso phases, A.D. 900–1300. This pattern is unique for the development of village agriculture, and we have just enough glimmerings of how it happened to allow us to speculate about why it happened. Let us, therefore, consider what may be the necessary conditions or prerequisites for our Jornada cultural development.

Obviously, among the prerequisites for development of the Chihuahua tradition in the Jornada region must be environmental factors. The facts that few domesticable plants exist in the Southwest and that this area has connections, albeit casual, with Mesoamerica—a center where much domestication occurred earlier—mean that the Southwest was an area where the development of agriculture occurred secondarily.

Also, this is a desert region with great seasonality, and the Jornada has major climatic fluctuations or unpredictable cycles, factors that made the development of agriculture difficult. Moreover, the various ecozones were not exploitable from a single base, a condition that promoted seasonal scheduling and a type of nomadism that favored collecting over sedentary agriculture. On the other hand, the fact that the Rio Grande created a lush natural ecozone that was not circumscribed meant that a successful or efficient foraging system was easier to accomplish than successful agriculture. All in all, the necessary conditions were not encouraging for the development of village agriculture, and only unique sufficient conditions or triggering causes brought it about, albeit very slowly indeed.

What factors or causes brought about the change? To understand them, we must analyze the sequence in terms of each phase, for the causes changed with each time period, although the environmental or ecological factors often remained the same.

Our story starts with the end of the Pleistocene, when Clovis, Folsom, and/or Angostura hunters roamed the Rio Grande–Chihuahua vegetational zone that was probably then a grassland where now-extinct buffalo, sloth, antelope, horse, camel, and the like grazed, while lower elevations in the mountains had an oak-pine forest. The shift from the Pleistocene to the Holocene was a momentous one for the hunters, for it brought climatic changes that caused a shift from grassland to desert, causing many herd animals to disappear. This change affected the hunters in a manner theretofore unknown. Previously, people had not developed numerous other subsistence options and had little knowledge of their environment, so they could not schedule their activities seasonally. Their response to previous glacial or climatic changes had been to change from rich hunters to poor ones. Now they had the possibility of changing from hunters to foragers and/or collectors with a broad-spectrum procurement system. This shift is the first one we see in the Hueco basin—from Clovis, Folsom, and Angostura to the initial Archaic stage of Gardner Springs, a trend that continued into the Keystone phase. In this second stage, people also developed a seasonally scheduled subsistence with resource specialization in seed collection, as attested by the rise of numerous grinding-tool types. The first change was therefore a positive-feedback system involving diminishing biomass, use of a broad-spectrum collecting system, specialization in seed collecting, and seasonal scheduling that further diminished the biomass, and so on, in a continuous cycle. By Keystone times, about 400 B.C., a new way of life had developed because of this specific set of causes.

During the Keystone phase the efficient desert foraging underwent further changes as a result of other causes. As climatic conditions worsened during the postglacial optimum and rainfall became less and less reliable, the foragers began to visit the Rio Grande's lush gallery forest, or thicket ecozone, more frequently. Here they established base camps, often with seasonal pit-house occupations. With this increasing sedentarism, the population slowly grew, making the foragers ready to adopt other subsistence options or specializations. This positive-feedback cycle led to the Fresnal way of life at about 2500 B.C.

Fresnal saw the shift from a calendar-round system of seasonal scheduling to one with pit-house base camps or seasonal villages from which small groups made forays to the various ecozones in certain seasons. As in Keystone times, there was increasing sedentarism and relatively rapid population growth, causing the people to adopt domesticates such as corn and pumpkins from Mesoamerica as supplements to their wet-season collecting. The growing reliance on corn in this arid environment led to the development of a new hybrid—Proto-Maiz de Ocho—which yielded more food, and in turn led to more sedentarism, more mouths to feed, and increasing use of bigger base camps with more people, and on and on, establishing a positive-feedback cycle. This development led to the Hueco and Mesilla phases, 1000 B.C. to A.D. 1000.

The next-to-last stage of horticulture reveals increasing sedentarism, resulting from longer and longer stays at the pit-house villages or base camps, with the result that population slowly increased, creating a demand for more food. This demand led to the adoption of more domesticates, such as beans, amaranth, and squash (*Cucurbita mixta*), production of new corn races (Maiz de Ocho and Pima-Papago), and more and more planting of those crops. Carbon 13-12 and nitrogen-15 studies, however, show that the Hueco people were still eating a great deal of wild plants and limited amounts of domesticates, so we surmise that they were practicing seasonal horticulture rather than agriculture.

Late in the Mesilla period another element was added to the positive-feedback cycle. A long drought, lasting from A.D. 700 to 1000 (Horowitz, Gerald, and Chaiffetz 1981), forced the horticulturists to do sufficient planting to support

Table 6.4. Necessary and sufficient conditions as a positive-feedback process for tertiary development in the Jornada region of the American Southwest

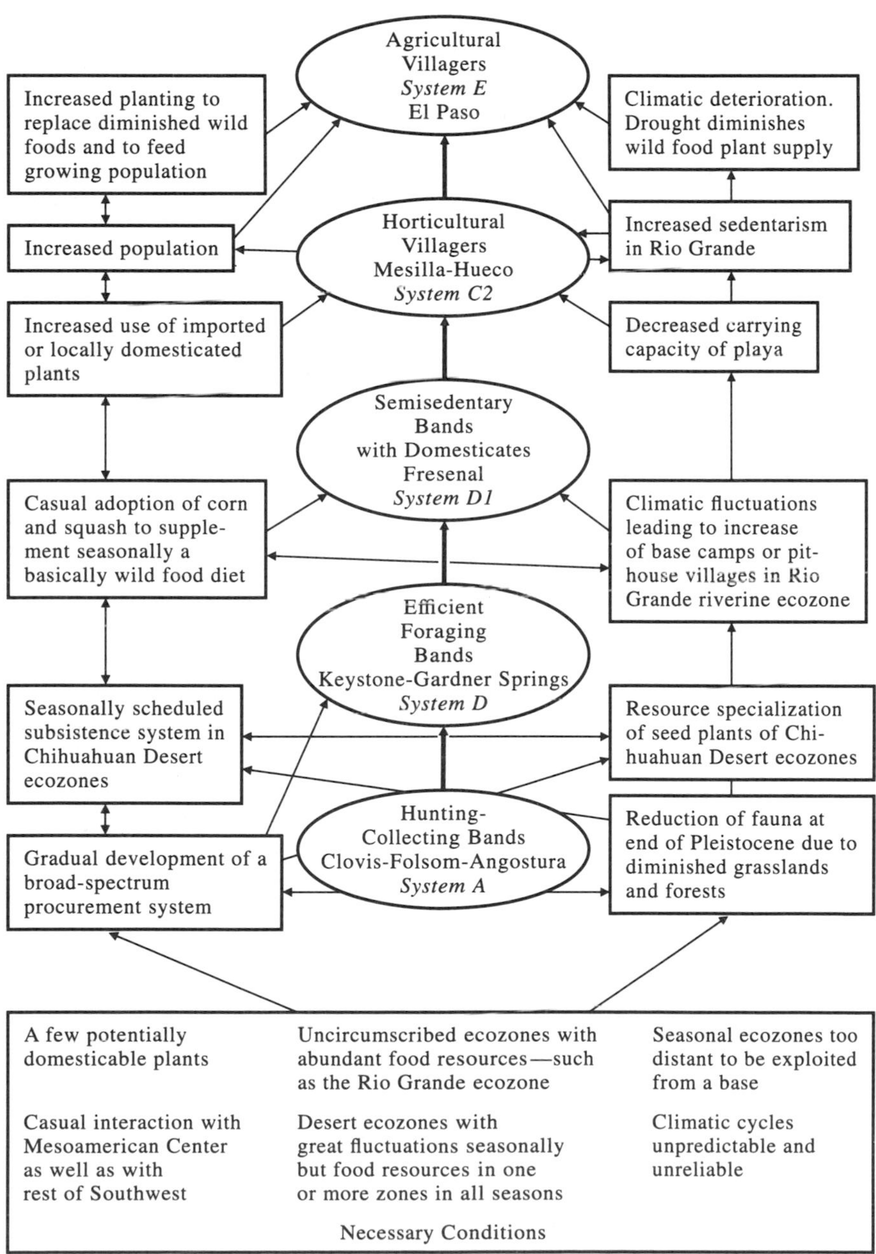

populations at sedentary year-round pueblos. By Doña Ana and El Paso times (A.D. 1100–1300) village agriculture was established. Other factors, such as a breakdown in the redistribution system and new kinds of social organization factors, may also have been part of a final positive-feedback cycle, but the archaeological evidence does not provide clear enough data to substantiate this idea.

We thus have an especially slow development from Hunter-Collectors to Village Agriculturists through specific evolutionary stages—Efficient Foragers to Foragers with Domesticates to Village and base-camp Horticulturists to Village Agriculturists—with each stage brought about by specific causes operating in four different positive-feedback cycles. In other words, the people became rescheduled in four different stages (Flannery 1968) rather than making one simple shift that resulted from one set of causes or a single cause (Cohen 1977a); this pattern is much like my tertiary development model.

Now we can ask, Did the other Southwestern Archaic developments leading to village agriculture operate in the same way? Our evidence from the Oshara tradition (Irwin-Williams 1973)—the best worked out sequence for the Southwest—seems to indicate that the answer is no. In fact, we believe that not only do the stages of the development from hunters to farmers differ but so do the causes of change and the kinds of positive-feedback cycles involved. One of the reasons the Oshara development is different from the Chihuahua is that many of the necessary conditions or prerequisites are significantly different. Obviously, both traditions are located in the great North American desert; both locations had few domesticable plants; and both had casual (indirect) contacts with Mesoamerica, which early became a Center of agriculture. There the similarity ends.

The Colorado Plateau and upper Rio Grande vegetational zones of the Oshara tradition are markedly different in terms of foodstuffs and ecozones from that of the Chihuahua Desert along the middle Rio Grande. Not only do the Colorado Plateau and upper Rio Grande have more animals and a definite winter season, but they do not seem to have been subject to the many unpredictable climatic fluctuations that made the Chihuahua region a more harsh and difficult one in which to live. The Colorado Plateau and upper Rio Grande also seem to have a number of oasislike ecozones—the lagoon canyon heads or Black Mesa lowland canyon heads—that are lusher than anything in the Chihuahua zone, even though the lush Colorado Plateau areas are circumscribed by harsh zones and can be outgrown by their population (Matson 1988). In fact, such lush zones may become bases for exploiting other ecozones in various seasons, creating a sort of centripetal effect rather than the centrifugal one we saw in the Chihuahua area. In the Jornada region the only really lush zone was the Rio Grande gallery forest or thicket, and it was certainly *not* circumscribed, nor could it be used as a base from which to exploit distant zones such as the oak-pine or ponderosa forest of the surrounding mountains.

Another possible factor, which is harder to define, concerns the interaction of the two areas. The Chihuahua zone, judging by trade materials recovered, seems to have had limited Archaic contacts with the other southwestern developments, whereas the Oshara development had more intense contact with other areas. The reason may have been Oshara's more central position, or its possible access to better passes and trade routes, or perhaps some unfathomable social preference. Whatever the cause, the Chihuahua development seems to have been on the fringe of the southwestern Archaic interaction sphere. All of these necessary conditions meant that rather different developments occurred in the Chihuahua and Oshara traditions.

Let us now examine the Oshara development to see what the sufficient conditions, or triggering causes, were that led to village agriculture there.

The first stage—from early hunters (Clovis-Folsom-Cody) to early foragers is similar to Gardner Springs, but the foragers of the Desha, Bajada, and Jay phases seem more affluent than those of Gardner Springs, perhaps because the Colorado Plateau was not as harsh an environment. People in both regions, however, reacted to the diminishing post-Pleistocene biomass by turning to broad-spectrum procurement foraging systems that involved seasonal scheduling and various newly acquired subsistence options. However, unlike Gardner Springs and Keystone, Oshara (Jay-Bajada-Desha) does not seem to have had seed-resource specialization, early pit houses, or base camps, or a calendar-round sort of seasonal scheduling. Rather, the Oshara peoples seem to have placed more emphasis on hunting in favored localities and "repeated re-occupations of favorable localities with access to fixed groups of macro-environments . . . ," causing them to adapt "to year-round exploitation of local resources" from well-watered base camps (C. Irwin 1973). Thus, even if the causes of change were similar, the two developments seem to have been headed along different paths from the outset.

During the next phase, San Jose, both the developmental stages and causes became noticeably different. While the Chihuahua tradition had a development toward Desert Foragers with Domesticates, the Oshara region groups were becoming even more Affluent Forager Bands or Foraging Villagers at the lush, circumscribed canyon heads. Factors leading to change in Oshara were increasing populations that overexploited or outgrew the lush base camp areas, and the development of more and better foraging subsistence strategies, not only at their permanently watered bases, but also in the surrounding zones into which bands could make forays to acquire specific resources in specific seasons.

This positive feedback cycle could go just so far, and by 800 B.C., the end of Armijo, a food and population crisis suddenly had to be met. To increase the food supply, the Oshara people quickly adopted corn, beans, pumpkins, gourds, and squash as a horticulture complex (En Medio and Trujillo phases, or on the Colorado Plateau White Dog Cave phase) and became villagers, setting up a new positive feedback cycle. More food from horticulture meant more people living in circumscribed zones, which led to more local domestication of plants (tepary beans, runner beans) and more importations of domesticates such as moschata and mixta squash. In this way horticulture rapidly became agriculture and led by Sky Village and Lolomai times to still more population increases. Rapid change was, therefore, inevitable. Overexploitation of resources, climatic fluctuations and drought, and trade connections were minor factors that only pushed along the two major ones. The Oshara development was thus no gradual drift into an un-

Table 6.5. Necessary and sufficient conditions as a positive-feedback process for secondary development in the upper Rio Grande and Colorado Plateau subarea of the American Southwest

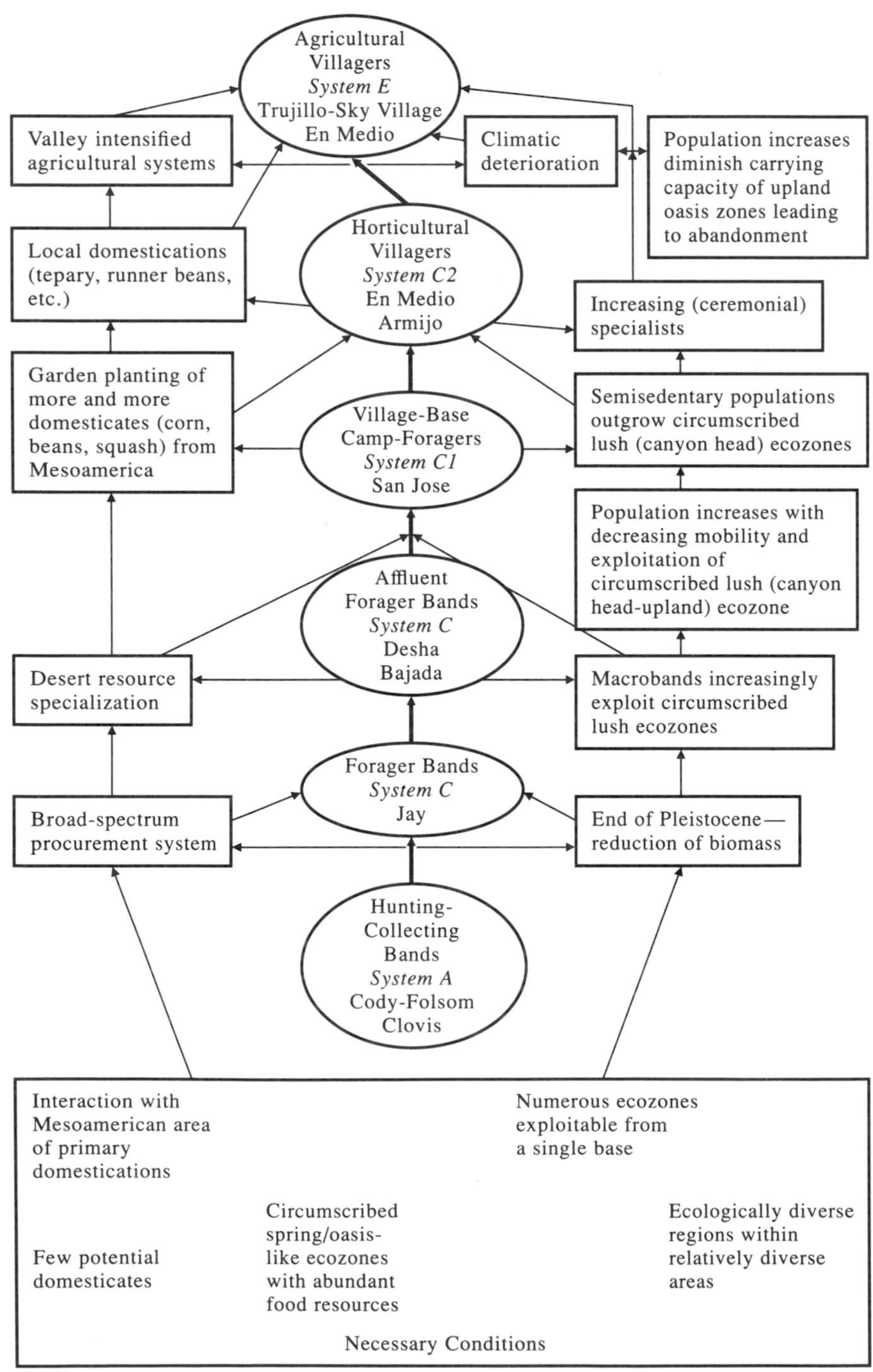

stable village agriculture like that of the Chihuahua tradition, but a rapid plunge into farming and village life. Table 6.5 shows the generalized stages of this secondary type of development.

Unfortunately, this scenario is both poorly documented and highly speculative. Many problems remain to be solved before it can be considered adequate proof of my secondary developmental hypothesis. First and foremost, in spite of the dictums of the New Archaeology, better chronological control is needed for each of the phases and subphases from 2000 B.C. to A.D. 1000 (Berry 1985). Only thus will we really be able to determine whether villages preceded agriculture in the Armijo phase or whether population increases were a cause or a result of agriculture. The chronological problem also plagues us when we try to determine exactly when and by what route the various domesticates arrived in the Southwest.

More careful excavation of older sites such as Bat Cave, as well as better digging of new sites (if they don't get looted by the amateurs first), in addition to more sophisticated studies of ancient food remains, like C13-12 or N15 isotopic analysis, should help us determine ancient diets and subsistence systems or patterns. Many of the broad statements made in this realm by archaeobotanists of the Southwest are completely undocumented (R. Ford 1981), and many reports do not even contain a list of the ecofacts in their archaeological contexts (chronological or otherwise). The whole demographic and population problem for these crucial late Archaic horizons needs to be investigated with more acute data-collection techniques and better analysis as well as more sophisticated methodological approaches. The same applies to the hints of ceremonial activities that seem to begin during this crucial period. Maybe the usually unconsidered nonmaterialistic ("emic") factors were just as important as the obvious materialistic ones.

However, there is a strong possibility that the sufficient conditions for secondary development were in operation in the Colorado Plateau and upper Rio Grande subareas of the U.S. Southwest between 1800 B.C. and A.D. 900. This is still only a possibility, albeit a strong one, and much good, solid data collection and analysis need to be done before the case can be proved. Furthermore, much research needs to be undertaken to determine if this Archaic Colorado Plateau and upper Rio Grande development also occurred in the other subareas of the Southwest, such as the Gila River drainage region. The Cochise data hint that similar developments pertained, but some regions—the Rio Grande or Mogollon Highlands—may have had a tertiary development. As of now, all we have are hints that the Southwest was a non-Center where the secondary and tertiary development of village agriculture had the same necessary and sufficient conditions as the lowland parts of the Centers. At this time we can conclude only that there is great hope that the data from the Southwest can be used to test my hypotheses.

CHAPTER 7

Europe

While the U.S. Southwest has the longest history of investigation on early village agriculture in the New World, Europe has the longest for the Old World, the Danes having become involved in "Mesolithic-Neolithic" archaeology as early as 1812. Like the Southwest, Europe is also a non-Center adjacent to a Center—the Near East—which gave this temperate area its first domesticates—emmer, einkorn, bread wheat, barley, and so on. As we shall see, both Europe and the Southwest have secondary and tertiary developments in different environments. But here the resemblances end.

Both in their rich, temperate ecology and in their many tertiary developments of village agriculture, the parts of Europe northwest of the Aegean more closely resemble the eastern United States (see chapter 8) than they do the Southwest. Obviously, the times at which these similar cultural processes occurred on their respective continents were different. In the eastern United States the development began in the Middle Archaic, with the first domesticates at 5000 B.C., continuing through the Woodland period of Foraging Villagers with horticulture, from about 500 B.C. through A.D. 1000, and the Mississippian period of Village Agriculturists, at A.D. 1000 to 1500. In Europe, the development was in the Mesolithic (8000–5000 or 6000 B.C.) and the Neolithic, ending about 2000 B.C. with true village agriculture. In both these temperate areas, the development of village agriculture was a long and slow process; it took between 3,000 and 5,000 years to shift from system A, Hunting-Collecting Bands, to system E, Village Agriculturists. The development of agriculture in the Aegean area, however, is a secondary one unlike anything so far discovered in the eastern United States, and there it occurred somewhat faster than in the New World.

The environments of Europe and the eastern United States have many features in common. Both have mainly temperate climates and deciduous broad-leaf forest vegetation. Both areas are included in or adjacent to steppe areas—the south Russian and Polish plains or the Great Plains of the United States—as well as areas of Mediterranean climate, such as the Mediterranean and Gulf coasts with their semitropical vegetation. Further, both Europe and the eastern United States are flanked by the Atlantic Ocean with its rich coastal marine resources and the ameliorating effects of the Gulf Stream. In fact, in terms of necessary conditions, the areas are relatively uniform overall, but within certain regions there may be great ecological diversity. Both areas have ecozones with abundant food resources that have made for relatively easy living for the ancient as well as the modern inhabitants, and in both areas archaeological investigations have been relatively easy. Research has been so full and of such long duration in Europe that my synopsis here must be brief and superficial, touching upon only a few of the high points of Neolithic and Mesolithic research.

A BRIEF HISTORY OF RESEARCH ON THE PROBLEM

Scientific archaeology in Europe began when the Danes established their National Museum of Ethnography and Archaeology by royal decree in 1812. Classification of specimens and investigations of mainly Mesolithic and Neolithic sites around the Baltic Sea soon followed. By 1830, Worsae, Nyerup, and others had developed the basic Stone, Bronze, and Iron Age sequences for that area as well as for all of human history. Through the middle of the nineteenth century the Danes pursued local investigations and continued to make further Neolithic and Mesolithic discoveries. As their research progressed, they developed better and better archaeological field techniques and methods and became more and more involved in such interdisciplinary studies as pollen analyses, clay varves, geomorphology, and changing sea levels.

At the same time, other things were happening that had considerable bearing on Neolithic and Mesolithic research. Of fundamental importance were the intellectual breakthroughs in 1859 of Charles Darwin and Alfred Russel Wallace in the field of biological evolution. These led indirectly to a true recognition of the Paleolithic and Neolithic by Lord Avebury in 1865 and stimulated further investigations of the Bronze and Iron ages of Europe. In France, a number of subdivisions of the Paleolithic were established, and differences in Neolithic pottery complexes were recognized. This pioneering period ended about 1880 with the recognition that some sort of transitional period was needed between the Paleolithic and the Neolithic.

The stage and term *Mesolithic* came into being with the work of Hedder Westropp. Needless to say, archaeological work in the Baltic and North Sea areas continued. Early finds were made of the "Swiss Lake Dwellers," and microliths were recognized as diagnostic lithic artifacts of this Mesolithic period (Heer 1866). In addition to this direct evidence of the Mesolithic as a distinct period and further ceramic subdivisions of the Neolithic, an important interdisciplinary field—philology and linguistics of Indo-European languages—had considerable effect on interpretation of the European Neolithic. Philological studies suggested to the linguists that waves of linguistic groups or families—Celtic, German, Latin, and so on—had swept into Europe at different times from their original Indo-European center in south-central Asia (India?). Since these "migrations" seem to have occurred before the Iron Age, the different waves unfortunately became tied to various Neolithic ceramic complexes. Many interpretations of the Mesolithic and Neolithic are still plagued by this "wave" concept.

Another concept that has caused trouble surfaced in the general period from

1880 to 1920, when the Marxists put forth their ideas about cultural evolution. Much of this difficulty arose out of the proposition that society evolved from savagery to barbarism to civilization. Barbarism was correlated with the Neolithic period, and gradually the idea developed that barbarism and/or the Neolithic meant sedentary village life with agriculture and pottery. This concept gradually gave rise to the basic axiom that if one finds an archaeological manifestation with pottery, then the peoples must have had sedentary life and agriculture—even though the ethnographic record is full of cases to the contrary. The postulate that pottery always equals agriculture, as well as the concept of economic determinism, has inhibited research on the theory of the origin and process leading to village agriculture. Often when archaeologists find a site with pottery, they assume these ancient people had agriculture rather than looking for evidence of such.

The rise of the Marxist dialectic in this period was not all bad, however, for it had one major redeeming virtue: it influenced the basic thoughts of the leader of the next period, V. Gordon Childe, who brilliantly dominated the field of European Neolithic studies for nearly half a century (Childe, 1925, 1931, 1942, 1951, 1958), from about 1920 to 1960. Childe produced an innovative study of the Danubian area and established the first solid chronology of the Neolithic for major regions of Europe, clearly indicating that agriculture had spread into Europe out of Asia Minor (Childe 1925). This significant study had many wide implications, for Childe recognized that a series of pottery complexes—Bell Beaker, Linear, and so on—characterized the spread of the Neolithic westward through Europe. On a much grander scale was his defining and elucidating the whole concept of the Neolithic Revolution and the oasis theory of agricultural origins, both strokes of genius in their time. Unfortunately, often through no fault of his own, these data were used to support the wave concept of the spread of village agriculture by specific ethnic and/or linguistic groups as well as to confirm the axiom that pottery automatically meant agriculture and sedentary life. Another minor flaw was that he did not fully consider the basic complex ecological factors, and the whole Mesolithic period was thus relegated to the status of a "dark period of extreme poverty" (Childe 1925, 1942).

The leader of the next period of archaeological research, from 1945 to 1965, was John Graham Douglas Clark. His basic studies of the Mesolithic led to novel considerations of the Neolithic, although like Childe, he too was a Marxist (J. G. D. Clark 1952), or in modern terms, a "cultural materialist" (M. Harris 1980). One of Clark's first ventures was a reconsideration of the Mesolithic of northern Europe (J. G. D. Clark 1936), and here he was most fortunate in having available the great mass of data meticulously and scientifically accumulated for over a century by the Danes and others in the Baltic Sea area. It was apparent to Clark that the Mesolithic was a significant transitional period during which people made a whole series of new adaptations to the changing ecological conditions following the Pleistocene. He saw this period as a prelude to civilization and agriculture rather than as some sort of dark age (J. G. D. Clark 1980). His important excavations of Star Carr and its interdisciplinary analysis led him to view the transition from Neolithic to Mesolithic less as a rapid revolution and more as a slow evolution. In fact, he later wrote that "the Mesolithic, so far from being a dead end, was in fact an essential prelude to fundamental changes in the development of culture" (J. G. D. Clark 1980, 7). One of his colleagues, G. I. Mathyushin (1976), saw the Mesolithic as "the most important epoch in history."

Clark and other scholars of this period increasingly recognized that throughout the Mesolithic people became more and more successful foragers, so that near the end of the period they had often settled down in villages, developed many resource specializations, and were in fact "ready for agriculture" (Braidwood 1951, 23). In northern and western Europe this meant that during the Neolithic people acquired domesticates only gradually, attaining true full-time village agriculture only near the end of the period. In the Classical Aegean area, however, early Neolithic and ceramic sites began to be uncovered that hinted at a quick development like that of the lowland Near East.

Much of the documentation of the acquisition of domesticates came in the next period, 1965 to the present (Price 1983). Hans Helbaek (1948), Willem Van Zeist (1967), Maria Hopf (1969), Jane Renfrew (1973), and others were leaders in this field of "palaeo-ethnobotany" for Europe. This was a busy group, for finds relevant to the origins of agriculture were being discovered everywhere. Although we shall deal with some of them in more detail later in this chapter, I will mention a few of particular significance here.

In the Aegean area a whole series of Neolithic and Proto-Neolithic sites was uncovered in Cyprus, Crete, and southern Greece. Upper Paleolithic remains with blades—examples of early hunters—were found at Kastritsa and Asprochaliko near Epirus and in the lower two zones of Franchthi Cave, which date at 25,000 to 15,000 B.C. and 11,000 to 8300 B.C. (T. Jacobsen 1981). Mesolithic remains were exhumed in zones IV and V that date at 8300 to 7300 B.C. and 7300 to 6300 B.C. There is an indication of a shift in the subsistence pattern for this final group, with signs of fishing becoming important. The people may have been Foraging Villagers (system C1) without agriculture and perhaps were the peoples of the lowest levels of Khirokitia in Cyprus and Argissa in Greece (Milojeic, Boessnick, and Hopf 1962). After 6300 B.C. agriculture was adopted rapidly. Village Horticulturists developed into Agriculturists from 6300 to 5500 B.C. at such places as Franchthi Cave (Zone VI), Argissa, Knossos in Crete, Khirokitia in Cyprus, and Gediki, Sesklo, Nea Nikomedeia, Achilleion, and others in Greece (J. Renfrew 1966). All of this is relevant to our problems of the origins of village agriculture via a secondary development. As research continues, our relevant data should get better and better.

Almost the equal of these revolutionary finds were many Neolithic excavations in the Balkans, particularly along the Danube River. Two sites near the Iron Gates on the Danube, Vlasac and Lepinski Vir, have yielded preceramic village remains and microliths indicating a Mesolithic heritage. Like them, but with pollen evidence indicating possible agriculture, is Icoana in Romania (Dolukhanov 1982). Earlier Mesolithic sites have also been discovered at Odmut Cave in Montenegro as well as at Cuina Turcului in Romania (Dennell 1983). Thus, we have a good deal of new evidence about the crucial stage leading to Neolithic village agriculture, and needless to say, a huge number of Neolithic sites (often stratified tells) have also been excavated in the past twenty years in southeastern Europe.

Perhaps as significant and as new as these discoveries were those of the Late Mesolithic and Early Neolithic along the shores of the western Mediterranean. Of particular importance have been the Mesolithic sites reported by Jean Guilaine in southern France as well as his reappraisals of all the new data coming in (Guilaine 1981).

While I have mentioned only three regions as yielding significant relevant material for Europe, obviously finds continued to be made in the regions where archaeological programs were initiated in earlier periods. The Danes carried on their fine Mesolithic-Neolithic research in the Baltic, and similar endeavors continued in the British Isles. Research also continued on the northwestern mainland of Europe, which had a long tradition of research on both the Mesolithic and Neolithic, particularly investigations concerned with megaliths. Last but not least, finds continued to be made of the Swiss Lake Dwellers. On every front, new data about the beginnings of village agriculture accumulated, and botanical data derived by new techniques were being analyzed by new palaeo-ethnobotanists.

Perhaps more significant has been the interpretation of data. The New Archaeology has caused data to be interpreted in new and radically different ways. A host of new scholars—Robin Dennell (1983), Jean Guilaine (1976), Colin Renfrew (1981), Ruth Tringham (1971), Hans Waterbolk (1968), Sarunas Milisauskas (1978), P. M. Dolukhanov (1982), M. Jarman (Jarman, Bailey and Jarman 1982), and many others—have made and still are making major conceptual contributions to our understanding of the European Mesolithic and Neolithic.

Perhaps one of the more important new concepts is that the "divide between Mesolithic and Neolithic . . . has been greatly overstated" and that the development of the Neolithic or village agriculture was one of Mesolithic assimilation rather than "waves" of Neolithic colonization (Dennell 1983, 43). The New Archaeology has prompted European archaeologists to take a new look at the Neolithic food data so that they may determine if groups had horticulture, incipient agriculture, or full-scale agriculture; they no longer automatically accept the dictum that pottery equals agriculture. At the same time, they are examining seasonal data more carefully to determine if sites were truly settled villages and not some sort of large base camps. Pollen studies are indicating that the cutting or clearing (by fire) of forested land to improve hunting and collecting often preceded agriculture in the Mesolithic and was not simply the result of land clearing for agriculture. In fact, the whole field of paleoecology and the relationship of Mesolithic and Neolithic peoples to their environments has moved on to a higher, more sophisticated conceptual plane in Europe (Dennell 1989).

To better understand such discoveries and their interpretations, a few brief statements should be made about the European environment itself as well as about the changes it experienced between the end of the Pleistocene (roughly 8000 B.C.) and the end of the Neolithic (about 2000–1500 B.C.)

THE ENVIRONMENT

In terms of topography, Europe has three general areas—coasts, lowlands, and uplands (see fig. 13). The coastal subareas comprise the Mediterranean shores, the Aegean Sea coast, the Atlantic Coast, and the two coastal subareas along the North and Baltic seas. The lowlands can be divided topographically into at least three main subareas: (1) the eastern plains, from southern Russia to northern Europe; (2) the western lowland subarea, the low, rolling country of the European interior (Germany, Denmark, Belgium, the Netherlands, and much of France); and (3) the dissected mesa-canyon country of the arid interior of Iberia. There are also three general subareas of uplands: (1) the Pyrenees, which separate the interior Spanish lowlands and the more fertile lowlands of France; (2) the Alpine subarea consisting of the Alps, intermediate mountains, and the Apennines of Italy; and (3) the Danubian drainage and basin in southeastern Europe, with its surrounding and inclusive mountainous country.

Major drainage systems cut many of these topographic subareas into segments, and of course, served as major routes of cultural migration and/or diffusion or acculturation. The Eastern plains subarea has the Dnieper, Bug, Dniester, and Siret rivers draining south into the Black Sea, while the Dvina, Neman, Vistula, and Oder flow north into the Baltic Sea. A series of rivers crosses the subarea of the low, rolling western lowlands and flows into the North Sea–Atlantic Ocean—the Rhine, Elbe, Seine, Loire, and Garonne—while the Rhone drains into the Mediterranean. The Tagus and Ebro cross Iberia and flow, respectively, into the Atlantic and Mediterranean. Although many streams cut through the Pyrenees and the Alps, it is the Danubian subarea of southeastern Europe that is extremely bisected by rivers—the Danube and its large tributaries such as the Sava, the Drava, the Morava, and the Tisza.

Temperatures are only slightly affected by the topographic and drainage features. Most of the area has a temperate climate except along the Mediterranean. Southern Spain and Portugal have less seasonality and rarely, if ever, suffer winter frosts. The topography and drainage do, however, affect the rainfall and wind currents that come in from the Atlantic and Mediterranean. The eastern plains and Iberian interior fall in the rain shadow of mountains and receive less rainfall, while the Mediterranean coast is well watered and mainly subtropical. The western seaboard has maritime seasonal rains, while the mountainous southeast has a variable but generally continental climate.

Perhaps more variable than the present-day climate of Europe was the climate following the Pleistocene and the retreat of the Alpine and continental Scandinavian ice sheets some eleven thousand years ago (see table 7.1). For northern Europe the changes were dramatic: a Younger Dryas period (8900–8300 B.C.), with arctic or glacial conditions that resulted in tundra and/or arctic forest vegetation, followed by a Pre-Boreal period (8300–7500 B.C.), when the climate became warmer and the Baltic area had birch and pine vegetation. The climate continued to get warmer in the Boreal period (7500–6200 B.C.), which had a more continental climate, and in the Atlantic period (6200–3000 B.C.) a still warmer maritime climate evolved. During the Boreal and Atlantic periods vegetation shifted

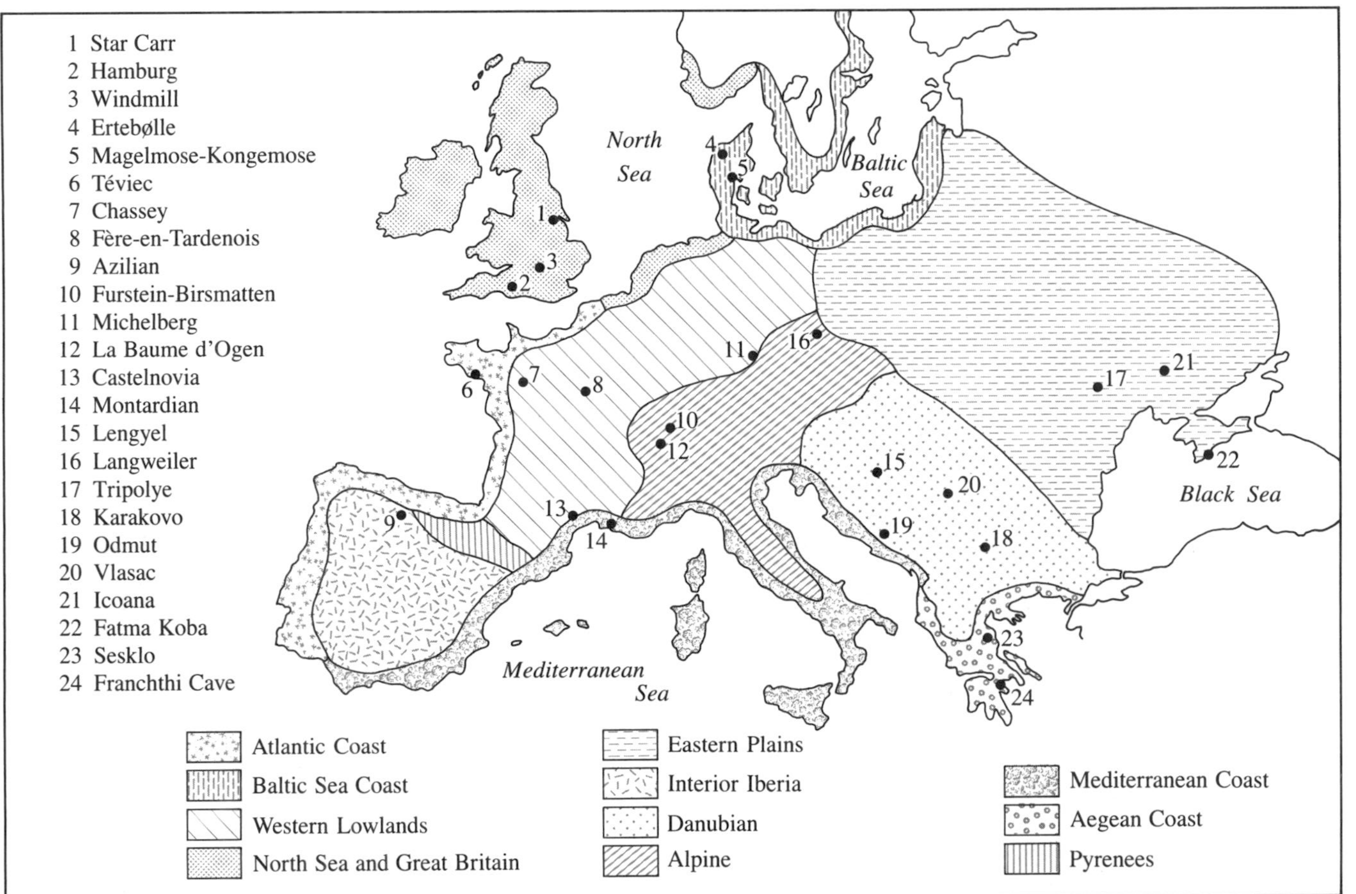

Figure 13. The environmental life zones and relevant sites in the European area

Table 7.1. The changing environment that affected Mesolithic and Neolithic developments in Europe

Dates B.C.	Cultural period	Climatic period	Dominant vegetation	Climate	Sea level stages of Baltic Sea	Dates
300						
	Iron Age				Present sea levels	
—	Bronze Age	Sub-Boreal	oak beech	More Continental		
—						2000 B.C.
3000	Neolithic					
					Littornia Sea	
—						
		Atlantic	oak elm	Warmer and Maritime		
—						
						5500 B.C.
	Mesolithic					
— 6200						
					Ancylus Lake	
—		Boreal	hazel pine oak	Warmer and Continental		
— 7500						
						7800 B.C.
— 8300		Pre-Boreal	birch pine	Warm Continental	Yoldia Lake	
						8300 B.C.
8900		Younger Dryas	forest tundra	Arctic	Baltic Ice Lake	
9800						

from open hazel, pine, and oak forests to dense oak and elm forests. The Sub-Boreal (3000–300 B.C.) brought a return to a more continental climate with less dense oak and beech forests. Much of this Sub-Boreal period falls after the Neolithic, as does the final period, which has continued up to the present and has brought a more maritime and slightly drier climate.

Obviously, the changing climate plus the topography and soils affected both flora and fauna. The Europe we are presently interested in has only five major zones of vegetation—the lush Mediterranean, the deciduous broad-leaf forest covering much of central Europe, the steppe area of the eastern plains, the xerophytic area of Iberia, and the Alpine coniferous forest areas of some of the uplands. These general zones shifted dramatically from 5000 to 2000 B.C. Of particular significance were the shifts from the Boreal period (7500–6200 B.C.), when pine-birch and mixed forest covered much of northern Europe, to the Atlantic period (6200–3000 B.C.), when these forests retreated northward and were replaced by broad-leaf deciduous forests and open woodland narrow-leaf deciduous forests; these, in turn, disappeared in the Sub-Boreal period, when the Late Neolithic and Bronze Age came into being. All of these changes affected Mesolithic foraging or plant-collecting activities as well as later Neolithic farming and herding or grazing practices.

Climatic changes as well as regional differences were also reflected in the fauna, but these changes may not have been quite so significant for people, with the exception of the extinction of the megafauna at the end of the Pleistocene. There can be little doubt that this radical change played a part in causing the Paleolithic hunter, who did little or no foraging, to change into a Mesolithic forager, who did many things in addition to hunting. Even after the loss of the large herds, however, abundant land animals still existed in the continental and Mediterranean forests and in the alpine areas; riverine and maritime resources were abundant as well. What is more, unlike our Centers, the western European regions that had abundant faunal food resources were neither restricted nor circumscribed, even though they may have varied regionally one from the other and differed from the eastern European regions, which had lusher, more circumscribed zones without the contrasting harsh areas.

While all of Europe is basically part of the major world zone of temperate-deciduous forest, the continent has great regional variation, and a number of native plants were susceptible to domestication (see table 7.2). Because of this great regional variation, Europe can be divided in numerous ways. I have chosen nine of the eleven major subareas or regions that take the Mesolithic and Neolithic developments into account.

There are five coastal subareas: (1) the western Mediterranean (from Albania to southern Spain); (2) the Atlantic (including the coasts of Portugal, Spain, France, and Belgium); (3) the North Sea (including the Netherlands and the British Isles); (4) the Baltic Sea (northern Europe); and (5) the Aegean coast (eastern Mediterranean). These subareas correlate well with the three basic Mesolithic types of culture, and each has slightly different kinds of climate, vegetation, and marine resources.

Three of the four inland subareas are also clear-cut. Interior Iberia is a good

Table 7.2. Some native cultivated and/or domesticated plants of Europe *

Scientific name	Popular name
**Agropyron sp.*	wheatgrass
**Agrostis sp.*	bent grass
**Allium cepa*	onion
**Allium sativum*	garlic
**Allium porrum*	leek
**Anethrum graveolens*	dill (herb)
Arbutas unedo	strawberry tree (evergreen)
**Atropa belladona*	belladonna (medicine)
**Avena sativa*	oats
**Avena strigosa*	fodder oats
**Beta vulgaris*	beet
**Brassica campestris*	rapeseed
**Brassica nigra*	mustard
**Brassica oleracea*	cabbage
**Brassica rapa*	turnip
**Brumus mermis*	smooth broomgrass
Carthamus tinctorius	safflower
**Carum carvi*	caraway
**Ceratonia siliqua*	carob
**Coriandrum sativum*	coriander
Cornus mas	cornelian cherry
Corylus avellana	hazelnut
Crataegus sp.	hawthorn
**Cucumis melo*	melon
**Cuminum cyminum*	cumin
**Cydonia oblonga*	quince
**Dactylis glomerata*	orchard grass
**Daucus carota*	carrot
**Digitalis purpurea*	digitalis
Echinochloa crus-galli	barnyard millet
Fagus sylvatica	beechnut
**Festuca arundinacea*	tall fescue
**Foeniculum vulgare*	fennel
Fragaria vesca	strawberry
**Glycyrrhiza glabra*	licorice
**Hyoscyamus muticus*	henbane
**Juglans regia*	English walnut
**Lactuca sativa*	lettuce
**Lepidium sativum*	garden cress
**Lolium sp.*	rye grass
**Olea europea*	olive
Panicum miliaceum	broomcorn millet
Papaver somniferum	codeine, morphine
**Petroselinum sativum*	parsley
**Phalaris arundinacea*	reed canary grass
**Phalaris tuberosa*	hardinggrass
**Phleum pratense*	timothy (grass)
**Pimpinella anisum*	anise
**Pisum sativum*	garden pea
Plantago psyllium	psyllium (medicine)
**Portulacea oleracea*	purselane
**Prunus avium*	cherry
**Prunus domestica*	plum

Table 7.2. (*continued*)

Scientific name	Popular name
Prunus instititia	ballice
Prunus spinosa	blackthorn
**Pyrus communis*	pear
**Pyrus malus*	apple
Rosa canina	rosehip
Rubus fruticosus	blackberry
Rubus idaeus	raspberry
Sambucus nigra	elderberry
**Secale cereale*	rye
**Setaria italica*	Italian millet
**Sorghum halepense*	Johnsongrass
Trapa natans	water chestnut
**Trifolium sp.*	true clover
Triticum polonicum	Polish wheat
Triticum spelta	spelt wheat
Triticum timopheevi	minor wheat (Russian)
Vaccinium myrtillus	bilberry
**Vicia sp.*	vetch
**Vitis vinifera*	grape

*After Harlan (1971); others, J. Renfrew (1973).

unit ecologically, although our Mesolithic and Neolithic knowledge of it is so poor that it will not be discussed. The other dry subarea with plains vegetation is the eastern plains—the steppes and/or open woodland of Poland and Russia. This subarea correlates with similar or related cultures in much of the Neolithic—Linear pottery in Early Neolithic, Funnel Beaker in Middle Neolithic, and Globular Amphora or Corded ware in Late Neolithic. The Alps or Alpine central subarea is equally distinctive; here the Swiss Lake Dweller sequence represents a distinctive cultural tradition for the Mesolithic and Neolithic. The Pyrenees subarea, on the other hand, has little archaeological data.

I felt fairly secure with these units, but determining the final ones caused much soul searching and consideration of the archaeological sequences. I finally decided one large, fairly unified subarea existed in southeastern Europe in the Danubian drainage and another subarea was in the lowlands of Western Europe. I am well aware that these subareas can be subdivided still further, but for the sake of clarity and unity I have not done so. In summary, then, I would divide Europe during the Mesolithic and Neolithic into the following eleven subareas of culture:

1. Coasts
   a. Western Mediterranean
   b. Atlantic
   c. North Sea
   d. Baltic
   e. Aegean
2. Lowlands
   a. Western lowlands

b. Eastern plains
c. Interior Iberia
3. Uplands
a. Alpine
b. Danubian
c. Pyrenees

RELEVANT ARCHAEOLOGICAL DATA

We shall consider the archaeological sequences in terms of the above cultural subareas—excluding the Pyrenees and inland Iberia, for which we have little relevant data—in the order of their relationship to the Near East Center. This geographic proximity often is reflected in how early a group acquired agriculture or domesticates and became acculturated to a Mesolithic or Neolithic way of life (see table 7.3).

We begin with the Aegean coastal subarea, which has a fine sequence from the Paleolithic to the Neolithic from Franchthi Cave in southern Greece (T. Jacobsen 1981). In the lower levels of this cave were two zones of the Upper Paleolithic dated at 25,000 to 15,000 B.C. and 11,000 to 8300 B.C. Bones as well as blade tools attest that these people were Hunting-Collecting Bands, and the size of occupation layers suggests that they were nomadic microbands that existed at the end of the Pleistocene. Unfortunately, our data from this cave as well as from the rest of the subarea are inadequate to tell how or why it developed into the Lower Mesolithic between roughly 8300 and 7300 B.C. Certainly the microlithic tool types and bone data of this period indicate these hunters had a broader spectrum subsistence system, and grinding stones suggest plant collecting. There is a suggestion that they had become Affluent Foragers—my system C. Further, some of the larger open sites—for example, Khirokitia in Cyprus and Argissa in Greece—have similar microlithic tools but no evidence of pottery or agriculture in their lower levels, suggesting that in this Lower Mesolithic time there may have been forager villages—system C1—with permanent architecture consisting of round houses. Unfortunately, reports do not adequately reinforce this supposition (J. Renfrew 1966). Thus, we really do not know if a system C1 existed in this area, although the following Upper Mesolithic remains in Franchthi Cave hint that such will eventually be found.

Above the Upper Mesolithic of zone V (7000–6300 B.C.) in Franchthi Cave are the aceramic Neolithic remains of zone VI (6300–5300 B.C.). These definitely had domesticated plants—emmer, hulled two-row barley, and lentils—all probably diffused from Anatolia (see table 7.4). Carbon-14 dates suggest that these remains are contemporaneous with a host of remains of aceramic villages—perhaps horticultural villages—system C2. They all have microlithic tools similar to those of the Upper Neolithic zone VII (5300–3400 B.C.) of Franchthi Cave, as well as the same domesticated plants plus other imports—naked two-row barley, oats, peas, vetch, bread wheat, and millet (see table 7.4). Certainly this indicates that village agriculture—system E—had arrived in the Aegean by 6500 to 5000 B.C., but our data are not concise enough to determine the sufficient conditions that brought it about. However, these Aegean data indi-

Table 7.3. Relevant archaeological sequences in Europe

Aegean Coast subarea	Danubian Uplands subarea	Western Mediterranean Coast subarea	Eastern Plains subarea	Alpine subarea	Western Lowlands subarea	Atlantic Coast subarea	British Isles North Sea Coast subarea	Baltic Sea Coast subarea	Dates
Classic	Baden	Bell Beaker	Bell Beaker	Horgen	Bell	Corded Ware	Peterborough		
			Globular		Beaker		Corded	Corded Wares	
		Almería	Corded				Windmill		2000 B.C.
		Megalithic	Funnel Beaker	Michelberg	Chassey	Seine	Megalithic		
Franchthi VII	Gumelnitsa		Tripolye		Megalithic	Passage		Megalithic	
						Grave	Hembury		
									4000 B.C.
	Lengyel								
	Karanovo IV		Linear		Linear	Linear			
		Cardial					Oronsay		
								Ertebølle	
									5000 B.C.
	Starcevo-			Birsmatten					
	Körös	Bug							
Franchthi VI					Tardenoisian				
Selko		Chateauneuf							
Argissa									
Nikomedeia						Téviec			6000 B.C.
		Baume de						Kongemose	
	Vlasac	l'Abeurador	Icoana						
									7000 B.C.
Franchthi V		Castelnovia							
				La Baume	Sauveterrian				
		Montardian	Fatma Koba	d'Ogens				Magelmose	
	Odmut						Mt. Sandel		8000 B.C.
Franchthi IV				Furstein			Star Carr		
					Azilian	Asturian			
	Cuina							Arhensberg	9000 B.C.
	Turcului								
Franchthi III									

cate a secondary development and have a Mesolithic and Neolithic sequence different from the other regions of western Europe.

Our next region is the Danubian subarea (table 7.3). While the exact transition from Paleolithic to Mesolithic is not well delineated, remains from Odmut Cave in northwestern Montenegro may date to between 7500 and 5500 B.C. Here triangular geometric microliths were replaced by trapezoidal ones about 6700 B.C. (Srejovic 1978). In addition to these microliths were small, flat bone harpoons with finely engraved geometric designs. Animal and fish bones as well as tool types suggest these people were Affluent Foragers—system C.

Somewhat later, 6000 B.C. ± 500, are the Vlasac stone-lined grave and midden material from the Iron Gates on the Danube as well as the material from Lepinski Vir (Srejovic and Letica-Vlasac 1978). Along with the continuation of microlithic stone tools and the unilateral multi-barbed bone harpoon or fish spear tradition, there are the remains of a semicircular (permanent?) village of seven or more triangular structures facing the river. The bones of both land animals and fish as well as bone and/or antler picks and mattocks for plant digging suggest these people were specialized Foraging Villagers—system C1. No evidence of horticulture or domesticates has yet been found, although the picks and mattocks could have been used for planting as well as for harvesting. If these foragers did not have domesticates or cultivars and were system C1, then they were ready for such.

Shortly thereafter the Starcevo-Körös complex (5500–4500 B.C.) developed, characterized by stamped or "impressed" (Impresso) pottery and evidence of agriculture based on emmer, einkorn, barley, and bread and club wheat (Milisauskas 1978). Remains from this period still contain large quantities of fish and land animal bones, so I suspect these people had reached system C2—horticultural villages, using domesticates imported from the Aegean or Asia Minor. In the upper levels of Lepinski Vir in Yugoslavia remains of trapezoidal houses on stone foundations comprised a village (Srejovic and Letica-Vlasac 1978), while in Bulgaria at Karanovo I-II on the Bosna River there was a large rectangular wattle-and-daub house inside an enclosure (Milisauskas 1978). In my opinion the horticultural village—system C2—had arrived by Early Neolithic times, 5500 to 4500 B.C.

Gradually, by Middle Neolithic times (4500–4000 B.C.), early village agriculture—system E—had evolved, as characterized by the upper levels of Karanovo III-IV in Yugoslavia and by Vadastra in Romania. This period has Linear decorated pottery and bigger villages; six-row hulled barley, peas, vetch, and perhaps other plants have been added to the larder (J. Renfrew 1973). The same trends continue into the Late Neolithic (4000–3000 B.C.), with such cultures as Gomolova (Dennell 1989), Lengyel, Gumelnitsa, Vinca Plocnik, and later Globular Amphora or Baden pottery-making people (Milisauskas 1978). New domesticates—lentils, vetch, peas, and others—appeared, sites were large and often fortified, and warfare seemed prevalent. Between 2500 and 2000 B.C. these cultures reached the Bronze Age, which is well past our period of interest. Despite the number of sites, our Mesolithic data are so poor it is difficult to deter-

Table 7.4. Sequence of cultivated and/or domesticated plants in Europe

Aegean	Danubian	Mediterranean coast	Eastern plains	Alpine	Atlantic mainland	Atlantic coast	British Isles	Baltic Sea
		Iron Age						
			Hradock					
					Nagyarpar			
			Langweiler 3					
							Windmill	
				Horgen				
							Vlaardiagen	
							Funnel Beaker	
							Hembury	
	Gumelnitsa							
				Michelsberg				
			Zarchowo					
			Tripolye					
			Lengyel					
			Langweiler 2					
				Cortalloid			Egolzwil	
	Karanovo IV							
			Coveta de l'Or					
			Langweiler 1					
Neolithic B								
	Koros-Karanovo 1							
Franchthi VII								
								Ertebølle
Neolithic A								
	Proto-Sesklo							
			Icoana					
			Baume de l'Abeurador					
Argissa								
Ghediki								
Sesklo								
Nikomedeia								
Knossos								
Franchthi VI								

mine the causes leading to village agriculture. Nor can we ascertain if a secondary or tertiary development occurred, although I suspect it was the former.

Not quite as well known is the Mesolithic and Neolithic of the Mediterranean Coast. The general sequence is similar, although significant differences include the emphasis on the exploitation of marine resources and a much longer Mesolithic sequence (Guilaine 1979). As yet, manifestations transitional from the Upper Paleolithic (perhaps Magdalenian) to the Epipaleolithic have not been well documented. However, burins and blades in Montardian Cave in South France as

Range of dates B.C.		Emmer	Einkorn	Hulled 6-row barley	Naked 6-row barley	Bread wheat	Club wheat	Millet	Lentils	Oats	Peas	Flax	Grapes	Broomcorn millet	Horse beans	Rye	Vetch	Spelt wheat	Poppies	Apples	Crabapple	Strawberry	Italian millet	Naked 2-row barley	Goosefoot	Barrenbroom
500	1000	x	x		x	x	x	x	x	x	x	x	x	x	x	x	x	x					x	x	x	x
1000	2000	x	x	x	x	x	x		x		x				x						x			x	x	
2000	2500	x	x	x	x											x	x		x					x	x	x
2500	3000	x		x	x					x					x	x		x							x	
2000	2500	x		x	x																					
2400	2800	x		x	x	x	x																			
1600	2600	x		x	x	x	x			x																
2400	3000	x		x	x	x	x			x																
3100	3400	x		x	x													x								
2600	3600	x	x	x	x		x	x	x		x						x									
2500	3000		x		x	x												x			x	x				
2500	3000	x	x	x	x	x	x										x									
2200	3800	x			x	x	x							x												
3000	3800	?	?	?	x	x			x	x	x	x		x	x	x										
3000	3800	x	x	x		x			x	x	x	x		x	?	?										
3300	4000	x	x		x	x	x				x	x		x					x	x			x			
3700	4200	x	x	x	x	x	x				x						x									
3800	4500	x	x	x		x																				
4000	4500	x	x	x												x										
4800	5160	x	x	x		x		x			x															
4500	5500	x	x	x		x	x																			
3400	5300	x	x					x																		
3500	5500		x																							
5210	5520	x	x			x	x	x			x															
5000	6000	x									x												x			
5900	6200	x																								
5000	6000								x									x								
5000	6000	x	x	x					x				x													
5000	6000	x	x						x		x						x						x	x		
5000	6000	x									x													x		
5000	6000	x	?	?		?					x													x		
5800	6300	x					x																	x		
5500	6400	x							x															x		

well as microblades and trapeze microliths in the Corsican Curacchiaghiu Cave site, which dates to about 6600 B.C., may be evidence of such a tradition; if not, they certainly are early in the Mesolithic development (J. G. D. Clark 1980). Both hint of a maritime adjustment, and the Corsican site indicates boats or seacraft were in use. This culture could belong to the Efficient Foraging Bands—system D.

The sporadic spatial distribution along the coast of both Mesolithic and Neolithic sites strongly suggests that cultural diffusion or acculturation was waterborne. This maritime adaptation of the Mesolithic, with its microblades and trapeze microliths, reached the western part of the subarea in Late Mesolithic times, 6500 to 5500 B.C. There are numerous coastal sites of this period at the Corruggi in Sicily and Grotto-del-Uzzo sites in Corsica as well as at the Curacchiaghiu sites of the Italian coast. Sites in France are Arene Candide and Chateauneuf-les-Martigues; in Spain are Castelnovian sites at Balma del Gai, L'Esplauga, and La Cocina (Guilaine et al. 1982); and Portugal has Mugem and Muge. Some of these were shell mounds and contain hints of sedentary occupations. At all these Mediterranean sites the people were presumably Efficient Foragers—system D. Further, Mesolithic foraging occupations, possibly using chickpeas (Dennell 1989), lentils, and vetch (table 7.4), have been found at Baume de l'Abeurador and Fontebregoua in southern France (Guilaine 1981). Occupants of these sites seem to have been Semisedentary Bands with Domesticates—system D1.

Between about 6000 and 5500 B.C., shell-mound sites with Impressed pottery appear in Italy in the southeast at Maddalene di Muccia, Leopardi, Coppa Nevigata, San Vito, and Grotta del Santuario della Madonna on the west coast. Shortly thereafter, Cardial (shell impressed) pottery sites occur on the south coast of France and Spain (Guilaine 1979). Jean Guilaine, J. Graham Clark, and others assume that these people also had agriculture and village life. However, only one cave site, Coveta de l'Or in Spain (4620–4315 B.C.), had remains of emmer, einkorn, six-row barley, and bread wheat (Hopf 1965). This evidence suggests Horticultural Villagers—system C2—rather than true agriculture. Perhaps true village agriculture arrived later—between 4600 and 3900 B.C.—at the end of Early Neolithic times (Milisauskas 1978), with the further spread of the Cardial pottery complex, or in Middle Neolithic times, with such cultures as Almería in Spain and the megaliths in France and Italy. Certainly such a way of life or cultural system is better documented for the Late Neolithic (with its Bell Beaker pottery) and for the Bronze Age.

There is a suggestion of a development from Magdalenian Hunting-Collecting Bands (system A) about 8000 B.C. to Efficient Foraging Bands (system D) during the Mesolithic (8000–7000 B.C.), then to Semisedentary Bands with Domesticates (system D1) developing to Early Neolithic horticultural villages (system C2) about 6000 to 4000 B.C., and finally reaching village agriculture (system E) between 4000 and 3000 B.C. This sequence seems to fit my tertiary development model of a slow evolution to village agriculture, but the data are still incomplete.

Adjacent to the Danubian subarea and seemingly acquiring village agriculture a short time later is the eastern plains subarea. Extensive and intensive excavations have been undertaken in Russia, Romania, and Poland, yet our data on the cultural process there are far from complete. Exact transitions from system A, Hunting-Collecting Bands, of the Magdalenian or the Swiderian (about 9000–7000 B.C.) of the Paleolithic to the Mesolithic are not well delineated, but there are a host of sites with microblades and trapeze microliths—for example, Petrovoorlovskaya and Fatma Koba in southwestern Russia and Janislawice in Poland—that must be classified as Mesolithic, between roughly 7000 and 5000 B.C. (J. G. D. Clark 1980).

Artifacts and bone materials suggest that these people lived in Efficient Foraging Bands—system D—or Affluent Foraging Bands—system C—but more analysis is needed to better describe their way of life. The continuity of microlithic tools suggests they developed into villagers with houses at Icoana in Romania (6000 B.C.). Two semisubterranean structures in Soroki II (5500–4500 B.C.) in the Dniester Valley of the south Russian plains indicate village development there. Both groups seemed to have done large amounts of foraging, but large-sized pollen of *Cerealia* in feces at Icoana hints that they might also have used domesticates and thus have been Horticultural Villagers—our system C2 (Dennell 1983). On the other hand, they could have been Semisedentary Bands with Domesticates—system D1—or Village Foragers—system C1.

Developing out of these were cultures using Linear pottery—Dniester-Bug, Tisza, and/or Soroki I—which were definitely villages in the period from roughly 4800 to 3900 B.C. These groups are often considered slash-and-burn agriculturists (Mathyushin 1976). However, let me hasten to point out that land animal and fish bones are abundant in such Linear sites as Stzelce and Olszanica in Poland as well as in the lowest Linear pottery level of Langweiler in nearby Germany, where emmer, einkorn, barley, and rye have been found. With this in mind, these peoples all might be better interpreted as Horticultural Villagers—system C2—than as Village Agriculturists—system E (Milisauskas 1978). The Lengyel cultures (3900–3300 B.C.) that followed them in central Europe and Langweiler II used more domesticates, including bread and club wheat, peas, lentils, oats, millet, and horsebeans, and may have been true Village Agriculturists. It is not yet certain whether the Tripolye people in southern Russia were still horticulturists or had become agriculturists (Dennell 1983). However, there are no doubts that the Funnel Beaker or Corded pottery-making peoples of the Late Neolithic (3300–2000 B.C.) were full-time Village Agriculturists—my system E. The eastern plains thus provide more evidence of a slow development (6000–3000 B.C.) of village agriculture possibly preceded by a relatively long period of village horticulture in the Late Mesolithic and Early Neolithic. Given the limited Mesolithic data, it is difficult to determine if this is a secondary or a tertiary development, let alone to discover the sufficient conditions or causes of its occurrence.

Having examined the developments in subareas with relatively easy access to the Aegean or Asia Minor, let us now consider the Alpine subarea, which is cut off from direct land or sea contacts with the Near Eastern Center of agriculture. Here the Mesolithic has been well studied. There is evidence of a transition from Magdalenian to Mesolithic at Oberrainknogel bei-Unken site, from roughly 10,000 to 9000 B.C. (J. G. D. Clark 1980). Needless to say, this earliest transition shows the peoples to have been basically Hunting-Collecting Bands utilizing blades and burins—system A. During the earliest Mesolithic period (roughly 9000–7000 B.C.), called variously Furstein or Fursteinerl-Seeberg, Efficient Foraging Bands—system D—were using microliths, nets, and harpoons. The subsequent period (7000–6000 or 5500 B.C.), called La Baume d'Ogens, had a similar subsistence system and microliths as well as triangular microblade projectile points were used. During the final period, Birsmatten (5500–4000 B.C.), which had five subdivisions, trapeze microliths came into use and people gradually settled down into villages or large base camps along the Swiss lake shores (J. G. D. Clark 1980).

As is obvious from the dates, this final period is contemporaneous with the Early Neolithic of southeastern Europe, when domesticated plants were being utilized, so these Lake Dwellers could have been Semisedentary Bands with Domesticates. On the basis of the careful work of R. Wyss, J. Graham Clark (Wyss, 1979) feels quite sure agriculture and Cortailloid pottery did not arrive until after 3000 B.C., and even then the use of domesticates was of minor importance (J. G. D. Clark 1980). At Egolzwil (3500–3000 B.C.), emmer, einkorn, bread and club wheats, barley, peas, and vetch were uncovered, while at Michelsberg (3000–2500 B.C.), only einkorn, bread wheat, spelt wheat, and barley occurred, so perhaps these people were Village Horticulturists—my system C2—rather than Village Agriculturists (Milisauskas 1978). In fact, according to J. Graham Clark, it is only in the Final Neolithic, during the period of Corded wares, or even in the Bronze Age, that true village agriculture developed in the Alpine subarea (Dennell 1989). This development thus differs from the areas to the east and also occurs much later.

A similar pattern occurs in the western lowlands of Europe: eastern France, western Germany, eastern Belgium, and the Netherlands (Scarre 1984). Here again the Mesolithic may be divided into three general periods. Azilian, in southern France and the Pyrenees of northern Portugal and Spain, represents the earliest period, roughly from 9000 to 7500 B.C. (J. G. D. Clark 1980). Antler and bone tools using the groove and splinter technique by burins and associated backed bladelets and harpoons indicate the Azilian culture probably developed out of the Magdalenian Hunting-Collecting Bands of Paleolithic times—my system A.

Recent investigation of Azilian, which has painted pebbles as well as microliths, suggests it was perhaps the inland and mountain (Pyrenees) facet of a seasonal subsistence system that had as its maritime facet the Asturian middens on the coast. Be that as it may, the Azilian peoples were certainly Efficient Foraging Bands—my system D.

Probably developing out of this phase was the Sauveterrian (about 7500–5800 B.C.) in central France. It had a similar subsistence system but utilized triangular and crescent microliths. Following it are the Tardenoisian periods I to III (roughly 5800–4400 B.C.), characterized by harpoons and trapeze microliths. While Sauveterrian and Tardenoisian populations and band size may have been a little larger and their occupations may have lasted a little longer than Azilian, it would appear that these two phases were basically Semisedentary Foraging Bands, and we do not know whether any domesticates were used. These people appear to have settled in villages and acquired Linear pottery in Early Neolithic times, roughly from 4400 to 3900 B.C. It has been assumed that they had horticulture, but exact proof is lacking (Scarre 1984).

Our data from Middle Neolithic times, roughly from 3800 to 2800 B.C., come mainly from various Megalithic chambered tombs, Chassey ceramic complexes in France (and Almería in Spain), and Funnel Beaker pottery sites and Michelsberg-type sites in western Germany, the Netherlands, and Belgium. The Michelsberg sites have yielded barley, einkorn, bread wheat, and spelt wheat; the Funnel Beaker sites have these grains plus club wheat and oats (Milisauskas 1978). There certainly is a strong suggestion that true agriculture had arrived, although more analysis is needed. In fact, it is only with the Bell Beaker pottery-making peoples of Late Neolithic times and the Bronze Age that we can be sure true village agriculture—system E—had taken hold in Europe's western lowlands. Late Neolithic sites such as Vlaardiagen in the Netherlands yielded abundant remains of emmer, bread and club wheat, six-row hulled and naked barley, and oats (J. Renfrew 1973). The pattern thus seems similar to the tertiary development in the subareas just discussed, albeit slightly more recent.

The pattern for the Atlantic Coast, which is adjacent to the subarea just described and often difficult to separate from it, is similar. Asturian (7000–5000 B.C.) is a definite coastal midden manifestation of northern Spain and perhaps southern France. Excavation at La Riera suggests it evolved out of the Magdalenian Paleolithic hunting cultures (J. G. D. Clark 1980). As mentioned previously, the microlithic Asturian culture could well be a coastal manifestation, with Azilian being the inland one, of a seasonal inland-coastal Efficient Foraging Band society—my system D.

Later Mesolithic manifestations, such as the burial site of Téviec in Morbihan, coastal France (roughly 5000–4000 B.C.), also had microliths as well as triangular and trapeze microlithic arrow points. There is little doubt that these people had a subsistence pattern of fishing and maritime collecting. Sites such as Téviec are large enough to suggest that they might have had year-round occupations or semisedentary bands, although no house or structure has been reported. Whether they had domesticates and belonged to system D1 cannot be determined at this point. In Téviec, where passage-graves were constructed, there are hints of burial ceremonialisms that developed further in Early and/or Middle Neolithic times. Certainly, in this period (4000–3000 B.C.), people who found it necessary to build such monuments probably belonged to villages with fair-sized populations and possibly used domesticates (Scarre 1984). Clark suggests that these peoples were basically Affluent Foragers (J. G. D. Clark 1980), perhaps with only a few domesticates—my system C2, Horticultural Villagers.

It would appear that true village agriculture did not arrive on the Atlantic Coast until those peoples making Bell Beaker pottery developed, or until this ceramic type was adopted by the locals in the period roughly from 3000 to 2000 B.C., a ceramic system that continued into the Bronze Age (Milisauskas 1978). Graham Clark and I think we can make an analogy of the Megalithic of Europe to the Hopewellian culture of the eastern United States (see chapter 8), for out of horticultural villages with ceremonialism, true village agriculture slowly developed in both areas (J. G. D. Clark 1980).

The final two subareas are the regions connected with the Baltic and North seas, that is, Denmark and the British Isles. From northern Germany and Denmark there is good documentation of the shift from Paleolithic Hunting-Collecting Bands (system A) to Mesolithic Efficient Foragers (system D). Such cultures as Ahrensberg, 10,000 to 8000 B.C. (Mellars 1978), Star Carr (J. G. D. Clark 1972), and Mount Sandel (8000–7000 B.C.) in England, and Maglemose in Denmark (7500–5700 B.C.) saw the early rise of Efficient Foraging Bands with microliths, triangular points, and harpoons—my system D (J. G. D. Clark 1980). In Denmark there is a further development of these foraging bands in such manifestations as Kongemose (5700–4600 B.C.), which had microblades and trapeze microliths.

At this Early Neolithic time there were domesticates and pottery in southeastern Europe, but to the west and north people were still Mesolithic Semisedentary Foragers. Pottery arrived in Denmark with the Ertebølle culture (4600–3300 B.C.), and there were a few possible house structures, but no evidence of the use of domesticates even in Early and/or Middle Neolithic times (S. Andersen 1974). In the British Isles, things were even further behind, and it is not until between 3400 and 2900 B.C. that we get pottery (Funnel Beaker-like) or hints of domesticates—a clay impression of spelt wheat at Hembury and emmer and barley impressions at the Windmill sites (Dennell 1983). These sites have enclosures and barrows, indicating the arrival of village life and megalithic ceremonialism. Basically, however, the people were still foragers who practiced a little horticulture—my system C2.

In Denmark in late Middle Neolithic times, 3000 to 2600 B.C., the situation was a little different, since the Funnel Beaker pottery and megalith-making peoples took over from Ertebølle. Although there is good evidence of villages, often containing large multiroom longhouses, the plant remains of barley, emmer, einkorn, club and bread wheat, and oats (table 7.4) seem to show that the people were basically Foragers who practiced a little horticulture but did not have agriculture (Helbaek 1954) or Semisedentary Bands with Domesticates—system D1. It must be emphasized that the Funnel Beaker remains occur earlier to the east, hence some groups in Germany and Poland may have been Village Agriculturists, but this is not true along the shores of the Baltic Sea (Dennell 1983).

Village agriculture (system E) seems to arrive in Late Neolithic times (roughly 2600–1500 B.C.), with the Bell Beaker pottery-making people in England and the Corded pottery-making people in Denmark. Even then, and continuing into the Bronze Age when villages were larger and more warlike, the number of plants cultivated was not numerous. On the European periphery, therefore, it took from roughly 3600 B.C. to almost 1500 B.C. before village agriculture became established. Between the earliest hunters and the first agriculturists there were foragers, who were followed by foragers with pottery and a ceremonial megalithic complex with only small amounts of horticulture and/or domesticates (J. G. D. Clark 1980). The story is much the same as that of the preceeding subareas, even if the times are slightly different.

Before we discuss why these developments took place, let me briefly summarize how they happened. At the end of the Pleistocene with its concomitant extinction of megafauna between 10,000 and 8300 B.C., we see a shift everywhere from Paleolithic Hunting-Collecting Bands to Mesolithic Efficient Foragers (see table 7.5).

In Pre-Boreal times—8300 to 7500 B.C.—the basic forager systems devel-

Table 7.5. Necessary and sufficient conditions as a positive-feedback process for secondary development in the Aegean area

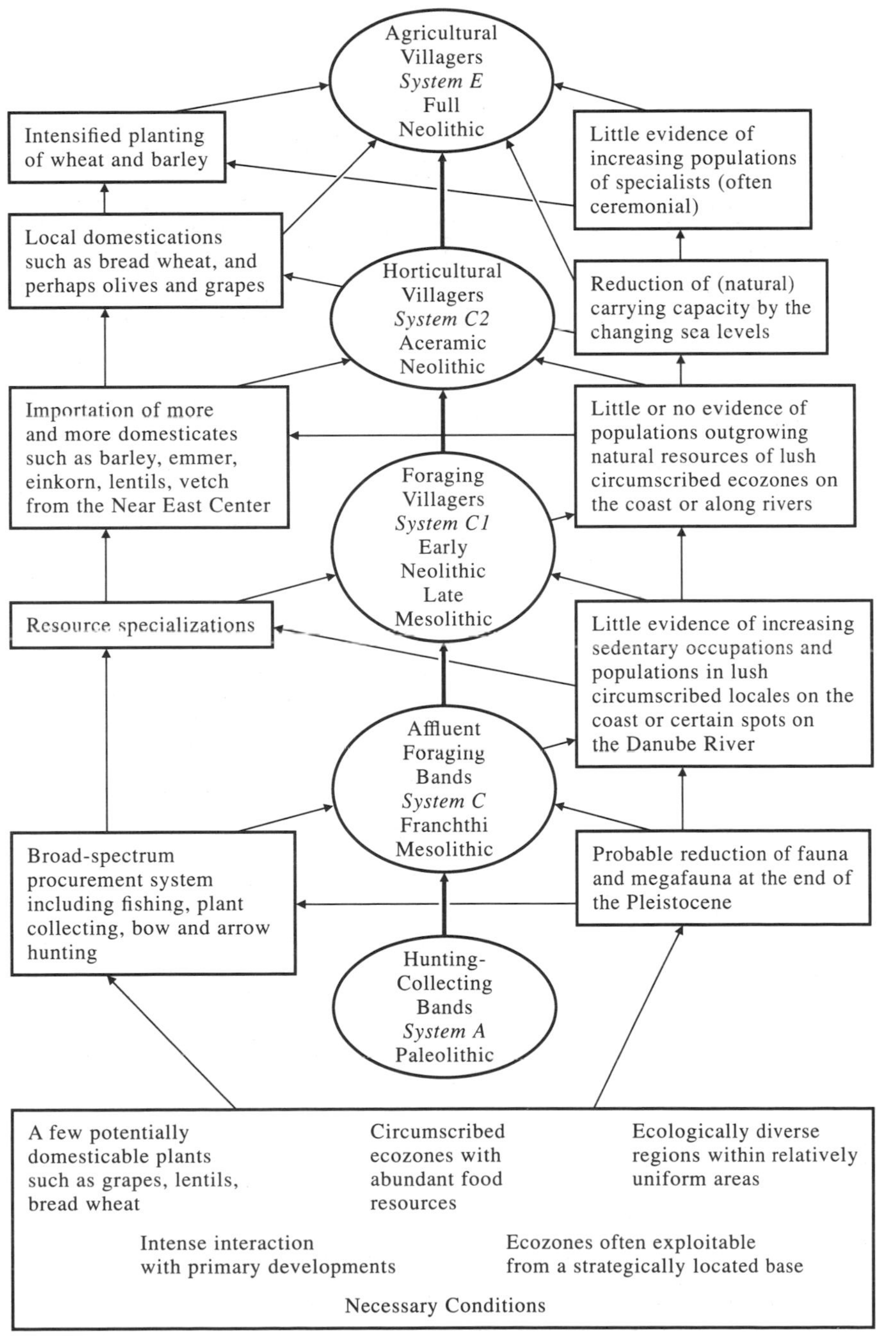

Table 7.6. Necessary and sufficient conditions as a positive-feedback process for tertiary development in Europe northwest of the Aegean

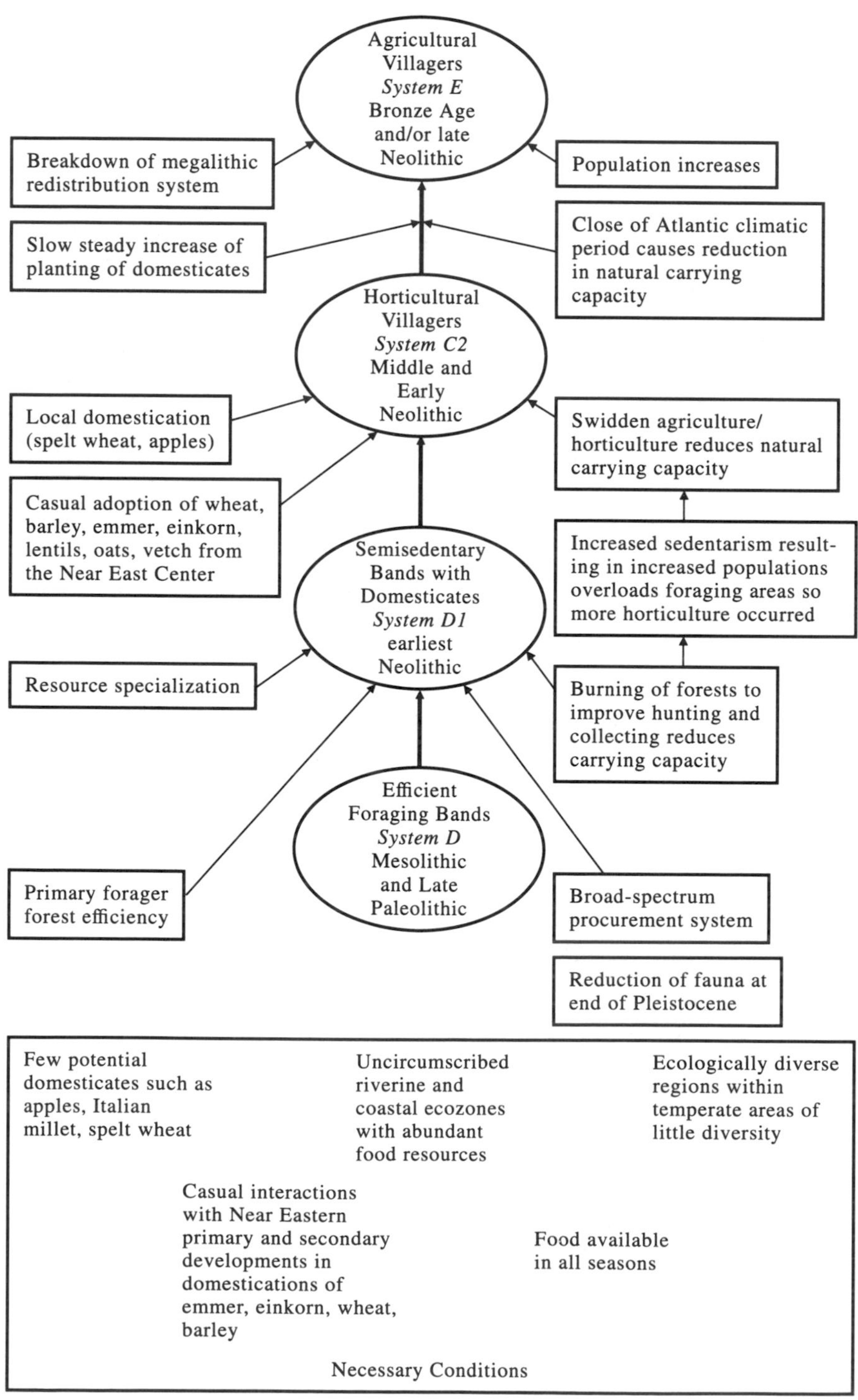

oped. Our data from western Europe are well documented, with Montardian occurring in the western Mediterranean, Furstein in the Alps, Azilian on the mainland, Asturian on the Atlantic Coast of southwest Europe, Arhensberg and Maglemose in the Baltic area, and Star Carr in the British Isles. Analysis of these remains indicates that the foragers had a series of scheduled subsistence options—hunting, trapping, picking, and food collecting—that were admirably adapted to the slightly inefficient environments in which they existed. In other words, Pre-Boreal peoples had developed a primary forager (forest) efficiency (Caldwell 1958) and exemplified my system D.

To the east we have fewer sites and less complete data (table 7.6) Odmut Cave (roughly 7000 B.C.) in the Adriatic area of Montenegro and Cuina Turcului Cave (10,750–8175 B.C.) in the Balkans both had limited microlithic and faunal materials, while Franchthi Cave zone IV (8300–7300 B.C.) had more Mesolithic artifacts and subsistence remains. The subsistence information suggests the people were foragers, but just what kind cannot be determined at present. However, the Franchthi Cave Mesolithic does have marine food resources, hinting that the people exploited the sea and that they were "shell-mound people"—Affluent Foragers, system C—although this speculation cannot be proved as yet.

In fact, it is not until the Boreal stage (7300–6200 B.C.) that one can discern that the east and west were developing along different paths. In the east, although the data are not as full as one would like, aceramic villages without agriculture existed at Vlasac (8200–6000 B.C.) and at Lepinski Vir (6500–5000 B.C.) in the Balkans and at Icoana (6000 B.C.) in Romania, apparently an indication that Foraging Villagers—system C1—had evolved (table 7.5). The data from the Aegean are less definite, but there are hints of pre-agricultural village remains in aceramic levels of Argissa and Nea Nikomedeia. The aceramic Neolithic levels of zone V (7300–6400 B.C.) of Franchthi Cave might be related to these, for evidence of agriculture only appears in zone VI (6300–5800 B.C.).

This evidence of the crucial middle stage of my secondary development model in the east contrasts sharply with what happened in the west. There, the basic Mesolithic efficient forager systems continued to develop, and as the bands became more and more efficient, they became more and more sedentary. La Baume d'Ogens and Birsmatten cultures developed in the Alps, Téviec on the Atlantic Coast, Kongemose and Ertebølle on the Baltic, Mount Sandel and Oronsay in the British Isles, and Sauveterrian and Tardenoisian in southern France. These sites provide good documentation of the slow evolution of Semisedentary Foraging Bands. Some examples in southern France even had a few domesticates—system D1—of my tertiary development (table 7.6).

Some of this sort of slow development continued well into Atlantic times (6200–3000 B.C.) in western Europe, but things were rapidly changing to the east. In Greece and the Aegean, village horticulture—system C2—began in the aceramic Neolithic (6200–5500 B.C.) and became village agriculture during the full Neolithic after 5000 B.C. Slightly later in the Balkans (5500–4500 B.C.), the Neolithic with villages and use of domesticates developed along with the use of Starcevo-Körös types of pottery. Similar developments occurred in the eastern plains, accompanied by the use of Linear-type pottery. Both eastern zones grew emmer and einkorn wheat, and barley, and both seem to have been Village Horticulturists—system C2. By Early Neolithic times (5500–4000 B.C.), with the further spread of Linear pottery, true village agriculture had been attained and secondary development completed.

It was not until early Atlantic times, when the forests were becoming thicker and were squeezing out the open meadows with their herd animals, that pottery, along with domesticates and villages, spread slowly into western Europe. Impresso and Cardial spread along the western Mediterranean coast, perhaps by boat, while Linear pottery moved through northern Europe, and Chassey-type and Cortailloid pottery appeared at Michelsberg in the Alpine ecozone. While people in these subareas had a variety of domesticates, they were all basically Village Horticulturists—system C2—perhaps with slash-and-burn agriculture rather than true agriculture.

Late in this time frame came the period of megalith building, which continued at least until 3000 B.C., when the sedentary villages of England may have seen their first domesticates. Village life with pottery but no domesticated plants continued, however, at Ertebølle in Denmark. During the Middle Neolithic (3900–2600 B.C.), the east saw village agriculture continuing with increasing populations, while village horticulture with increasing ceremonialism continued in the west as the Atlantic climatic period gave way to the Sub-Boreal. The Late Neolithic period, from 2600 to 1500 B.C. (and perhaps the Bronze Age), brought the development of more specialized pottery types, more regional or subareal cultural differentiation, and village agriculture everywhere.

Having given the overview of the periods crucial to the development of village agriculture in Europe and having seen that the east had a secondary development while the west fits the model of tertiary development, let us see if we can determine the necessary and sufficient conditions (or causes) that brought about such developments.

THEORETICAL CONSIDERATIONS

Since the eastern sequence is less well documented, particularly the Mesolithic portion of it, let us analyze it first (table 7.5). Specifically, I shall deal with the sequence of the Aegean and Danubian subareas, although the latter is a little later than the former. These two regions had a number of necessary conditions pertaining to my secondary development model, and these conditions caused them to differ from the west. Obviously, both the Aegean and Danubian subareas are adjacent to the Near East, a Center where village agriculture developed early.

Certainly, the Aegean Mesolithic indicates an intense interaction with this region of primary developments, as evidenced by the similarity of the microlithic tools of Franchthi Cave to those of Belbasi and Beldibi of southern coastal Anatolia. Later in the Neolithic, the Aegean also received all of its domesticates—emmer, einkorn, lentils, vetch, peas, and others—from the Near East Center, along with possible concepts about pottery, architecture, and so on. While the Danubian area does not have evidence of such intense interaction with the Near East Center, it has ample evidence of intense interaction with the secondary development of the Aegean non-Center. This type of relationship between eastern

Europe and the Near East Center is clearly different from that of western Europe, whose regions had casual and/or indirect interactions with the Near East and/or eastern Europe. This is one of the reasons western Europe experienced a tertiary development rather than a secondary one.

Other differing necessary conditions further distinguish the two regions. The eastern regions are less well defined because of the limited number of Mesolithic sites whose ecological attributes can be studied. Both the Aegean and Danubian sites of early Foraging Villagers are quite similar in terms of necessary conditions. Franchthi Cave, then near the coast, was situated against the flanking hills, both zones of rich resources. The Danubian sites include Lepinski Vir, "in a small inaccessible cove immediately off the Danube" (Dennell 1983, 43), Vlasac, and Icoana, which are also in the Iron Gate gorge region. All were surrounded by steep, rocky environs with numerous ecozones rich in game such as chamois and fox, and with acorns in the forests and aquatic foods in the oceans or river. All of these zones were exploitable from a strategically located base. In addition, all seem to have been in circumscribed ecozones with abundant food resources.

The location of sites to the west is much more varied and, relatively speaking, these conditions do not pertain. It remains to be seen if future Mesolithic-Neolithic sites conform to these conditions.

On a more general level, our eastern sites have local ecological diversity but are areally uniform, while our Mesolithic sites to the west occur in diverse subareas. Both subareas have a few potentially domesticable plants (J. Renfrew 1973). As one can see, the necessary conditions are different for east and west and are seemingly interacting with different sufficient conditions, or triggering causes.

I use the word *seemingly,* because our Mesolithic data for the east are meager, and except for Franchthi Cave the transition from hunters to foragers is poorly understood. As Dennell and others have indicated, at the close of the Pleistocene, reindeer, horse, and mammoth were replaced in the Balkans by roe deer, elk, red deer, and pig. Interacting with this reduction of fauna was the development of a broad-spectrum procurement system (Dennell 1983). These changes could have been the causes for the shift from a Paleolithic hunting-collecting way of life in zone II to Mesolithic foraging in zone III of Franchthi Cave. It remains to be seen if this shift holds true for the whole subarea.

The next stage, the shift from Affluent Foragers to Foraging Villagers, is not well documented, but Franchthi Cave again suggests more resource specializations were occurring. Whether these were interacting in a positive-feedback situation with increasingly sedentary occupations and populations in lush circumscribed locales is difficult to prove either in the Aegean or elsewhere. It is equally difficult to determine if the people outgrew those lush circumscribed areas or if the natural carrying capacity was reduced, leading to the importation of domesticates from the Near East. By full Neolithic times the Aegean people were domesticating such local plants as bread wheat and perhaps olives and planting all intensively. Eastern European data indicate that much research is still needed before adequate testing of my secondary developmental hypothesis is possible (table 7.5).

In western Europe, on the other hand, the Mesolithic is well known, and many environmental studies have been undertaken (table 7.6). Western Europe is undeniably a region of uncircumscribed ecozones with abundant food resources, of which one or more are available year-round (J. G. D. Clark 1952). Aside from these ecozones of rich resources, other ecozones may vary considerably because of temperature, elevation, soils, precipitation, forest cover, and so on, although the few lush spots mean people could settle down and exploit the other local ecozones from a base rather than engaging in calendar-round seasonal movements. Besides leading to decreased mobility, these environmental conditions also resulted in increased logistic mobility in terms of natural food resources. Further, western Europe, a non-Center, was not adjacent to the Near East Center, and therefore it lacked easy access to a number of basic plants—einkorn, emmer, and other wheats, barley, lentils, vetch, oats—domesticated in that Center. Contacts were neither direct nor intense. Unless certain sufficient conditions pertained, there was a tendency in Mesolithic western Europe to increase wild-food procurement rather than to acquire domesticates, let alone agriculture.

Intensification of wild-food procurement and resource specialization leading to efficient foraging systems certainly were taking place throughout the European Mesolithic as people learned to fish and developed better use of nets, boats, and harpoons; collected more and more marine resources; engaged in more and better plant collecting; and improved their hunting technique with such items as better bows and arrows and atlatls (throwing sticks). These processes led to intensive foraging that eventually upset the carrying capacity of the lush ecozones. Pollen profiles and soil studies provide evidence that in the Mesolithic (before agriculture) large tracts of forest were being cut down or burned to make hunting and plant collecting more efficient. The tendency toward larger base camps for bands and/or sedentary life perhaps meant that a larger population was exploiting the deteriorating natural environs, making a supplement to foraging techniques readily acceptable.

Domesticated plants came into use as a horticultural supplement to foraging during the Late Mesolithic in southern France (Guilaine 1981), during the Early Neolithic in the east, and during the Middle to Late Neolithic in the west. This incipient agriculture—perhaps initially of the slash-and-burn type that ruined the soils—plus the onset of the Atlantic climatic and vegetational period (6200–3000 B.C.), when forests became denser and made hunting and plant collecting more difficult, further reduced the carrying capacity of the land. In addition, a growing population in Middle to Late Neolithic times meant increasing use of domesticates and cultivars as well as domestication of a few local plants such as spelt wheat and broomcorn millet. Added to these factors was the possibility that the extensive exchange and redistribution systems of foraged foods connected with megalithic ceremonialism gradually broke down. In this way the relatively successful foragers with horticulture slowly became Village Agriculturists by the end of the Neolithic or Bronze Age almost everywhere in Europe (table 7.6).

Thus, the basic necessary and sufficient conditions do seem to pertain in western Europe and to confirm my model of tertiary development. The more dynamic testing of the model by a consideration of these factors or conditions as part of a positive-feedback system makes the confirmation even stronger. In terms of the European Mesolithic-Neolithic data, this cycle might be expressed as follows: the Eurpean Mesolithic bands had relatively efficient foraging systems in an easily exploitable temperate environment that had many uncircumscribed and relatively lush ecozones. These bands gradually settled in villages and continued to improve their efficient foraging system even though they had exchanges (of obsidian and the like) with an area (the Near East) in which domestication and/or horticulture and agriculture had already developed.

It was only as their system began to overextend the local European environment as a result of burning or clearing forests in the Late Mesolithic, or as that normally bountiful area slowly changed for the worse duc to natural causes in the Atlantic climatic time when plants and animals become harder to obtain in the dense forests, that people began to practice horticulture (Table 7.5), at first adding only a few domesticates from the Near East (emmer, lentils, vetch, einkorn, other wheats, barley, and the like) to the more southwesterly Aegean and Balkan foraging subsistence that was one of the bases for Neolithic village life. Slowly, as the Middle to Late Neolithic population gradually increased and the carrying capacity of the various local environments slowly diminished due to slash-and-burn agriculture, changing climate in Atlantic times, and larger populations in the villages—perhaps combined with the breakdown of the exchange or redistribution systems connected with megalithic ceremonialism—the people acquired more and more domesticates (oats, vetch, lentils) and domesticated some local plants (spelt wheat, broomcorn millet, apples, crab apples, and so on). Following this pattern, horticulture gradually became agriculture, and by Late Neolithic or Bronze Age times the people were true Village Agriculturists.

Thus, my tertiary model of the development of village agriculture has been tested by European Mesolithic and Neolithic data and found valid (Table 7.6). This confirmation of the model is different in time and kind from the previous test done with Japanese data. As we shall see, however, it bears some resemblance to the next case considered—the eastern United States, from Archaic to Mississippian times. Differences in time and specific data, however, are of minor importance. What is significant is that the cultural process is the same and came about for the same reasons or causes. Thus my tertiary development hypothesis is gradually proving to be a generalization or "law" about cultural process.

A major difference between the European data and those from the eastern United States is that there is no evidence of secondary development in the latter. Of course, even in eastern Europe where such might have occurred, the data are not yet adequate to test this hypothetical development. More research is also needed in the eastern United States, but some tentative conclusions can now be reached.

CHAPTER 8

The Eastern United States

The eastern United States, like Europe, was a non-Center that received its first domesticates (squash and gourds) from a Center—Mesoamerica—and then later, like western Europe, cultivated a few of its own native plants. Only after a long interlude of village horticulture or incipient agriculture did village agriculture finally develop. As in western Europe, these processes in the peripheral areas lagged a number of millennia behind the Centers. Unlike the secondary development, either in the Centers or non-Centers, these tertiary developments were a slow and uneven process.

Development of agriculture in the New World lagged far behind the process in the Old World. New World domestication began in the Mesoamerican Center about 5000 B.C. and agriculture about 2000 B.C., while such processes occurred about 3000 B.C. and A.D. 1000, respectively, in eastern North America. Domestication in the Old World, on the other hand, may have begun in the Near East well before 9000 B.C. and agriculture before 5000 B.C., while there is little evidence of any plant or animal domestication or cultivation in western Europe much before 6000 to 5000 B.C., and in some areas Neolithic village agriculture did not begin until as late as 1500 B.C.

That the sequential timing was so similar is somewhat surprising considering that the distances and type of diffusion routes were different. Obviously, the distance even from the Fertile Crescent of the Near East to Ireland is thousands of miles less than that from Central Mexico to New England. This is not the main difference, however. The seemingly significant difference lies in the routes of diffusion. In Europe the spread of domesticates and agriculture by land went from one contiguous group to another, and by sea, from one relatively contiguous part to another. In the United States, however, there is little or no evidence of diffusion by sea—except perhaps the gourd dated 5300 B.C. found at the Windover site in Florida (Watson and Cowan 1989)—and the overland diffusion seems to have been mainly a spread from Mesoamerica to the Southwest through an area of northern Mexico that has no evidence of the use of domesticates or agriculture. A further diffusion took place from the Southwest to the eastern United States by way of the southern Plains, another area that lacks evidence of early plant domesticates or agriculture.

Obviously, some of these differences in routes were due to geographic and physiographic features. In Europe, for example, no north-south cordilleran chain or area of desert or grassland plain separates the Center from the non-Centers. Yet there are some significant similarities, for the plain in southern Russia and eastern Poland is analogous to the western Great Plains of the United States. Also, south of a boreal forest zone both areas have northern subareas of hardwood forest with a cool, humid continental microthermal climate. In addition both have a major southern band of softwood forest with a warm, humid subtropical mesothermal climate. Further, both areas have long coastlines with abundant marine resources and both warm Atlantic and subtropical facets (namely, the northern Mediterranean and the Gulf of Mexico and the Carribean shorelines as well as the major river drainages cutting across them). Both areas lie in northern latitudes and have humid temperate climates, and both have experienced a similar sequence of climate change since the end of the Pleistocene (Dryas, Pre-Boreal, Boreal, Atlantic, and Sub-Boreal periods). Needless to say, these climatic regimes have made the archaeological results somewhat similar, for there is little preservation of the crucial archaeological plant remains, and good factual data about the origin and spread of domesticates or agriculture are hard to come by.

A BRIEF HISTORY OF RESEARCH ON THE PROBLEM

The kind of crucial vegetal archaeological data that can be acquired from bogs, pollen, soils, and flotation of burned materials is far more numerous for Europe than for the eastern United States. Our information about the origin and spread of agriculture as well as the archaeology of the eastern United States therefore tends to lag far behind that of Europe, but with the "flotation revolution" of the 1980s, research in the United States has begun to catch up (B. Smith 1989). In the early part of the nineteenth century the Danes were already digging up archaeological data and analyzing them to arrive at the basic European periods, while in the eastern United States and the Great Plains, the Native Americans were still forming proto-historic sites (and we have yet to find or dig up, let alone analyze, many of them).

In fact, in the United States the first real evidence of any interest in ethnobotany or preserved plant remains from archaeological contexts was Melvin Gilmore's study (1931) of the botanical materials that Mark Raymond Harrington had "uncovered" in 1922 and 1923 from the Bluff Rock Shelters in the Ozarks of northwestern Arkansas and southwestern Missouri during an "exploring" expedition for the Museum of the American Indian of New York. The period roughly from 1918 to 1934 was also the epoch in which the first "careful" diggings were undertaken in the eastern United States, out of which came the first meager archaeological regional sequence. These investigations were accomplished by Fay-Cooper Cole and Thorne Deuel in Fulton County, Illinois (Cole and Deuel 1937); William ("Doc") McKern in Wisconsin (McKern 1939); Arthur Parker in New

York (Parker 1922); Duncan Strong in the Plains (Strong 1935); and others, although "mound digging" (and I use the word in the loosest sense) had preceded their efforts for almost a century (Moorehead 1892). Before moving to the next stage, it should be noted that in 1924, Ralph Linton wrote a most intuitive paper, "The Significance of Certain Traits in North American Maize Culture," for the *American Anthropologist* and defined the eastern agricultural complex, identifying a host of local "small grain" domesticates.

The next period, from 1935 to 1950, was one of great progress and saw the eruption of Works Progress Administration (WPA) archaeology as well as salvage archaeology for the Tennessee Valley Authority (TVA) and for the Missouri Valley Authority (MVA). Literally hundreds of people worked on crews doing archaeology all over the eastern United States, particularly in the southeast. The first Archaic sequences for many regions were formed at this time, and a few attempts were made by James Ford, Gordon Willey, James Griffin, and others to synthesize this chronological data (Ford and Willey 1941; Griffin 1978).

During this period incredible amounts of archaeological data were exhumed, usually by newly devised "scientific" techniques. Some of it (not enough) was analyzed, and the results (far too few) were published, and some (even less) was in the realm of ethnobotany. To my mind the new leader was Volney Jones, who undertook to analyze the botanical materials coming in, including the all-important ones from Newt Kash Hollow, Red Eye Cave, and other caves in Kentucky (Jones 1936). He also elaborated on the idea of the eastern agricultural complex. Also working in the realm of theory was Alfred Whiting, who wrote on corn (Whiting 1944), and Edgar Anderson of the Missouri Botanical Garden of Saint Louis, who hypothesized about the origin of agriculture itself (Anderson 1952). Hugh Cutler of the same institution carried on this interest in U.S. archaeobotany (Cutler and Agogino 1960; Cutler and Blake, 1973), and Paul Mangelsdorf was becoming interested in the problem, if not in the area. Although this period was one of great archaeological data collection and little analysis, things were about to change.

The period from 1951 to 1966 saw the rise of the interdisciplinary approach, as botanists and archaeologists consciously worked together. Volney Jones (Jones 1968) and Hugh Cutler (Cutler and Blake 1973) were still leaders in the field of ethnobotany in the eastern United States. By the end of the period they had been joined by Richard Yarnell (1964, 1972, 1976, 1977, and 1978), Lawrence Kaplan (Kaplan 1970; Kaplan and Maina 1977), Charles Heiser (Heiser and Nelson 1974), Walton Galinat (1970), and others. As changes occurred in ethnobotanical thinking, new speculations and hypotheses about prehistoric agriculture and archaeological data were developed. Archaeological investigation in the eastern United States was also shifting. Although the great crews of the 1930s were gone, salvage archaeology continued, mainly in the Plains and Texas, with many of the archaeologists of the previous period still undertaking investigations. Among the many new faces were Joseph Caldwell (Caldwell 1958), Michael Fowler (Fowler 1973), and Lewis Binford (Binford et al. 1970). Meanwhile, the University of Michigan had inherited the mantle of the University of Chicago, and we had a new leader, James Griffin. He, in his benevolent despotic way, led the field to bigger and better chronologies—in large part thanks to the advent of carbon-14 dating (Griffin 1978). Already there had been a subtle shift from data collection to more sophisticated analysis for chronology and the integrating of ethnobotanical data into the eastern U.S. archaeological sequences by the interdisciplinary method, but further changes, in the guise of the New Archaeology, were in the wings.

As the New Archaeology took effect, the emphasis shifted from chronology and cultural process to cultural contexts. Computer analysis, ecology, and systems theory became catchwords, and new techniques—flotation, phytoliths, pollen analysis, and carbon 12-13 and nitrogen-15 isotopic studies—were brought to bear on collecting more and better botanical or nutritional data from archaeological contexts (Struever and Vickery 1973). More thought was given to reconstructing ancient subsistence systems, in part through the use of more and better ethnographic analogies as well as through actual experiments or replications of ancient subsistence techniques using replicated tools (Asch and Asch 1977, 1985). Paleo-nutrition studies became more sophisticated, and older opinions and postulates about subsistence were reanalyzed, reexamined, and restated (Waselkov 1975). Joining the older leaders—Jones, Yarnell, Heiser, Cutler, Kaplan, and others—were such new faces as Richard Ford (Ford 1985), Nancy and David Asch (Asch, Ford, and Asch 1972), Deborah Pearsall, Wilma Wetterstrom, Bruce Smith, C. Wesley Cowan, Francis King, James P. Gallagher, Barbara Bender, Mary J. Adair, Mark Lynott, Jon Muller, and William Woods.

While many new ethnobotanists were being trained, archaeologists were becoming familiar with ethnobotanical techniques and methods. Perhaps most important from our standpoint was the fact that specific programs were oriented to attacking the problem of the origin and spread of agriculture in the eastern United States. The most notable of these was the Koster site near Kampsville, Illinois, under the direction of Stuart Struever. Equally important were the projects of Patty Jo Watson (Watson 1969, 1974, 1985) and William Marquardt (Marquardt and Watson 1977) on Archaic sites in Kentucky, and the University of Michigan's many new investigations all over the east and southeast (R. Ford 1981, 1985). As data piled up on every front, our knowledge of the all-important Archaic stage greatly increased, as did information about the Woodland and Mississippian stages. Now the jargon is beginning to change again, and the New Archaeology has mellowed, but the mass of new data relevant to the problem of prehistoric agriculture in the eastern United States may help in testing my models—particularly the one of tertiary development.

THE ENVIRONMENT

A discussion about the environment of the area and its various ecological zones and their subareas is in order. We will use this ecological orientation to describe the pertinent archaeological sequences (see fig. 14).

The first area to be considered is the Great Plains, which is not really in the eastern United States but is a buffer zone between the Southwest and the Eastern Woodland areas. This vast steppe area has a middle-latitude dry climate, with hot summers and cold, windy winters. Its rainfall is limited because it lies in the rain

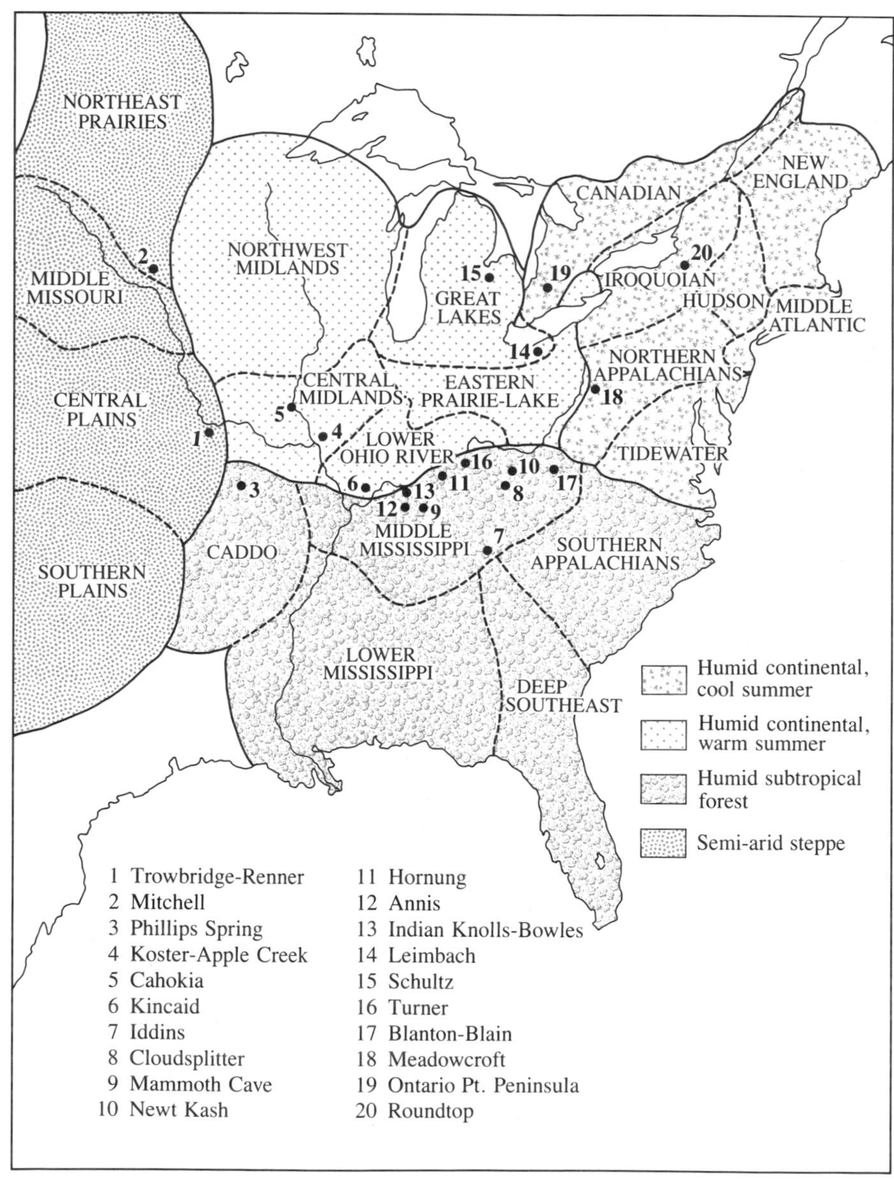

Figure 14. The environmental life zones and relevant sites in the eastern United States

shadow of the Rocky Mountains, which block the westerlies coming from the Pacific Ocean, although some storms do sweep down to the plains from the boreal forests of central Canada. In spite of its name, the Great Plains is not as flat as one might expect, and there are two major north-south divisions—the high plains and the lower prairie. Both have some low rolling hills, buttes, and wide river drainage systems, many of which connect with the Missouri River and have high terraces (examples are the Red River, the Arkansas River, and the Platte River). Other variations in geography are the Black Hills in North and South Dakota and the Llano Estacado (Staked Plain) in Texas. The difference in rainfall from east to west correlates with soil differences and results in major vegetational domains, running in bands from west to east—a western belt of short grasses, a central belt with mixed grasses, and finally the tallgrass-bluestem prairie to the east. The humidity and rich soils along the middle Missouri result in a gallery forest of cottonwood, cypress, and scrubs, while the plains of Oklahoma and northern Texas have vegetation that is almost xerophytic. In spite of this floral variation, the fauna was dominated by one animal, the buffalo, or American Bison, vast herds of which once roamed the plains. Other large animals included pronghorn antelope, mule and white-tailed deer, elk, and the formidable Plains grizzly bear, making the Plains a hunter's paradise. This abundance of game is reflected in the archaeological record, which shows agriculture arriving late and even then mainly along the middle Missouri.

In addition to the two major north-south subdivisions, other subareas of the Great Plains are the dry southern plains, the more forested central plains, the scrub-grass northwestern high plains or middle Missouri, and the northeastern tallgrass prairies. We have fairly complete archaeological sequences for most of these four subareas, with some limited information on their ancient subsistence systems.

The eastern United States, which is characterized by woodlands and not steppe, is the region east of the Great Plains. Ecologically, it may be divided into three major zones on the basis of its major climate zones: the Northeast, which has cool, humid continental summers; the Midwest, which has warm, humid continental summers; and the Southeast, which includes the Ozark-Ouachita hill country, and has hot, humid subtropical summers. Each of the areas has its own vegetation: in the Northeast, there are hardwoods as well as conifers in the Appalachian Highlands; the Midwest has hardwood forest and prairie vegetation; and the Southeast has softwood forest. Obviously, each of these three areas can be further divided, and these subdivisions often have significant and unique archaeological sequences. Let us therefore take a closer look at the three major divisions of the eastern United States.

We shall start with the Southeast since it is nearest to the Southwest, a key area of prehistoric agricultural development, but separated from that region by the southern and central Plains. Generally speaking, the Southeast is a humid area with heavy winter and spring rains coming from the Gulf of Mexico and the South Atlantic, and occasional fall hurricanes. Although frost and occasional snow occur along its northern borders on the Ohio River and in the northern Ozarks and southern Appalachians, frosts are infrequent and the climate is generally warm with short winter seasons and hot sticky summers. Much of the topography of the Southeast consists of coastal plains covered by softwoods, and there are large swampy regions with cypress and gumwood, while hardwood forests occur in the uplands of the Ozarks and southern Appalachians. The area is drained by many large waterways, from the Mississippi, Ohio, and Arkansas to a host of small streams, resulting in wide river bottoms, fertile soils, and lush vegetation. In addition to many edible wild plants, game is fairly abundant, although except for deer it consists mainly of small mammals and reptiles. Furthermore, the rivers abound with fish and the coastal regions have a wealth of marine resources. Like southern Europe, the Southeast was and is a warm, temperate forested land of plenty but with few domesticable plants (table 8.1).

I have made five subareas in the Southeast. Nearest the Southwest is the subarea comprised of east Texas, Oklahoma, Arkansas, and southern Missouri. It is characterized by the forested Ozark-Ouachita hills and the drainage areas of the Red and Arkansas rivers. Because of its dominant ethnic group, I refer to this region as the Caddo subarea. Farther east, along the Mississippi south of Memphis on the Tennessee River, and including the adjacent states of Mississippi and Alabama, is a subarea I call the Lower Mississippi. For many regions of this subarea we have long, comprehensive archaeological sequences, but our subsistence information is limited, and few local plants were domesticated (table 8.1).

Two subareas of the Southeast front on the Atlantic Ocean. The northern one, or Southern Appalachians, includes the Appalachians in Virginia south of the Roanoke River, and North and South Carolina. To the south is the Deep Southeast, the subarea that includes Georgia and Florida with their coastal plains and extensive swampland. These subareas had their archaeological heyday during WPA times, but the tradition of good archaeology continues there.

The subarea I call the Middle Mississippi is the area south of the Mississippi and Ohio rivers but north of the Ozarks and the Gulf drainages of the Alabama and Mississippi rivers as well as the Appalachians to the east. This is softwood

Table 8.1. Some native cultivated and/or domesticated plants of the eastern United States*

Scientific name	Popular name
Amaranthus hypochondriacus	amaranth
Ambrosia trificia	giant ragweed
Chenopodium album	grass seed
Chenopodium bushianum	lamb's-quarters, goosefoot
Helianthus annuus L.	sunflower
Helianthus tuberosus	Jerusalem artichoke
Hordeum pusillum	little barley
Iva annua L.	swampweed, sumpweed
Iva annua or *macrocarpa*	marsh elder
Nicotiana guadrivalvus or *N. rustica*	Indian tobacco
Phalaris caroliniana	maygrass
Polygonum erectum	knotweed

*After R. Ford (1985).

hill country with the Tennessee and Cumberland rivers, along with the middle part of the Mississippi, as its major drainage area. (It is often called the Tennessee-Cumberland, but because it extends westward into the Mississippi drainage, I prefer the term *Middle Mississippi.*) Patty Jo Watson's program in the Kentucky region should yield much significant data about early agriculture there, with implications for the whole subarea.

A similarly influential program is the Koster program in the Midwest zone. This differs from its southern neighbor in being noticeably colder with longer winters and often fierce blizzards. It is still relatively humid, with weather feeding in from both the Gulf of Mexico and central Canada. Not only does it have many river systems—the Mississippi, Missouri, Illinois, Rock, Wabash, and others—but also its north-central border is formed by four of the Great Lakes: Michigan, Superior, Huron, and Erie. These waterways have created fertile flood plains and terrace systems. As a whole the zone is relatively undulating and hilly.

Its northern part has many glacial features, resulting in slightly different soils. This feature, coupled with colder temperatures, has resulted in both conifers and hardwoods. Farther to the south a mixed hardwood forest–prairie region covers northern Missouri, Illinois, Indiana, and Ohio. South of it, where soils are more alluvial and the climate warmer, hardwoods are mixed with some softwoods along the lower Ohio River.

Like the Southeast, this zone has many edible plants (see table 8.1), and game abounds. The northern east-west prairie strip of the Midwest probably had buffalo in prehistoric and early historic times, while the most northerly part had elk and moose. Riverine and aquatic resources, particularly in the Great Lakes, provided yet another food resource.

Needless to say, the great environmental variation of the Midwest zone or area allows it to be divided into a number of subareas. The first of these, the Lower Ohio River subarea, lies north of the Middle Mississippi subarea of the Southeast. It stretches from about Cincinnati, Ohio, to Cape Girardeau, Missouri, across the fertile lowlands of southern Illinois, Indiana, and southeastern Ohio and includes the lower part of the Wabash River Valley. Northwest of it is the Central Midlands subarea, reaching from western Illinois north to the Rock River. This subarea, which includes the Illinois River Valley and northern Missouri, is where the pioneering efforts in archaeology occurred and where the innovative Koster research continues. To the north, in the colder and more heavily forested regions, is the subarea of the Northwest Midlands, which include much of Iowa, Minnesota, Wisconsin, and northwestern Illinois. Much like this subarea is the region to the east, the Great Lakes, which includes southeastern Wisconsin, northeastern Illinois, northwestern Indiana along Lake Michigan, the state of Michigan itself, and northwestern Ohio. This subarea and that following have had less relevant archaeological investigation than the Central Midlands. South of the Great Lakes area is the prairie-like region of northern Indiana and Ohio, which I call the Eastern Prairie Lake subarea. This subarea had the great Hopewell Mound cultures, which unfortunately were mostly excavated long before there were interests in early agriculture or sophisticated methods for the reconstruction of ancient subsistence systems.

This brings us to the final major area or zone, the Northeast, which has similar problems of archaeological investigations. The Northeast is a humid climatic zone with much of its weather affected by the Gulf Stream of the Atlantic. While some parts have milder temperatures because of this current, the more mountainous interior portions have cold winters and usually heavy snows. Physiographically, the northern Appalachian chain bisects the area from southeast to northeast, with the Alleghenies, Catskills, Adirondacks, and Green and White mountains being part of it. Conifers and pines abound, while hardwoods—maple, beech, oak, and the like—grow in the coastal strip to the east. In the northwestern part of the area the hardwoods blend into the boreal forest. In addition, there is a whole series of major river valleys—the Saint Lawrence, Delaware, Hudson, Connecticut, Merrimack, and so on—which have specific vegetational regions and rich aquatic resources. Along the coast and lakes Ontario, Erie, and Huron are aquatic and lacustrine resources. Also plentiful is game, including deer, moose, elk, and a host of smaller, fur-bearing animals. Although the agricultural potential of the Northeast was far below that of the Southeast or Midwest, its wild resources were just as plentiful.

Like the other major areas or zones, the Northeast may be subdivided in various ways. Because of the kinds of archaeological sequences, I divide it as follows. Nearest the Midwest is the Canadian subarea, which includes Ontario and the Saint Lawrence River. The northeastern part of Ohio, most of central New York to the Hudson River, and northern Pennsylvania along the Susquehanna River consistute the Iroquoian subarea. East of the Susquehanna is the Hudson subarea, reaching from Lake Champlain to Peekskill. This would abut the Middle Atlantic subarea, which includes New Jersey, New York City and Long Island, and Connecticut. North of it is the New England subarea, while south to the Roanoke (or James) River is the Tidewater subarea, which abuts the Southern Appalachian subarea of the Southeast. To the west, in West Virginia, Virginia, and Pennsylvania, is the Northern Appalachian subarea.

The quality and quantity of the archaeological sequences of these subareas vary considerably, and the early archaeological data have not been studied in a sophisticated manner. Nevertheless, there are archaeological materials useful for investigating the problem of origins of agriculture. Let us begin with the most poorly known zone or area, the Northeast.

RELEVANT ARCHAEOLOGICAL DATA

Eastern Fluted and Clovis points (9000–7000 B.C.) have now been uncovered or excavated from all subareas of the Northeast, and a component that may be even earlier has been exhumed in the lower levels of Meadowcroft Rock shelter in the western Pennsylvania region of the north Appalachian subarea (Adovasio et al. 1975). None of these sites or the Early Archaic ones (roughly 7000–5000 B.C.), characterized by Kirk, Palmer, and bifurcated-base points, has as yet yielded any plant remains (Griffin 1965). A few pieces of charcoal representing possible plant remains have come from Middle Archaic sites (5000–3000 B.C.) such as Neville in the New England subarea, Vergennes and Vosburg in the Hudson River subarea, and Brewerton in the Iroquoian subarea of Late Archaic times

(3500–1000 B.C.), but none has been identified as domesticated plant remains (Ritchie 1969). The culture, however, seems to have been Efficient Foraging Bands—my system D (table 8.2).

In fact, although most of the Early Woodland sites have burial cult remains, such as the Middlesex remains of central New York, Orient on Long Island (Ritchie 1969), Baskill (Adena) on the Delaware River, or Saugeen in Ontario (Wright and Anderson 1963), they are without plant remains except for the middle levels of Meadowcroft in the western Pennsylvania region of the northern Appalachian subarea (about 870 B.C.), which supposedly had pumpkin (*Cucurbita pepo*) remains (Adovasio et al. 1975). This Meadowcroft carbon-14 date should be viewed with suspicion, however, since corn remains from slightly higher levels in the same cave were radiocarbon dated at 340 and 375 B.C., earlier than any in the Midwest or the Southeast.

If the identification is correct and if the radiocarbon date is valid, then the Meadowcroft remains would be the earliest evidence of domesticates in the northeastern United States. Even so, this specimen occurs three or four millennia after the similar Koster and Napoleon Hollow remains in Illinois (Asch and Asch 1985) and five or six thousand years after the earliest specimens in Tamaulipas (MacNeish 1958). Perhaps the earliest reliable evidence of domesticates comes from a few charred corncob fragments from the late Point Pennisula sites of the Princess Point complex in the southwestern Ontario region of the Canadian subarea (Stothers 1977), which should fall in the general time period from A.D. 500 to 900 (see tables 8.2 and 8.3). Slightly less reliable evidence of corn comes from the roughly contemporaneous Hunter Home and Kipp Island No. 4 sites in western New York (Richie 1969).

Preliminary studies suggest some if not all of the sites in this general time period were permanent occupations, and various posthole patterns may be evidence of village life at this time, although they could be signs of Semisedentary Bands with Domesticates. Unfortunately, adequate samples of food remains are rare for this time period, and sophisticated studies of the few food remains we do have leave much to be desired. The limited sample seems to indicate that the subsistence at this time came basically from wild plants and animals, not from domesticates. In other words, the people were successful Village Foragers, with or without horticulture, or more probably they were Semisedentary Bands with Domesticates—my system D1.

In fact, it is not until Glen Meyer, Pickering, or Uren times (A.D. 1000–1250) in Ontario (J. Wright 1966), or Owasco in upper New York State and along the Hudson River, or Monongahela or McFate in Pennsylvania that charred remains of corn, squash, and beans have been recovered in significant amounts in villages that have longhouses, often inside palisades (Ritchie 1969). Such remains are often associated with numerous charred wild-plant remains as well as with fish and other animal bones and shells. As yet, studies are too unsophisticated to determine if the inhabitants were Incipient Village Agriculturists, Horticulturists, or Agriculturists, but I suspect they probably were Village Horticulturists—my system C2.

Whether groups in New Jersey—such as Overpeck of Clearview and Bowman

Table 8.2. Relevant archaeological sequences from the eastern United States

Great Plains subarea	Caddo subarea	Lower Mississippi subarea	Southern Appalachians and Deep Southeast subareas	Lower Ohio and Middle Mississippi subareas	Eastern Prairie–Middle Mississippi subareas	Central Midlands subarea	Iroquoian and Northern Appalachian subarea	Dates
Middle Missouri					Monongahela	Cahokia	Iroquois	
Oneonta	Fulton	Natchez	Moundville	Kincaid		Sand Prairie	Owasco	
Upper Reb.-Neb.	Gibson	Plaquemine	Lamar		Ft. Ancient	Moorehead		1000 A.D.
				Douglas	Blaine	Stirling		
Sterns Creek	Alto	Coles Creek	Weeden Island			Fairmount		
	Christensen	Troyville	Kolomoki	Lewis	Turpin	Weaver	Point	
Kansas City	Cave	Marksville	Swift Creek	Baumer	Harness	Ogden-Fette	Peninsula	0
Hopewell					Temper	Havana	Middlesex	
		Tchefuncte	Deptford	Salt Cave	Adena	Red Ochre	Meadowcroft 4	
					Newt Kash	Black Sand		
	Fourche	Poverty Pt.						
Chalk Hollow	Maline					Titterington		1000 B.C.
					Cloudsplitter			
				Iddins				
			Stalling's					
Lo Dais Ka C			Island				Laurentian	2000 B.C.

	Phillips Spring	Amite	Kirk	Faulkner	Bacon Bend	Helden Kahlman		3000 B.C.
Lo Dais Ka D					Annis	Napoleon Hollow	Lamoka	
		Jones Creek	Hardaway			Koster 8	Neville	4000 B.C.
				Eva		Koster 11	Meadowcroft 3	5000 B.C.
Logan Creek	Dalton		Windover			Modoc Starved Rock		6000 B.C.
								7000 B.C.
Yuma							Meadowcroft 2	8000 B.C.
Folsom Clovis							Meadowcroft 1	more than 10,000 B.C.

Brook in the Long Island region—or contemporaneous groups in Delaware or New England had a similar subsistence system is at present unknown. It could well be that such groups did not attain that status until the final Woodland period, A.D. 1250 to 1650. To the west, however, the Iroquois or Iroquois-dominated groups did have agriculture in this final period, utilizing not only corn, beans, and squash, but also sunflower and tobacco (Ritchie 1969). Village agriculture—system E—thus arrived late in the Northeast, perhaps after both Horticultural Villagers (system C2) and Semisedentary Bands with Domesticates (system D1), with village agriculture slowly becoming the dominant way of life only just before historic times. The data generally tend to confirm my tertiary development model, but unfortunately, research in this area is just too incomplete to test the model adequately.

Much the same may be said of the second most poorly documented area in the eastern United States—the Great Plains (see table 8.2, left column). There are Early Man and Archaic remains from every one of its subareas, but no plant remains (Wedel 1961). In fact, the earliest plant remains come from just outside the Kansas City region, barely in the Great Plains, at two Havana-type Illinois Hopewell sites—Renner and Trowbridge—which date roughly from 0 to A.D. 300. The Renner site had maize and reputedly beans, while Trowbridge had pumpkin (*Cucurbita pepo*) and maize (A. Johnson 1976). Another possible early occurrence of maize is from the Lo Dais Ka site in the high plains of eastern Colorado, but accurately dating the associated Complex B is impossible (Irwin-Williams and Irwin 1966). One other record of plant remains from Middle Woodland times is the mention of squash—probably *Cucurbita pepo*—from the Sterns Creek site (A.D. 500–1000) in the central Great Plains subarea (Strong 1935), so the people there might have been Semisedentary Bands with Domesticates—system D1.

Whether other Great Plains Woodland sites of roughly this time period utilized domesticated plants is unknown. In fact, at this time period, any domesticated plants seem to have been passing through the plains into the eastern Woodland villages. It was not until A.D. 1000 that the Mississippi Horticultural Villagers—system C2—spread up the Missouri River Valley, and even then concrete examples of that complex are rare. Many of the Middle Missouri sites from A.D. 1000 onward have corn, squash, and beans, and the Mitchell site is also mentioned as having beans and tobacco (Cutler and Blake 1973). Supposedly the contemporaneous Central Plains–Upper Republican and Pawnee complexes had the same type of village horticulture or agriculture as did the Washita River and Custer foci, but good contextual evidence is lacking (Lehmer 1971). It is therefore most difficult to test any hypothesis about the origins of village agriculture with the limited Plains data. There are hints it followed a tertiary development: groups evolved from Efficient buffalo-hunting Foraging Bands (system D) in the Archaic to Semisedentary Bands with Domesticates (system D1) in Middle Woodland times to Horticultural Villagers (system C2) with Chapalote corn, squash, and sumpweed at about A.D. 1000, becoming Village Agriculturists (system E), with the addition of beans, sunflower, and gourds, in a few places along the Missouri River by historic times (Adair 1988).

The evidence is slightly better for the Southeast area, although still far from really adequate (see table 8.3). As noted previously, the climate of the Southeast is warmer and more humid than that of the Great Plains, which may have made life a little easier for the Indians; however, it makes matters worse for the archaeologists, since ancient plant remains are poorly preserved. Yet a tremendous amount of archaeological investigation has been undertaken in the Southeast, and relatively complete sequences since Paleo-Indian times have been established for all our subareas—the Caddo region, the Lower Mississippi, the Middle Mississippi, the Southern Appalachian, and the Deep Southeast.

Although there are scattered finds of Paleo-indian points—Clovis, Plainview, or Dalton in the Caddo area, and Dalton in the Lower Mississippi—it is the Archaic remains, with evidence of Efficient Foraging Bands (system D), that are pertinent to the problems of early domesticates. Remains of several varieties of Archaic culture occur in each subdivision (Muller 1983). In the Caddo subarea are the Late Archaic remains of Fourche Maline, as well as remains from the Ozark Bluff Dwellers (Harrington 1924). Of relevance are the Phillip Springs *Cucurbita pepo* remains dated at 2350 B.C., as well as Ozark Mississippian remains with two varieties of corn, *Cucurbita mixta,* and amaranth (Fritz 1984). Moreover, carbon 12-13 analysis of skeletons from this area indicates agriculture came late in Mississippian times, and earlier people consumed mainly wild foods (Lynoff et al. 1986). Although the Lower Mississippi subarea had Early to Middle Archaic remains at Jones Creek and Amile River, only one squash seed was uncovered at the Teoc Creek site in Mississippi (R. Ford 1981), and the late Poverty Point and early Woodland Tchefuncte sites have yielded no plant remains. Russell Cave had *Chenopodium* seeds that could have been cultivars, but the case is debatable (B. Smith 1984—see table 8.3). Much the same may be said for the Deep Southeast with its Stanfield and Stalling Island remains and the magnificent Archaic sequence Joffre Coe established for the Southern Appalachian area: from Hardaway to Palmer to Kirk to Stanley, ending with Quilform (Coe 1964). But we do have one gourd remain from the Windover site in Florida (Watson 1988), dated at 5300 B.C.

In fact, the most relevant Archaic remains come from the Middle Mississippi subarea, where Patty Jo Watson and her team (Marquardt and Watson 1977) established an Archaic sequence. Following the Dalton phase of Early Archaic times (before 5000 B.C.), we have Eva, Faulkner, Three Mile, and Indian Knolls phases. The Annis site of the Three Mile phase and the Hayes site (roughly 2339 to 3199 B.C.) yielded squash (*Cucurbita pepo*) remains as did the Bacon Bend site at roughly 2400 to 2100 B.C. (Ferguson 1978). While there are few Middle Archaic components with plant remains, Late Archaic ones (such as the Indian Knolls phase) are more numerous. These include the Iddins site (roughly 1500 B.C.) with gourds and squash (R. Ford 1981); the Bowles site (1440 to 1490 B.C.) with gourds (Marquardt and Watson 1977); the Cloudsplitter site (1500 to 500 B.C.) with squash, gourds, and *Chenopodia* (R. Ford 1981); the Sparks site (1500–500 B.C.) with squash (Applegarth 1977); the Higgs site (1020–280 B.C.) with *Chenopodia* and squash (Ford 1981); as well as Newt Kash Hollow with sumpweed, sunflower, squash, and gourds (Jones 1936) and Mammoth Cave

Table 8.3. Sequence of cultivated and/or domesticated plants in the eastern United States

Great Plains Northern subarea	Great Plains Central subarea	Caddo subarea	Lower Mississippi subarea	Middle Mississippi subarea	Central Midlands subarea	Lower Ohio River subarea	Northeast subarea	Deep Southeast subarea	Range of dates	gourd	pumpkin (*C. pepo*)	sunflower	swampweed-marshelder	lamb's-quarters, goosefoot	knotweed	maygrass	ragweed	amaranth	corn	beans	tobacco	warty squash (*C. moschata*)
			Historic						1500 1900 A.D.	x	x	x	x	x		x	x	x	x	x	x	x
								Moundville	1100 1500 A.D.	x	x	x							x	x		x
		Fatherland (Natchez)							1300 1600 A.D.	x	x								x	x	x	x
							Iroquois		1200 1700 A.D.	x	x	x							x	x	x	x
					Range				1200 1600 A.D.										x	x	x	
Rock Village									1200 1600 A.D.										x	x		x
							Round top-Owasco		900 1100 A.D.	x	x								x	x		
Mitchell									1400 1700 A.D.										x	x	x	
Mahhaha									1400 1700 A.D.										x		x	
				Kincaid					1300 1600 A.D.		x	x		x			x	x	x	x		
					Cahokia				1000 1600 A.D.		x	x							x	x		
						Blanton			900 1200 A.D.		x								x	?		
						Ft. Ancient Blaine			900 1100 A.D.										x	x		
	Alto								800 1100 A.D.		x								x	x		
Sterns Creek									700 1100 A.D.		x											
						Turner			500 900 A.D.										x			
						Newbridge			260 1060 A.D.					x								
						Sand Ridge			500 900 A.D.			x										
							Pt. Pennisula (Ont)		500 900 A.D.										x			

Site															
Newman	100	600	A.D.										x		
Schultz	200	600	A.D.		x								?		
Harness	200	600	A.D.										x		
Brandenberg	200	600	A.D.	x											
Macoupin	200	600	A.D.	x	x										x
Apple Creek	100	500	A.D.			x	x	x		x	x	x			
Synder	100	500	A.D.			x	x	x		x	x	x			
Renner	100	300	A.D.									x	x	x	
Trowbridge	0	250	A.D.		x								x?		
Russell Cave	0							x							
Hornug	0	200	A.D.		x								x		
Meadowcroft 4	0	400	B.C.			x							x?		
Cowan Creek	0	500	B.C.	x				x							
Leimbach	0	500	B.C.		x										
Meadowcroft 3	500	1000	B.C.		x										
Salts Cave	300	1000	B.C.	x	x	x	x								
Bones Cave	300	1000	B.C.		x										
Mammoth Cave	500	1000	B.C.			x	x	x							
Newt Kash	400	800	B.C.	x	x	x	x		x						
Higgs	780	1020	B.C.			x									
Sparks	800	1200	B.C.		x										
Riverton	400	1200	B.C.			x									
Cloudsplitter	500	2000	B.C.	x	x		?	x							
Bowles	500	2100	B.C.		x										
Iddins	1000	2000	B.C.	x	x										
Kahlman	1000	2000	B.C.	x											
Koster 8	1500	2000	B.C.			x									
Napoleon Hollow 2	2000	2200	B.C.		x	x									
Bacon Bend	2100	2400	B.C.		x										
Phillips Spring	2100	2400	B.C.	x	x										
Annis	2400	3700	B.C.		x										
Koster 11	4000	5000	B.C.	?	x										
Napoleon Hollow 1	3100	5000	B.C.		x										
Windover	5300		B.C.	x											

with squash, gourds, and sunflower (Watson 1969), the last two sites being in the time period roughly from 1000 to 300 B.C. Although the picture is far from clear, it would appear that in Middle Archaic times *Cucurbita pepo* and gourds were imported from northeastern Mexico, but even this is debated (Heiser 1985; Decker 1988; B. Smith 1989). In later Archaic times sunflower, sumpweed, and *Chenopodia* were locally domesticated or cultivated (see table 8.3). Let me hasten to add that all these domesticates comprised an extremely small portion of the total diet, and horticulture or incipient agriculture was a minor occupation. All of these groups could be considered Semisedentary Bands with Domesticates—system D1.

The Early Woodland (500 B.C.–0) and Middle Woodland (0–A.D. 700) periods are well represented in many regions of the Central Midlands, Lower Ohio River, and Eastern Prairie Lake subareas, but only the Middle Mississippi groups have many domesticated plant remains. There, squash, gourds, sunflower, sumpweed, and perhaps a few others appear (table 8.3). We do, however, have a few examples of maize at the Adena Hornung site (200 B.C.–0) in northern Kentucky, and at the Hopewellian-like Peter site (A.D. 300–900) in Tennessee (R. Ford 1981). Many authors assume that Middle Woodland cultures such as Havana, Hopewell, and the like had a subsistence based on maize agriculture because of their village life (Sears 1968), but the little evidence we have—mainly from carbon 12-13 isotopic analysis (Buikstra et al. 1987; Ambrose 1987; Bender et al. 1981)—suggests that at this time these sedentary peoples of the Midwest had a successful foraging subsistence system with only a little horticulture, providing at best a meager supplement to their rich "wild" diet. In fact, long ago Joe Caldwell dubbed the system "primary forest efficiency" (Caldwell 1958), and I would classify these people as just barely Horticultural Villagers—my system C2.

In fact, it is not until Mississippian times—after A.D. 900—that we have real evidence of the use of corn agriculture (Muller 1987). Yet even this is rare, occurring only at Kincaid in Illinois, Alto in Texas, Moundville in Alabama, Fatherland in Mississippi, Hardin in Kentucky, and a few other sites. However, carbon 12-13 isotopic studies do indicate full-time corn agriculture (Bender 1981). Also, beans have been found at some of these large town or village temple-mound sites, as have some of the older plants, such as squash, gourds, marsh elder, and sunflower (Muller 1983), so village agriculture—system E—at last seems to have arrived. Obviously, much more data on plant remains are needed for the southwestern Midwest subareas at all periods before we can reach any valid conclusions or adequately test my tertiary development hypothesis, even though we have many almost complete archaeological sequences.

Much the same may be said of our final subarea, the Central Midlands, although it has a few more plant remains—mainly from the inadequately reported Koster site (Asch, Ford, and Asch 1972). Stray finds of Paleo-indians occur; in Missouri, Clovis points were found at Shriver and Kimmswich sites, and Clovis or Clovis-like points occur as surface finds increase in all parts of the Midwest. These distinctive points infer an emphasis on hunting that lasted into Early Archaic times (7000–5000 B.C.). Dalton points occur in Missouri and Southern Illinois; Starved Rock in central Illinois and Flambeau in Wisconsin have Agate

Basin-like Plains affiliations; and Holcombe in Michigan and Brohm in western Ontario have Plainview points with similar affiliation. So far we have no evidence of domesticates or cultivars in this period (Griffin 1978). The people seem to have been Early Hunting-Collecting Bands (system A), becoming Efficient Foragers (system D) in Early Archaic times.

Our first evidence of domesticates comes in the Middle Archaic. Squash seeds were found in level 11 (estimated 5000–4000 B.C.) at the Koster site as well as at the nearby Napoleon Hollow site and Phillips Spring site, Missouri, dating roughly from 2400 to 2000 B.C. (Chomko and Crawford 1978); they were also found at Modoc in southern Illinois, the Markee site in Wisconsin, and perhaps at some of the earlier Shield Archaic components in Michigan.

As might be expected, the Late Archaic is even better represented. Among the sites are Faulkner in southern Illinois (MacNeish 1948), Heldon and Titterington at the Koster site, Old Copper in Wisconsin, McCain in southern Ohio, and Shield Archaic and/or Glacial Kame in northern Ohio and Michigan (Griffin 1978). The only plant remains, however, were sumpweed seeds found at Koster level 8 (Asch, Ford, and Asch 1972) and the Collins site (R. Ford 1981); squash seeds at Riverton in southern Indiana (Yarnell 1976); and gourds at Bony Creek, Missouri (King and McMillan 1975). In a vague way this mirrors the sequence in the Middle Mississippi, but the evidence that horticulture was practiced is even less convincing. In fact, all these sites could be better classified as having Semisedentary Bands with Domesticates—system D1.

The evidence is only slightly better for the Early Woodland period with its burial mounds and first pottery, although gourds and perhaps *Chenopodia* occurred at the Adena and Cowan Creek sites in the Lower Ohio, while Leimback had squash (Shane 1967), and maize was found at the Daines Mound in Ohio (Murphy 1971) and at Newman in Illinois (Struever and Vickery 1973).

It is also doubtful that the well-represented Hopewellian burial mound horizons of Middle Woodland times (0–A.D. 900) had agriculture. In their intensive flotation efforts in central Illinois, David and Nancy Asch and Michael Ford found abundant evidence of the use of many wild plants as well as of land animals, fish, and mollusks but few remains of domesticates or cultivars. The McGraw, Hopeton Square, Horner, and Turner mounds in Ohio had corn; in addition, Macoupin and Scovill in Illinois had gourds and squash (R. Ford 1981), while the Schultz site in Michigan had gourds, but this is meager evidence for domesticated plants (Ford 1981). Even slimmer is the evidence that sunflower, sumpweed, and marsh elder were cultivated at this time, although that has been reported for the Apple Creek (Kaplan and Maina 1977), Snyder, Koster, and Stillwell sites in central Illinois (R. Ford 1981). Sunflower is also reported from one Late Woodland component (A.D. 700–1000) at the Sand Ridge site in Ohio, and a gourd fragment came out of the Brandenberg mound of roughly the same period in central Missouri (Ford 1981). Again, Caldwell's primary forestry efficiency pattern of subsistence seems to pertain to Middle Woodland times (Caldwell 1958), and Fowler's idea that people were practicing a sort of "dump heap" horticulture (Fowler 1973) may well turn out to characterize their subsistence system, which seems to have been system C2—Horticultural Villagers.

Also following Fowler (1974), it would seem that only in Mississippian times (after A.D. 900) do we have sure evidence from carbon 12-13 isotopic studies (Buikstra et al. 1987; Lynoff et al. 1986) of village agriculture—my system E. This system occurs mainly at such Middle Mississippi manifestations as Cahokia in central Illinois and Kincaid in southern Illinois. The people there definitely were supplying their huge towns or villages with agricultural produce from the rich river bottomlands. In addition to corn and beans, Kincaid gives some evidence that sunflower, squash, *Chenopodia,* ragweed, and marsh elder were probably grown (in Muller 1983, 1987). Farther east, in Ohio, the Upper Mississippi site of Fort Ancient Blain has yielded maize remains (Galinat 1970) and evidence of village agriculture (Watson 1988; B. D. Smith 1989). Evidence of agriculture for other Upper Mississippi manifestations—the Oneonta phase in Wisconsin, Iowa, and Missouri; and Fisher and Danvers from Starved Rock in northern Illinois—is more suggestive than convincing, but there is a good possibility that these people too had village agriculture (Bender, Baerreis, and Stevenson 1981). Whether the contemporaneous Woodland groups such as Monongahela in northern Ohio and Juntunen in Michigan practiced agriculture is problematic. As stated before, the sequential data for the material culture since Middle Archaic times are full and well documented, but reliable subsistence information is woefully inadequate. This becomes clear when we compare my tertiary model of agricultural development with the archaeological data for this area.

On the basis of slim evidence it would seem that the sequence of development for the eastern United States is roughly as follows. Following the period of Paleoindians during the Pleistocene, Hunting-Collecting Bands—system A—in some areas of the east gradually evolved into Efficient Foraging Bands (system D) during the Early and Middle Archaic. Caldwell long ago speculated that they probably had a seasonally scheduled settlement pattern, a broad-spectrum forager economy with seasonal scheduling, and had made the first steps toward a primary forest efficiency way of life (Caldwell 1958). During this period, while they were developing a number of resource specializations, they took on the planting and use of squash and gourds (see table 8.3) as a further subsistence option (or perhaps used these plants as containers or fishnet floats).

During the Late Archaic and perhaps even into the Early Woodland period (2000 B.C. to the time of Christ), technical innovations as well as increasing resource-procurement specialization led to greater logistic mobility; at the same time, people lived longer in one place—decreasing residential mobility. They had become Foraging Bands with Domesticates—system D1. During this general period, perhaps via the "dump heap" mechanism or some imitation, people began to plant *Cucurbita,* and this led them to start planting and using a number of local cultivars: sunflower, marsh elder, sumpgrass, *Chenopodia,* amaranth, and others. This was a further type of resource specialization, and it increased logistic mobility by improving security of food supply. Slightly later these people probably also added the planting of maize (and perhaps beans) to their already affluent foraging or primary forest efficiency subsistence system. These changes allowed them to live in larger groups and to stay longer in one place. By Middle

Woodland times (0–A.D. 900) there were large Hopewellian villages with spectacular burial mounds and a chiefdom type of social organization; the people had become Horticultural Villagers—system C2.

At this time the primary forest efficiency system had reached its zenith. Through the long period of development, however, various things had happened that caused the people gradually to place more and more reliance on their planted produce. By A.D. 1000 in the Mississippi period they had become Agricultural Villagers—my system E. Thus, development in the eastern United States (or Woodland) fits well with my hypothetical tertiary developmental model. In terms of evolution and cultural process, the development is similar to that described for Japan (chapter 5) and Europe (chapter 7), although the temporal periods and domesticated plants are different.

THEORETICAL CONSIDERATIONS

Now come the critical questions: What were the conditions or causes of this development? Does this United States development allow further testing of my hypothetical tertiary model? The inadequacy of the Eastern Woodlands prehistoric subsistence data means that discerning the causes or conditions of change is difficult and often based more on speculation than on fact. I believe such speculation worthwhile, however, because it causes us to focus on what the problems are and what directions future investigation ought to take to solve these problems.

Certainly the necessary conditions for tertiary development pertain to the eastern United States (table 8.4). There are few plants that were potentially domesticable. In fact, perhaps only the sunflower went through a genetic change and was truly domesticated. All the other plants used—sumpweed, ragweed, marsh elder, and the like—were basically only cultivated; that is, rather than undergoing genetic changes, they were changed morphologically by human selection and planting.

Further, domesticates were hard to come by. The Eastern Woodlands had only casual and/or indirect contacts with a Center—Mesoamerica—from which the earlier pumpkins and gourds probably came, and similar contacts with the Southwest non-Center, from which the easterners received corn, beans, squash, and perhaps other plants. This necessary condition thus affected the Eastern United States in much the same manner as did Japan, western Europe, and perhaps parts of the Southwest.

Perhaps more important, the third and fourth conditions of the tertiary model, which have to do with the availability of "natural" foods, also pertain. Unlike the environments of either the primary or the secondary developments (uncircumscribed riverine and coastal ecozones), the Eastern Woodlands of the United States had abundant resources, and food was available in all seasons. Even during cold winters when plant food was not available in the woodlands, there were still fish in the lakes and rivers, and large fur-bearing animals to be hunted. What this means is that, unlike the areas of primary and secondary developments, there was relatively less pressure on groups to change from their efficient food-collecting system to a food-producing one.

Table 8.4. Necessary and sufficient conditions as a positive-feedback process for tertiary development in the eastern United States

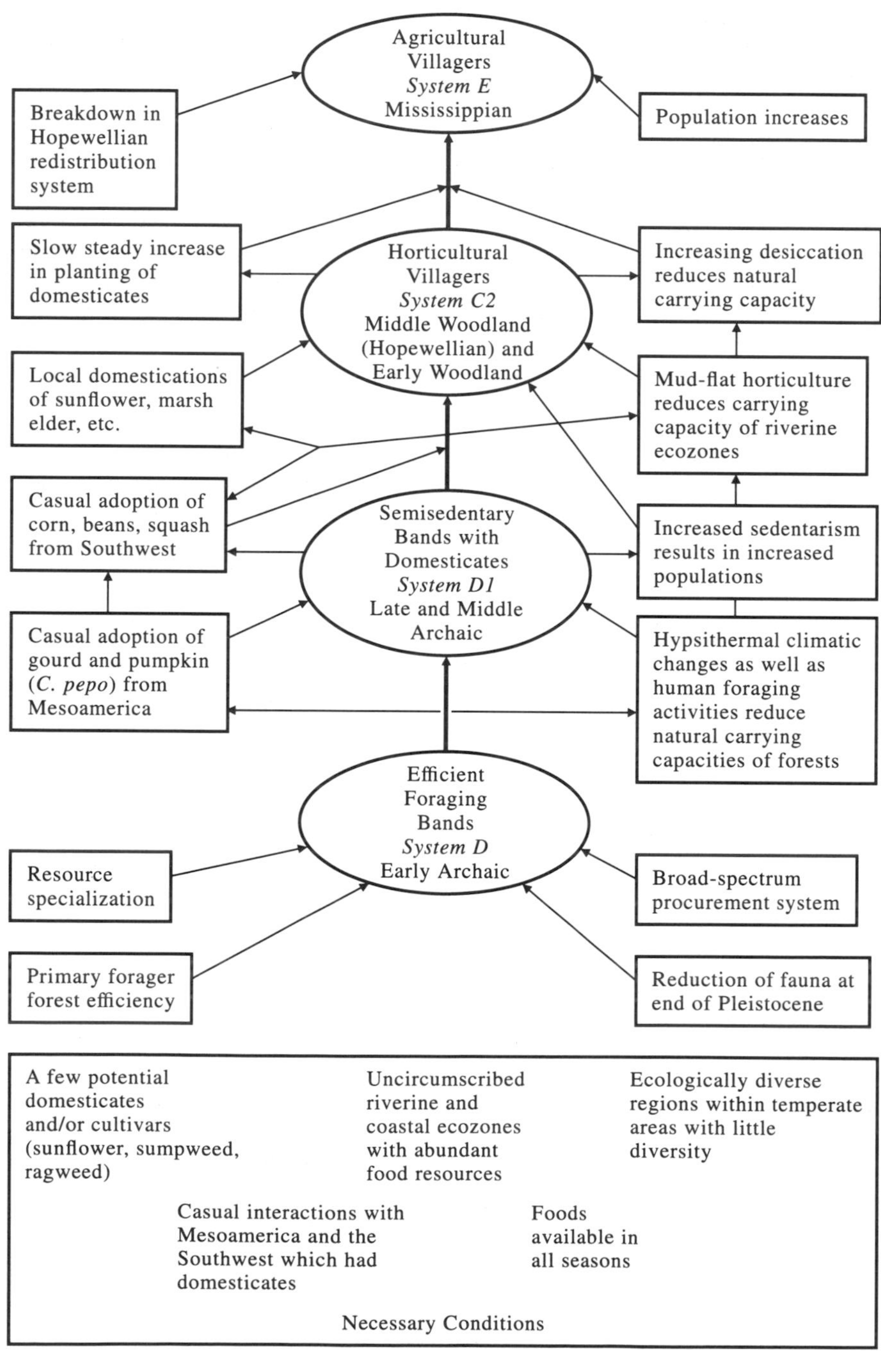

Also, the general environmental aspects show that each region of the eastern United States had a series of differing ecozones and that each region was slightly different from the others. Thus the necessary condition of regional or subareal diversity as well as areal diversity was met. This condition, of course, differs from the necessary condition for secondary development of diversity in subareas or subregions but relative uniformity in regions. Superficially, this ecological diversity seems like that of the primary developments, but it is different. In the eastern United States food resources are relatively abundant in all or most of the ecozones in each region in most seasons. The regional ecozones in areas of primary development, on the other hand, lack food resources in one or more seasons and rarely have one or more seasons with abundant food resources. We therefore see that the necessary conditions of my tertiary model pertain to the Eastern Woodlands of North America. This aspect of the hypothesis seems to have been adequately tested. Now, what about the all-important triggering causes, or sufficient conditions?

Although not well documented in the archaeological record of the Eastern Woodlands (or, for that matter, of Europe or Japan), the post-Pleistocene environmental changes and changing fauna and/or faunal distributions seem to have led to the development of a broad-spectrum procurement system in Early Archaic times. The environmental conditions encouraged resource specialization, such as hook fishing, net fishing, spear fishing, trapping, shell collecting, and so on, and this increased specialization acted as a positive-feedback system that led to the development of a primary forager forest efficiency. In other words, as resource specialization and development increased, the foragers became more efficient at maintaining a stable food supply, which led to more specialization, and so on. One of the resource specializations that occurred during this time was the planting of pumpkins and gourds acquired indirectly from northeastern Mexico (i.e., from Tamaulipas). All this gradually caused Efficient Foraging Bands—system D—to become Semisedentary Bands with Domesticates—system D1.

The planting of such local domesticates as sunflowers and of cultivars such as marsh elder, sumpweed, and ragweed began to act in a positive-feedback cycle with the reduced carrying capacity brought about by further foraging practices such as the burning of forests, overfishing, and so on. These conditions interacted with the decreasing mobility or growing sedentarism of groups and the concomitant population increases and concentrations. All these interactions took place very slowly throughout Late Archaic and/or Early Woodland times, possibly causing the evolution of horticultural villages—system C2—by Middle Woodland times. Although some of these conditions, such as population increases, continued to operate in a positive-feedback system, further desiccation of the environment and increasing plantings brought about new conditions that led to agriculture. One of these new conditions was the breakdown of the Middle Woodland–Hopewellian redistribution system. Accompanying this change was the general drying up of the environment about A.D. 1000, which further reduced the carrying capacity of the land. As a result, larger amounts of food had to be produced by more and more planting to feed the increasingly sedentary population. Thus, by A.D. 1000 or slightly later (depending on how far east one was) village agriculture—system E—had developed.

The sufficient conditions for tertiary development therefore seem to have occurred in the eastern United States during the long and slow cultural development from 7000 B.C. to A.D. 1000. My hypothesis seems to have been adequately tested and also seems to pertain to Japan and perhaps to the Jornada region of the Southwest and to western Europe. Unfortunately, the archaeology of the eastern United States is not so complete as that of western Europe nor have as extensive amounts of prehistoric foodstuffs been uncovered or as good subsistence reconstruction been attempted. As of now, therefore, the hypothesis is obviously not completely confirmed. I sincerely hope my speculations will stimulate scholars to attack this problem so that the tertiary development model will be confirmed or, better yet, more acutely phrased hypotheses will be developed. While there is more hope than light in our speculations, perhaps this hope will lead to a better light.

Let us now turn to the final set of developments of agriculture and settled life—those in tropical areas. As we shall see, the evidence for the testing of hypotheses about the causes leading to village agriculture there are even more meager than for the eastern United States, and of course, much archaeological research is needed. However, the areas to be tested are challenging, and the data collected so far are provocative.

PART 3

The Tropical Non-Centers

CHAPTER 9

India, Southeast Asia, and Oceania

While data from the temperate non-Centers—the U.S. Southwest, Europe, and the eastern United States—were not completely satisfactory for testing my hypothesis, information from the tropical non-Centers—southeastern Asia, the New World Tropics, and Africa—is even less so. The first of these areas, southeast Asia, has received a great deal of attention in literature on the origins of agriculture. Along with this area I include, as part of a more general non-Center, the Indian subcontinent and Oceania. India and Pakistan have closer connections to the Near East than to the Far East, but since they acquired rice and India acquired potato yams from the latter and have yielded too little data to warrant separate consideration, I have placed them in the general southeastern Asia region.

Southeastern Asia has much in common with the temperate non-Centers. It has relatively rich food resources, and even more of its wild plants were susceptible to domestication than were those in temperate climates. There are many lush riverine or coastal areas that are not geographically circumscribed, and within certain regions there may be a number of ecozones that can be exploited from a well-chosen base. Also like the temperate non-Centers, this tropical area is very humid. Preservation is therefore poor, and remains of ancient plants are even less frequent. Furthermore, southeast Asia is peripheral to a primary Center—the Far East—although, as noted earlier, that Center has the least reliable data pertaining to origins of village agriculture. Given these similarites to the temperate non-Centers, one expects tropical Asia to have experienced a secondary or tertiary development. As we shall see, however, the relevant data are too meager to give the area a definitive place, or even to decide if its development belongs to still another type. Ironically, the dearth of data, rather than inhibiting speculation and theorizing, seems to have stimulated it.

A BRIEF HISTORY OF SPECULATIONS ON THE PROBLEM

Actual speculation that southeastern Asia was a Center of plant domestication and agricultural innovation has prevailed since the 1920s, with Oswald Menghin (1931), N. I. Vavilov (1951), Ralph Linton (1955), and others taking the lead.

The basis for much of their hypothesis was the assumption (perhaps invalid) that the area's wide range of wild plants susceptible to domestication meant that those plants were domesticated there. Following this lead were later hypotheses by A. G. Haudricourt and Louis Hedin (1944), E. Werth (1954), H. Wissmann (1957), and, last and most important, Carl Sauer (1952). Their scenario was more or less as follows: Because of the lush environment, collectors settled down. Once settled down and having leisure time because of abundant food resources, they experimented with the local plants, and thus root crops (taro?) were first domesticated. Later, perhaps in terraced taro fields where rice (*Oryza sativa*) was a weed—as suggested by Jacques Barrau (1965, 1974)—they started dry-rice agriculture and then finally turned to wet-rice cultivation.

This scenario was widely accepted until discoveries by Chester Gorman (1970), William Solheim (1969), and others in Spirit Cave, Thailand, in the 1960s. This remarkable cave, which yielded early (10,000–7000 B.C.) radiocarbon determinations, and additional materials excavated from nearby sites in the 1970s, gave rise to a new set of hypotheses further indicating that Southeast Asia was a Center and not a non-Center. Based upon some shaky plant identifications and equivocal carbon-14 dates, it was hypothesized that at the end of the Pleistocene (14,000–7000 B.C.) the Hoabinhians had developed a broad-spectrum tropical subsistence system (Gorman 1969) and stable village life, which in turn led to taro domestication in the lowlands and dry-rice farming in the piedmont areas in the period from 7000 to 4500 B.C. After this agricultural complex spread, wet-rice farming and irrigation developed in the alluvial plains about 3000 to 2000 B.C. According to this hypothesis, Southeast Asia was not only a Center for tropical root crops and agriculture but also the place of origin for rice, the basic seed crop of the Far East (Gorman 1977).

Needless to say, this hypothesis has inspired numerous reactions. Karl Hutterer (1983), D. E. Yen (1980), and others have pointed out that the early domesticated plant identifications in Thailand are not reliable. K. C. Chang (1979) and others (e.g., Pearson 1979) have shown that rice in South China was earlier, although now rice at Koldihawa, India, has been dated to 6970 B.C. (Lone, Khan, and Buth 1988). As a further reaction to this hypothesis, H. D. Sankalia (1972), Vishnu-Mittre (1977), and Jerome Jacobson (1976) indicated that this scenario did not fit with data uncovered for India. Southeast Asia was once again relegated to the status of a non-Center (Harlan 1971). But perhaps more important, Harold Conklin (1957), Peter Kunstader (1978), Karl Hutterer (1983), and perhaps Jacques Barrau (1974) thought that Gorman and Solheim's concepts of the ecology of domestication and their interpretation of the environment of the tropics did not make their hypothesis tenable. This point is well taken, and I believe it justifies a brief discussion of the environmental aspects of the area.

THE ENVIRONMENT

Tropical southeastern Asia is the vast area lying between the tropics of Cancer and Capricorn and extending eastward from the Indian subcontinent and across Southeast Asia to Indonesia and the Philippines and on to the hundreds of islands of Oceania (see fig. 15). It is basically hot and completely frost free with moder-

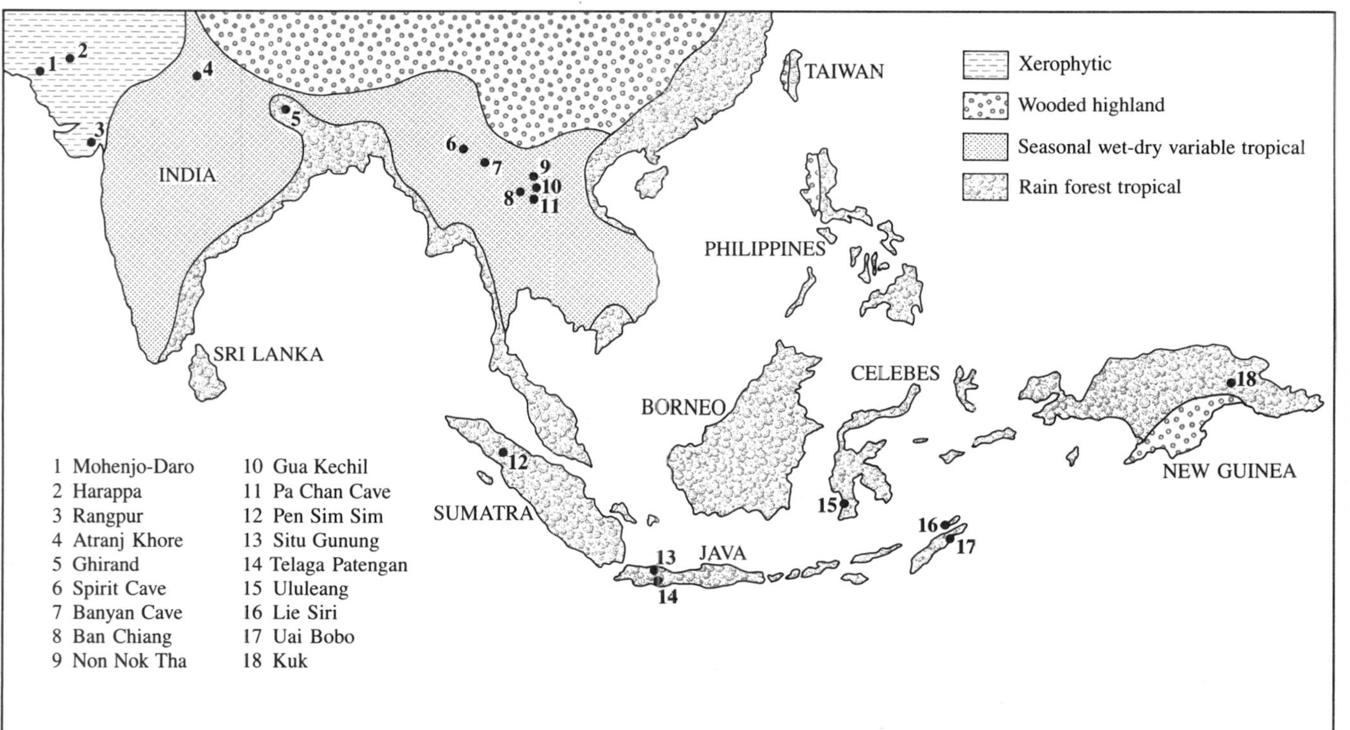

Figure 15. The environmental life zones and relevant archaeological sites in the Southeast Asia and Oceania area

ate to heavy rainfall in every month, which results in lush evergreen vegetation and abundant animal life. In terms of necessary conditions it has luxuriant vegetation in uncircumscribed ecozones that can be exploited from many bases; it also has numerous domesticable plants and relatively diverse ecozones, including large subareas of wooded highlands, tropical rain forest, and seasonally wet and dry tropics, all of which contain diverse ecozones. Obviously, such a complex and huge area—including both continental and insular regions—has considerable variation. The dry northwestern part of India, including the Indus River Valley, is markedly different (and perhaps should not be included in the study area), and the huge areas of swamp in Thailand, Vietnam, Sumatra, and Borneo are also exceptional subareas. Generally speaking, the major area can be broken into two broad subareas: the humid tropics of evergreen rain forest with short or no dry seasons, and the seasonal wet-dry tropics with one or more dry seasons and a varying evergreen forest (Barrau 1965).

The seasonal wet-dry variable tropics, the larger of the two subareas, includes most of India and the southeastern mainland of Asia as well as a small section of the eastern Philippines, eastern Taiwan, southern New Guinea, and a few other island spots. Yearly temperatures are high, but daily and seasonal fluctuations are more pronounced than in the humid tropics, and rainfall is slightly lower. More important is yearly distribution of rainfall. The seasonal wet-dry tropics usually have one or two dry seasons, which can vary from a month to six or eight months in length, when evaporation exceeds rainfall. Obviously, wind currents (the northeast trades), distance from the sea, topography, and other factors contribute to this great variation (Hutterer 1983).

Although large portions of southern Thailand and Cambodia are relatively level basin, much of this country is hilly, with the hills increasing in ruggedness as one goes northward towards the east-west–running Himalayas. A series of small north-south mountain chains divide the subarea: the Western Ghats along the west coast of India, the Arakan Yoma along western Burma, the Tenasserim along eastern Burma and Malaysia, and a series of ridges along western Vietnam. Thus the many long rivers and their drainages run in roughly the same southerly direction. Moving from west to east these rivers are the Indus, Normada, Ganges, Brahmaputra, Irawaddy, Yom, and Mekong.

The variations in climate cause differences in length of the dry season and considerable diversity in vegetation (during the dry season, plant growth often slows or halts). Where the dry season is short (less than three months), vegetation is tropical evergreen rain forest with high canopies, but as the length of the dry season increases, the height and density of the forests decrease, and the plant communities range from scrub forest to tree savanna to lush savanna to grassland and/or xerophytic scrub and form the seasonal wet-dry tropical subarea. Wide variation in plant kingdoms means that many plants accumulate energy in the form of starches and fat in the wet season to last them through the dry season. These plant products provide a rich diet for human consumption. Matching this plant variety are the many land animals, both small game and large herd animals, and aquatic resources from the oceans and rivers. Many of the region's plants are potentially domesticable (see table 9.1). Definitely coming from this subarea are

Table 9.1. Some native cultivated and/or domesticated plants of southeastern Asia*

Scientific name	Popular name
Alocasia macrorhiza	ape elephant ear
Amaranthus mangostanus	amaranth
Amorphophallus sp.	aroid tuber
Areca catechu	betel nut
Arenga saccharifera	sugar plum
Artocarpus communis	breadfruit
Artocarpus integrifolia	jack fruit
Averrhoa bilimbi	camias or bilimbi
Averrhoa carambola	carambole
Bambusa sp.	bamboo
Benincasa cerifera	white gourd
Borassus flabellifer	palmyra palm
Brassica juncea	sarson
Cajanus cajan	pigeon pea
Canavalia gladiata	jackbean
Cassia angustifolia	senna
Cinnamomum cassia	cassia bark
Citrus aurantiifolia	lime
Citrus aurantium	sour orange
Citrus decumanus	shaddock
Citrus grandis	pummelo
Citrus limon	lemon
Citrus medica	citron
Citrus nobilis	tangerine
Citrus paradisi	grapefruit
Citrus sinensis	sweet orange
Cocos nucifera	coconut
Coix lachryma-jobi	Job's tears
Colocasia esculenta	taro
Corchorus capsularis	jute
Cordyline fructicosa	cordyline
Crotaliaria juncea	sun hemp
Croton tiglium	croton oil
Curcuma longa	tumeric
Cyemopsis tetragonolobus	guar
Digitaria cruciata	millet
Dioscorea alata	winged yam
Dioscorea esculenta	yam
Dolichos biflorus	hyacinth bean
Durio zibethinus	durian
Echinochloa frumentacea	Japanese millet
Eleocharis tuberosa	water chestnut
Elettaria cardamomum	cardomon
Eugenia caryophylla	clove
Eugenia sp.	jambos
Garcinia mangostana	mangosteen
Gossypium arboreum	cotton
Hibiscus cannabinus	kenaf (fiber)
Lageneria siceraria	gourd
Lansium domesticum	lansone
Lawsonia inermis	henna
Luffa acutangula	loofa

Table 9.1. (*continued*)

Scientific name	Popular name
**Mangifera indica*	mango
Momordica charantia	balsam pear
**Metroxylon sagus*	sago palm
**Musa acuminta*	banana
Musa australomusa	Feii banana
**Musa balbisiana*	plantain
Musa sapientum	banana
**Musa textilis*	manila hemp
**Myristica fragrans*	nutmeg
**Nephelium lappaceum*	rambutan
**Nephelium longana*	longan
**Oryza sativa*	rice
Pachyrhizus erosus	yam bean
**Panicum miliare*	slender millet
**Paspalum scrobicalatum*	slender millet
**Piper betle*	betel leaf
**Piper nigrum*	black pepper
**Piper methysticum*	kara
**Psophocarpus tetragonolobus*	winged bean
Pueraria sic.	pueronia
**Pueraria lobata*	yam bean
**Saccharum officinarum*	sugar cane
**Sesamum indicum*	sesame
**Solanum melongena*	eggplant
**Syzygium aromaticum*	clove
**Tacca leontopetaloides*	arrowroot
Tamarindus indica	sweet pod tree
Terminalia catappa	kotamba
Trichosanthes anguina	serpent gourd
**Vigna aconitifolia*	mat bean
**Vigna calcarata*	rice bean
**Vigna mungo*	urd, black gram
**Vigna radiata*	mung bean
Zingiber officinale	ginger root

*Domesticates according to Harlan (1971).

arrowroot, millet, yam-bean, sugar cane, yams, betel nut and leaf, and perhaps gourds, cloves, nutmeg, and tumeric, as well as rice. As yet, however, no intensive studies of the ecozones of this subarea have been undertaken, although we do have archaeological sequences from eastern India, Burma, Thailand, Vietnam, and Malaysia.

The seasonally dry tropics contrast with the humid tropics, the second major subarea (Hutterer 1983), which comprises the east coasts of China, Vietnam, the Philippines, Malaysia, and the islands of Melanesia and Polynesia. Temperatures are high, and rainfall (averaging 1,000 millimeters per month) exceeds evaporation throughout the year, so there is no real dry season. Topography varies from coastal strips to relatively high mountains, but there are no great river valleys as there are in the drier areas. Because of the heavy rainfall, the vegetation is exceedingly lush evergreen rain forest or jungle, often with a high canopy that is rich in animal species.

Since a large percentage of nuts and fruit as well as many small animals are found in the canopy, the humid tropics or rainfall tropical subareas do not have the great food potential of the drier subarea; however, root crops abound, as do marine resources. Among the host of potentially domesticable plants (see table 9.1) the mainland has aroid tuber, banana, Job's tears, rice, sago, taro, ape elephant ear, bamboo, black pepper, cardamon, cloves, ginger root, nutmeg, water chestnut, and tumeric. The islands have Tahitian chestnut, kava, sweet potato, breadfruit, many species of citrus, coconut palm, durian, and rambutan. In addition, numerous plants are domesticated for their fibers, such as cordyline and manila hemp. Archaeological sequences come from Malaysia and its islands of Melanesia and Polynesia.

A subdivision associated with the humid tropics is one best described as swamps, and some of those in Borneo, Ncw Guinea, Melanesia, and the mainland are huge. While rainfall and temperature in the swamps are much the same as in the humid tropics, swamp topography is flat and watery. Although lush, the vegetation is not as dense as in the humid tropics; it has fewer species, and rarely a high canopy. Yet some of the root crops categorized as potentially domesticable for the humid tropics may, in fact, be native to the swamps. As might be expected, however, little archaeology has been done in this soggy ecozone.

A different situation exists in northwest India and Pakistan, a subarea that probably has the most complete archaeological sequence in all of southeastern Asia. The subarea is north of the Tropic of Cancer; it is also very dry, and the vegetation is basically xerophytic. The land is hilly and dominated by river valleys, including that of the Indus. This subarea could easily be considered either an eastern subdivision of the Near East Center or a western extension of the Southeast Asia non-Center. The occurrence of rice in northwest India and Pakistan and the lack of local domesticates, however, have caused me to include this subarea in the southeastern Asia non-Center.

RELEVANT ARCHAEOLOGICAL DATA

Now let us examine the pertinent archaeological sequences in southeastern Asia, starting with this northwestern subarea in India and Pakistan (see table 9.2). The northwest has probably the longest and most complete sequence of the southeastern Asia non-Center, due in large part to its long history of archaeological investigation, including the famous excavations of Sir Mortimer Wheeler at Mohenjo-Daro (Wheeler 1959, 1968). Some Early and Middle Paleolithic remains, either choppers and/or Acheulian hand axes, have been reported for western India, and a definite Upper Paleolithic "blade-and-burin complex" comes from Rajasthan and Gujarat, which B. Allchin, A. Goudie, and K. Hedge (1978) believe could be put in the period roughly from twenty thousand to ten thousand years ago. In terms of my model these remains might represent early Hunting-Collecting Bands—system A.

Following these remains would be microlithic tools, dated about 4500 B.C., at the Bagor site in Rajasthan. This so-called Mesolithic site is distinctive in having

Table 9.2. Relevant archaeological sequences in southeastern Asia

Northwest India and Pakistan	Tropical India	Southeast Asia	Malaysia	Melanesia	Polynesia	Dates
					Vaitootia	
						1000 A.D.
					Tonga	
						0
						1000 B.C.
Pirak	Gondhara					
				Lapita Pottery		
		Dong Dau				
		Non Nok Tha	Lie Siri			
						2000 B.C.
Harappan	Chirand	Banyan Cave				
			Uai Bobo			
		Ban Chiang				
						3000 B.C.
	Koldihawa		Ulu Leang			
Nehrgarh						
						4000 B.C.
				Balif		
		Tham Pa Chan				
Bagor	Sarai					
						5000 B.C.
	Nahar					
						6000 B.C.
	Rai			Kuk		7000 B.C.
		Spirit Cave				
		Hoabinhian				more than 8000 B.C.
	Blade-and-burin complex					
Gujarat						

floors of schist slabs bracing the walls of some sort of "circular earthen structure" (J. Jacobson 1979). Thus, we have the remains of a hamlet or village, and its grinding stones and microlithic tools suggest its people were Foraging Villagers—system C1 of my model. Although no plant remains have been found, the possibility exists that the people were Horticultural Villagers—system C2. Reputedly there is evidence for a "herding economy," suggesting that the latter may be the more correct reconstruction, but the evidence is far from secure (Allchin 1969). Just to the north, in Baluchistan, there are "aceramic Neolithic" sites at Gumla and Rana Ghundai, dating roughly from 4000 to 3000 B.C., as well as ceramic Neolithic sites, roughly from 3500 or 3000 B.C. to 2300 B.C. (J. Jacobson 1979). On the basis of the size of villages and some tools, these people are assumed to have had agriculture. If so, they represent village agriculture—system E—but let me caution that there are no plant remains that yield direct evidence of domesticated plants, let alone of agriculture. These people, then, were probably Horticultural Villagers—system C2. Possible evidence that the Harappa-phase Neolithic peoples did have agriculture is the definite plant remains of six-row barley, six-row wheat, field peas, sorghum, mustard, millet, sesame, lentils, date palms, and chick peas found at Chirand and other sites. Rice also occurs at about 1800 B.C. at Ahar and Atranjikhera. As has been noted by Vishnu-Mittre (1977), all of these plants were domesticated and imported into this subarea from elsewhere. In terms of cultural evolution, it also should be noted that these Harappan remains occur well after the village agriculture stage, for there is considerable evidence that Mohenjo-Daro and similar sites were cities and possibly capitals of states (Wheeler 1968).

One might speculate that this region was a non-Center of secondary development with the blade-and-burin Upper Paleolithic remains representing Hunting-Collecting Bands—system A—and the microlithic cultures, Affluent Foraging Bands—system C. These in turn became the Bagor Foraging Villagers—system C1—who evolved into the aceramic Neolithic Horticultural Villagers—system C2—and thence to the Neolithic Agricultural Villagers—system E—with most of the plants being imported. A case could be made that this region has the necessary conditions for my secondary developments. Its oasis and Indus River Valley locations are lush and have various ecological zones that can be exploited from a strategically located base. They seem circumscribed by deserts, and the region is next to a Center—the Near East—with which it obviously interacted. I must caution, however, that not only are there no plant remains to confirm this hypothetical development, but also the basic ecological studies have not been undertaken to determine if the necessary conditions of northwest India either exist or pertain to the model. Needless to say, any consideration of sufficient conditions at this stage is out of the question until much well-planned archaeological research is undertaken. Future evidence might well show there was a tertiary, rather than secondary, development or even a whole other kind of development. About all we can be sure of now is that primary development did not occur here.

From this discouraging region let us turn to the rest of India, which H. D. Sankalia, the doyen of India's prehistory, characterized as being where European prehistory "was in 1860" (Sankalia 1974).

This more easterly part of India is (or was) in the seasonal wet-dry tropics and once had a varied dry tropical jungle vegetation as well as major rivers—including the Ganges—running through it. There were then all the abundant food resources of such a habitat.

Early and Middle Paleolithic remains in the subarea are again characterized by pebble chopping tools and (late?) Acheulian hand axes. Late Paleolithic blade-and-burin complexes dating between twenty thousand and ten thousand years ago come from two caves in Bhimbetka and in Muchchatla as well as Chintamanu Gavi Cave in Andhra Pradesh state (J. Jacobson 1979). Whether these people were Hunting-Collecting Bands—system A—or Foraging Bands—system C or D—is difficult to determine, but M. L. K. Murty (1979) and others differentiate three groups in this complex, suggesting that it might be one of the latter.

The microlithic tools of the Mesolithic, 8000 to 5000 B.C., are well represented at rock-shelters and open sites in central India and coastal sites in Tamil. The most important is Sarai Nahar Rai on an oxbow lake of the Ganges in Uttar Pradesh state in north-central India (J. Jacobson 1979). This site, dated at roughly 8300 B.C., was a large village of beehive-type pole structures. Tools and wild goat and/or sheep bones indicate that the people were probably sedentary foragers. If so, they were Foraging Villagers—my system C1—but they may also have had domesticates and practiced horticulture. At present it appears that both foraging bands and villages existed side by side in central India during the Mesolithic.

Exactly when village horticulture—system C2—or agriculture began is difficult to determine. The Neolithic village of Koldihawa on the Belan River, a branch of the Ganges, had impressions of possible domesticated and wild rice on some pottery dated to 6570 to 4500 B.C. (Lone, Khan, and Buth 1986; Buth, Lone, and Khan 1986). Koldihawa is apparently related to other Belan sites—pottery from Mahagara and Panchoh—that may date from 4500 to 3500 B.C. Whether these represent village agriculture or village horticulture has not been demonstrated, but the nearby Chirand site, dated from 3500 to 2600 B.C., definitely represents village agriculture—system E. In addition to house structures and pottery, this site had remains of wheat, barley, peas, rice, lentils, and sorghum.

The question is, Do we have secondary development, tertiary development, or some other kind? The ecological factors that might be necessary conditions suggest a tertiary development, for many of the site locations along the Ganges drainage are lush and not circumscribed, with food resources abundant in all seasons. The relative uniformity of the riverine region, however, suggests a secondary development. As we have shown, the archaeological sequences can be interpreted in numerous ways, and it is difficult to tell whether the people were Efficient or Affluent Foragers or Semisedentary Bands with Domesticates or Foraging Villagers. It will take years of solid research before we clarify this situation and can either test my models or make new ones.

At a similar stage of investigation (Jennings 1979; Bellwood 1979), but at the opposite end of southeastern Asia, is Polynesia. Although Polynesia is in the humid tropics, its island climate gives it many unique features, including great ecological diversity, a number of potentially domesticable native plants, and abundant marine food resources. Its cultural history is also unique, since humans and all other life forms invaded these island regions from elsewhere in southeastern Asia (see table 9.2). These migrations were relatively late, thus the subarea has neither Paleolithic nor Mesolithic remains—system A and perhaps C and D as well. The earliest remains occur about 1500 B.C. in the western part of the subarea under the guise of invaders with Lapita pottery, a village way of life, and possibly agriculture or horticulture—either system E or system C2 (Green 1976).

Although it is often hypothesized that these people brought with them from Melanesia banana, sugar cane, pandanus, plantain, and breadfruit—the basis for an agricultural way of life (system E)—not a single plant remain has been uncovered to back up this speculation. In fact, recent research in the Marquesan Islands (which may have been invaded first) suggests that the invaders may have been Foraging Villagers with or without horticulture. This way of life then could have developed into village agriculture in Polynesia. At the Vaitootia site in the Society Islands we do have the remains of coconut, pandanus, kava, and gourd (table 9.3) in deposits from A.D. 600 to 1200 (Kirch 1979). However, the crucial sequential archaeological data on agriculture are too meager to attempt to determine whether we have secondary, tertiary, or some other kind of development. Determining necessary or sufficient conditions or testing models about the evolution of village agriculture must await the results of future investigations.

Much the same may be said of Melanesia, although some new and startling finds have been made in the highlands of New Guinea by the Australians. Stone tools, often choppers, suggest people entered this subarea between fifty thousand and thirty thousand years ago, probably when much of it was connected to the mainland of southeastern Asia. The early "Paleolithic" complexes last up until 11,000 to 8000 B.C. and may well represent Hunting-Collecting Bands—system A. Between 4000 and 5000 B.C. and lasting up until the first pottery, about 3000 to 2000 B.C., there were large sites—possibly villages—or semisedentary forager base camps. The ridged fields at sites such as Kuk (Golson and Hughes 1976) suggest that horticulture or root crops and/or the use of domesticates, perhaps taro, came from southeastern Asia.

The bones of domesticated pigs, occuring near the end of this time period, somewhat confirm this tentative evidence of domestication of root crops and horticulture by villagers. It has been speculated that expanding population in this period of Neolithic village horticulture á la Ester Boserup (Boserup 1965) led to village agriculture in the next period (Bellwood 1979). It is thought that the first agricultural village began about 2000 to 1500 B.C. with Lapita-like pottery and agriculture based on bananas, sugar cane, pandanus, plantain, and breadfruit.

This scenario fits my tertiary development model extremely well: Hunting-Collecting Bands—system A—in the Paleolithic before 11,000 B.C. led to Efficient Foraging Bands—System D—from 11,000 to 5000 B.C., which developed into Semisedentary Bands with Domesticates—system D1—about 4000 B.C. By 3000 B.C. the people of New Guinea had become Horticultural Villagers—system C2—using taro (and pigs) imported from southeastern Asia. They evolved into Village Agriculturists—system E—at about 2000 B.C. Unfortunately, there is not one single preserved plant remain to confirm this elaborate hypothesis. Obviously this speculative data cannot be used to confirm my model, for doing so would be testing a hypothesis with a hypothesis.

It is equally difficult to test my causative model of tertiary developments with this shaky data, although some of the necessary conditions were certainly present in New Guinea. Its many mountain valleys give it (1) great regional diversity, although as a whole it has the areal uniformity of the humid tropics. Its mountain valleys are (2) extremely lush and exploitable from a base; the wide varieties of plants and animals are certainly not in circumscribed locations. New Guinea also has (3) evidence of contact with southeastern Asia and ultimately China, which earlier possessed cultivars and domesticates as well as practices of horticulture and agriculture. The archaeological sequence from 8000 to 2000 B.C., furthermore, suggests that some of the sufficient conditions were also present: (1) intensification of wild-food procurement and (2) resource specialization, indicated by domesticated pigs and ridged fields and possible evidence of population pressures. Thus far, intensive studies of the subarea have not been sufficient to change this interesting fiction into fact so that we can determine whether the other conditions pertain.

Moving westward to Malaysia, we find that Ian Glover (1979) and others are turning up more and more relevant materials. As we are all well aware, this region has a long Paleolithic history going back to the Java Ape Man, and before that, it is characterized by early chopping tools. The later Paleolithic periods (20,000 to 10,000 years ago), also characterized by choppers, could well represent my system A, Hunting-Collecting Bands. Recently, preceramic remains much like Hoabinhian have been found in the cave excavation of Ulu Leang in the Celebes as well as at the Lie Siri and Uai Bobo sites in Timor (Glover 1979). Dating between ten thousand and five thousand years old, these are a continuation of the chopper tradition. The two shelters on Timor are of particular interest, for with the preceramic were reputed remains of domesticated betel nut and betel leaf, seeds of Job's tears, and such wild plants as hackberry and candlenut (table 9.3). These could well be indications of some sort of foragers with domesticates, but whether they were my system D1 or Village Horticulturists—system C2—cannot be determined. In addition to ceramics in the upper levels of these caves there were still other domesticated plant remains—gourds, bamboo, Polynesia water chestnut, and millet. Rice is also reported for the upper levels of Ulu Leang. These remains fit a general Neolithic pattern and could represent either Village Agriculturists (system E) or Village Horticulturists (system C2). While these intriguing finds indicate the great potential of Malaysia, it is presently premature to attempt to connect it with my models.

Perhaps this qualification should also be made of the final region: the mainland of southeastern Asia—Burma, Thailand, Cambodia, Laos, and Vietnam—although much more work has been undertaken in this subarea (see table 9.3). The finds from Thailand, particularly Spirit Cave, have received much publicity relevant to early agriculture (Solheim 1969, 1972; Gorman 1969, 1970, 1977; Bayard 1970). Like Malaysia, the southeastern Asian mainland has a long Paleolithic sequence of chopping-chopper tools. The final part of this tradition, roughly from 25,000 to 18,000 years ago, may represent Hunting-Collecting Bands—system A.

Between 16,000 and 7000 B.C. existed the Hoabinhian culture, characterized by flake artifacts struck from pebbles, pebble choppers, pebble grinding tools, and slate knives (Gorman 1969). These tools, plus the wide variety of bones of large and small animals and some plant remains, suggest to Solheim and Gorman that these people were "broad-spectrum hunting and gathering foragers" (Gorman 1977). A number of open sites—Gua Kechil, Sai Yok, Laang Spean, Padah Lin (dated about 11,450 B.C.), and others—as well as Ongba Cave (about 7310 B.C.)—suggest they lived in seasonal bands and had a broad-spectrum procurement system (Hutterer 1983). In terms of my model these people could have been Efficient Foragers—system D. In the upper levels of Ongba Cave, above the Hoabinhian remains, were stratified deposits with cord-marked ceramics dated roughly 7000 to 5500 B.C. Even more important were the excavations in Spirit Cave, which had cord-marked pottery in its lower levels dating from 9500 to 5500 B.C. (Gorman, 1970). Hutterer (1983, 190) lists plant remains associated with this ceramic complex (table 9.3):

> candle nut (*Aleurites*), hackberry (*Celtis*), castor oil (*Ricinus*), canarium nut (*Canarium*), butter nut (*Madhuca*), almond (*Prunus*), terminalia nut (*Terminalia*), and a chestnut-like nut (*Castanopsis*): several vegetables, among them cucumber (*Cucumis*), bottle gourd (*Lagenaria sic.*), water chestnut (*Trapa*), possibly pea (*Pisum*), bean (*Phaseolus*) or soy bean (*Glycine*), [bamboo] broad bean (*Vicia* or *Phaseolus*), and bitter melon (*Momordica*), lotus (*Nelumbium*), [and] melon (*Trichosanthes*) or gourd (*Luffa*): as well as mild narcotics or condiments such as betel nut (*Areca*) and betel leaf (*Piper*).

Many of these may be remains of wild plants (Yen 1977), but the gourd, betel nut, betel leaf, and water chestnut were probably domesticates, hence the occupants of Spirit Cave, who had pottery, can be classified as Semisedentary Bands with Domesticates—my system D1. Furthermore, from Banyan Valley Cave, dated at 3500 B.C., and Tham Pa Chan Cave, dated 5500 to 3500 B.C., mango and rice were found with similar pottery (Yen 1977). Remains from some open sites, such as the upper levels of Gua Kechil (II and III), suggest that at this stage or shortly thereafter (at about 4000 B.C.) some people had settled down in villages, which would make them Horticultural Villagers—my system C2.

Following this stage would be that represented by materials from Non Nok Tha hamlet and Ban Chiang village, roughly 4000 to 2000 B.C., which had painted pottery, bronzes, and—most important—definite evidence of rice (Yen 1982), although it was perhaps "intermediate between the wild rice and the weed rice" (T. T. Chang 1976, 146). Other evidence of rice domestication comes from caves in the Banyan Valley, which has deposits that date from 3500 B.C. to 700 A.D. as well as from Khok Phanom DI dating from 4000 to 1350 B.C. (Higham 1988). This evidence probably represents the beginning of village agriculture—my system E—in Thailand. Some, however, think that this stage represents village hor-

Table 9.3. Sequence of cultivated and/or domesticated plants in southeastern Asia

Northwest India	Tropical India	Southeast Asia	Malaysia	Melanesia	Polynesia	Range of dates	breadfruit	plantain	sugarcane	6-row wheat	barley	peas	rice	chickpeas	sorghum	millet (*penisetum*)	millet (*eleusine*)	brassica	sesame	cotton
					Historic		x	x	x											
					Vaitootia	600 1200 A.D.														
					Tonga	0														
				Lapita		1500 B.C. 0	?	?	?											
			Dong Dau			1000 2000 B.C.						x								
		Banyan Cave				1000 3500 B.C.						x								
	Koldihawa					1500 4500 B.C.							x							
Harappan						1600 2300 B.C.				x	x	x			x			x	x	x
	Oriyup					1600 2000 B.C.							x							
	Baldipur					1600 2000 B.C.				?			x							
	Singhblan					1600 2000 B.C.				?			x							
	Hallir					1800 2000 B.C.											x			
Ahar						1800 2000 B.C.							x		x	x				
Atranjikhera						1500 2000 B.C.				x	x		x							
Chenbu Daro						1750 2500 B.C.				x	x									
Mohenjo Daro						1750 2500 B.C.				x	x									
Harappa						1750 2500 B.C.				x	x									
	Chirand					1800 3500 B.C.				x	x	x	x	x	x					
Burzahom						1800 2500 B.C.				x	x									
Kalibogan						2000 3000 B.C.					x									
Rangpur						2000 3000 B.C.							x							
Lothal						2300 3000 B.C.							x							
			Lie Siri			2500 3500 B.C.														
			Uai Bobo			2500 3500 B.C.														
			Uai Bobo			3000 4000 B.C.														
		Tham Pa Chan				3500 5500 B.C.							x							
			Ulu Leang			3500 4500 B.C.							x							
	Belan River - Holdihawa					4000 5000 B.C.							x							
		Spirit Cave				5500 9500 B.C.							x							

ticulture, with village agriculture of wet rice not beginning until the period from 2000 B.C. to the time of Christ (T. T. Chang 1976). Others place rice cultivation even later, in the Iron Age, starting at 700 B.C. at the Lopburi Artillery site, when people moved out of the piedmont down into the plains and river valleys.

Needless to say, these materials have created considerable controversy, particularly concerning the possibility that domesticates such as rice appear earlier in Thailand than in China. As indicated in chapter 5, however, the dates for rice are older in South China than in Thailand. In my opinion, therefore, Thailand is a non-Center rather than a Center, and the diffusion into southeastern Asia was from north to south. Lest these conclusions appear too definite, let me point out that analysis of the materials is not complete, and that field work is continuing. Never-

date palm	hackberry	candle nut	canorium nut	butter nut	almond	tominalia nut	chestnut	cucumber	gourd	water chestnut	oca bean	soybean/castor bean	bitter melon	sweet melon	lotus	mango	betel nut	pepper betel	job's tears	polychestnut	bamboo	foxtail millet	coconut	pandamus	kava	taro	banana
																							x	x	x	x	x
																							x	x	x		
																							?	?	?	x	
																							?	?	?	?	?
																x											
x																											
									x											x	x	x					
									x											x	x	x					
	x	x															x	x	x								
	x	x	x	x	x	x	x	x	x		x	x	x	x	x	x											

theless, the Thai sequences conform closely to my tertiary development model: i.e., from Hunting-Collecting Bands to Efficient Foraging Bands to Semisedentary Bands with Domesticates to Horticultural Villagers to Agricultural Villagers.

THEORETICAL CONSIDERATIONS

Preliminary studies of the ecology of southeastern Asia (Solheim and Bayard 1973) suggest that many of the necessary conditions of the tertiary developmental model pertain. There was considerable regional diversity of ecozones: lush river bottom, rolling plains, open bush or savanna, high knolls with trees, narrow foothills with dense vegetation, slopes with dense bamboo vegetation, and mountaintops with tropical jungle canopy. While the seasonal dry tropics were relatively similar throughout the region, they did include several different ecozones—humid tropics with little seasonality, swamps, the tropical coasts, and the more arid zones of Pakistan. Also, the wet season in these tropics does not diminish the food resources, which are available in all seasons. This area also had some lush ecozones (the river valleys) that were not circumscribed. The presence of rice, water chestnuts, soybeans, gourds, and other plants indicates casual and/or direct exchanges with an adjacent Center (China) that possessed cultivars and domesticates and had agriculture earlier.

One necessary condition for tertiary development is slightly different from that of the temperate culture area. Like other tropical regions, southeastern Asia has many more plants that were domesticated than cultivated, although—like the temperate zones—these plants were more cultivated than domesticated. At the present stage of research, it is difficult to comprehend the significance of this difference. Should I modify the necessary condition for tertiary developments regarding potential domesticates to read: "a few potentially domesticable plants but many potential cultivars"? Or is this a necessary condition for still another model that explains agricultural development in the tropics? As of now I cannot answer these questions, but I suspect the model needs to be modified rather than replaced.

Now, what about the sufficient conditions, the triggering causes for change? (See table 9.4.) Although I cannot give any final answers for southeastern Asia, some sufficient conditions of the tertiary development model do seem to pertain. The development of agriculture begins near the end of the Pleistocene, some fifteen thousand years ago, with the final chopping-chopper complexes and/or earliest Hoabinhian Hunting-Collecting Bands—system A. The first period of development occurs at the time of the extinction of the Pleistocene megafauna and general climatic amelioration. "The southeastern Asian climatic regimen, however, seems to have been only slightly affected; a more serious environmental change—keyed to the terminal Pleistocene/Early Recent eustatic rise in sea level—was the reduction of the land mass of southeastern Asia to approximately one-half of its late Pleistocene size" (Gorman 1977, 349).

Obviously, this change not only brought about some extinctions but also redistributed fauna and upset the equilibrium of the biomass. Contemporaneous with this change were new subsistence patterns: "These patterns can best be described as broad-spectrum hunting, fishing, and gathering of flora and fauna that indicate the [seasonal] exploitation of a number of vertically stratified ecozones" (Gorman 1977, 341). Whether these activities acted in a positive-feedback relationship "remains to be researched" (Gorman 1977, 349), but it seems extremely likely that they were the causes of the development of an efficient foraging system—my system D—in middle and/or late Hoabinhian times, about twelve thousand years ago. This primary tropical forest efficiency continued and led to more resource specialization, as evidenced by hoes for root digging, ground-stone knives for cutting cereal grasses, and axes for felling trees. Accompanying resource specialization was the importation of a few domesticates: water chestnuts, gourds, soybeans, and, later, rice, indirectly and/or casually from the Chinese Center. These new crops, in turn, required clearing of the forest for gardens.

Table 9.4. Necessary and sufficient conditions as a positive-feedback process for tertiary development in Southeast Asia

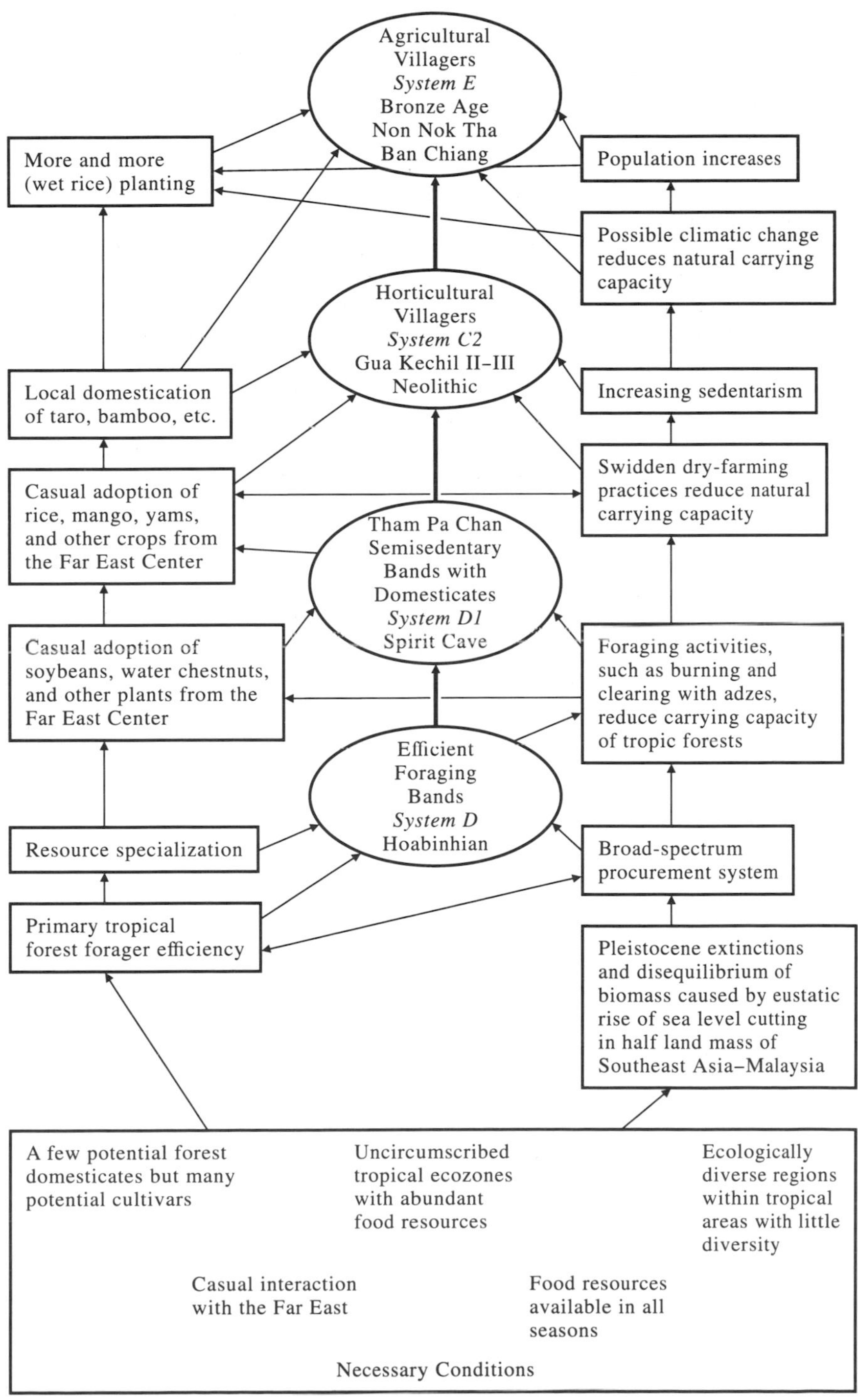

Some of these subsistence activities may have reduced the carrying capacities of the tropical forest, which has a delicate ecological balance. All these factors, perhaps acting in a positive-feedback cycle, led to the development of Semisedentary Bands with Domesticates—system D1—by about 5000 to 3500 B.C. As more domesticates were being imported, such as rice for dry planting, some local domestication of taro, bamboo, and other plants occurred. Meanwhile, the carrying capacity of the forest steadily diminished as forests were cleared for collecting and then for swidden horticulture with its slash-and-burn techniques. All this led to the slow development of horticultural villages—system C2—at Gua Kechil and Non Nok Tha by about 3000 to 2000 B.C.

These developments, however, are not documented in the archaeological records and much "remains to be researched" (Gorman 1977, 349). This is equally true for the next stage, the development to village agriculture. However, signs appear of population increases and more intensive wet-rice planting, along with the continuation of some of the other causes that seem to have been driving forces. Whether these changes were natural ones that reduced the carrying capacity, as in some other tertiary developments, cannot presently be determined, but the remainder, with varying degrees of archaeological documentation, do seem to pertain.

Thus my tertiary development model can be partially tested in the Southeast Asia area of the southeastern Asia non-Center. There are hints that similar developments occurred in Malaysia and Melanesia, but the data are insufficient for testing hypotheses about causation. To the west, in India and Pakistan, hints of secondary developments appear, but again, the data are insufficient to test hypotheses. The same may be said for the next area: the New World tropics.

CHAPTER 10

The New World Tropics

Studies of the New World tropics suffer from many of the problems that plague endeavors in southeastern Asia. Obviously, the most troublesome is that in neither area is there good preservation of plant remains, especially of root crops. Consequently, information about domesticated crops has been derived indirectly, for example, from studies of tool types. One problem with this procedure is that tools—griddles in the New World or hoes in Asia—can be used for a variety of tasks other than cultivating specific plants such as manioc or taro. Another problem is that only recently have investigations in either area employed interdisciplinary techniques—for example, flotation analysis, studies of carbon 12-13 and nitrogen-15 isotopes, raciation studies to determine amino acids in diets, research on phytoliths and pollen—to derive data about ancient diets or plant remains. Fortunately, some recent interdisciplinary studies (Piperno 1988; Roosevelt 1980; Stothert 1985; Linares and Ranere 1980) in this tropical area of the New World may yield reliable data from plants that can be used to reconstruct subsistence systems more accurately.

A BRIEF HISTORY OF SPECULATIONS ON THE PROBLEM

Unfortunately, research on the problems of origins of agriculture in the tropics has long been colored by speculation based upon the geographical distributions of plants. In the period roughly from 1920 to 1930, the most intensive use of this approach was by Ivan Vavilov and his team of Russian scientists (Vavilov 1951). On the basis of the numbers and kinds of cultivated plants and their geographical frequencies, Vavilov and his colleagues assumed that plant domestication developed in areas with a wide diversity of plants. They decided that one major center of diversity was in South America, which could be divided into Andean and Brazilian-Paraguayan subcenters. (I include the latter in the non-Center of the New World tropics.) Their assumption was that, because of the numerous domesticated plants (about forty-five), domestication had taken place first in the tropics and then spread to other regions. Further, since a number of the crops of this "Center" were root crops—manioc, sweet potatoes, and the like—the Brazil-Paraguay area became the "New World Root Crop Center."

The American geographer Carl Sauer derived a similar hypothesis. His opinion was that, while southeastern Asia was the primary cradle of agriculture, a "secondary center" occurred in northwestern South America. This theory was based upon the following assumptions (Sauer 1936, 19–28):

1. Early people in the lush tropical forest of northwestern South America "lived at a comfortable margin about the level of wants" and thus had the leisure time for the "improvement of plants by selecting for better utility to man."
2. Agriculture began in wooded lands like the South American tropics.
3. The "inventors of agriculture had previously acquired special skills in other directions that predisposed them to agricultural experiments."
4. "The hearths of domestication are to be sought in areas of marked diversity of plants or animals, where there were varied and good raw materials to experiment with, or in other words, where there was a large reservoir of genes to be sorted out and recombined" (this condition implies well-diversified ecozones and perhaps a variety of climates).
5. Finally, "above all, the founders of agriculture were sedentary folk," for "growing crops requires constant attention. . . . what is food for man is feasts for beasts; and therefore, by day and night someone must drive off the unbidden wild guests. Planting a field and then leaving it until the harvest would mean loss of harvest." Sauer believed that such conditions existed in northwestern South America in ancient times and that was why agriculture first began there.

In terms of my hypothesis we may say that Sauer believed that the first farmers were originally sedentary affluent foragers and fishermen living in the tropics in wooded areas of marked plant diversity (some of which, such as root crops, were susceptible to domestication). The foragers' subsistence pattern was so successful that they had leisure time to experiment with plants, which led to plant domestications and agriculture. In archaeological terms this development would mean that in the tropical forest we should find villages of Affluent Foragers (system C) who quickly became sedentary root crop Agriculturists (system E), the earliest farmers on the South American continent. From this tropical agricultural Center, seed crop agriculture then spread to other areas such as the Andes and Mesoamerica. No archaeological data supported this senario when Sauer conceived it in the 1930s, and little concrete evidence has been uncovered since to confirm it, whereas much has been found to refute it.

Nevertheless, Sauer's hypothesis remains popular (Lathrap 1970), although many strenuously object to it (Mangelsdorf 1952). Perhaps one of the first to object was Julian Steward, who led the field in the period from about 1936 to the 1950s (Steward 1955). Steward's idea of cultural ecology provided the basic premise upon which the South American archaeological research projects of the 1930s and 1940s were founded, and the *Handbook of South American Indians* was built upon his idea (Steward 1946). Steward's core concept relevant to New World agriculture was that seed crops and irrigated agriculture evolved in the desert areas of Peru and diffused to the rest of South America. Andean potato root cropping thus gave rise to tropical manioc root cropping, not vice versa. Based on this were several great archaeological programs such as the Virú Valley program in Peru. From 1935 to 1950 many famous archaeologists, among them Julio Tello, Duncan Strong, Wendell Bennett, Jim Ford, Cliff Evans, John Murra, and Gordon Willey, dug up material in the Andes that tended to confirm Steward's hypothesis. During this period, however, little was happening in the tropics, and neither Sauer's nor Steward's hypothesis was being tested by data from that area, although some preliminary excavations were made in Venezuela by Alfred Kidder II (1944), George Howard (1943), Mario Sanoja (1963), Cornelius Osgood (1943), and Wendell Bennett (1937).

Research in the New World tropics began in earnest in the period from 1950 to about 1965. Two of the most energetic investigators were Cliff Evans and Betty Meggers. They undertook major excavation programs—five sites in Venezuela's Orinoco region (Evans, Meggers, and Cruxent 1959), the Guiana region (Evans and Meggers 1960), two regions in Brazil (Meggers and Evans 1957, 1973), and Valdivia in Ecuador (Meggers, Evans and Estrada 1965; Evans and Meggers 1968). In addition, they trained many students and sponsored and assisted anthropological programs all over South America. During these years they amassed great amounts of data, established a number of fine basic cultural sequences (mainly of the ceramic period), and made some interesting speculations about trans-Pacific contacts. Their theoretical contributions to the problems of origins of agriculture in the tropics, however, differed little from those of Julian Steward (Steward 1946). Following an environmental determinist approach, they looked on the tropics as areas of low agricultural potential with little chance of technological improvement. To them, the initial agriculture and later chiefdoms at Marajó Island in Brazil were basically importations from the nuclear Andean region (Meggers and Evans 1957). As we shall see, major reactions against these views came from R. L. Carneiro (1974), Donald Lathrap (1970), A. J. Ranere and P. Hansell (1978), and others.

In the final period—from 1965 to the present—the main focus was on collection of archaeological data, and many contributed to this effort. One indefatigable data collector and builder of regional sequences was Ben Rouse, who worked with José Cruxent in Venezuela (Rouse and Cruxent 1963). Their endeavors overlapped with those in the related Antilles subarea (Rouse 1964) by Ricardo Alegria (1965), Ripley Bullen (1962), Froleich Rainey, Gordon Willey, Henry Nicholson (Alegria, Nicholson, and Willey, 1955), Cornelius Osgood (1942), John Goggin, and others. While no new theories came out of these endeavors, the long, well-documented sequences of ceramic traditions, which proved to be as early as any in Peru, suggested that agriculture and village life in Venezuela had a history of their own. Further, early griddles suggested early manioc agriculture, and the huge size of some of the chiefdoms hinted that the potential of the tropics was not as limited as previously thought.

Before going further, I must interject a cautionary comment about an assumption that became popular at this time and that continues to plague the interpretation of early agriculture in the New World tropics. Simply stated, this assumption is that the presence of griddles automatically means that the ancients grated manioc, with the concomitant conclusion that they grew only domesticated manioc. Lathrap's conclusions about subsistence in the tropics of eastern Peru (Lathrap 1958) were based upon this false assumption and led in the next period to a revival of Sauer's original hypothesis as well as other "novel" ideas (Lathrap 1977). As Olga Linares (1979) notes, the matter is not that simple. In an exhaustive comparison of artifacts classified in museums as manioc griddles (*budares*) or as maize-cooking griddles (*comales*), De Boer (1975) concludes that "it is not possible to distinguish between the two categories on any criteria such as shape, size or form" (Linares 1979, 33). The presence of griddles could indicate the grinding of corn or other plants in addition to indicating manioc grinding; griddles therefore do not equal manioc agriculture.

The building of basic ceramic sequences in Venezuela was duplicated in Colombia by Gerado Reichel-Dolmatoff (1965), Jim Ford (1944), Wendell Bennett (1944), and others, while Gordon Willey and Bob McGimsey (1954), John Ladd (1964), Bob Greengo, and others were doing the same in Panama. In Panama they established a four-thousand-year ceramic sequence and found preceramic remains dated at about 4800 B.C. at Cerro Mangote (McGimsey 1958). These endeavors established the chronological framework for the ceramic periods of the New World tropics.

The final period has brought new interpretations and new controversies (Binford 1962). A fundamental principle of this period grew out of the tropical African studies of the geographer Ester Boserup, who pointed out that "population growth is . . . the independent variable which in turn is a major factor determining agricultural development" (Boserup 1965, 11). Mark Cohen elaborated this principle to explain the beginning of agriculture in Peru (Cohen 1977a), and Marvin Harris, a cultural materialist, was in full agreement (M. Harris 1980). Further, Robert Carneiro (1960)—in opposition to Betty Meggers, Clifford Evans, Julian Steward, and other cultural ecologists—saw the tropics as having some areas of great agricultural potential, which allowed population pressures in circumscribed areas to be prime causes or movers of cultural change. These themes, often unstated and unrecognized, were woven into the thinking of many of the archaeologists of this final period and were compatible with the resurrection of many of Sauer's earlier ideas.

Perhaps the most vociferous of these investigators was Donald Lathrap, who gradually shifted his field endeavors from the selva of eastern Peru to tropical coastal Ecuador in the same area of Valdivia where Betty Meggers and Clifford Evans had worked earlier. In addition to attempting to refute their findings, he added many new data about the period of the first villages with pottery and agriculture (4000–1000 B.C.), and ironically came up with good evidence that beans and corn had preceded root crops in the South American tropics (Lathrap 1967). In spite of this evidence, he and his many students remain firm believers that root crops preceded seed crops in the tropics and that many of Sauer's assumptions pertain to their findings (see Hill 1972; Marcos 1978; Stothert 1977, 1985; Zevallos-Menendez 1971).

Also in agreement with some of these dicta are Anthony Ranere, Olga Linares,

Payson Sheets, Richard McCarty, Patricia Hensell, Richard Cooks, Deloros Piperno, and others in Panama, where major research programs were undertaken during this period (see, e.g., Linares and Ranere 1980). Studies by Alexandra Bartlett, Elsa Barghoorn, and Rainer Berger in the area of the Gatun Locks yielded dated maize pollen (Bartlett, Barghoorn, and Berger 1969), and plant remains from excavation identified by Walton Galinat, C. Earle Smith, and Lawrence Kaplan (in Linares and Ranere 1980) had bearing on the problems of the origin of agriculture in the tropics. These data, when taken in conjunction with early works of Robert McGimsey, Gordon Willey, Samuel Lathrap, Robert Greengo, Cornelius Osgood, Frederick Johnson, Matthew Stirling, and others make Panama one of the best-studied subareas in the New World tropics.

Running a close second is Venezuela, where Cruxent and Rouse were joined by a whole new group, including Erika Wagner, Anna Roosevelt, Mario Sanoja, Elsa Redman, Chuck Spencer, Luisa Vargas, and Arenos Vargas. Of particular significance were Anna Roosevelt's endeavors in the Parmana region of central Venezuela; her monograph (1980) contains an accurate analysis of the problems of agricultural archaeology in the tropics.

An increasing number of locally directed archaeological programs also began in the New World tropics. In Colombia local projects were carried out by Henning Bishoff, Gerado Reichel-Dolmotoff, Urrego Correal, William Bray, and Wesly Hurt. Among those working in Brazil were Clifford Evans, Betty Meggers, the late Annette Laming Empaire, Wesly Hurt, Ruth Gruhn, and Allan Bryan. All in all, the period from 1965 through 1988 saw great progress in uncovering data relevant to the problems of the origins of village agriculture in this area.

Before discussing those data, a word or two is needed about the area itself, for its environment is far from uniform, and developments may not always occur by the same process in all parts of this vast region that travels under the overall rubric of "the tropics."

THE ENVIRONMENT

Generally speaking, the tropics have a humid climate that can be divided into two zones, wet (rain forest) and seasonal wet and dry (savanna). South America also has a few areas of drier tropical steppe—the northwest Pacific Coast of Ecuador, the North gulf coast of Venezuela, and the eastern Brazilian Uplands (see fig. 16). Within the tropical wet zone, a further division into subareas can be made between the small narrow strip on the coast of Colombia and Ecuador, which receives its rainfall and temperature from the Pacific Ocean, and the huge Amazon drainage in Brazil, which receives its weather from the Atlantic. Despite these differences, all this vast area has a hot, frost-free climate and receives considerable rain (over 1,000 millimeters a year) even in the drier tropical steppes.

The topography and physiography further subdivide the region into eight culture subareas. In the west the Andes divide the northwest Pacific coastal plain subareas of Columbia and Ecuador from the tropical rain forest. Central America can be divided into two subareas—rain forest on the Caribbean and savanna along the Pacific. The North gulf coast subarea of Colombia and Venezuela, like Central America, has both wet and seasonally dry areas. The Antilles of the

Figure 16. The environmental life zones and relevant sites in the New World tropics

Caribbean, their tropical climate modified by their island location, form another subarea. The northern savanna subarea includes the Orinoco lowlands, the Guiana Highlands, and the adjacent coasts, all with a seasonally tropical climate. The Amazon Basin, stretching from the foothills of the Andes eastward to the river's huge delta, forms another subarea. East and south of it lie two subareas—the seasonally wet-and-dry climate of the eastern Brazilian Uplands and, farther southeast, the southeastern Brazilian tropical coast, where climate ranges from tropical and wet to a slightly cooler marine subtropical zone. I have further divided the eight subareas into a number of regions because of archaeological investigations. These are:

1. Central American subarea
    a. wet tropical Gulf Coast
    b. seasonal tropical Pacific Coast
2. Northwest Pacific Coast subarea
    a. Colombian coast
    b. Ecuadorean coast
3. North Gulf Coast subarea
    a. Colombian wet tropical coast
    b. Venezuelan dry tropical coast
4. Antilles subarea
5. Northern savanna subarea
    a. Lower Orinoco lowlands
    b. Upper Orinoco lowlands
    c. Guiana Highlands and coasts
6. Amazonia subarea
    a. Western Andean lowland
    b. Middle Amazon
    c. Deltaic Amazon
7. Eastern Brazilian Uplands subarea
8. Southeastern Brazilian tropical coast subarea

As table 10.1 indicates, the variety of ecozones is reflected in the number of plants native to the New World tropics that were susceptible to domestication. I have not attempted to assign any of them to a specific subarea. Many of these plants spread from the tropics into the Andes and Mesoamerica and became staples in those Centers.

RELEVANT ARCHAEOLOGICAL DATA

Although not enough good contextual archaeological information exists to determine the necessary or sufficient conditions that led to village agriculture in the New World tropics, there are some data relevant to our problem. Let me hasten to add that any interpretation of the archaeological information from the New World tropics is fraught with contention, for this realm is filled with controversies, feuds, recriminations, animosities, and strongly held opposing opinions.

We shall start with the areas or subareas with the least significant data, those farthest from where agriculture first occurred in the area, and end up discussing the subareas where we have some solid data for testing my models.

Table 10.1. Some native cultivated and/or domesticated plants of the New World tropics*

Scientific name	Popular name
Acrocomia vinifera	palm
**Anacardium occidentale*	cashew
**Ananas comosus*	pineapple
Annona cherimolia	cherimoya
**Arachis hypogaea*	peanut
**Bertholletia excelsa*	Brazil nut
**Bixa orrellana*	achiote
**Byrsonima crassifolia*	nance
**Calathea allouia*	lairen
Canavalia ensiformis	sword bean
Canavalia plagiosperma	jack bean
Canna edulis	achira
**Capsicum annuum*	chile pepper
**Capsicum baccatum*	pepper
**Capsicum frutescens*	pepper
Carica candicans	papaya
Carica sp.	papaya
**Centrosema pubescens*	centro
Cocos sp.	coconut
**Crescentia cujute*	tree gourd
Cucurbita maxima	squash
**Datura sp.*	stramonium, Jimsonweed
**Desmodium sp.*	tick clover
**Dicscorea trifida*	yam
**Erythroxylon coca*	coca
**Gossypium barbadense*	see island cotton
Hymenaea courbarit	algarrobo (West Indian locust)
**Ilex paraguariensis*	maté
Ilex vomitoria	yaupon
**Inga feuillei*	pacae
Ipomoea batatas	sweet potato
Lageneria siceraria	bottle gourd
**Manihot esculenta*	manioc
**Nicotiana rustica*	tobacco
**Nicotiana tabacum*	tobacco
Pachyrhizus erosus	jícama, yam bean
**Paspalum dilatatum*	dallisgrass
**Paullinia cupana*	guaraná
**Paullinia yoco*	yoco
**Persea americana*	avocado
**Physalis peruvianum*	uchuba (berry)
Scheelia zenensis	palm
**Stylosanthes gracilis*	stylo
**Tripsacum laxam*	Guademala grass
Xanthosoma sagittifolium	xanthosoma

*Domesticates according to Harlan (1971).

Starting with the southeastern Brazilian tropical coast subarea, we have a long sequence—about seven thousand years of the Sambaqui tradition (see table 10.2). It ends at about A.D. 900 with the Una and Aratu sites in the Rio de Janeiro region (Meggers and Evans 1978). These coastal sites were adapted to a marine subsistence pattern. Although pottery developed about 1000 B.C. (Periperi phase) and some of these sites may have been occupied by sedentary groups, there is no evidence of agriculture. Following this Sambaqui tradition (about A.D. 500) Tupiguaraní sites appear with distinctive black and red or white corrugated and brushed pottery (Meggers and Evans 1978). Because of the presence of this pottery, griddles, and other traits affiliated with the Amazon, it is assumed that people with manioc agriculture moved into the southeast. Although there is no hard evidence for this assumption, the sequence might be from early Sambaqui, marine Affluent Foraging Bands—system C—to Periperi, Foraging Villagers—system C1—to Tupiguaraní, Village Horticulturists or Agriculturists. As indicated, the subsistence data pertaining to domesticated plants have not been found nor has the Hunter-Collector period been defined.

The adjacent inland subarea, eastern Brazilian uplands, is drier and has more savanna vegetation, especially with increasing distance from the marine resources (see fig. 16). It has an even longer sequence (see table 10.2), with sites at Rio Claro and the Serranapolis dating before 8000 B.C. and fitting the Hunting-Collecting Bands—system A—of my model (Bryan and Beltrao 1978). Following are the Humita and Umbu lithic traditions that last from at least 5000 B.C. to at least 1000 B.C. Their bola stones, projectile points, anvil stones, and other tools suggest these people were some sort of hunter-gatherers and/or foragers (Meggers and Evans 1978).

These traditions are followed by the Lagoa Santa sites of preceramic foragers, whose way of life may have persisted until A.D. 500, when three ceramic phases appeared: Taquero in the state of Rio Grande do Sul, the Una tradition in Minas Gerais, and the Papeba tradition in the Papeba region. All of these places have large sites, possibly villages, and many have urn burials, but there is no evidence of agriculture until A.D. 1000, during the later Averias phase, which has large village sites said to contain evidence of maize. This suggests that village horticulture and/or agriculture did not arrive in the eastern Brazilian Uplands until that time. Our information is so meager, however, that these data cannot be used to test any of my models.

Much the same can be said of the Amazonian subarea to the north of the subareas just discussed, for although it has some fairly complete sequences for the last four thousand years, not a plant remain has been uncovered. There are no preceramic remains, although Meggers and Evans (1978) mention some sites without ceramics in the southwest portion of the region. On the other hand, some authors, in particular Donald Lathrap, believe this is the major center for manioc domestication, and in fact view it as the initial area of all plant domestication after ocean currents many thousands of years ago carried gourds from Africa that reputedly stimulated domestication (Lathrap 1977). In the light of these speculations, let us examine the Amazonian data, starting with Lathrap's sequences for eastern Peru along the Ucayli River in the upper Amazon region (see table 10.2).

Table 10.2. Relevant archaeological sequences from the New World tropical lowlands

CENTRAL AMERICAN SUBAREA			NORTH GULF COAST SUBAREA		ANTILLES SUBAREA
Northwest Pacific Coast subarea	Panama Pacific Coast region	Panama Caribbean Coast region	Colombia Coast region	Venezuela dry Coast region	
Inca	Veraguas	Chiriquí	Sinu	Valencoid	Chicoid
					Meillacoid
		Aguacate-Cerro Punta		Guayabtoid	
Milagro	Coclé		Zambrano	Terroid	
	St. María Bugaba			Ocumaroid	Ostionoid
					Saladoid
Guangala					
Jambelli	Pueblo Nuevo	Concepción	Momil 2	Tocuyanoid	
				Saladoid	
			Momil 1		
Chorrera			Malambo	Dabajuroid	
Machalilla	Sarique				
			Barlovento		Banwari
			Canapote	Ranch Peludo	
	Monagrillo				
San Pablo					
Real Alto					
Valdivia			Bucareli		
San Pedro					
		Talamanca	Puerto Hormiga	Manicuarie	Ortoire
	Cueva de los Ladrones			El Heneal	
					Loiza
	Cerro Mangote		San Nicolas		
Vegas					
					Couri
				Casitas	
Manantial	Madden Lake				El Jobo

Here Lathrap defined an early Tutishcainyo ceramic complex. Its resemblances to Wairajicra in Huánuco, Peru (which had corn and not manioc agriculture), and radiocarbon dates from 1750 to 1200 B.C., led him to conclude that his complex dated from roughly 2000 to 1000 B.C. (Lathrap 1970). On the basis of some reputed stone axes and a few possible griddle fragments, he also concluded that these people were Villagers with manioc agriculture. Since this stage would be the end of the developmental sequence, none of the earlier stages would be known. Lathrap, however, may not be inviolable, so we shall briefly outline the rest of his sequence.

NORTH SAVANNA SUBAREA		BRAZILIAN AMAZON SUBAREA			EASTERN BRAZIL UPLAND SUBAREA	SOUTHWESTERN BRAZIL TROPICAL COAST SUBAREA	
Venezuela Lower Orinoco region	Venezuela Upper Orinoco region	Brazil Upper Amazon region	Brazil Middle Amazon region	Brazil Lower Amazon region			Dates
	Camoruco III	Shipibo-Catacocha-Arua				Curimatau-Ipaca	
					Calunda		
		Caimito	Napo	Marajoara			A.D. 1000
	Camoruco II						
Arauquinoid		Cashibocana		Formiga		Gratu	
	Camoruco I		Tivacundo		Una		
	Corozal III	Yarinacocha			Taquero		
Barrancoid							
	Corozal II	Hupa-iya		Mangueirus			
	Corozal I						
Saladoid						Periperi	1000 B.C.
	Ronquin	Shakimu		Ananatuba			
			Yasuni				
		Tutishcainyo		Mina		Macedo	
							2000 B.C.
	La Gruta						
					Marchioki		
						Gomes	3000 B.C.
Canaima							
						Maratua	4000 B.C.
					Vierra		
						Umbu	
							5000 B.C.
					Humita		6000 B.C.
					Rio Claro		7000 B.C.

The next phase, Shakimu, has a radiocarbon determination of 650 B.C. (Lathrap 1958). Lathrap believes these people were Villagers with manioc agriculture who were the recipients of influence from the Peruvian highlands, while the earlier Tutishcainyo phase was a donor to Peru. No direct evidence exists for either of these conclusions nor for the following Hupa-iya culture, dating from 200 B.C. to the time of Christ.

Stratigraphically following Hupa-iya is Yarinacocha with a date of A.D. 90. This culture is thought of as Barrancoid, derived from Venezuelan migrants with manioc-based village agriculture. Others believe that village manioc and perhaps corn agriculture were introduced into the region at this time because Barrancoid and related Correal phases of Venezuela really do have concrete evidence of corn if not manioc agriculture (Roosevelt 1980).

Following Yarinacocha are the ceramic phases of Pacacocha, Cashibocana, Nueva Esperanza, Cumancaya, and Caimito, with the last perhaps representing the ceramic complex of the Shipibo tribe of historic times. These people are thought of as manioc Village Agriculturists who moved in and out of the region from 2000 B.C. to A.D. 1500 like the reputed waves of migrants in Mesolithic and Neolithic Europe. There is, however, no exact evidence of their subsistence system or of a series of migrations.

The sequential material from nearby Rio Napo of the Upper Amazon region of the Amazonian subarea is not much better for testing my hypothesis, but it reflects a different interpretation from that of Lathrap (Evans and Meggers 1968). The earliest ceramics, called Yasuni, come from a village site and perhaps date around 1000 B.C. Although the people may have been Foragers, it is not until the Tivacundo phase, about A.D. 510, that griddles occur, leading to the common assumption that they represent manioc preparation. The Napo phase, just after A.D. 1000, had manos and metates, as well as griddles, so it is concluded that its people were Village Agriculturists growing both manioc and maize. The same is believed for the final phase, Catacocha, which may last into historic times (Evans and Meggers 1968).

The other Amazon regional sequence comes from the delta region and is as long as or longer than that from the Ucayli, and new investigations by Anna Roosevelt may radically change it (Roosevelt 1989). Mina, the earliest phase of shell-tempered pottery, has a radiocarbon date earlier than 2000 B.C. (Meggers and Evans 1978), and there are hints of earlier pottery at 4000 B.C. (Roosevelt, personal communication). While the Mina-phase sites are large (perhaps villages), there is no evidence that they had any sort of agriculture either in the delta or in the equally early ceramic Alaka phase site from the northwest coast of Guyana.

Evans and Meggers interpret the Ananatuba phase, about 1400 B.C., as villages with incipient agriculture, supposedly domesticated manioc. However, the pottery does not include griddles, so there is no evidence that the people were anything but Affluent Foragers living in large sedentary villages—my system C. In the early centuries of the Christian era were the Mangueirus, Aeauan, and Formiga ceramic phases, which seem to represent villages or hamlets comprised of communal longhouses. Once again, no indications (not even griddles) exist of the subsistence pattern (Willey 1971).

Marajoara, the final black and red-on-white polychrome pottery phase, is represented by huge sites with habitation mounds and urn burials indicating class differences. It is assumed that these people had intensive manioc and maize agriculture, as did the following Arua phase representing the tribal groups of historic times. Meggers and Evans (1957) believe they were Village Agriculturists, perhaps chiefdoms. If their reconstruction is correct, then the acquisition of full-time agriculture took a long time, and my system E did not reach the deltaic region of the Amazon until A.D. 1000 to 1500, an interpretation diametrically opposed to that of Lathrap.

Since we have drifted down the Amazon to the ocean, let us move on to the cultural development in the Antilles, where changes in subsistence systems are equally poorly understood because of the same lack of preserved plant remains (Alegria 1965). Furthermore, village agriculture did not develop *in situ* in the Antilles but was introduced by migrations of Saladoid peoples from Venezuela.

The earliest dated remains (about 4000 B.C.) come from the Aguas Verdes and other manifestations in Cuba (Kozlowski 1977). These and the Couri-type remains of Haiti have large macroblades, scraper planes, and contracting-stem points that seem related to the earlier Sand Hill and Orange Walk phases of Belize. Mesoamerica may be the region from which people first migrated to the Antilles (MacNeish, in press). Late preceramic remains such as Loiza, Ortoire, and Krum Bay, that may date up until the time of Christ, are characterized by shell and bone tools as well as edge-ground manos (Willey 1971). These traits suggest the Antillean peoples were related to such traditions as El Heneal and Manicuari of coastal Venezuela. It has been suggested that these peoples were seagoing foraging bands who migrated to the Antilles by boat from Venezuela (Rouse and Cruxent 1963).

The next maritime migration to the Antilles, occurring at the beginning of the Christian era, was that of the Saladoid peoples who brought their village agriculture from the mouth of the Orinoco River of Venezuela. In Puerto Rico the Ostinoid subtradition later evolved (A.D. 700) from the Saladoid and spread out to nearby islands. In Haiti and/or Cuba it developed into the Meillacoid subtradition, while to the east it developed after A.D. 1000 into the Chicoid subtradition. Finally, in Trinidad and the Lesser Antilles, the earlier traditions were replaced between A.D. 1000 and 1500 by the Guayabitoid peoples, who shipped into the area from the eastern coast of Venezuela. All these migrants were probably Village Agriculturists—system E—and we have no evidence of how their agriculture developed. To learn this, we must turn to the mainland.

On the coast of Colombia, literally north of the Andes, there is a savanna and xerophytic region with a tropical wet-dry climate (see fig. 16) in the gulf subareas. Although fishtail points have been found here, perhaps representing Hunting-Collecting Bands—my system A—some large nonceramic sites like San Nicolas have scraper planes, turtle-back scrapers, and the heavy lithic materials that perhaps represent the Archaic. The earliest reliable excavated materials, however, dating about 3000 B.C. (Reichel-Dolmatoff 1965), come from Puerto Hormiga (Reichel-Dolmatoff 1961). This stratified shell mound, circular in shape, is assumed to represent a Forager village. Associated with its fiber-temper and sand-temper pottery—some of the earliest dated pottery in the New World—are grinding slabs, large anvil stones or mortars, pebble pestles and/or hammers, as well as crude choppers and scraping tools. It is assumed that these tools were "used for grinding or mashing roots or seeds" (Lathrap 1970), but since domesticated corn was already present in both Panama to the north and Ecuador to the south, the tools may, in fact, have been used to prepare corn instead. New dates of 5300 B.C. on corn pollen from the nearby Calima Valley and of 3200 B.C. in the El Dorado Valley (Rue 1988) tend to confirm this hypothesis.

About 2000 B.C., Canapote probably developed out of this earlier manifestation and continued the same way of life—either Foraging Villagers (system C1) or Foraging Bands without domesticates (system D), or with them (system D1). Also, the presence of grinding stones suggests the possibility of Horticultural Villagers with corn (system C2). Seemingly developing out of this phase was the Malambo ceramic phase (1120 B.C.–A.D. 70), which according to Gerado Reichel-Dolmatoff (1965) had a new way of life—"early horticulturist." The presence of griddles suggests that these sedentary villagers had manioc horticulture (my system C2) or village agriculture (my system E). The following phase, Momil, had both griddles and manos and metates, suggesting a combination of manioc and corn agriculture. Probably a true village agricultural way of life—system E—had evolved in this part of Colombia by the time of Christ. If not, then it certainly had occurred by the Zambrano complex, about A.D. 700, which had house mounds, gold metallurgy, agricultural terrraces, and a number of other traits that suggested to Reichel-Dolmatoff (1965) that the "chiefdom" way of life had developed.

The sequence from possible early Affluent Foraging Bands (system C) to Foraging Villagers (system C1) to Horticultural Villagers (system C2) to Agricultural Villagers (system E) indicates that my secondary development model occurred in this part of Colombia. However, no preserved plant remains have been collected by which adequate estimates could be made of subsistence systems.

Data are only slightly better for the even drier Venezuelan coast east of the mouth of the Orinoco, the north Gulf Coast subarea, which is a tropical or subtropical steppe or desert (see fig. 16). First of all, some definite preceramic remains from this region have been found just east of Lake Maracaibo, yielding El Jobo-type remains at Muaco, Taima-Taima, and Cucurucyu and dating from roughly 14,000 to 7000 B.C. (Cruxent 1967). Projectile points associated with the bones of mastodon, horse, giant sloth, and other extinct animals certainly indicate these peoples were Hunting-Collecting Bands—system A. Hunting continued into Las Casitas times, with contracting-stem points that might date about 5000 B.C. Next came a series of complexes indicating Efficient Foraging Bands—my system D—oriented to exploiting marine resources.

On the central coast, El Heneal, dating between 3800 and 1550 B.C., had crude celts, edge grinders, and bone tools, while the Manicuari complex, on the east coast, with dates from 2325 to 1190 B.C., had milling stones, bone points, and a host of shell tools (Rouse and Cruxent 1963). Following these (and possibly developing out of them) were materials from Rancho Peludo, dating between 1880 and 445 B.C. Among the village materials that were apparently eroding out of the bank of the Rio Guesere were urn burials, pottery of Dabajuroid series, and parts of two griddles. This suggested to Rouse and Cruxent that the people "had begun to cultivate manioc" to a small extent and were perhaps Horticultural Villagers—my system C2. More data are needed for real proof, for the manioc could have been wild and not domesticated, and the griddles could have been used as comales for toasting corn and not manioc. This tradition seems to continue on to historic times with its subsistence system little changed. Metates found with the Guasave phase at the time of Christ may indicate that maize then replaced manioc or that maize was being prepared in a new manner.

Appearing in the area from about 400 B.C. to A.D. 300 was the Tocuyanoid series of various kinds of painted pottery. Like the preceding tradition, these people seem to have been Villagers, and a few metates suggest they used maize and perhaps were Village Agriculturists or Village Horticulturists. The latter stage may not have occurred, however, until the time of Ocumaroid-style pottery from A.D. 300 to 1500. While these materials suggest a secondary development, we lack the crucial direct subsistence data to confirm it (see table 10.2).

Data are better for the seasonally wet-dry savanna subarea of Venezuela (see fig. 16), which is dominated by the drainage of the Orinoco River. We have two good sequences here—one for the delta and the other for the middle Orinoco. The delta area lacks preceramic remains representing Hunting-Collecting Bands and Foragers. The earliest remains are of Saladoid red-on-white pottery, which in this area date from about 1010 to 700 B.C. (Rouse and Cruxent 1963). The presence of metates and griddles indicates that these peoples were already Village Agriculturists, as were their successors—the Barrancoid peoples who appeared about 700 B.C. and the Arauquinoid peoples whose culture developed about A.D. 1000. If this reconstruction is valid, then the whole sequence occurs after the development of village agriculture and thus can tell us little about its evolution.

We do have some direct subsistence evidence for the middle Orinoco (Roosevelt 1980) but lack the crucial preceramic remains, although the contracting-stem points and scrapers of early hunter type from the Canaima workshop remains at Bolívar site may pertain to this stage (Meggers and Evans 1978). The earliest definite remains come from La Gruta, which has radiocarbon dates from 2100 to 1600 B.C. (Roosevelt 1980). It and the following phases—Ronquin (1600 to 1100 B.C.) and Ronquin Sombra (1100 to 800 B.C.)—belong to the Saladoid white-on-red pottery tradition. Among the lithics were stemmed projectile points. There were also remains of a few poorly preserved mammal and fish bones and charcoal remains of wild plants. Thick ceramic griddles and chips without discernible use-wear might pertain to the subsistence system.

About 800 B.C. appeared the Corozal tradition characterized by pottery, which was often red and decorated with incisions and appliqués. This tradition has been divided into three periods: Corozal I, 800 to 400 B.C.; Corozal II, 400 B.C. to A.D. 100; and Corozal III, A.D. 100 to 400. Settlement pattern analysis reveals a noticeable increase in population, from roughly 112 people per village to 1,744 people by Corozal III times. Corozal III had both chipped-stone and ground-stone tools, including metates. A few carbonized grains of maize were found, along with many wild plant remains, including those of unidentified legumes. There were also animal bones, possibly of wild species. Carbon 12-13 isotope studies of human bone from these periods gave readings that could indicate consumption of a large amount of aquatic (fish) resources and/or manioc, but this possibility cannot be determined without analyzing the nitrogen-15 or performing raciation for amino acid analysis.

The Camoruco phase that followed is likewise divided into three periods: Camoruco I, A.D. 400 to 700; Camoruco II, A.D. 700 to 1100; and Camoruco III, A.D. 1100 to 1500. It belongs to the Arauquinoid ceramic series characterized by appliqué and punctuated decoration. During this phase the population increased further, perhaps to 2,354 people, and the larger villages had dwellings on mounds or platforms. Unlike previous horizons, many kernels of maize of the Pollo or Chandelle races were uncovered, along with fragments of sword beans (*Canavalia ensiformis*) and burned cakes of manioc on griddles. Little mention is made of remains of wild plants or animals.

This Orinoco sequence, which is quite adequate in spite of the lack of the preceramic, has been well analyzed and described by Anna Roosevelt, Mario Sanoja, Juan Vargas, Irving Rouse, Alberto Zucchi, and others. They have suggested three varying reconstructions. The most popular hypothesis is that the La Gruta phase represents manioc-using Horticultural Villagers who develop into manioc-based chiefdoms with some maize agriculture in Corozal times, changing to villagers with mainly maize agriculture in Camoruco times (Roosevelt 1980). The emphasis on population changes suggests this hypothesis might fit the secondary development model, but we need evidence of the earlier developmental periods before we can make a definite classification.

The other two interpretations suggest the development was either secondary or tertiary. Zucchi (1978) sees the La Gruta tradition as Foraging Villagers (system C1) developing into Horticultural Villagers with maize (system C2) in Corozal times, while the Camoruco phases represent chiefdoms with true villages and full-time corn, bean, and manioc agriculture—system E. The third interpretation views La Gruta people as Semisedentary Bands with Domesticates (manioc horticulture)—system D1—developing into Horticultural Villagers—system C2—by Corozal times, and Agricultural Villagers—system E—during the Camoruco phase. It is hoped that data from further excavation or new kinds of innovative analysis of the type Anna Roosevelt is now undertaking will yield more solid reconstructions.

Although more research is also needed for Central America, this subarea has a long archaeological history, beginning in the early 1930s with work by Samuel Lothrop, Frank Roberts, and Frederick Johnson at Coclé (see Lothrop 1966). The archaeological sequences found by Gordon Willey, Charles McGimsey, Robert Greengo, and others (see Willey and McGimsey 1954) on the south Pacific Coast with its tropical savanna vegetation seem quite different from those found inland in the northerly tropical rain forest along the Caribbean Coast by Olga Linares, Anthony Ranere, Payson Sheets, and others in 1960s and 1970s (Linares and Ranere 1980). Finally, a significant series of dated pollen profiles from the Gatun Locks of the Panama Canal Zone have bearing on the beginning of domesticates in this subarea (table 10.3). These profiles, analyzed by Alexandra Bartlett and Else Barghoorn (Bartlett, Barghoorn, and Berger 1969) include large-size maize pollen in levels dated at 5350 B.C. ± 130 and/or 4280 B.C. ± 80. This pollen cannot come from wild plants or teosinte, since neither was native to the canal zone region. Both maize and manioc pollen have been identified in levels dated at 1200 B.C. ± 60, and abundant maize pollen and possible sweet potatoes have been found in more recent levels. Any archaeological sequence for the region must therefore take into account this basic dated sequence of domesticated plants (see table 10.3).

In early times the Pacific and Caribbean culture areas were probably similar

Table 10.3. Sequence of cultivated and/or domesticated plants in the New World tropical lowlands

Honduras / Panama / Colombia / Ecuador Pacific Coast / Venezuela Upper Orinoco / Southern Brazil	Range of dates			gourds	corn	common beans	canavalia beans	achira	manioc	coca	sweet potato	pineapple	coconut	peanut
Tupiguaraní	1200	1500	A.D.						x	x				
Avenias	1000	1500	A.D.		x									
Camoruco	400	1500	A.D.		x	x			x	?				
Sitio Conté	800	1200	A.D.		x									
Palenque core	400	800	A.D.		P				P			x	x	x
Cerro Punto	400	600	A.D.		x	x					x			
Gatun core-upper section	0	500	A.D.		P				P		x			
La Betania	0	300	A.D.		x									
Corozal III	0	400	A.D.		x									
Concepción	200	300	A.D.		x									
Chiriquí	0	300	B.C.		?				?		x			
Corozal II	0	400	B.C.		x									
Saladoid	0	1000	B.C.		?				?					
Aramquail core	0	1000	B.C.		P									
Corozal I	400	1000	B.C.						C 12 - 13					
Chorrera	600	1100	B.C.		x				?					
Gatun core-middle section	1000	1500	B.C.		P				P					
Aguadulce	1000	2000	B.C.		P									
San Pedro	2000	2500	B.C.		x									
Real Alto	2100	2500	B.C.		ph	x	x	x	?					
Monagrillo	1300	2600	B.C.		x									
El Dorado Valley	3020	3380	B.C.		P									
Cueva de los Ladrones	1820	4910	B.C.		ph									
Lake Yojoa	2435	3205	B.C.		P									
Gatun core-lower section	4000	5000	B.C.		P									
Las Vegas	390	6100?	B.C.	x	ph									

P = pollen
ph = phytolith
C = isotopic analysis

and had Hunting-Collecting Bands (system A) with fishtail points, blades, and scrapers like those found by Richard Cooke and Junius Bird near Madden Lake and elsewhere (Bird and Cooke 1978). The later sequences, however, have noticeable differences.

Let us first consider the sequence along the drier tropical steppe lands off the Pacific Coast, specifically in or near Parita Bay (see fig. 16). The earliest remains found were at the shell mound of Cerro Mangote, a small midden about thirty meters in diameter, on the Santa María River some ten kilometers from the present coast. Faunal studies indicate that, when occupied, it was along the active lagoons on the Pacific Coast (McGimsey 1956). The tools are mostly crude choppers, hammers, bone awls, shell beads and pendants, and heavy pebble-edge grinders that were used in shallow "metates" or milling stones. The lower levels of Aguadulce shelter, a few kilometers inland, had a similar artifact complex (Ranere and Hansell 1978). The size of the sites and the seasonality of the shells found at Cerro Mangote indicate that both these occupations were seasonal camps of micro- or macrobands. Charles McGimsey's analysis (1956) of the faunal remains and tools suggest that the occupants were also seasonal or Efficient Foraging Bands—my System D.

The next phase comes from the preceramic levels at Cueva de los Ladrones, 4910 to 1820 B.C., where maize phytoliths appear, making these people Semisedentary Bands with domesticated maize—system D1 (Piperno 1984). Possible corn phytoliths also appear at about the same time in nearby Cueva de Vampiros (Piperno 1988). Maize also first appears in the Gatun pollen profile at this time. It would seem that corn spread rapidly south from Mesoamerica at an early date (table 10.3).

Following this phase and probably developing from it is the ceramic Monagrillo phase dating from roughly 2455 to 1295 B.C. (Willey and McGimsey 1954). This phase is characterized by monochrome red-on-buff decorated pottery. Included in the phase are three sites near Monagrillo and Zapotal a few kilometers inland. Evidence suggests that during the period of occupation these sites were at the edge of the water. Other components appear to be the upper level of Aguadulce shelter, 1680 B.C. ± 95, and some of the undated levels of Cueva de los Ladrones (Ranere and Hansell 1978). The edge grinders, milling stones, and choppers at these sites suggest they evolved out of the earlier Cerro Mangote and Ladrones phases. The four coastal components, however, are larger, and postholes and other features are signs of permanent occupation, while faunal remains also indicate year-round occupations—signs that these people were basically villagers. Moreover, both at Monagrillo and Aguadulce shelters, corn pollen as well as chemical residues of maize occur, so these people or their neighbors were practicing some maize horticulture. In terms of our typology, then, Monagrillo had a horticultural village culture with maize—system C2. This conclusion agrees with the pollen date of 1200 B.C. from the Gatun Locks.

Exactly when village agriculture—system E—arrived in this subarea is difficult to determine. The culture of the Sarique phase, 1500 to 500 B.C., seems to have been similar to Monagrillo, so system E may not have occurred until Pueblo Nuevo (500 B.C. to the time of Christ) or Santa María (0–A.D. 500), and if not then, certainly by Coclé times, which had chiefdoms with villages and/or towns, elaborate polychrome pottery, gold work, and so on (Willey 1971). Ethnographic studies in the region and some archaeological dates from the Chiriquí and Sitio Sierra sites as well as the Palenque pollen core, suggest agriculture in this final phase included cultivation not only of corn but also of beans, manioc, and sweet potatoes. Certainly a strong case can be made that this subarea is an example of tertiary development.

Some evidence of a similar development is found in the nearby northern highlands with its gulf-oriented rain forest or our Panamanian Caribbean Coast subarea. The basic adaptations and culture phases, however, were quite different. Excavations by Anthony Ranere and company have uncovered a series of preceramic stratified rock-shelters—Casita de Piedra, Trapiche, Zarsa Dero, and Horacio Gonzales. While ceramic occupations overlay these shelters, the fuller components of the ceramic phases came mainly from nearby open sites or "ruins" excavated by Olga Linares and others (Linares and Ranere 1980). These sites have solid stratigraphy buttressed by a series of radiocarbon determinations and interdisciplinary studies aimed at obtaining reliable information about ancient subsistence patterns.

Talamanca, the earliest phase, is defined by a series of nine layers from the bottom of these shelters. Its six radiocarbon determinations range from 4610 to 1920 B.C. and are the basis for Ranere's estimate that the phase existed from 4600 to 2300 B.C.—roughly just after Cerro Mangote in the adjacent region (Ranere 1980). Like that site, the Talamanca occupations seem to have been by relatively small groups for brief seasonal periods. Also similar are the edge-ground cobbles, anvils, and milling stones. Artifacts unlike those of Cerro Mangote include wedges, choppers, scraper planes, and chipped axes or adzes. This tool assemblage, plus some charred remains of wild plants, suggests that the people were Efficient Foraging Bands—system D.

Evolving out of Talamanca is the Boquete phase, which occurs in upper or middle layers of all four rock-shelters as well as at some open sites (Ranere 1980). Its four radiocarbon determinations (from 2135 B.C. ± 75 to 350 B.C. ± 75) are the basis for estimating that the phase existed from 2300 to 300 B.C. Occupations still seem to have been seasonal, albeit longer, although Ranere believes, on not very good evidence, that these were sedentary occupations like Monagrillo and Sarique. I think they could better be considered semisedentary. Unlike Monagrillo and Sarique, however, the Boquete complex has no pottery and is defined on the basis of lithic tools—edge-ground cobbles, muller stones, hammerstones, anvils, irregular and bifacial cores, scraper planes, small tabular chisels, and ground and polished stone axes, celts, and chisels. Abundant plant remains, all wild, and some faunal remains definitely indicate these peoples were efficient foragers. Ranere speculated that they also had manioc. If so, they were Semisedentary Bands with Domesticates (manioc)—system D1.

Evidence for the subsistence system is on somewhat firmer footing for the following phases—Concepción and/or Barriles—when a few corn fragments occurred (Galinat 1980). These people could be classified as Horticultural Villagers—system C2. Definite evidence of village agriculture—system E—appears with the ceramic Bugaba phase, 0 to A.D. 800, judging from the corn, canavalia beans, and sweet potatoes uncovered and identified by C. Earle Smith at the Cerro Punto site dating A.D. 400 to 600 (C. E. Smith 1980a).

Although the evidence is far from perfect, there are indications of a slow tertiary development in this region of Central America. The sequence would be from Efficient Foraging Bands to Semisedentary Bands with Domesticates to Horticultural Villagers to Agricultural Villagers. Whether we have a similar tertiary development in the final tropical subarea—the Pacific Coast of Colombia and Ecuador—is debatable.

As elswhere in the tropics, Hunting-Collecting Bands—system A—are not well represented, although the Exacto complex, which has unifaces and burins, might be such. The following Manantial complex from the Santa Elena peninsula of southern Ecuador, about 9000 to 7000 B.C., has leaf bifacial points, choppers, denticulates, and large end scrapers and might represent Affluent Foraging Bands—system C (Lanning 1967a). Both the above complexes, however, are represented mainly by surface collections rather than by excavated contexts. This is not true of the following Vegas complex, dated at roughly 7000 to 4000 B.C., where Karen Stothert has excavated a large number of burials as well as midden sites (Stothert 1977—see table 10.4).

The Vegas sites have few projectile points, but the various pebble choppers, grinders, and scrapers as well as site locations and refuse that includes animal remains suggest the presence of seasonal foraging bands and/or villagers, perhaps Foraging Villagers—my system C1. Stothert and others have speculated that these people might have used manioc (Stothert 1985), but there is no supporting evidence, and the hints of corn agriculture are not convincing. Further doubt is cast on this hypothesis by the absence of any traces of manioc in the following complex, which is called Valdivia (Meggers, Evans, and Estrada 1965). Stothert has, however, found a "ghost feature" of a possible gourd that she believes is perhaps ten thousand years old (early Vegas). At site 80 in late Vegas, roughly 6000 to 3900 B.C., there are phytoliths of maize (Piperno 1984). If these identifications are confirmed with better contextual data and better dates, perhaps village horticulture—system C2—began in latest Vegas times. Stothert also believes that, although the following Valdivia phase developed out of Vegas, there is a gap before Valdivia (Stothert 1985).

Dating between 3000 (or 2700) and 2000 B.C., Valdivia has the first pottery and is represented by large village sites. The Real Alto site, for example, includes low mounds, a plaza, and oval houses, as well as complex burials and other evidence of ceremonialism (Lathrap, Marcos, and Zeidler 1977). Most of the sites are near the sea and show evidence of marine exploitation, hunting, and collecting. The Loma Alta site (Norton 1977), dated at about 2600 B.C., lies inland on an arable flood plain and has numerous grinding stones that suggest intensive plant collecting or agriculture, while carbonized corn kernels were uncovered at the large San Pablo site, roughly 2000 B.C. (Zevallos-Menendez 1971). At Real Alto, dated at about 3000 B.C., Deborah Pearsall (1978) identified phytoliths of corn as well as evidence of the use of beans (perhaps canavalia as well as common beans) and perhaps cacao. These people were certainly villagers, but it is difficult to determine whether they were, as I suspect, Horticultural Villagers—system C2—or Agricultural Villagers—system E.

Following the Valdivia complex in Ecuador are the Machalilla and Chorrera phases (roughly 1500 to 500 B.C.) as well as the Tumaca phase in Colombia, all of which were assuredly agricultural villages—system E—and perhaps even chiefdoms with influences from Mesoamerica. Inland Chorrera has evidence of corn and perhaps manioc (Meggers and Evans 1978).

A case could be made for a slow tertiary development from the Semisedentary Bands with Domesticates—system D1—of Vegas to the Horticultural Villagers—system C2—of Valdivia to the true Agricultural Villagers—system E—at Chorrera. However, the present weight of evidence from the somewhat inadequate site reports seems to favor a relatively quick secondary development from Foraging Villagers at 4000 B.C. during the Vegas phase to Agricultural Villagers in Valdivia times between 3000 and 2000 B.C. (table 10.4).

THEORETICAL CONSIDERATIONS

The above reconstruction might mean that a relatively rapid secondary development on the tropical Pacific Coast of Ecuador followed an evolution similar to that of India in southeastern Asia, the Aegean in Europe, or Egypt in Africa. Like those locations, Ecuador was adjacent to secondary developments in a Center. This development would make the Ecuadorean secondary development a sort of appendage development of the Andean Center, following a process that occurred roughly contemporaneously with that of the desert coast of adjacent Peru. Now the question becomes, Do the same sorts of necessary and sufficient conditions pertain to this secondary development of the tropical Pacific Coast of Ecuador? Although the site reports with the necessary interdisciplinary ecological studies are at present relatively inadequate, I believe we can adapt and reword the causes or conditions of secondary development to fit the Ecuadorean archaeological data in the general period from 4500 to 1500 B.C. The basic necessary conditions of secondary development would be as follows.

As indicated in table 10.4, this region had a number of potential domesticates, and the presence of such Andean domesticates as common beans, canavalia beans, and achira in Valdivia was a sign of intense interaction with a major Center. The maize in Valdivia and possibly in late Vegas also may indicate interaction with a Center, although maize, of course, was originally domesticated much earlier in Mesoamerica. The brief description of the environment of Valdivia (Meggers, Evans, and Estrada 1965) indicates that sites of this phase often occur on *salitres*—barren raised areas on beaches, often located at the end of peninsulas and between inlets or river mouths. These certainly are circumscribed coastal ecozones yielding both abundant marine and terrestrial food resources, and numerous ecozones are exploitable from these bases.

Since sophisticated ecological studies have not been undertaken for the region, we cannot know if these conditions pertain to all Valdivia sites as well as to other phases, particularly the riverine Vegas sites. It is also difficult to determine if there is regional ecological diversity but subareal uniformity, although the peninsular locations of Valdivia sites suggest this necessary condition might pertain. The prerequisite causes for the secondary development thus do seem to pertain to coastal Ecuador, and this part of that hypothesis seems tested. Now what about the all-important triggering causes—the sufficient conditions?

As indicated earlier, it is difficult to discover the causes for the development

Table 10.4. Necessary and sufficient conditions as a positive-feedback process for secondary development in coastal Ecuador

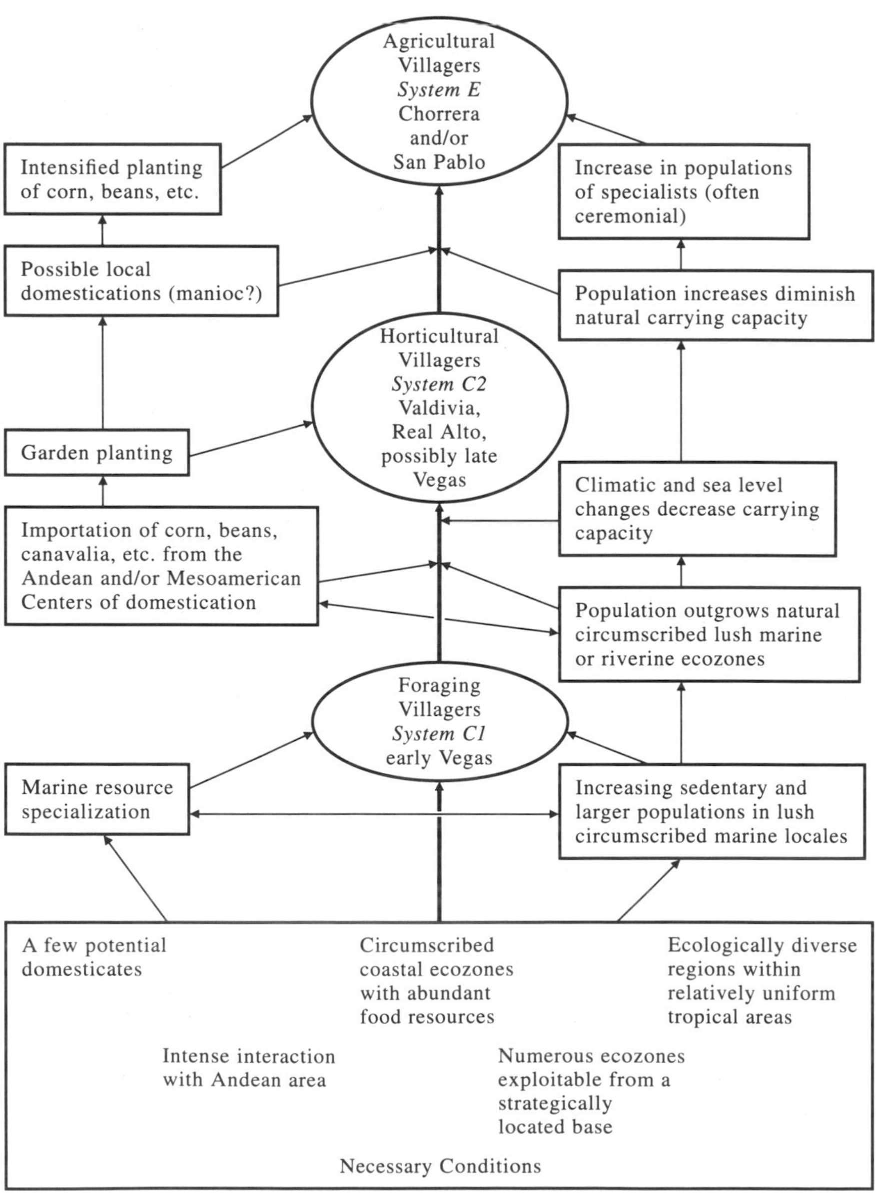

from Exacto or pre-Vegas, possibly Hunting-Collecting Bands (system A), to possible Archaic Foragers of Manantial times, and the transition to early Vegas Foraging Villagers (system C1). These early stages have not been well documented by data from excavation. It should be noted, however, that evidence of the extinction of megafauna in this general period can be found in the tar pits of nearby Talara in Peru. In Vegas times there certainly was a broad-spectrum procurement system, so future research may show that these sufficient conditions did occur.

In both the Vegas and Valdivia phases evidence occurs of increasingly specialized exploitation of marine resources. The location of Vegas riverine sites to the *salitre* would seem to indicate that the people were becoming increasingly sedentary, populations were increasing, and people were moving to lush circumscribed areas along the river or on the coast. Also, between late Vegas and Valdivia times, domesticates were being imported from the Andean Center. Whether the above causes, as well as the Vegas population's outgrowing the lush riverine circumscribed village locations and climatic deterioration and/or changing sea levels, are valid is difficult to document.

In Valdivia times, populations clearly increase, and Real Alto evidence indicates increasing numbers of ceremonial specialists; intensified planting of domesticates also occurred during this period. Whether Valdivia people were interacting in a positive-feedback cycle with diminished coastal carrying capacity or whether local domestication pertained cannot be demonstrated. The data we have serve more to test my secondary development model than actually to prove it. However, continuing research should help us better understand the causes of this secondary development in this region and in other parts of the tropics such as Venezuela, Colombia, and Brazil.

More research is also occurring in other tropical areas, where tertiary developments might have taken place. Our data from Panama, for example, are sufficient to confirm this hypothetical model, although the early stages of development are still not well documented. The necessary conditions certainly seem to have occurred in the coastal region, but it is difficult to determine whether they also occurred in the inland areas. As table 10.5 indicates, there were a few potential domesticates, and during the Panama Archaic and Formative, interaction with either Mesoamerica or the Andes took place. In Panama's tropical zones, food resources were available in all seasons, and most ecozones—even the coastal shell heaps—had abundant food resources and were not circumscribed. In addition, while both the inland and coastal regions have a number of ecozones, the two subareas are rather different from each other ecologically.

As stated previously, the earliest (pre-5000 B.C.) stages of this tertiary development in Panama are so poorly understood that we cannot discern the sufficient conditions causing the changes from Hunting-Collecting Bands using fishtail points—system A—to the Efficient Foragers of Cerro Mangote. We do know that extinction of megafauna occurred at the end of the Pleistocene and that a broad-spectrum procurement system had developed. Fortunately, starting with Cerro Mangote at about 5000 B.C., we do have some data for hypothesis testing. These people had developed an efficient foraging way of life based on marine resources,

Table 10.5. Necessary and sufficient conditions as a positive-feedback process for tertiary development in Panama

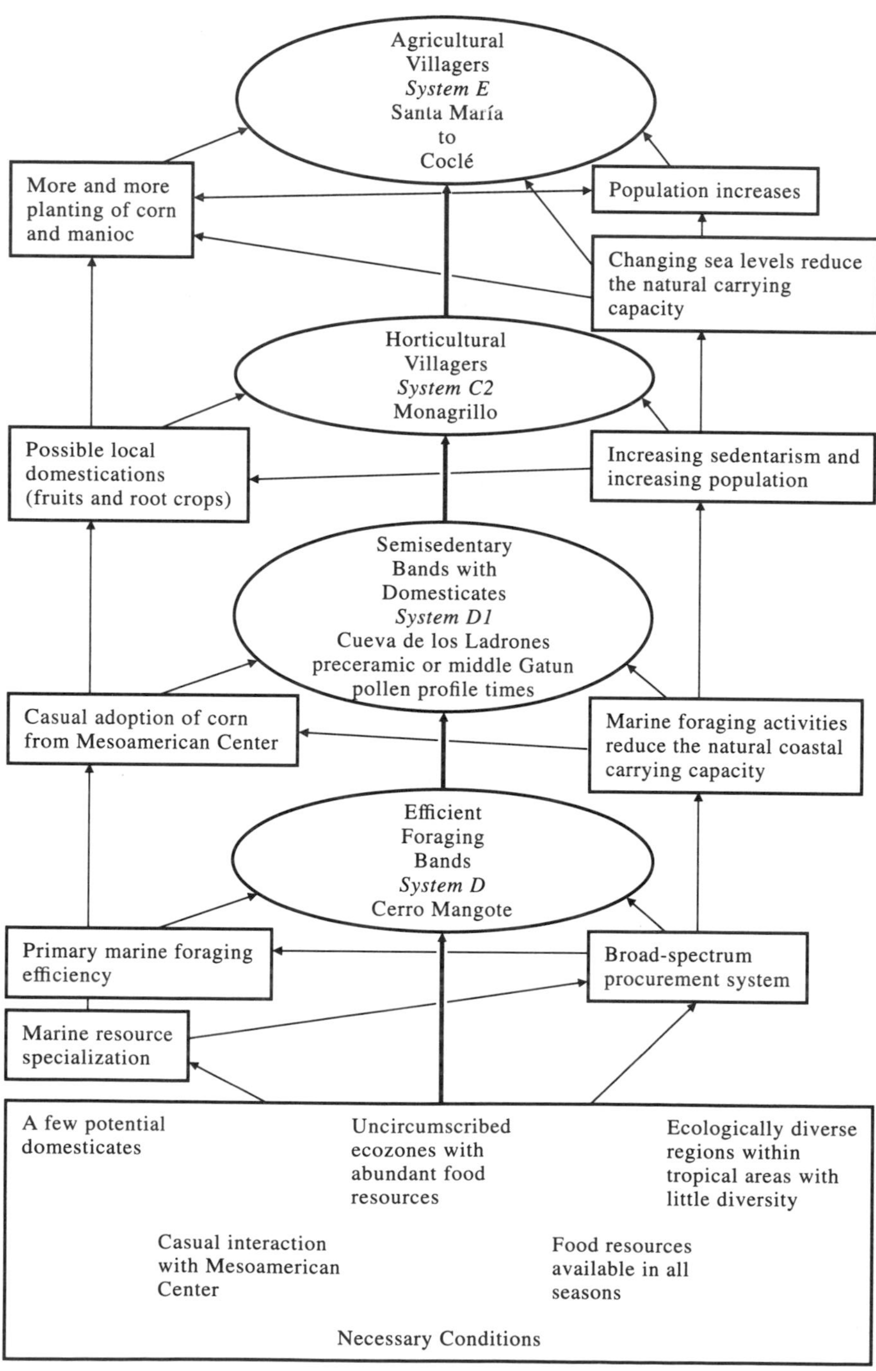

and the carrying capacity of the coastal locales had possibly been reduced. This positive-feedback cycle made the people amenable to other subsistence options, and both the Gatun core as well as the phytoliths from Cueva de los Ladrones and Cueva de Vampiros suggest they began to use corn (Piperno 1988). Ranere, Linares, Hansell, and others think they also used manioc (Linares and Ranere 1980), but proof is lacking.

The use of domesticates, including possible local domestications of plants such as fruit trees, combined with marine specializations may have meant the foragers became semisedentary, with a concomitant increase in population. While I cannot prove this theory, I believe these were the causes of the shift from the Semisedentary Bands with Domesticates in the preceramic times of Cueva de Los Ladrones to Horticultural Villagers in Monagrillo times, about 2000 B.C. Throughout this period, as well as later, the receding sea levels probably further reduced the carrying capacity of Parita Bay and pushed the villagers to do more and more planting. Combined with a sedentary way of life, the change in sea level led to population increases in Monagrillo and Sarique times, which encouraged more planting of corn and manioc, resulting in further population growth, and so on, until by the time of Christ (Santa María or Coclé phases) the people were Agricultural Villagers. This long, gradual development on the Pacific Coast of Panama seems to fit my model of tertiary development. Whether this scenario could also be applied to the inland rain forest zones in Panama remains to be seen. This scenario also might pertain to development along the middle Orinoco, but, as in inland Panama, we simply do not have enough solid data to confirm my tertiary model. Before I can prove or even adequately test my hypotheses for the tropical lowlands of the New World, data need to be derived from well-designed field programs with clearly specified objectives and the use of numerous interdisciplinary approaches. Despite these shortcomings, archaeological investigation on the origins of agriculture in the New World tropics is still better than in our final area: Africa.

CHAPTER 11

Africa

Our final area—Africa—is one of the least well known and also the largest and most diverse ecologically, ranging from Mediterranean coastal vegetation to tropical rain forest to desert to savanna. Although I have included it with the other tropical areas, it really is more than that. One might, therefore, expect it to yield large amounts of materials. Yet we have so few pertinent materials from Africa that almost none of the data can be used to test the validity of any hypothesis. This lack is rather surprising, for parts of Africa—particularly Egypt—have been under archaeological investigation as long as any place in the world. Further, preservation in the Egypt-Sudan region is excellent, and preserved plants have been exhumed there; yet still there are few findings that have direct bearing on the origins of agriculture. Perhaps that is because most research on Africa has been concerned either with "fossil man," the African Iron Age, or the "great civilization" (i.e., Egypt), most of which does not concern problems of origins of agriculture. In this section, therefore, I shall not attempt an extensive history of the archaeological research for the area. Instead I shall briefly consider the history of theories and speculations about how and why village agriculture came into being on the African continent.

A BRIEF HISTORY OF RESEARCH ON THE PROBLEM

Once again we start with the endeavors of Vavilov and his colleagues in the 1920s (Vavilov 1951). On the basis of plant distribution and genetic variability, they decided that one of the major areas of plant domestication was Abyssinia (Ethiopia) and concluded that yams, sorghum, the millets, and the like were first domesticated here. These conclusions were reached in spite of the predynastic Egyptian early domesticated plant remains found by Sir Flinders Petrie at Amratian and Gerzean sites (Petrie and Quibell 1895) and George Caton-Thompson's discoveries at Fayum in the early twenties and at Badaria (Caton-Thompson and Gardner 1934).

Vavilov's conclusions, as well as the work of the French in the Sahara, led Andre Chevalier (1938) and others to think of the Sahara as a Center. Paul Munson (1968), Roland Portères (1950, 1970), and others continued to amass evidence that millets and sorghum were first domesticated in the Sahara, albeit at a relatively late period. The 1940s and 1950s saw the rise of Peter Murdoch's hypothesis that the Gold Coast of West Africa (i.e., the "Bend on the Niger") was the center of plant domestication in Africa—mainly for root crops such as yams (Murdoch 1959). This theory agreed with Sauer's popular theories of the time (Sauer 1952). Later, David Harris, with his theories about vegiculture in the tropics (Harris 1969), as well as David Coursey (1976), Oliver Davies (1967), and others joined this team. In the late 1950s and 1960s, Jacques Barrau and Roland Portères (Portères 1976) proposed that there were three principal Centers or foci (or cradles): (1) North Africa from Egypt to Morocco; (2) the belt of savannah and steppe peripheral to the forest heart of Africa; and (3) the rain forest of the Gold Coast and Central Africa.

Although these three "clusterings" were still recognized as important, from the late 1960s to the present, Jack Harlan, John de Wet, and Ann Stemler (Harlan, de Wet, Stemler 1976), Thurston Shaw, Creighton Gabel, D. Davies, and others began to talk of Africa as a non-Center or "as basically noncentric" (Harlan 1982, 178). Further, Harlan speculated that "there were a number of factors that might have triggered off the process," mainly "the desiccation of the Sahara" roughly 4000–2000 years ago, "the expansion of the Near Eastern agricultural system," and the "gradual increase in population density to the point that adjustments were required" (Harlan 1977, 41). Although I agree fundamentally with these causes, or sufficient conditions, I would phrase them slightly differently and come to other conclusions on the basis of rather different kinds of direct sequential archaeological evidence. Because climatic change is so obvious in Africa, let us consider the African environment and geography as a whole.

THE ENVIRONMENT

Basically, Africa has four bands of east-west climatic zones (Fig. 17). Those north of the equator are duplicated to the south in a mirror image. Along the equator is the equinoctial zone or rain forest subarea, which has the highest temperatures and heaviest rains, which fall throughout the year (Butzer 1972). This subarea would include the modern countries of northwest Zaïre, Congo, Gabon, Cameroon, the Central African Republic, and the Guinea Coast countries of Nigeria, Benin, Togo, Ghana, Guinea, Sierra Leone, Ivory Coast, and Liberia (see fig. 17). Generally speaking, elevations near the equator are low; rivers are large and sluggish (including the massive Congo, or Zaïre); and the vegetation is lush tropical rain forest with a thick canopy (Butzer and Cooke 1982).

North and south of this zone, as well as east, limited rainfall results in savanna—hot zones with seasonal summer rains and woodland and grass savanna vegetation. The northern savanna-woodland strip subarea north of the equator includes most of present-day Sudan, southern Chad, Niger, Upper Volta, Senegal, Gambia, southern Mali, Mauritania, and southern Morocco (formerly Western Sahara). Actually, this east-west strip of savanna comprises two parts: a narrow, more southerly strip, which is slightly more humid and wooded, and a northern savanna. Because this humid strip is so narrow, we consider the whole subarea a single zone.

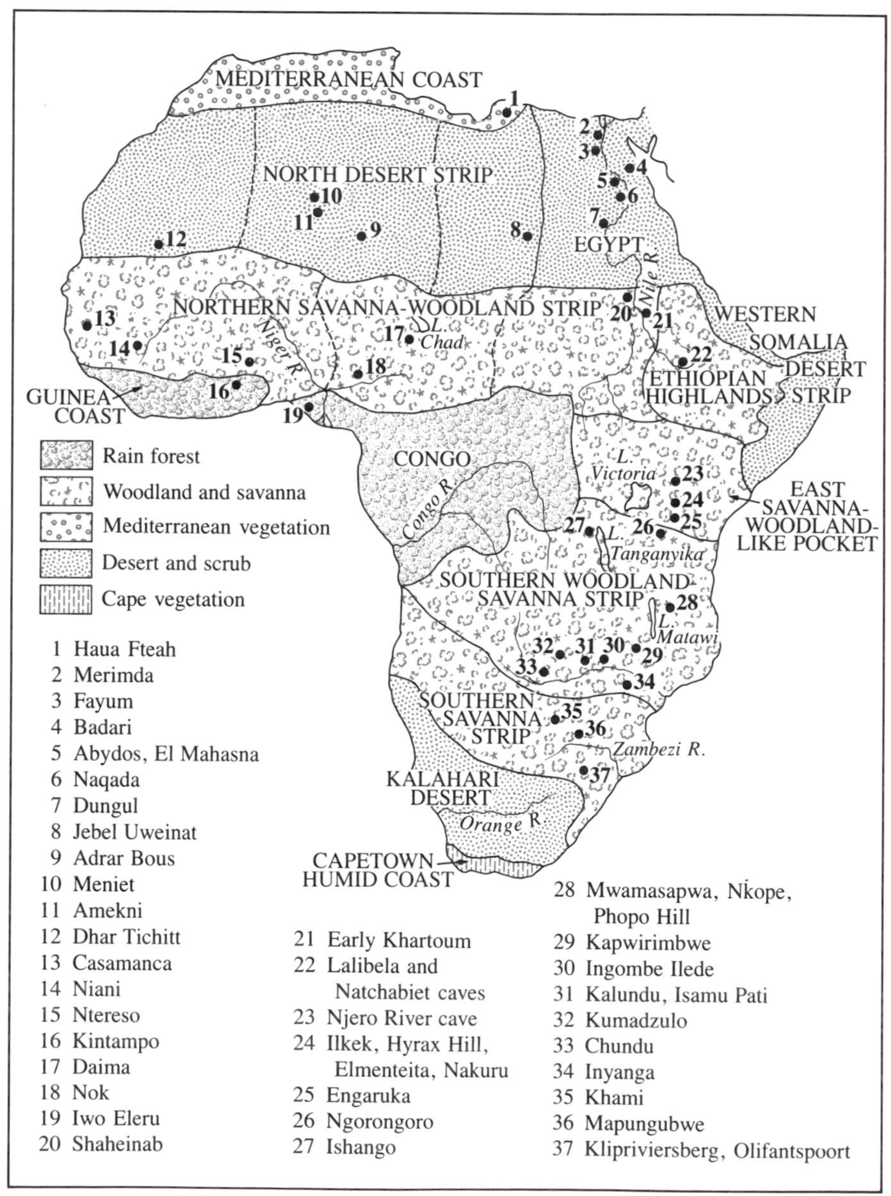

Figure 17. The environmental life zones and relevant sites in Africa

As one moves north the climate becomes drier, and the savanna is replaced by the north desert strip subarea, which has little vegetation and many sand dunes. This desert contains occasional oases and, of course, the crucial strip of arable land along the Nile River. This desert zone has hot temperatures and little rainfall, limited mainly to the winter months. Within this zone lie northern Sudan, Egypt, Libya, most of Algeria, and the northern parts of Chad, Niger, Mali, and Mauritania.

Finally, at the far north is the Mediterranean coastal subarea, which has lush Mediterranean vegetation. The climate is warm and rains are heavier and fall in winter.

Roughly equivalent to this northernmost strip is its opposite, the zone along the southern coast of South Africa, called the Capetown humid coast subarea. The only difference is that rain falls there in the summer months (January through March), rather than in winter. North of this lush zone is the Kalahari Desert subarea and its extension to the west, the Namib. This desert zone includes northern South Africa, western Botswana, and Namibia. Desert blends into the southern savanna and woodland-savanna subareas as one moves north into Mozambique, Zimbabwe, southernmost Zambia, and Angola. The central rainforest subarea is the east woodland savanna pocket, which includes Tanzania, Malawi, Zambia, Rwanda, Burundi, and the southern part of Zaïre.

East Africa has three subareas that are slightly different from the pattern found in the rest of the continent because they are at higher elevations and receive moisture from the Indian Ocean rather than from the Atlantic. One of these subareas covers much of the Ethiopian Highlands and has high mountains with evergreen vegetation, deep valleys, and heavy summer rains. To the east, in western Somalia and along the Red Sea Coast of Ethiopia, is a strip of desert. The third subarea, the eastern savanna woodland-like pocket is more mountainous and includes a number of lakes that have winter rains as well as wooded areas and savanna. This subarea covers Uganda, part of Rwanda, Kenya, and much of northern Tanzania. The subareas discussed above add up to eleven major subareas. For cultural reasons, I believe Egypt and its fertile Nile River Valley should be considered separately, as an eleventh zone.

Egypt is distinct from the rest of Africa because it (and much of the Sudan) utilized a series of domesticated plants that were imported from the Near East at an early time. These plants included emmer, barley, club wheat, flax, chickpeas, peas, lentils, broad beans, cabbage, and oats (see table 11.1).

Another area that used imported plants was eastern Africa—Madagascar, Kenya, and Ethiopia. Among the plants imported there from southeastern Asia were banana, mango, Asian yams, coconuts, sugar cane, and sesame. Later domesticates in this subarea came from local plants susceptible to domestication (Portères 1970): Ethiopian oats, chat drugs, coffee, finger millet, enset bananas (*Ensete ventricosum*), teff, noog oil, castor oil, guinea grass, Rhodes grass, and stargrass.

The savanna zones (mainly north of the equator) had a number of other plants susceptible to domestication (Purseglove 1977). These include baobab, sheabutter tree, watermelon, potherb, hungry rice, black hungry rice, kenef,

Table 11.1. Some native cultivated and/or domesticated plants from various subareas of Africa*

Scientific name	Popular name
East Africa	
**Avena abyssinica*	Ethiopian oats
**Catha edulis*	chat drugs
**Chloris gayana*	Rhodes grass
**Cofea arabica*	coffee
**Cynodon aethiopicum*	stargrass
**Cynodon dactylon*	Bermuda grass
**Eleusine coracana*	finger millet
**Ensete ventricosum*	enset beans
**Eragrostis tef*	teff (cereal)
**Gossypium herbaceum*	African cotton
Guizotia abyssinica	noog oil
**Panicum maximum*	guinea grass
**Ricinus communis*	castor oil
North Savanna	
**Adansonia digitata*	baobab
**Butyrospermum paradoxum*	shea-butter tree
**Ceratotheca sesamoides*	savanna seed
**Colocynthis citrullus*	watermelon
**Corchorus olitorius*	potherb
**Digitaria exilis*	hungry rice
**Digitaria iburua*	black hungry rice
**Hibiscus cannabinus*	kenaf
**Hibiscus sabdariffa*	roselle
**Lagenaria siceraria*	bottle-gourd
**Oryza glaberrima*	African rice
**Parkia biglobosa*	sweet pod
**Pennisetum americanum*	pearl millet
Polygala butyracea	black beniseed oil
**Sesamum alatum*	sesame
**Sesamum radiatum (?)*	sesame
**Solanum aethiopicum*	African tomato
**Solanum incanum*	bitter tomato
**Solanum macrocarpon*	nightshade
**Sorhum bicolor*	sorghum
Tropical Central Africa	
**Digitaria decumbens*	pangolagrass
Cajanas cajan	pigeon pea
**Eragrostis curvula (?)*	weeping lovegrass
**Eragrostis lehmanniana*	Lehmann's lovegrass
Gossypium herbaceum	cotton
**Pennisetum clandestinum*	kikuyu grass
Pennisetum purpureum	elephant grass
Sesamum indicum	sesame
**Tamarindus indica*	tamarind
Tropical West Africa and Congo Basin	
**Abelmoschus esculentus*	gumbo, okra
**Afromomum melegueta*	grains of paradise, spice
**Blighia sapida*	akee apple
**Brachiaria deflexa*	guinea millet
**Coffea canephora*	robusta coffee

Table 11.1. (*continued*)

Scientific name	Popular name
**Cola acuminata*	cola nut
**Cola nitida*	cola nut
**Cucumeropsis edulis*	African seed
Dioscorea bulbifera	air potato
**Dioscorea cayenensis*	yellow Guinea yam
Dioscorea dumetorum	bitter yam
Dioscorea praehensilis	bush yam
**Dioscorea rotundata*	white Guinea yam
**Elaeis guineensis*	oil palm
Hibiscus esculentus	okra
**Hyparrhenia sirufa*	jaragua grass
**Kerstingiella geocarpa*	groundnut
**Lablab niger*	hyacinth bean
**Piper guineense*	piper seed
**Plectranthus esculentus*	kafir potato
**Solenostemon rotundifolius*	piasa
**Sphenostylis stenocarpa*	yam pea
**Telfairia occidentalis*	fluted gourd
**Vigna unguiculata*	cowpea
Voandzeia subterranea	groundnut

*After Harlan (1982).

cannabinus, roselle, bottle-gourd, African rice, sweet pod, pearl millet, black beniseed oil, African tomato, nightshade, bitter tomato, and sorghum. In the savanna, but of uncertain origin, were pigeon pea, African cotton, sesame seed, and tamarind.

The other main region with domesticable plants is West Africa, or Harlan's Swamplands (Harlan 1988), mainly in the rain-forest subarea. Some of these plants also are native to the rain-forest Congo to the east as well as to the tropical forest–savanna woodland margin in West Africa. Plants coming mainly from West Africa include a number of root crops: gumbo, grains of paradise, akee apple, guinea millet, robusta coffee, cola nut, air potato, yellow guinea yam, bitter yam, bush yam, white guinea yam, oil palm, okra, jaragua grass, sirufa, groundnut, hyacinth bean, piper seed, kafir potato, piasa, yam pea, fluted gourd, and cowpea.

Thus Africa has a number of possible culture subareas with potential domesticates. Most of these subareas were relatively lush, had foods in all seasons, and were quite uniform, although regional ecological diversity sometimes existed within certain sections. Environmental uniformity is especially true of the deserts, but even there oases and the fertile Nile create lush areas.

Although the specifics are different, Africa bears considerable ecological resemblance to the other tropical areas and is not significantly different from the temperate zones. With these ecological characteristics in mind, I divide the African continent into twelve culture subareas, as follows:

1. Egypt
2. Mediterranean Coast

3. North desert strip (Sahara)
4. Ethiopian Highlands
5. Northern savanna–woodland strip
6. Eastern savanna–woodland pocket
7. Rain forest: Guinea Coast and Congo
8. Western Somalia desert strip
9. Southern woodland–savanna strip
10. Southern savanna strip
11. Kalahari-Namib Desert strip
12. Capetown humid coast

RELEVANT ARCHAEOLOGICAL DATA

Although my classification of Africa into cultural subareas does take into account both geographic and cultural factors, its main purpose is to create convenient units for describing archaeological sequences. As we shall see, there has been considerable cultural uniformity through time in each of these culture subareas (see table 11.2). As with other regions, I shall consider these subareas in an ascending sequence of importance, looking first at those with the least amount of archaeological data relevant to our problems of the origins of agriculture and settled life. Selecting the subarea with the least data was difficult, for many had few or none of the preserved plant remains crucial for discussing ancient agricultural developments.

Let us begin with the lush Capetown humid coast, where even in historic times agricultural village life had not developed among the Strandlopers (Bushman-Hottentot—Phillipson 1977). Paleolithic cultural sequences for this subarea run well back into the Pleistocene; the earliest one relevant to our problem is the Oakhurst complex, which lasted from about 12,000 B.C. until about 8000 B.C. While the preceding Paleolithic culture may have been Hunting-Collecting Bands (system A), the Oakhurst peoples were probably Efficient Foraging Bands (system D), for they used numerous chipped end scrapers, grinding stones, ivory and rib-bone mattocks, and bone adzes. Their way of life seems to have continued for millennia as the Wilton complex. This microlithic complex evolved through a series of stages: Early (ca. 8000–4000 B.C.), Classic (4000–1000 B.C.), Developed (1000 B.C. to the time of Christ), and Ceramic, which might have lasted up until A.D. 1000. Even in their latest phase these people seem to have been Foraging Bands with no domesticates. The Strandlopers of the Ceramic Wilton had flake tools and perhaps iron ones. They may have lived in shell-midden villages that were decimated by white colonists long before the people acquired domesticates.

This situation is about the same for the Kalahari-Namib Desert strip, where early Paleolithic remains were followed by the Lockshook complex that lasted at least until 6000 B.C. (Phillipson 1977; J. D. Clark 1981). Scrapers, bone spears, and bone harpoons suggest that these people were Foragers. Following the Lockshook complex was a long sequence of the Wilton microlithic ending with ceramics in the first centuries after Christ. This was followed by the Smithfield complex, which had dentate stamp pottery and microliths, but there was no evidence of sedentary culture, let alone agriculture. Although these are good sequences, no domesticated plant remains have been recovered.

The same situation is more or less true of the opposite end of the continent—the Mediterranean Coast. From historic records we know that peoples of the Classic period and modern times had a Near Eastern-type agriculture with wheat, barley, and so on (Balout 1981). Archaeological research has uncovered well-defined Paleolithic complexes, and the latest Aterian one certainly was Hunting-Collecting Bands—system A. Starting at about 12,000 B.C. was the Capsian tradition, which had microliths and probably a foraging economy. Certainly the peoples making trapezoids and crescentic microliths at La Mouillah and Tebessa from 8000 to 5000 B.C. were Efficient Foragers, as were those using Cardial pottery from 4700 to 1000 B.C. These people may have reached only the sedentary foraging stage, for we find evidence of domesticated plant remains only in the upper levels of Capeletti Grotto, where C. Thurston Shaw (1972) found sorghum and pearl millet, and in the Neolithic levels (4000 B.C.) of Haua Fteah Cave in Cyrenaica, where C. B. McBurney (1967) found emmer. The Neolithic peoples of the Capsian tradition may also have been Foraging Villagers with inland savanna horticulture, but it was not until the Iron Age that real village agriculture based on barley, wheat, and so on arrived from the Near East. Concrete evidence of that development, however, is lacking. Although there are at least five complete sequences from the Paleolithic through the Neolithic, rarely, if ever, have plant remains been found (see table 11.3).

The meagerness of plant remains also characterizes our other two major subareas south of the equator in Africa. The southern savanna strip has remains of Paleolithic Hunters and an early Forager lithic complex called Pomongwe that lasted until about 6000 B.C. Then the usual Wilton microlithic sequence of Foragers continues until about A.D. 700, when Iron Age (Bantu) invaders arrive. At some of the large fortified ruins like Inyanga (700 B.C.), remains of kaffir melon, groundnut, finger and pearl millet, cowpeas, and sorghum were found, while sorghum and cowpeas were found at Mapungubwe, Bambandydanalo, Klipriviersberg, and Olifantspoort (Phillipson 1982; Shaw 1972).

To the north, in the southern savanna–woodland strip, the general situation is much the same. A long sequence of the Paleolithic is followed by a Forager microlithic sequence—Nachikufam I, IIA, and IIB—lasting until about the time of Christ, when the people settled down and acquired dentate stamp pottery (J. D. Clark 1981). At Nkope, Phopo Hill, and Mwamasapwa there is evidence the people used cowpeas, perhaps indicating that they were Horticultural Villagers. Again, evidence of village agriculture does not occur until the Iron Age (Bantu) invasion after A.D. 600, and here the evidence is a few grains of sorghum at the Isamu Pati and Ingombe Iledi sites (Shaw 1972). Again, the archaeology is good, but the documentation for agriculture is extremely poor.

This lack of documentation is not too surprising considering that none of these areas had many local plants susceptible to domestication, but it is most disappointing for the next subarea—the tropical rain forest. This area has often been mentioned as a Center of domestication and does have many wild plants that be-

Table 11.2. Relevant archaeological sequences from Africa

MEDITERRANEAN COAST SUBAREA	EGYPT SUBAREA	ETHIOPIAN HIGHLANDS SUBAREA	NORTH DESERT STRIP SUBAREA			GUINEA COAST SUBAREA	EAST SAVANNA WOODLAND-LIKE POCKET SUBAREA	SOUTHERN (WOODLAND) SAVANNA STRIPS SUBAREAS	
			Western Sahara region	Central Sahara region	Sudan region				Dates
Iron Age		Iron Age		Nok	Iron Age		Hyrax Hill	Smithfields	A.D. 1000
	Ptolemaic		Berbers	Daima		Niani			0
					Nubian				
		Axum				Niger Delta			
			Akjinjeir Arriane		Jebel Moya		Engaruka	Ceramic Wilton	1000 B.C.
		Natchabiet							
			Chebka Naghez Nkhal			Kintampo	Njero Cave	Developed Wilton	
Grotto Capeletti		Lalibela		Arriean					
	Dynastic		Goungon Khimiya						
				Naghar					
			Akreijit						2000 B.C.
				Adrar Bous			North Horr		
						Iwo Eleru			

Neolithic of Capsian tradition	Pre-Dynastic	Gobedra			Afyeh A			Classic Wilton	
									3000 B.C.
				Meniet					
Cardial	Merimda				Shaheinab				
							Ishango		
		Wilton							4000 B.C.
Haua Fteah	Badarian				Khartoum Neolithic			Early Wilton	
							Gamble's Cave		5000 B.C.
	Fayum					Micro-lithic			
					Khartoum Mesolithic				6000 B.C.
			Capsian				Capsian Nderit		
	Sebilian				Qadan				7000 B.C.

Table 11.3. Sequence of cultivated and/or domesticated plants in Africa

Mediterranean Coast	Egypt	Sudan	Ethiopia	Central Sahara	Western Sahara	Guinea Coast	Central Eastern	South savanna	Range of dates		flax	emmer	barley	club wheat
								Klipriviersberg	1500	1800 A.D.				
								Mapungubwe	1250	1350 A.D.				
								Isamu Pati	800	1000 A.D.				
								Kalundu	800	1000 A.D.				
								Mwanasapawa	700	900 A.D.				
						Niani			650	850 A.D.				
								Ingombe Ilede	640	720 A.D.				
				Daima					800	900 A.D.				
								Inyanga	500	700 A.D.				
								Engaruka	300	1800 A.D.				
						Niger Delta			0	500 A.D.				
								Nkope	0	215 A.D.				
		Jebel Moya							A.D. 200	200 B.C.			?	?
			Natchabiet						200	400 B.C.			x	
	Senacherid								500	900 B.C.				
						Hajer bin Humeid			900	1100 B.C.				
				Ariane					700	900 B.C.				
						Njero Cave			1020	1345 B.C.				
Capeletti Grotto									1000	1200 B.C.				
					Chebka				900	1000 B.C.				
					Naghez				1000	1100 B.C.				
			Lalibela						1000	1200 B.C.			x	
	Dynastic					Kintampo			1400	1700 B.C.				
				Adrar Bous 2					1500	2500 B.C.				
	Pre-Dynastic								2000	3000 B.C.	?	x	x	x
			Gobedra						2000	3000 B.C.				
		Afreh A							2340	2430 B.C.			x	
		Shaheinab							3100	3500 B.C.			x	
				Adrar Bous 1					3500	4500 B.C.			x	
	Abydos								3600	4400 B.C.				
		Kadero							3600	4600 B.C.			x	
	Merimde								3500	5200 B.C.		x	x	x
	Badarian								3700	5580 B.C.		x	x	
				Meniet					3100	3800 B.C.		x	?	?
Haua Fteah									4500	5000 B.C.		x		
	Fayum								4000	5200 B.C.	x	x	x	?
	Nabta Playa								6000	6300 B.C.			x	

came domesticated. Archaeological excavations in the tropical rain forest of the Congo and eastern Cameroon have produced no plant remains whatsoever, and the archaeological sequence is not well defined from the Paleolithic Lupemban and Tschitolian complexes, which have Levallois-like flake tools, through a Wilton-like microlithic complex and the so-called Leopoldion Neolithic (J. D.

castor (oil) bean	bread wheat	guinea millet	oil palm	finger millet	pearl millet	sorghum-guinea corn	garlic	cetta	cowpeas	groundnut	teff	chickpeas	gourd	rice	yam	foxtail millet	kaffir melon	coffee
						x												
x					x	x			x								x	
						x												
						x												
					x	x												
						x												
						x												
						x												
				x	x	x			x	x							x	
						x												
														x	x	x		
									x									
						x												
												x						
					x													?
											x							
						x												
													x					
					x	x												
		x		?	?							x						
		x			x							?						
													x					
			x						x									
					?	x												
							x	x										
				x														
	x																	
			x															
		x																
	x																	
				?	?													
x																		

Clark 1981). This sequence contrasts with the Guinea Coast area, where there is a better-defined sequence. Only one site—Kintampo, situated inland on the edge of the wet tropics—has yielded a few specimens of cowpeas and husks of the oil palm (Shaw 1981). In the archaeological sequence there are early Paleolithic flake-tool remains called Asejire that may date before 10,000 B.C. and Guinea microlithic tools at Adwuku and Iwo Eleru that date roughly from 10,000 to 2000 B.C. This is followed by the Punpun phase in Kintampo shelter, which has dentate stamp pottery, celts, and so on associated with the domesticated plant remains just mentioned.

The nearby and closely related northern savanna–woodland strip subarea is equally poorly known. There is a long microlithic Capsian-like preceramic with the first pottery arriving between 2000 and 1700 B.C. in the upper Yengema complex. Ntereso pottery follows at 1300 B.C., but true evidence of village agriculture does not occur until the Iron Age, at which time sorghum is found with ceramics at the sites of Daima (A.D. 800–900), Niani (A.D. 1000), and later Nok (Shaw 1972, 1981).

Given this poor evidence in one so-called "cradle," let us turn to another reputed center, the Ethiopian Highlands (Sutton 1981). This area has a long sequence of Paleolithic remains, as well as such famous fossils as Lucy, indicating the presence of people, some of whom undoubtedly were great hunters. A similar economy is perhaps suggested by the Hargesian industry with its backed blades, burins, and other tools, which may have lasted until about 6000 B.C. Following these remains are microlithic Wilton-like complexes, such as those at Melka Knotoure and Mandera in Somalia. Some of the earliest Ethiopian pottery occurs in the upper levels of Gobedra rock-shelter along with specimens of finger millet that date roughly from 4000 to 3000 B.C. Later ceramic deposits from Lalibela (roughly 1000 B.C.) and Natchabiet shelters (time of Christ) yielded barley and chickpeas. Whether these early ceramic complexes pertained to horticultural villages or agricultural villages cannot be determined adequately at present, but the limited evidence favors the former. True village agriculture, probably of wheat and barley as well as of millets, sorghum, and other plants, is present by the time of the great ruins of Axum (constructed about the time of Christ) and the following Ethiopian Empire periods. Future research should contribute much crucial data on how and why village agriculture came into being in this relatively unexplored subarea.

Better explored, of course, is the adjacent eastern savanna pocket subarea (including Kenya) where numerous finds of early hominids and Paleolithic Hunting-Collecting Bands have been made by the Leakeys and others. The subarea has also yielded good solid sequential artifacts of periods relevant to the problems of the origins of agriculture. Perhaps most important are the materials often designated Lower Kenya Capsian and thought to date to about 10,000 B.C. These include a complex characterized by backed blades, burins, and other tools from sites like Nderit. Somewhat later is the development called Upper Kenya Capsian, which has backed blades, crescents, burins and scrapers, ostrich-shell beads, and bone points from such places as Gamble's Cave.

These remains suggest we are dealing with Foragers whose way of life lasted perhaps until 4000 B.C. (J. D. Clark 1981). Some of these people may have occupied shell mounds. The main fishing complexes, with wavy-line pottery, net-sinkers, harpoons and so on, were on the shores of Lake Victoria, Lake Rudolf, and the Nile in the southern Sudan from 4000 to 3000 B.C. These fishing complexes were certainly either Village Foragers or Semisedentary Bands with Domesticates. Later, pottery and stone-bowl complexes appeared at nearby North Horr and Nderit (2000 B.C.); in addition, Njero also had a fragment of domesticated gourd (*Lageneria siceraria*—Shaw 1972). The people at this time, as well as during the later Narosura axe complex, were perhaps Horticultural Villagers or Foraging Bands with Domesticates. True village agriculture does not seem to occur until the Iron Age, at the time of Christ, at such sites as Engaruka and Hyrax Hill. The sequence for this region is long and well-documented but needs more evidence of horticulture and agriculture. Changing lake levels between 4000 and 2000 B.C. as well as rising populations do, however, hint that the people slowly adopted domesticates and eventually become Village Agriculturists. There is tantalizing evidence suggesting secondary and/or tertiary developments, but unfortunately it is not sufficient to test my models adequately.

About the same may be said of the desert area northwest of Kenya, which also experienced a drying of its lakes and a shift from savanna to desert roughly between 5500 and 2000 B.C. (Hugot 1981). Here we have two good sequences following earlier remains of Paleolithic Aterian Hunters and Capsian-like microlithic Foragers that may have lasted up until five or six thousand years ago. Amekni, on Lake Chad, seems to have been a large fishing village that lasted up until ceramic times, while Meniet, with rocker-stamp pottery at about 3500 B.C., has evidence of domesticated emmer. These were possibly sites of early Semisedentary Bands with Domesticates. A later development is the Adrar Bous site, which had Tenerean-type pottery and barley, guinea millet, and, later (2000 B.C.), sorghum. Similar complexes seem to occur at later sites such as Naghar and Arriean, but whether the people were Horticultural Villagers or Agricultural Villagers is difficult to determine. The Iron Age complexes in the early part of the Christian Era, found at such sites as Daima and Niani, certainly were agricultural villages.

The other northern desert (Sahara) subarea that has produced relevant results is the Dhar Tichitt region of south-central Mauritania. Here are finds, lasting until about 6000 B.C., of Aterian Paleolithic and Capsian microlithic Foragers (Munson 1968). Akreijit-phase sites on ancient beach levels that date about 2000 B.C. (1750 B.C. ± 130) have pottery and seem to reveal the presence of Efficient Foragers—system D. After a gap between 1500 and 1400 B.C. comes the Khimiya phase, in which land animal and fish bones as well as plant remains suggest Foraging Villagers, but impressions of pearl millet (*Pennisetum sp.*) on pottery suggest the people were either picking wild millet or planting small amounts of it in gardens. This pattern continued through the Goungon and Nkhal phases until about 1100 to 1000 B.C., the Naghez phase, when the encampments were larger and had stone masonry construction, and there were seeds of both pearl millet and guinea millet (*Brachiaria deflexa*). Whether these seeds were wild or in the first stages of domestication is unknown, but the people might well be classified as Semisedentary Bands with Domesticates—system D1.

The shift in proportion of millet in the following Chebka phase, 1000 to 900 B.C., indicates planting of millet and is evidence of village horticulture—system C2. Through the Arriane phase to the Akjinjeir phase, ending about 400 B.C., there seems to have been a slow evolution of village agriculture—system E.

Table 11.4. Correlation of cultural, ecological, and subsistence changes in Neolithic Mauritania (based on Munson, 1968)

Dates B.C.	Developmental Systems	Archaeological Phase	Type of Settlements	Probable Subsistence Pattern	Possible Percentages of Plants: Wild plants	*Pennisetum* sp.	*Brachiaria deflexa*	*Panicum lactum*	Sorghum bicolor	*Panicum turgida*	Environmental changes
400–800	Village Agriculture (*System E*)	Akjinjeir	Villages and/or hamlets	Cultivation, herding, limited collecting and hunting	7	90	1	1	?	1	Invasion of desert species
800–900		Arriane			16	82	1	1	x		Rainfall less
900–1000	Village Horticulture (*System C2*)	Chebka	Fortified hamlets	Herding, cultivation, collecting, and limited hunting	34	61	2	3	x		Lakes gone
1000–1100	Foraging Bands with Domesticates (*System D1*)	Naghez	Semisedentary macroband camps	Herding, collecting, limited hunting and fishing, incipient cultivation	75	4	12	9			Shrinking lakes, decreasing rainfall
1100–1200		Nkhal	Temporary microband camps	Herding, fishing, limited collecting and hunting		?	?	?			
1200–1400		Goungon			99	1					Large lakes plentiful
1400–1500		Khimiya			x						
1500–2000		Akreijit		Hunting, fishing, limited collecting							Very dry

After this, "mobile, horse riding" Libyan Berbers with iron tools invaded the area, ending the Neolithic.

Mauritania thus provides the best evidence for a tertiary development in Africa, and there is enough ecological and population data for a tentative discussion of necessary and sufficient conditions. A revision of Munson's chart (table 11.4) graphically portrays the correlations of these significant data.

The final two zones in Africa, the Sudan and the Egyptian subarea have, of course, been the scene of a huge amount of research for over two centuries. Nevertheless, the data relevant to the beginning of village agriculture are not much better than they are for Mauritania (DeBono 1981). In the Sudan, the less well known of the two subareas, controversial remains from Khartoum fit in the major period of transition from food collecting to food production. Fred Wendorf, Peter Shinnie (1971), and others have found numerous Paleolithic remains. Qadan remains, 11,000 to 8000 B.C., and remains with Sebilian flake tools and grinding tools may represent early Foragers. Next come the Khartoum hospital-site remains, which include abundant microlithic tools and harpoons. There is evidence of pottery (and perhaps a pre-pottery Mesolithic) and of village life from 7000 to 4000 B.C. Although the data are subject to debate, a case could be made that these were fishing Village Foragers—system C1 (Arkell 1975). Near the end of this period comes early Khartoum, indicated by wavy-line pottery and barley at Kadero and sites near Nabta Playa, and by a great increase in village size. These people could be classified as Horticultural Villagers—system C2 (see table 11.4).

Sites grew larger in the following Khartoum periods, and at Esh Shaheinab (3500–3100 B.C.) there was evidence of wheat and barley. Whether this represents village agriculture is unknown, but agriculture clearly had occurred in this area by the time of Christ, when the huge structures at Jebel Moya were built. This quick development of village agriculture seems to be a secondary one and rather different from developments in the central or western desert or from the slow tertiary development in Kenya.

Like the Sudan, Egypt has abundant Paleolithic remains of early hunters. Our story begins with the Sebilian at about 10,000 B.C. (see table 11.5). Evidence of Affluent Foragers is abundant at such sites as Kom Ombo, where people probably lived in large villages or base camps along the Upper Nile (Reed 1965). Although Fayum A (5200–3000 B.C.) and Badarian (5588–3700 B.C.) bear little evidence of cultural continuity to the Sebilian, there are traces of agricultural or horticultural villages using emmer, barley, and club wheat (Caton-Thompson and Gardner 1934). This sequence looks like a faster secondary development, but our data are inadequate for clearly determining necessary and sufficient conditions (F. A. Hassan 1984).

THEORETICAL CONSIDERATIONS

Because there are not enough data to test my hypothesis about secondary development in Egypt, let us turn back to the Sudan (see table 11.6). During the Khartoum Mesolithic (about 5000 B.C.) the sedentary Foragers dependent on fishing seem to have had a positive-feedback cycle between increasing population and

Table 11.5. Relevant archaeological sequences in Egypt and northern Sudan

Sudan	Lower Nubia	Upper Egypt	Lower Egypt	Dates
		Old Kingdom		
Omdurman Bridge	Jebel Moya	Second Dynasty		
		First Dynasty		
				3000 B.C.
		Late Gerzean	Ma'adi	
	Abkan			
Khartoum Neolithic			El Omari	
	Shamarkian	Early Gerzean		3500 B.C.
		Amratian	Merimde	
Early Khartoum				4000 B.C.
	Early Shamarkian	Badarian	Fayum A	
				4500 B.C.
Khartoum Mesolithic				

specialized food collection from circumscribed riverine resources. Apparently there was also some sort of exchange with the Near East or Egypt, where domestication and/or horticulture and/or agriculture had already occurred. As tensions increased, the Khartoum foragers supplemented or replaced their wild resources with barley horticulture from the adjacent area to the north. Horticulture of barley in Nabta Playa led relatively quickly to population increases.

By Early Khartoum (4000 B.C.), villages may have exhausted the riverine territory into which they could expand and may have destroyed the carrying capacity of the circumscribed fishing locales, thereby diminishing food production from wild resources. This would have necessitated more planting of borrowed domesticates—such cultivars as barley and emmer—as well as of local domesticates, for example, oil palm. The Horticultural Villagers of Middle Khartoum (Esh Shaheinab, 3000 B.C.) became Agricultural Villagers by late Afreh (2400 B.C.) or later Jebel Moya. This development might be stated in the following terms of necessary conditions (p. 316):

Table 11.6. Necessary and sufficient conditions as a positive-feedback process for secondary development in prehistoric Sudan

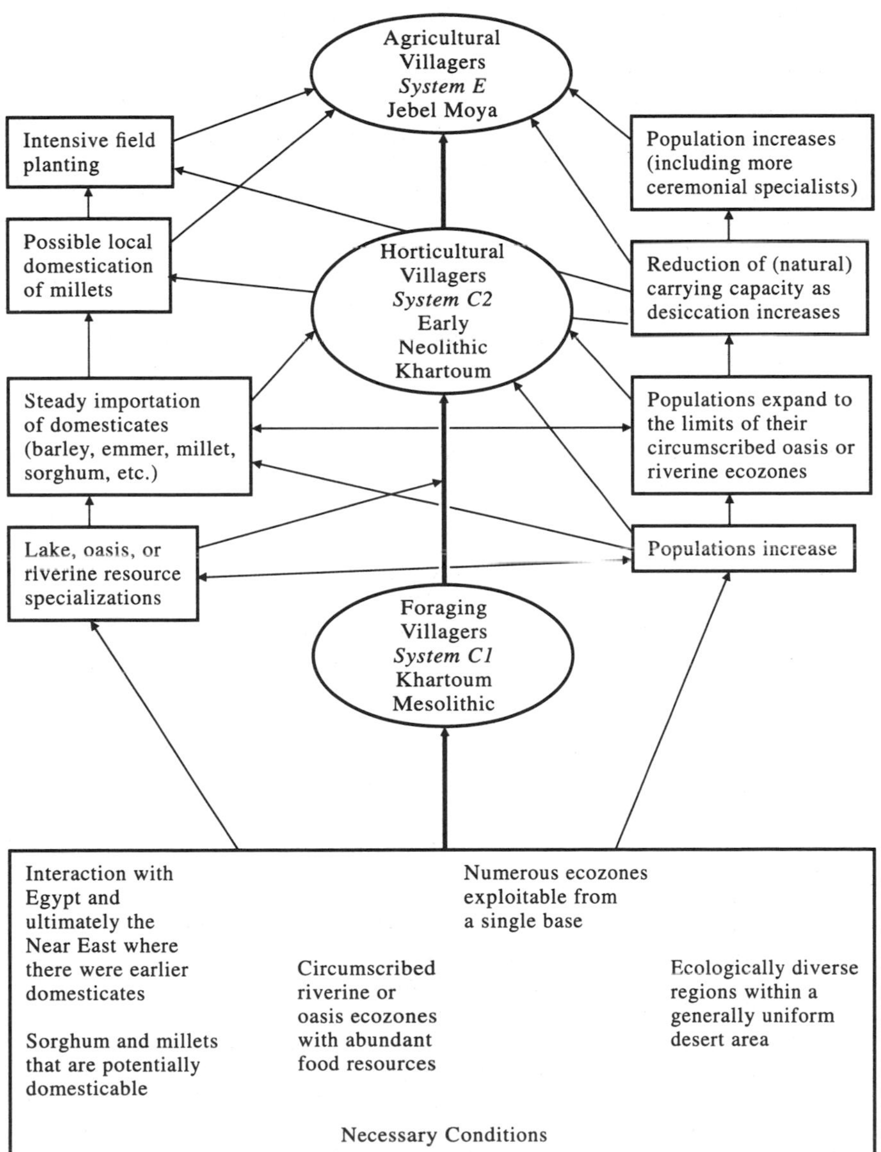

During Mesolithic Khartoum, sedentary Village Foragers specializing in the aquatic resources of riverine or lake ecozones lived in the Sudan. The regions in which they lived had regional and areal diversity, could be exploited from riverine or oasis bases, and had one or more circumscribed ecozones with easily exploitable, multiple resources (fish, inland game, gallery forest plants). The regions also had a number of local potentially domesticable millets and sorghum and were part of an interaction sphere, with access to the Near East or Egypt, where earlier domestications of wheat, barley, and other plants had taken place.

Given these necessary conditions, development of village agriculture may have occurred if the following sufficient conditions took place. Populations increased, resulting in large part from sedentarism in the Khartoum Mesolithic and Nabta Playa Foraging Villager phases. These increases caused decreases in the carrying capacity of one or more of the rich aquatic ecozones and resources became insufficient to support the populations in the lush circumscribed ecozones. Failure to develop or invent more resource specializations led to adoption of domesticates (barley) and incipient agriculture from Egyptian neighbors, which made the Khartoum people become Village Horticulturists. This change resulted in population increases that required more intensive planting as well as domestication of local plants, causing the people to become Village Agriculturists in Late Khartoum times.

What about tertiary development in Africa? The closest example seems to be the development at Dhar Tichitt in Mauritania (Table 11.7), where seasonally nomadic Capsian Foragers evolved into Efficient Foraging Bands—System D—at Akreijit (2000 B.C.), developing by Nkhal times (1200 B.C.) into Semisedentary Bands with Domesticates. In the Chebka phase (1000 B.C.) these people became Horticultural Villagers—system C2—who developed into Agricultural Villagers—system E—in Akjinjeir times (400 B.C.). This area contains evidence of population increases, environmental exploitation, and environmental changes that seemed to cause tertiary development. Similar developments seem to have occurred at a slightly earlier date at Lake Chad in the Mesolithic-Neolithic sequence and in the region of Lake Victoria. Neither of these, however, provides the fine causative data P. J. Munson has discussed for Mauritania (Munson 1968). Nevertheless, I believe all three developments may be portrayed as arising from a positive-feedback cycle that may be stated as follows.

If Capsian-like bands have relatively efficient foraging systems in an easily exploitable and uncircumscribed lake environment in the desert or savanna, they will continue to improve this efficient fishing or aquatic foraging system until forager villages develop, even if these foragers have exchanges with an area such as Egypt and/or the Sudan, in which domestication and/or horticulture and/or agriculture have developed. Only as the foraging system begins to overextend its local lake ecozones and the lush environment slowly changes for the worse during the general desiccation between 4000 and 1000 B.C., which shrinks the lakes and changes the savanna to desert, will the foragers begin to use domesticates. At first they would add only a few domesticates—sorghum or barley from Egypt or the Sudan—to supplement their efficient semisedentary subsistence system of

Table 11.7. Necessary and sufficient conditions as a positive-feedback process for tertiary development in Mauritania

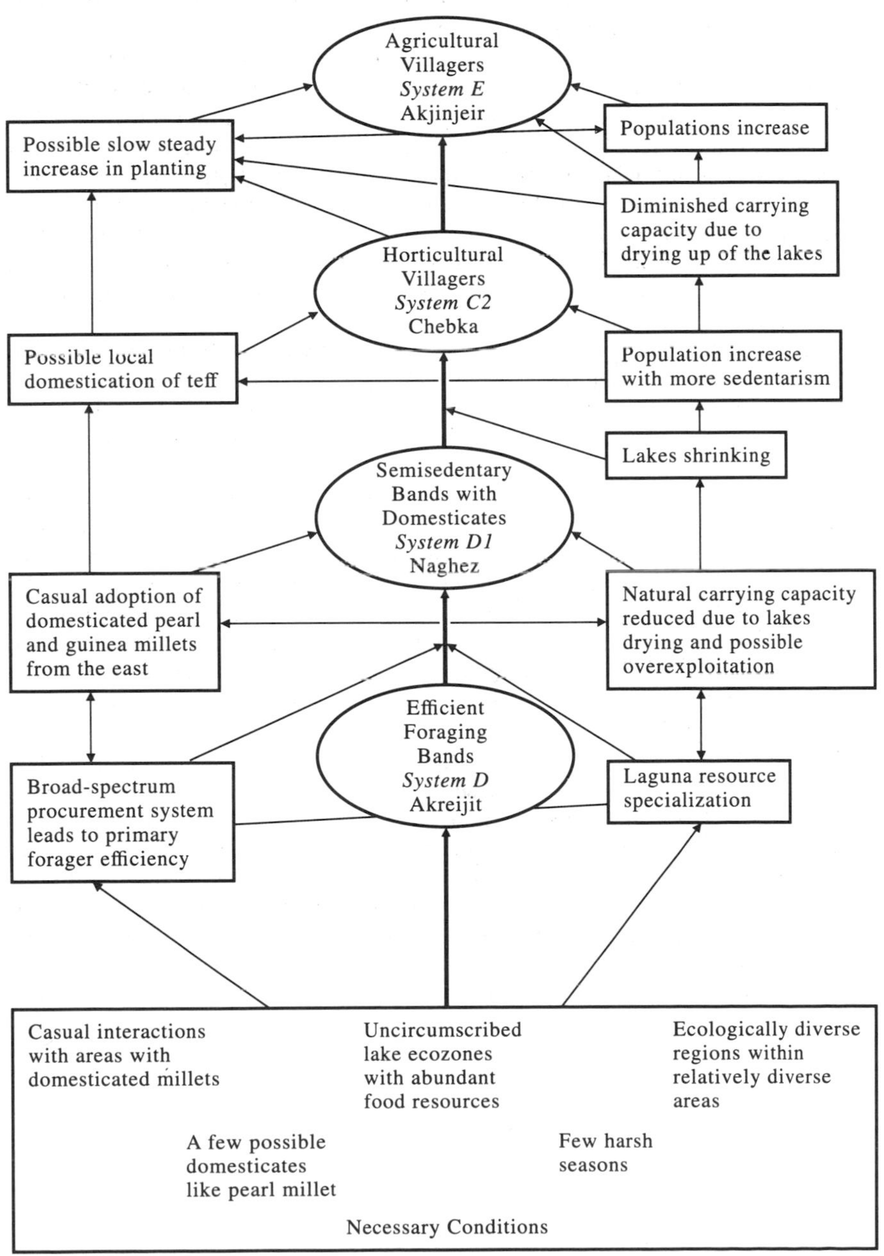

foraging. As population gradually increased in Neolithic times and the environment's carrying capacity slowly diminished due to climate change, the people would adopt more and more plants—millets, wheat, and so on—until horticultural villages developed. By the Iron Age, agricultural villages would exist.

Following are the necessary and sufficient conditions that seem to apply to this feedback cycle in the Dhar Tichitt region. Future research may show that the Lake Chad and Lake Victoria areas developed through a similar process. Dhar Tichitt Capsian forager bands had the following necessary conditions: Regional diversity around the lakes in Mauritania as well as xerophytic diversity for the area as a whole, combined with lush uncircumscribed ecozones such as oases, food resources available in all seasons, some local plants (e.g., pearl millet) susceptible to domestication, and exchanges with adjacent areas such as Egypt and the Sudan that possessed cultivars and domesticates and were practicing horticulture and/or agriculture.

When these necessary conditions were combined with the following sufficient conditions, village agriculture developed. Wild-food procurement in Akreijit times intensified, as did resource specialization with energy-efficient fishing and collecting foraging systems. By Naghez times, intensive foraging upset the carrying capacity of the lush ecozones. At the same time a natural reduction of the carrying capacity of local foraging areas was taking place as a result of shrinking lakes and decreasing rainfall, so that the relatively successful foragers slowly adopted barley and millet horticulture in Chebka times and thereby became increasingly sedentary, which resulted in population increases while the environment's carrying capacity was shrinking due to more desiccation and climatic change. This made it necessary for the people to adopt more and more domesticated plants, which caused further increases in population, creating a positive-feedback system that caused the people to become Village Agriculturists by Akjinjeir times.

Even though the data are insufficient, we do seem to have a valid case for tertiary development in Mauritania. Research in the Lake Chad and Lake Victoria regions will perhaps provide sequential data to test the model further.

The pattern of tertiary development in western Africa resembles that of the New World tropics and parts of southeastern Asia and (in a nonspecific way) Europe and the Eastern Woodlands of the United States. Further testing of the model is needed to confirm or refute the various parts of my hypothetical reconstructions. This preliminary attempt at synthesis and analysis should indicate the directions future research should take.

CHAPTER 12

Conclusions

In the preceding chapters I have attempted to test my three hypotheses about the development of village agriculture with data from various regions of the major areas of the prehistoric world. Available data were uneven, and the testing in all the culture areas is still inadequate. The tests of my primary developmental hypothesis, which concerns the initial domestication of plants and then later the relatively slow evolution of village agriculture, by way of routes 1 and 2 (see fig. 18), was attempted using sequential archaeological information from three (highland) regions of the Andean culture area, from four regions in Mesoamerica, and from a single area of the Near East. The other major primary culture area—the Far East—has not yet yielded sufficient relevant sequences to test my hypothesis.

Testing of the secondary developmental hypothesis (by way of routes 6 and 3) was more adequate because we have relevant sequences from all the above-mentioned Centers as well as from some non-Centers. From the Andes there were four regional sequences; from Mesoamerica, one; from the Near East, four lowland sequences and a possible one from the highlands; and from the south-central coast of China, one possible sequence. The non-Centers yielded two sequences from the U.S. Southwest, perhaps two from India in the Ganges and Indus river valleys, two from the Nile region of Africa in Egypt and Sudan, one from the Ecuadorean Coast of the New World tropics, and at least one from the Aegean area of Europe.

Tertiary developments (by way of routes 7, 5, and 4) all came from non-Centers, except for one in Japan. Although more numerous than other developments, they had fewer relevant sequential remains (preserved domesticated plants) and varied considerably one from the other. The best data on tertiary development came from eight sequences in Europe northwest of the Aegean. Some relevant information also came from six regional sequences of the eastern United States, possibly from one or two of the U.S. Southwest sequences, and from four possible sequences from southeastern Asia and Oceania, from three or four relatively poor ones from Africa, and from two or three poorly documented ones from the New World tropics.

All in all, I have made about fifty tests of my models—eight of the primary hypothesis, about seventeen of the secondary, and possibly twenty-six of the tertiary. There is thus some data to substantiate my trilinear theory of the development of village agriculture.

Each of the three hypothetical models tested had its own set of almost mutually exclusive necessary and sufficient conditions that caused it to evolve. I say "almost mutually exclusive" because two necessary conditions—potentially domesticable plants and an environment susceptible to planting—are common to all three models. I might add that these common conditions are often absent from areas whose people remained Hunter-Collectors (stage 1) or who developed from Hunter-Collectors to Foragers (step 2) but never became Village Agriculturists (stage 3). Another factor common to my three sequences is the sufficient condition of the post-Pleistocene reduction of food resources due to the extinction of fauna and megafauna. Yet reactions to even this condition differed in the various sequences. Obviously, the mutually exclusive conditions are the essential characteristics of my theory.

By way of summary, let us reexamine the conditions or causes of each of the three developments, not just to assess their validity in terms of the previously described sequences but also to examine their similarities to or differences from each other. As a means of further enhancing their validity, I shall briefly mention some similar sequences that lack some or all of these conditions and therefore did not reach village agriculture.

THE PRIMARY DEVELOPMENTAL HYPOTHESIS AS THEORY

Let us first examine the necessary conditions for primary development. Originally we conceived this as the development from Hunting-Collecting Bands (system A) via route 1 to Incipient Agricultural Bands (system B1), then via route 2 to Village Agriculturists (system E). Our comparative data suggested, however, that these three stages should be modified with transitional substages: a Destitute Forager stage (system B) between stages 1 and 2, and an Agricultural Band stage (system B2) between stages 2 and 3.

Region-by-region hypothesis testing also modified the necessary and sufficient conditions of primary development as follows (see Tables 1.2 and 12.1):

Necessary Conditions

1. Environments with layered ecozones that have regional and areal diversity
2. Multiple food resources that cannot be exploited from a single base
3. Seasonality with "harsh" (often dry) seasons in which food resources are limited
4. Intense interaction with ecologically similar regions (that may have developed domesticates in a similar manner) as well as with regions having secondary development
5. Potentially domesticable plants in one or more of the ecozones

Sufficient Conditions

1. End of the Pleistocene with concomitant climate change, making harsh seasons harsher, and reduction of fauna and megafauna and/or food supplies
2. Development of new subsistence options such as intensive seed collection or food (seed) storage
3. Development of seasonally scheduled subsistence systems, which often upset the local ecosystems and the genetics of potentially domesticable plants (because of selection of mutants or more productive plants from the random variation of the species)
4. The development of micro- macroband scheduled settlement patterns
5. A gradual increase in population (causing greater demand for food)
6. Planting to ensure seasonal security and allow longer stays and decreased residential mobility
7. Importation of more domesticates
8. Specialized plant processing (including hybridizations)
9. Still larger, more sedentary populations
10. Further decrease in residential mobility, leading eventually to sedentarism and village agriculture

Let us now consider each of these basic conditions in more detail, emphasizing their uniqueness. The necessary condition of the presence of domesticable plants (that is, of "genetically unstable plants, capable of improving their food yields and that could be planted" [Dimbleby 1967, 80]) is a pre-condition for all three kinds of development. Domesticable plants also existed in some other areas such as the Owens Valley of California, where the people developed from Hunters to Incipient Agriculturists but did not become Village Agriculturists, because certain sufficient conditions, which will be discussed shortly, were lacking (Thomas 1983, 1986). Needless to say, the lack of domesticable plants or environments where planting was impossible kept many peoples from advancing beyond the Forager stage—among them the Basin Plateau peoples of the western United States, the Eskimos of the New World Arctic, and the Australian Aborigines. Other groups who lacked these conditions remained Hunters—the Alacalaf and the Araucanians of South America, for example.

A similar situation applies to another necessary condition: the intense interaction with ecologically similar regions that had also developed domesticates. Obviously, there are all sorts of places in the world that did not interact intensively with the highland regions of the Andes, the Near East, Mesoamerica, or the Yellow River region of the Far East. These places failed to develop incipient agriculture (system B1) via route 1, and consequently they did not go on to village agriculture via route 2.

Areas such as southern Bolivia, northern Chile, and northwestern Argentina, for example, had environments quite similar to highland Peru (Willey 1971) but went through a secondary or tertiary development rather than a primary one because their interactions with the primary development of Peru were not sufficiently intense. Further, they seemed to lack certain sufficient conditions, such as

Table 12.1. Comparison of the necessary and sufficient conditions of primary development with the cultural sequences of areas that followed the differing routes of development

	Route 1					Route 2				Route 3				
	Far East, Yellow River	Andean highlands	Mesoamerican highlands	Near East highlands	Owens Valley, California	Andean highlands	Mesoamerican highlands	Near East highlands	Far East, Yellow River	Coastal Ecuador	Coastal Peru	Mesoamerican lowlands	Levant (Near East)	Near East riverine
Necessary conditions														
1. Environments with layered ecozones that have regional and areal diversity	x	x	x	x	x	x	x	x	x					
2. Multiple food resources *not* exploitable from a base	x	x	x	x	x	x	x	x	x					
3. Seasonality with "harsh" (dry) seasons with limited food resources	x	x	x	x	x	x	x	x	x					
4. Intense interaction with similar areas and/or regions with similar developments (including primary domestications)	?	x	x	x		x	x	x						
5. Potential domesticates	x	x	x	x	x	x	x	x	x	x	x	x	x	x
Sufficient conditions														
1. The end of Pleistocene causes														
(a) reduction in fauna and megafauna	x	x	x	x	x	x	x	x	x	x	x	x	x	x
(b) harsher seasons	x	x	x	x	x	x	x	x	x	x	x	x	x	?
2. Development of new subsistence options:														
(a) seed collection	?	x	x	x	x	x	x	x	x	?	x	x	x	x
(b) seed storage	?	x	x	x	x	x	x	x	?	x	x	x	x	x
3. Seasonal scheduling, which upsets														
(a) the local ecosystem(s)		?	x	x	?	?	x	x						
(b) genetics of the domesticable plants		x	x	x		x	x	x						
4. Micro-macroband settlement pattern		x	x	x		x	x	x		?	x	x	x	?
5. Longer stays by larger populations who need more food		x	x	?		x	x	?			x	?	x	x
6. Plant for														
(a) seasonal security		x	x	x	?	x	x	x						
(b) longer residence		x	x	x		x	x	x			x		x	?
7. Import of more domesticates for planting		x	x	x		x	x	x		x	x	x	x	x
8. Specialized planting (often of hybrids)		x	x	x		x	x	x						
9. Larger populations		x	x	x		x	x	x		x	x	x	x	x
10. Living in one spot longer and longer until sedentary life is attained		x	x	x		x	x	x	?					

a seasonal settlement pattern or a subsistence system that included seed collection and storage. This lack of the sufficient conditions needed to stimulate early collectors or foragers to take routes 1 or 2 to village agriculture is also the reason many other groups that had the necessary conditions for primary development did not evolve in this manner.

Many groups that went through secondary or tertiary development lacked the

Route 3						Routes 4-5-12								Rte 9	Rtes 6-10-11			Route 7-8					Route 13		
Coastal China	Egypt riverine	Sudan riverine	Aegean coastal	India riverine	U.S. Southwest	Europe	Eastern U.S.	Arid Africa	Tropical Africa	Tropical New World	Southeastern Asia	Oceania	Japan	Owens Valley, California	Northwest Coast, Canada	California coast	Chile north coast	Monitor Valley, Nevada	Danger Cave, Utah	Eskimo	Araucanians, Chile	Boreal tribes, Canada	Bushman, South Africa	West Australians	Alacaluf, Chile
					x									x				x	x	?	?		?	x	
														x				x	x		x		x	x	
					x									x				x	x	x	x	?	x	x	x
					?																				
x			x	x	x	x	x	x	x	x	x	x	?	x											
x	x	x	x	x	x	x	x	x	x	x	x	x	x	x	x	x	x	x	x	x	x	x	x	x	x
?	x	x	?	x	x	x	x	x	?					x				x	x		?		x	x	x
x	x	x	x	x	x	x	x	x	?	?	x			x		x	?	x	x		x		?	?	
x	x	x	x	x	x	x	x	x			?	x	x	x		x	x	x	x		x		x	x	
					?	?	?	x						x	?			?	?						
						x	?	?																	
	x	x			x																				
					x																				
					x																				
					x																				
x	x	x	x	x	x	x	x	x	?	x	?	?	x												
x	x	x	x	x	x	x	x	x	x	x	x	x	x												

necessary condition of seasonality with harsh (often dry) seasons that limited food resources. A few areas of secondary development did have seasonality—the American Southwest, Egypt-Sudan, and the Levant—as did a few areas of tertiary development, such as the African regions around Lake Chad and Mauritania. Also, many desert peoples (Australian Aborigines, Bushmen, etc.) who never left stage I also experienced these environmental conditions (see table 12.1).

Many of the nondeveloping desert groups lived in lagoon or oasis areas, such as the Owens Valley (Thomas 1969), where many of their basic resources could be exploited from a strategically located base. In contrast, a necessary condition of primary development is that the basic food resources are *not* exploitable from a single base, which causes people on their way to becoming Incipient Agriculturists to be seasonally nomadic and exploit a variety of ecozones, each of which has different plants, some of them potential domesticates.

The final necessary condition is of a more general environmental nature, but it clearly differentiates primary development from secondary and tertiary. Areas of primary development appear to have both regional and areal diversity. Each consists of a series of regions containing different, often layered, ecozones. In other words, the six different ecozones of the Tehuacán Valley in Mexico—El Riego oasis, travertine slopes, alluvial thorn forest, dissected slopes, valley center steppe, and humid river bottoms—differ from each other (MacNeish, Fowler et al. 1975). These ecozones are often layered, cannot be exploited from a single base, have different food resources, and often have resources that are only exploitable in certain seasons. In other words, there is regional diversity. Furthermore, the ecozones of the Tehuacán Valley differ from those of neighboring areas—the Valley of Mexico, Puebla, and Oaxaca (Schoenwetter 1974, fig. 21)—whose microenvironments also differ internally from each other.

In addition, each of the series of regions making up the Tehuacán subarea has a different complex of ecozones (often because of differing elevations), giving areal diversity to the whole subarea.

A few highland areas such as Scandinavia, the Canadian Rockies, and the Himalayas of northern India may have this kind of ecology but never developed agriculture. Usually they lack the necessary possibilities for exchange systems or for planting potential domesticates, or they have different sufficient conditions that prevented them from evolving agriculture. However, one of the prime factors differentiating areas of secondary or tertiary development from areas of primary development is that the areas of primary development have contrasting, almost dichotomous, general ecologies rather than similar ones. These necessary conditions have limited, if not governed, the direction by which the three developments evolved toward village agriculture.

More important than necessary conditions in making primary developments unique are sufficient conditions (see table 12.1). The exception is the first condition—the end of the Pleistocene with its concomitant diminished (faunal) food resources and increased seasonality, making the harsh and often dry seasons become harsher or drier and leading to more seasonal contrast in climates. With few exceptions, this change occurred worldwide and affected most cultural developments—from the Paleolithic to the Mesolithic or Epipaleolithic, Paleoindian or Lithic to the Archaic, and so on.

As mentioned previously, this drastic change often resulted in the development of a broad-spectrum subsistence strategy and an increase in subsistence options (Binford 1968; Flannery 1966). In areas of primary development, one option is always collecting and storing seeds for the harsh season(s). This option was adopted by many of the Mesolithic or Epipaleolithic cultures in the Levant, the

Archaic of California and/or the Great Basin, and Saharan Africa. Areas of primary developments were unique, however, in that seed collection and storage operated in conjunction with several other conditions—a seasonally scheduled subsistence system that exploited a number of environments (some with potentially domesticable seed plants), a micro- macroband settlement pattern, and the selection of seeds with larger size or other beneficial characteristics for food. Annual reoccupations of the seed collection zone caused changes in the natural environment of the seeds and in the genetic structure of the seed plants. All this eventually led to planting of genetically different seeds, that is, domesticates. The combination of these features, well-documented for Tehuacán-Oaxaca and Tamaulipas in Mesoamerica (see chapter 3), seems unique to primary development. There is a strong suggestion that this process also occurred in the Shanidar area of Iraq (chapter 4), as well as in the Junín and Ayacucho regions of the Andes (chapter 2).

Now the question becomes, Why did this development not occur earlier, for the necessary conditions certainly existed in the primary Centers before the end of the Pleistocene and the climatic fluctuations that reduced the food supplies and made the harsh seasons harsher. I think the answer is that the Hunters had not yet developed a wide variety of subsistence options (such as seed storage) or the tools and technology for them (baskets, nets, and so on), nor had they acquired the knowledge of the environment to allow them to create a seasonally scheduled subsistence system. This meant that the first part of the positive-feedback cycle, which involved reduction of food staples and development of a broad-spectrum procurement system with its necessary special tools and seasonal scheduling, could not occur. In other words, two of the three elements necessary for development—broad-spectrum procurement and seasonal scheduling—had not yet evolved. The upsetting of the ecosystem by seasonal scheduling accompanied by seed selection therefore could not take place. Both the right natural causes and the appropriate cultural conditions or causes had to be present.

At roughly the same time—the end of the Pleistocene—the combination of necessary and sufficient conditions found in the Centers did not occur in any of the non-Centers, and this prevented them from developing along route 1. The unique combination of necessary and sufficient conditions found in the primary Centers did not occur in any of the sixteen or so secondary developments or in the thirty or so tertiary developments, nor has such a combination been documented for any of the places that did not develop village agriculture or for any Foraging Bands with Incipient Agriculture such as those of the southwest coast of Australia (Hallam 1988). A possible exception is the Owens Valley of California (Lawton et al. 1976), but that area lacks one necessary condition. Most of the rich resources of the valley were exploited from a single base camp, and that condition prevented the evolution from system B1 to E.

Basically, broad-spectrum procurement, seasonal scheduling, and a reduction of food staples are the conditions leading from system A to B and B1 via route 1. A driving force was the need to ensure sustenance by obtaining enough food from storable and grown domesticates so the people could survive the harsh seasons. Once this security was attained, the trend toward village agriculture via

Table 12.2. Necessary and sufficient conditions as a positive-feedback process in primary developments

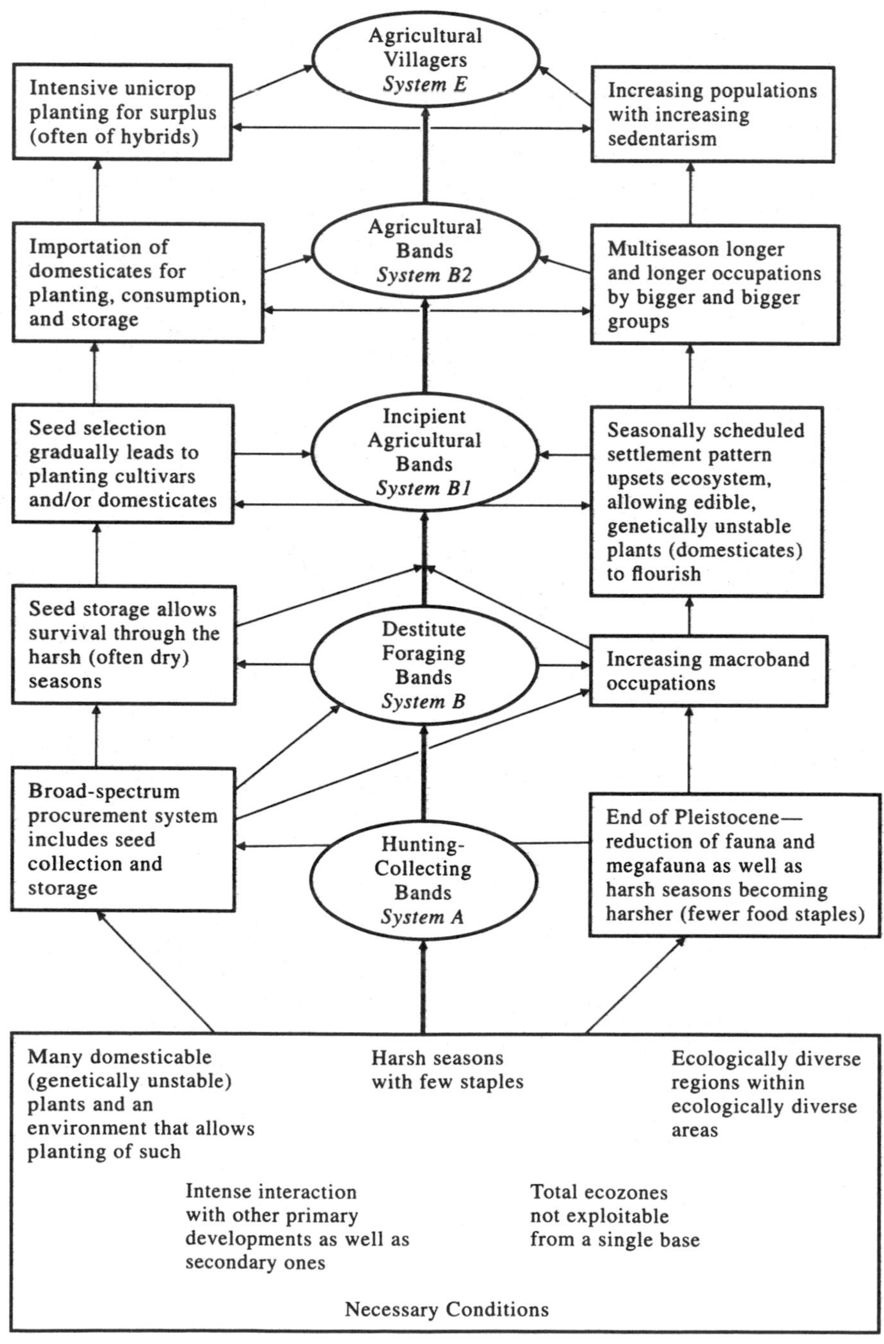

route 2 slowly and imperceptibly became a drive not just for food resource security but for a more stable, sedentary way of life. The latter was attained by a slow, steady increase in agricultural production.

Other sufficient conditions that triggered this gradual development of village agriculture include the planting and use of more domesticated plants (including hybrids) that the Foraging Bands with Incipient Agriculture often received in exchanges with similarly developing neighbors. Increased planting caused people to stay longer in one spot, which resulted in population growth. This, in turn, led to more specialized planting, often of hybrids or newly developed, more productive seed plants that further increased populations. All these conditions seem well-documented for the Tehuacán, Tamaulipas, and Ayacucho sequences and may also pertain to the Junín area of Peru and the Jarmo-Shanidar highland regions of Iraq. Yet not all of them seem to apply to secondary or tertiary developments, where other, rather different, triggering causes occur. Although the lack of all the sufficient conditions for primary development prevented many Foragers or Collectors from evolving into Village Agriculturists, it was not just the necessary or sufficient conditions in and of themselves that brought about the development, but rather the interaction of these conditions in a positive-feedback cycle or in "deviation amplifying processes" (Flannery 1968). Let me restate my modified scenario of this process (see table 12.2).

When the Pleistocene ended, drastically reducing faunal food resources and making harsh seasons harsher, some groups survived in certain distinctive cultural areas that had ecological, regional, and areal diversity by developing a broad-spectrum subsistence system. That system exploited the diverse ecozones, which could not be exploited from a single base, through seasonal scheduling. This annual schedule included collection of potentially domesticable seeds, which the collectors stored for use during the harsh season. This behavior resulted in an annual return to the seed-collection areas and tended to alter the ecozone by the growing of seeds in disturbed habitats; at the same time, seed selection changed the genetics of some of the seeds. Often the result was the planting of some seeds, now domesticates, "for more and larger seeds emerge from deep planting" in disturbed habitats (Harlan, de Wet, and Stemler 1976, 43; Bell 1965). Planting resulted in increased food production, which often led microbands to congregate into macrobands. Again, the larger number of people needed still more food. Planting of local domesticates and importing of more domesticates from similarly developing interacting ecozones resulted in still larger populations and a tendency, for reasons of food security, for longer and longer seasonal occupations by macrobands. As more food was needed because of this decreased residential mobility, more intensive planting, selection of domesticates (often hybrids), and specialization in the growing of certain productive plants evolved. Larger and larger groups continued to live in one spot longer and longer until the Agricultural Bands gradually came to reside all year in one place as sedentary Agricultural Villagers.

Documenting this dynamic cultural process with convincing archaeological "facts" presents major problems. Yet in the regions of primary development in Peru, Mesoamerica, and perhaps the Near East, certain recognizable archaeological elements or traits seem to reflect the salient features of this positive-feedback process (see table 12.3). These elements are roughly equivalent, as noted below, to the sufficient conditions shown in Table 12.1. In brief, the elements are as follows:

1. The disappearance of the megafauna and decreasing numbers of fauna at the end of the Pleistocene, which can be documented by paleontological studies as well as by proportions of bones to other ecofacts and artifacts (equivalent to sufficient condition 1a in table 12.1).
2. Greater annual seasonality and indications that harsh (dry) seasons became harsher or drier, often evidenced by study of pollen profiles (equivalent to sufficient condition 1b).
3. Increase in foraging or subsistence options other than hunting, which may be documented by various complexes of artifacts and/or use-wear studies that reflect subsistence practices (equivalent to sufficient condition 2a).
4. Increased seed storage, which may be evidenced by the presence of pits or by a greater number of pits (equivalent to sufficient condition 2b).
5. A seasonal settlement pattern, evidenced by ecofactual seasonal indicators in various types of occupations (equivalent to sufficient condition 3).
6. Seed selection, often evidenced by increasing proportions of certain seeds, seed size, pollen grain size, etc. (roughly equivalent to sufficient conditions 2a and 2b).
7. Increased numbers of macroband occupations, as documented by increasing site or occupation size as well as by increased number of fireplaces and the like (equivalent to sufficient conditions 4 and 5).
8. Importation of domesticates, which may be indicated by plant remains of domesticates not native to the local region or by earlier dates on the same domesticates found in other regions (equivalent to sufficient condition 7).
9. Increase in length of occupations or multiseason occupations, which may be indicated by analysis of seasonal indicators in various occupations (roughly equivalent to sufficient conditions 5, 6, and 10).
10. Planting for storage, which may be documented by study of plant remains (equivalent to sufficient condition 2b and 6a).
11. Increased use of hybrids or derived races, as indicated by botanical study of domesticated plant remains (equivalent to sufficient condition 8).
12. Increase in population, as may be evidenced by study of site size and settlement pattern analysis (equivalent to sufficient conditions 9 and 10).
13. Intensive planting of domesticates, as may be documented by plant remains and/or pollen analysis (equivalent to sufficient condition 8).

Not only should the above traits appear in the relevant archaeological sequences, but they should appear as complexes in roughly the above order. The existence of such a pattern may indicate that a specific positive-feedback process or deviation-amplifying process occurred. As table 12.3 indicates, this sequence of traits or elements in the expected order does seem to reveal that such a process occurred in various regions of primary development.

The first two elements in table 12.3—disappearance of megafauna and increased seasonality—occurred to the early (Pleistocene) Hunters in all levels of development in many parts of the world about ten thousand years ago. However, due in large part to various necessary (environmental) conditions, the results were unique in areas of primary development and set the stage for some fundamental changes.

In the Ayacucho region the extinction of mastodon, horse, sloth, and other animals (element 1, table 12.3) between Huanta and Puente times—roughly ten to eleven thousand years ago—as well as the shrinking of the tundra while the humid forest expanded and seasonality increased (element 2), as indicated by pollen profiles (MacNeish 1981b), seems to have led to an increase in subsistence options (element 3) in late Puente and Jaywa times as well as to an emphasis on seed collecting in Jaywa times (MacNeish, Vierra et al. 1983). The numerous storage pits in Jaywa times suggest increased storage (element 4), and settlement-pattern studies suggest a seasonally scheduled subsistence system (element 5—MacNeish et al. 1983). This positive-feedback process caused the development from Hunting-Collecting Bands to Foraging Bands with Incipient Agriculture—from stage 1 to stage 2—at Ayacucho roughly between 9000 and 5800 B.C.

Elements 1 and 2 of the primary development sequence are not as well documented for the Junín–Cerro de Pasco region of the central Andes, but again horses seem to have become extinct between Huargo to Panalauca times. The Lake Junín pollen profile indicates major desiccation and perhaps more seasonality at about 8000 B.C. (Pearsall 1988). Whether the following Pachamachay phase 1 had more subsistence options (element 3) than the Panalauca phase is not so well documented, but storage pits (element 4) and evidence of seed selection (element 6) were uncovered. John Rick's analysis of individual floors at Pachamachay suggests that a seasonally scheduled settlement pattern and subsistence system (element 5) had developed by Pachamachay phase 2, but more reconnaissance is still needed (Rick 1980).

The third major investigated region of the Peruvian highlands—the Callejón de Huaylas—has the first part of development, stage 1, the Hunting-Collecting Bands. Extinction of megafauna (element 1) is perhaps not well documented in Guitarrero I, and the pollen profile is equivocal (Lynch 1980). Guitarrero II, however, does seem to show increasing subsistence options (element 3), evidence of storage (element 4) and seed selection (element 6), and hints of a seasonally scheduled settlement pattern (element 5). Again, more survey is needed to document this final feature. Nevertheless, in spite of these few gaps, the shift from stage 1 to stage 2 and the elements possibly bringing it about are quite well documented for the central highlands of Peru.

The evidence for the development from Hunting-Collecting Bands to Incipient Agricultural Bands is even better for four of the Mesoamerican regions. Extinction of horse, giant turtle, jackrabbit, and probably mammoth and mastodon as well as other animals (element 1) occurred between early and late Ajuereado times (8000–2000 B.C.) in both Tehuacán (MacNeish, Fowler et al. 1975) and Oaxaca (Flannery et al. 1981; Flannery 1986). A similar phenomenon seems to have occurred in the Valsequillo region of Puebla (Irwin-Williams 1967c) and in the Valley of Mexico (Niederberger 1976), all of which also show evidence of greater seasonality (element 2).

Table 12.3. Comparison of the traits of regional sequences that may represent the positive-feedback processes of primary developments, stages 1 to 3

Areas

- Ayacucho region
- Junín region
- Callejón de Huaylas region (Andes)
- Oaxaca region
- Tehuacán region
- Southwest Tamaulipas region (Mesoamerica)
- Sierra de Tamaulipas region
- Iraq-Iran Hilly Flanks (Near East)

Cultural phases		Range of dates	**Hunting-Collecting Bands**	1. Pleistocene faunal extinctions	2. Climatic change – harsher conditions	3. Increased subsistence options	4. Seed storage pits	5. Seasonal scheduling	6. Seed selection	7. Increasing macroband occupations	**Incipient Agricultural Bands**	8. Importation of domesticates	9. Longer occupations	10. Planting for storage	11. Increased use of hybrids	12. Increase in population	13. Intensive specialized planting	**Agricultural Villagers**
	Jarmo	6000 – 6500 B.C.												x	x	x	x	x
STAGE 3	Laguna	0 – 500 B.C.												x	x	x	x	x
	Mesa de Guaje	800 – 1200 B.C.												x	x	x	x	x
	Purrón-Ajalpan	900 – 2300 B.C.												x	x	x	x	x
	Espiradon	1400 – 1600 B.C.												x	?	x	x	x
	Toril	1000 – 1750 B.C.												x	?	x	x	x
	San Blas-Mitos	1000 – 2000 B.C.												x	?	x	x	x
	Cachi-Andamarka	1250 – 3100 B.C.												x	x	x	x	x
	Çayönü	6500 – 8000 B.C.						x		?	x		x	x		x	?	
STAGES 2–3	Almagre	1500 – 2500 B.C.						x	x	x	x		x	x	x	x		
	Flacco-Guerra	1200 – 2500 B.C.						x	?	x	x	x	x	x	x	x		
	Abejas	2300 – 3500 B.C.						x	x	x	x	x	x	x	x	x		
	Martinez	2000 – 3500 B.C.						x		x	x	?	x	x		x		

	Quishqui Punco	? ?						x		?	x	?	?	?		x
	Pachamachay phases 4-5	2800 – 4400 B.C.						x	x	x	x	?	?			x
Chihua		3100 – 4400 B.C.						x	x	x	x	x	x	x	x	x
	Zawi Chemi-Shanidar B1	8500 – 9500 B.C.				?	x	x	x	x	x	?	x	?		
STAGE 2	La Perra	2500 – 3000 B.C.						x	x	x	x	x	x	?		
	Ocampo	2800 – 4000 B.C.				x	x	x	x	x	x	x	?	x		
	Coxcatlán	3500 – 5000 B.C.				?	x	x	x	x	x	x	x	x		
	Blanca	3500 – 5000 B.C.				?		x		?	x					
	Guitarrero IIe	5400 – 6000 B.C.				?	?	x	x	?	x	x		?		
	Pachamachay phases 2-3	4400 – 6175 B.C.				x	?	x	x	?	x	x				
Piki		4400 – 5800 B.C.					x	x	x	x	x	x	?			
	Shanidar B2-Zarzi	9500 – 1100 B.C.			x	x	x	x	?	?						
STAGES 1–2	Nogales	5000 – 6000 B.C.				x	x	x	?	x						
	Infiernillo	5000 – 7000 B.C.				x	x	x	x	x	?					
	El Riego-L. Ajuereado	5000 – 9000 B.C.				x	x	x	x	x	?	x				
	Jicaras-L. Naquitz	5000 – 7000 B.C.			?	x	x	?	?	x						
	Guitarrero IIa-IId	6000 – 7400 B.C.				x	x	x	x	?	?	x				
	Pachamachay phase 1	6175 – 7000 B.C.			x	x	x	?								
Jaywa		5800 – 7100 B.C.				x	x			x						
	Palegawra	11,000 –13,000 B.C.	x	x	x	x										
STAGE 1	Lerma	7000 – 9000 B.C.	x	x	x											
	Early Ajuereado	9000 –20,000 B.C.	x	x	x											
	Early Naquitz	7000 – 9000 B.C.	x	x	x											
	Guitarrero I	7400 – 9000 B.C.	x	?	?											
	Lauricocha 1-Huargo	7000 –10,000 B.C.	x	x	x											
Huanta-Puente		7100 –11,000 B.C.	x	x	x											

Earlier studies at Canyon Diablo terraces, and soils and plant and animal remains from two regions of Tamaulipas (MacNeish 1958) indicate a similar climatic shift (element 2). Terraces associated with the Diablo remains in the Sierra de Tamaulipas also had bones of extinct horse and beaver (element 1), so the same initial positive-feedback process could have been in operation.

The second process, involving increased subsistence options (element 3), seed storage (element 4), collection of a wide variety of seeds (element 6), and the development of a seasonally scheduled settlement pattern and subsistence system (element 5), is just as well documented for the same regions of Mesoamerica. From Naquitz to Jicaras in Oaxaca (Flannery et al. 1981; Flannery 1986), from late Ajuereado through El Riego in Tehuacán (MacNeish, Fowler et al. 1975), and from Lerma to Infiernillo in Tamaulipas (MacNeish 1958), all of the above elements seem to have been in operation. By about 6000 to 5000 B.C. we have Foraging Bands with Incipient Agriculture and the first evidence of plant domestication: pumpkins and probably corn in Oaxaca; *mixta* squash, avocado, and amaranth in Tehuacán; and pumpkins and perhaps chile in Tamaulipas.

The case for the Near East is the least clear of all the primary developments. Zone B2 of Shanidar Cave, however, does seem to have some evidence of more subsistence options than Palegawra, as indicated by more varied geometric microliths (element 3), while more grinding stones of querns appeared at the Zarzi site. The latter, along with increased *Cerealia* pollen, may indicate seed selection (element 6). Shanidar B2 has an increased number of storage pits (element 4), and its hypothesized winter occupation (R. L. Solecki 1964) may show a seasonally scheduled settlement pattern (element 5).

In all three areas the evidence from various regional sequences—three in the Andes, three or four in Mesoamerica, and perhaps one in the Near East—indicates the presence of the basic elements for the positive-feedback process in the shift from Hunters (stage 1) to Incipient Agriculturists (stage 2).

Elements of a new positive-feedback system appear in stage 2. Our best data come from the Mesoamerican highland areas with their great ecological diversity. In Tehuacán, system B1 is represented by the Coxcatlán phase, which had a seasonally scheduled subsistence system (element 5) and increasing use of domesticates from other regions (element 8). The following Abejas phase had increasingly longer and bigger occupations (elements 7 and 9) and more use of hybrid corn (element 11). Pits filled with domesticates (mainly corn) suggest specialized planting for storage (element 10) and longer occupations (element 9). All the above elements, working as part of a positive-feedback system, seem to have caused Abejas to develop gradually into Purrón and Ajalpan, whose people were Village Agriculturists, system E of stage 3 (MacNeish 1975).

The case for the Sierra de Tamaulipas and the Sierra Madre of southwest Tamaulipas is equally good. The La Perra phase had seed storage (element 4), increased subsistence options (element 3), seasonally scheduled subsistence occupations (element 5), and the use of foreign domesticates (element 8)—corn and beans—as well as locally domesticated pumpkins (element 6). During the following Flacco-Guerra and Almagre phases, there were longer and bigger occupations (elements 9 and 12) and pits full of corn (often hybrid, element 11), indicating intensive, specialized planting for storage (element 10). These elements combined to cause the people to become sedentary Villagers by Mesa de Guaje times, about 1500 B.C. (MacNeish 1958).

The documentation of the process in Oaxaca, from Blanca to Martinez to the Village Agriculturists of Espiradon, is not quite so good, but some of the elements—such as pits with domesticates, hybrid corn, and perhaps longer occupations at Yanhuitlán—do occur earlier, so the same positive-feedback process may have existed (Flannery, Marcus, and Kowalewski 1981; Flannery 1986). Zohapilco, in the Valley of Mexico, has a few of the above traits (Niederberger 1976) and may have experienced the same process.

Two regions of highland Peru also seem to have followed the same positive-feedback process from stage 2 to stage 3. The best case is in the Ayacucho Valley (MacNeish, Vierra et al. 1983). Here the Piki phase, with increased subsistence options (element 3), storage pits for seeds (element 4), a seasonally scheduled subsistence system (element 5), and incipient agriculture led in Chihua and Cachi times to a shift from macroband camps to some villages (element 7), increasing use of foreign domesticates (element 8), use of corn hybrids (element 11), longer and bigger occupations (elements 9 and 12), and the development of an intensive specialized vertical economy (element 13) by Cachi-Andamarka times (MacNeish, Vierra et al. 1983).

This process is less well documented for the other two regions in Peru. In the Junín region, Pachamachay phases 4 and 5 had evidence of increasing quinoa seed size as well as suggestions of specialized planting for storage (elements 6 and 10), longer and larger occupations (elements 9 and 12), and a gradual evolution of a sedentary way of life (Rick 1980). Unfortunately, this later process is not documented in the Callejón de Huaylas, although the possible late preceramic Quisque Punca remains may fit in this stage (Lynch 1980).

Equally difficult to document is the process in the Near East. Zawi Chemi, with its large open-site occupation, may be evidence of the development of base camps (elements 7 and/or 9 and 12), while its pollen profiles (Leroi-Gourhan 1969) may indicate seed selection (element 3) and perhaps domestication (element 6). The following stage, which is represented by Karim Shahir, Çayönü (Redman 1978), and the lower levels of Ganj Dareh (P. Smith 1975), had larger and longer occupations (elements 9 and 12), more pits for the storage of domesticates that were perhaps planted for that purpose (element 10), and imported grains such as emmer, einkorn, and/or barley (element 8). These elements may represent the transition to village life.

In summary, as tables 12.2 and 12.3 indicate, we do have some validation of my primary developmental hypothesis. The positive-feedback process resulting in primary development seems to be represented in three parts of the world. It is present in four areas of Mesoamerica (i.e., in two places in Tamaulipas, in Tehuacán, and perhaps in Oaxaca). In Peru, it is best documented for Ayacucho and perhaps for Junín, and part of the process seems to occur in the Callejón de Huaylas. In the Hilly Flanks of the Near East the Çayönü–Shanidar–Zawi Chemi–Jarmo complex suggests the process may also have occurred there. As yet we have no relevant data from the Far East. Obviously, we need more and better information, not only from all the sequences just mentioned, but also from each one of the four major cultural areas or Centers.

It is clear from these limited reports that we need basic data collection on all fronts, from the highlands of Peru and Mexico to the Hilly Flanks of the Near East and the whole of the Far East. What is more, we need more sophisticated scientific studies, such as nitrogen 15 and carbon 13 and 12 isotope analysis and amino acid and strontium studies of human bone to determine the presence (or absence) and proportions of domesticated plants in ancient diets as well as to determine whether the people were incipient or full-time agriculturists.

THE SECONDARY DEVELOPMENTAL HYPOTHESIS AS THEORY

From many standpoints the primary developmental hypothesis is not as well tested as the secondary one. Through secondary development, Hunting-Collecting Bands (system A) evolved relatively rapidly via route 6 into Foraging Villagers (system C1)—and then rapidly evolved into Agricultural Villagers (system E) via route 3 and/or routes 12 and 15. Often this development took place by substages. Between stages 1 and 2, an Affluent Foraging Band substage (system C) led to Foraging Villagers (system C1). Then, between stages 2 and 3, the Horticultural Villagers substage (system C2) intervened before Village Agriculture (system E). (See tables 12.4 and 12.6.) The systems and routes of secondary development differ from those of primary development, as do most of the necessary and sufficient conditions and the positive-feedback sequence for that development.

Primary and secondary developments both usually occurred in areas with potentially domesticable plants and environments susceptible to planting, so they have this necessary condition in common. In areas of secondary development the initial introduction of domesticates or cultivars came from areas that were ecologically dissimilar. In other words, the intense interaction was with the opposite end of the dichotomous interaction sphere rather than with similar areas of that sphere, as was the case in the primary development.

Other necessary conditions for these two developments exhibit even greater differences. A fundamental difference concerns the general aspects of the two environments. Secondary developments seem to have occurred where microenvironments differed within each subarea but where congeries of microenvironments were similar if not the same from one subarea to another. For example, the environments of each river valley along the Peruvian coast may include a variety of ecozones: rocky points, ocean bays, river deltas, and lower, middle, and upper riverine zones with gallery forests surrounded by lomas. Yet when several coastal valleys are compared—for example the Lurín, the Chilca, and the Pisco valleys—the same complexes of ecozones occur over and over again. A combination of relative areal uniformity with regional diversity thus exists in areas of secondary development.

Even more fundamental as a necessary condition for secondary development is a combination of lush ecozones with less productive areas. Usually one or more of the regional ecozones may be called "lush," in terms of having multiple and/or numerous food resources. However, these lush ecozones—for example, oases, rocky coastal points, deltas, or locations at rapids or the confluence of two or more rivers—are usually circumscribed by zones of less abundance. Usually all the lush zones or oases can be exploited from a single base, for they are located more or less radially from that base. This is a striking difference from areas of primary development, where productive ecozones cannot be exploited from one base.

Although areas of secondary development experience seasonality, none of the seasons is harsh enough to yield few or no food staples. Food may be gathered in all seasons, unlike the case in areas of primary development, where food resources were limited or unavailable in certain seasons. It is thus clear that secondary developments, whether in Centers or non-Centers, occurred in environments fundamentally different from those in which primary developments took place. An even greater contrast was occasioned by the sufficient conditions that brought about the secondary development of village agriculture.

Secondary developments, like primary, were triggered by the end of the Pleistocene with the reduction in faunal resources and greater seasonality that forced people to use a broad-spectrum subsistence system with more options and seasonal scheduling. Because of the different necessary (ecological) conditions in areas of secondary development, the reasons for change were somewhat different (see table 12.4). The sufficient conditions included intensification of the wild-food procurement system and resource specialization that caused people to become Sedentary Foragers. Their increased sedentarism, often at circumscribed "lush" base camps or villages, contributed to a population increase. Eventually the Sedentary Foragers were forced to accept domesticates from other groups living in areas of primary or secondary development and with whom they were interacting. The gradual use of horticulture of both imported and local domesticates led relatively rapidly to further increases in population, which in turn meant more use of domesticates. Foraging with horticulture then rapidly shifted to true agriculture.

Now let us examine each of these necessary and sufficient conditions in the areas in which secondary development occurred as well as where such conditions did not occur. The primary developments seem to have occurred in layered ecozones (not radial regions), where the areas or subareas differ. In contrast, secondary developments are characterized by areal uniformity and regional radial diversity. The Peruvian coast is the prime example of this type of environment. Within each river valley is regional diversity, but all river valleys in the coastal area or subarea are much the same. We do not as yet have enough Archaic sequences from coastal Mesoamerica to generalize, but the one relatively complete sequence from coastal Belize has a complex of ecozones that are relatively similar to those of Chantuto and Puerto Marquez on the Mexican Pacific Coast as well as to the Tecolutla River Valley of Veracruz, Mexico (see chapter 3).

All the Peruvian and Mesoamerican coastal river valley locations consist of a few circumscribed lush spots surrounded by a complex of ecozones or catchment basins. This pattern bears a general likeness to similar riverine zones in other parts of the world—Mureybit and Ali Kosh in the Near East, Khartoum on the Nile, Mesolithic sites on the Indus and Ganges rivers of India, and perhaps South American tropical coastal sites such as Valdivia in Ecuador. Vita-Finzi and

Table 12.4. Comparison of necessary and sufficient conditions of secondary developments with the cultural sequences of various areas that followed the differing routes of development

	Route 1					Route 2				Route 3				
	Far East, Yellow River	Andean highlands	Mesoamerican highlands	Near East highlands	Owens Valley, California	Andean highlands	Mesoamerican highlands	Near East highlands	Far East, Yellow River	Coastal Ecuador	Coastal Peru	Mesoamerican lowlands	Levant (Near East)	Near East riverine
Necessary conditions														
5. Potential domesticates	x	x	x	x	x	x	x	x	x	x	x	x	x	x
6. Intense interaction with dissimilar areas and/or regions with domesticates										x	x	x	x	x
7. Environment with (radial or catchment) regional diversity but areal uniformity										x	x	x	x	x
8. Multiple food resources exploitable from strategically located base										x	x	x	x	x
9. Some "lush" circumscribed ecozones with multiple food resources										x	x	x	x	x
10. Seasonality with relatively harsh seasons that rarely limit food resources										x	x	x	x	x
Sufficient conditions														
1. The end of Pleistocene causes														
(a) reduction in fauna and megafauna	x	x	x	x	x	x	x	x	x	x	x	x	x	x
(b) harsher seasons	x	x	x	x	x	x	x	x	x	x	x	x	x	?
2. Development of new subsistence options:														
(a) seed collection	?	x	x	x	x	x	x	x	?	?	x	x	x	x
(b) seed storage	?	x	x		x	x	x	x	?	x	x	x	x	x
3. Seasonal scheduling	?	x	x	x	x	x	x	x	?	x	x	x	x	x
11. Intensification of wild-food procurement systems										x	x	x	?	x
12. Resource wild-food specialization										x	x	?	x	
13. Sedentary life of skilled foragers										x	x	?	x	?
5. Larger populations										x	x	x	x	x
14. Populations outgrow circumscribed lush ecozones										x	x	x	x	x
15. Carrying capacity of this and other ecozones is diminished										x	x	x	x	x
16. Domesticates imported from areas of primary developments										x	x	x	x	x
17. Horticulture of imported and local domesticates										x	x	x	x	x
18. Further population increases result in further manipulation of domesticates and development of agricultural specialization										x	x	x	x	x

Route 3						Routes 4-5-12								Rte 9	Rtes 6-10-11			Route 7-8					Route 13		
Coastal China	Egypt riverine	Sudan riverine	Aegean coastal	India riverine	U.S. Southwest-Oshara/Gila	Europe	Eastern U.S.	Arid Africa	Tropical Africa	Tropical New World	Southeast Asia	Oceania	Japan	Owens Valley, California	Northwest Coast, Canada	California coast	Chile north coast	Monitor Valley, Nevada	Danger Cave, Utah	Eskimo	Araucanians, Chile	Boreal tribes, Canada	Bushman, South Africa	West Australians	Alacaluf, Chile
x	x	x	x	x	x	x	x	x	x	x	x	x	x	x											
x	x	x	x	x	x						?		?												
x	x	x	x	x	x								?		x	x	x					?			?
x	x	x	x	x	x	x	x	x	x	x	x	x	x		x	x	x			x					?
x	x	x	x	x	x		?								x	x	x								
x	x	x	x	x	x	x	x		x	x	x	x	x		x	x	x					x			x
x	x	x	x	x	x	x	x	x	x	x	x	x	x	x	x	x	x	x	x	x	x	x	x	x	x
?	x	x	?	x	x	x	x	x	?					x				x	x		?		x	x	x
x	x	x	x	x	x	x	x	x	?	?	x			x	x			x	x		x		x	x	
x	x	x	x	x	x	x	x	x			x	x	x	x				x	x		x		x	x	
x	x	x	x	x	x	x	x	x	?	?	x	x	x	x	x	x	x	x	x	?	x	?	?	?	?
x	x	x	x	x	x	x	x	x	?	?					x	x	x			x	x	x			
x	x	x	x	x	x	x	x	x				x	x		x	x	x			x					
x	x	x	x	x	x	x	x	x			?	x	x		x	x	x								
x	x	x	x	x	x	x	x	x			x	x	x		x	x	x								
x	x	x	x	x	x			x																	
x	x	x	x	x	x			x																	
x	x	x	x	x	x																				
x	x	x	x	x	x	x	x	x	?	?	?	?	?												
x	x	x	x	x	x																				

Higgs pointed out (1970) that a similar situation applies to the coastal oasis locations of the Levant, and Cynthia Irwin-Williams (1979) states that similar conditions prevail in the canyon-head environments of the Oshara-tradition sites in the U.S. Southwest. The area of secondary development in Peru has both riverine and coastal sites—Ancón, Playa Hermosa, Paracas, Asia, Trujillo—at lush circumscribed locations with radial ecozones around them. This same pattern seems true for Mesolithic sites in the Aegean, the Jomon sites in Japan, and the late preceramic and early ceramic (Valdivia) of the Ecuadorian Coast.

Many other coastal foragers—such as the people of the Northwest Coast of the United States, British Columbia, and Alaska and the Californian and Chilean coastal shell-mound people—also had regional diversity combined with areal uniformity and the possibility of exploiting a series of ecozones from a strategically located base. Yet their lush ecozones were less circumscribed and, unlike people in areas of secondary development, they did not have potential domesticates or intense interaction with a dissimilar area that had domesticates and/or cultivars. Thus, areas of secondary development do have some necessary conditions that not only distinguish them from areas of primary and tertiary development but also from areas of Efficient Foragers who did not evolve to village agriculture.

What really sets the secondary developments apart, however, is the sufficient conditions, or triggering causes. Most areas were affected by the diminution of faunal food resources at the end of the Pleistocene, with seasonal scheduling and a broader spectrum of subsistence options developing in response. Areas of secondary development, however, evolved other responses as well. They greatly intensified their wild-food procurement systems (often in the lush areas or catchment basins) and increased their resource specialization, gathering riverine as well as marine resources in coastal regions and possibly roots in more tropical environs. These options were not possible in areas of primary development, in part because of environmental conditions. Nor were such options feasible in areas of harsh climate where village agriculture never developed. Increased procurement of wild foods and resource specialization in areas of secondary development generally resulted in Affluent Foraging Bands becoming Foraging Villagers, with concomitant tendencies for increases in population and concentration into hamlets and/or villages. Since secondary developments were concentrated in circumscribed environs, larger populations meant that groups outgrew their natural food resources and often diminished the carrying capacity of these crucial environmental niches.

Neither areas of tertiary development nor the Affluent or Efficient Foraging Bands of many of the bountiful coastal regions had such circumscription. They could therefore expand into other areas where food resources were sufficient or could develop still more efficient food-procurement systems or even continue successful specialized procurement patterns. The groups undergoing secondary development, however, had reached a "food crisis" (Cohen 1977a).

Their response to this crisis was to adopt domesticates from adjacent areas of primary development. Consequently, the Foraging Bands relatively rapidly became Horticultural Villagers. This acceptance of horticulture probably meant planting both imported domesticates and some locally domesticated plants. This activity, in turn, led to further population increases, which resulted in more and more planting, until people quickly became Village Agriculturists. In other words, unlike primary development, where agriculturists slowly became villagers, in secondary development villagers rapidly became agriculturists. The main driving force of this "revolutionary" secondary development was the relationship between relatively abundant food resources in circumscribed areas and population increases that upset this ecological equilibrium. Later, the need for food surpluses to feed full-time specialists (many with religious functions) also became a driving force. This unique development may be envisioned in a more dynamic way by considering it as a positive-feedback process (see table 12.5).

The worsening environmental conditions at the end of the Pleistocene probably forced bands to develop seasonal scheduling as well as a broad-spectrum procurement system. Those living in an environment having (1) areal uniformity but regional radial diversity, (2) at least one lush circumscribed zone, and (3) many other ecozones that could be exploited from a single base developed foraging that involved resource specialization and decreased mobility (use of base camps). As these foragers became increasingly sedentary, population increases led to more resource-procurement specialization. Eventually, the people settled in villages, and concomitant population increases caused still more dependence on resource specialization, steadily increasing pressure on the food resources of the lush circumscribed ecozone.

Exchanges with an area in which domestication and/or horticulture and/or agriculture had already occurred caused the Foraging Villagers to adopt domesticates as a substitute for or supplement to further wild-resource specialization. The use of imported domesticates led to horticulture, which relatively quickly resulted in further population increases and often destroyed the carrying capacity of the local circumscribed environments. As food production from wild resources diminished, it became necessary to plant more of the borrowed domesticates or cultivars as well as plants domesticated locally. Relatively quickly, the foragers became Agricultural Villagers.

This positive-feedback process produces a sequence of twelve elements or traits that are similar to the sufficient conditions for secondary development (see table 12.4). Although the first two elements are also similar to those of primary developments (see table 12.1), most are significantly different. The elements of secondary development are as follows:

1. The end of the Pleistocene brought worsening environmental conditions, as evidenced by pollen profiles, and loss of megafauna, as evidenced by paleontological data (equivalent to sufficient condition 1 in table 12.4).
2. A broad-spectrum procurement system developed, as evidenced by ecofacts and artifacts indicating new subsistence options (equivalent to sufficient condition 2).
3. Base camps were established in lush circumscribed locales (roughly equivalent to sufficient condition 14).
4. Resource specialization occurred (equivalent to sufficient conditions 11 and 12).

Table 12.5. Necessary and sufficient conditions as a positive-feedback process in secondary developments

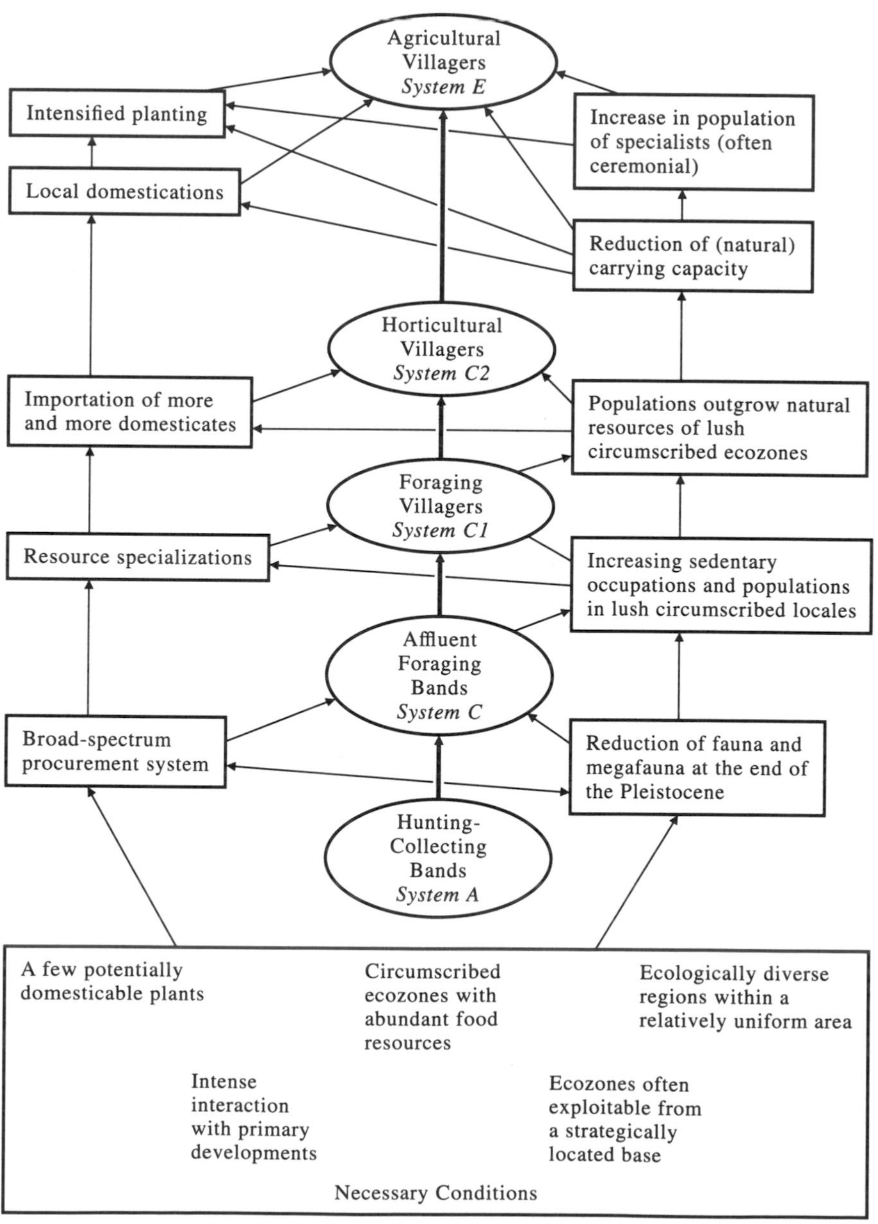

5. Foragers adopted sedentary lifestyles in certain lush circumscribed locales (roughly equivalent to sufficient condition 13).
6. Population increased (equivalent to sufficient condition 5).
7. The sedentary foragers accepted domesticates from areas of primary development (equivalent to sufficient condition 16).
8. Acceptance of domesticates rapidly brought about a shift to horticulture, with populations increasing to the limits of their circumscribed zones (equivalent to sufficient condition 14).
9. The carrying capacity of the area was reduced (equivalent to sufficient condition 15).
10. The people turned to local domesticates and more planting (equivalent to sufficient condition 17).
11. The need for more food increased with the growing number of full-time specialists (included in sufficient condition 18).
12. Population growth resulted in more intensification of planting, resulting in village agriculture (equivalent to sufficient condition 18).

Now let us see how well this positive-feedback process of secondary development applies to each of the areas of secondary development described in earlier chapters. The feedback process seems to have occurred in at least four of the five or six coastal regions of Peru. All twelve steps seem to have occurred in the Lurín-Chilca central coast region; most of the later elements are well documented for the north-central coast of Gavilanes and elsewhere as well as for the south coast around Paracas; the later elements are documented for the north coast near Trujillo. The nearby tropical coast of Ecuador, although not as thoroughly reported, seems to have gone through a similar evolution, and a case can be made that at least the last eight elements of the positive-feedback cycle occurred.

Of the four lowland regions of the Near East, two regions—the Levant and the Syrian—seem to have undergone this same sort of evolution from Hunting-Collecting Bands (about eighteen thousand years ago) to Village Foragers (about ten thousand years ago) to Village Agriculturists (about nine thousand years ago). All twelve elements of the positive-feedback cycle seem to pertain there. Three riverine sites—Ali Kosh on the Tigris as well as Abu Hureyra and Mureybit on the Euphrates—seem to have experienced only the final elements of this evolution.

The U.S. Southwest is somewhat similar to the Levant in its xerophytic vegetation and relatively lush oasislike or canyon-head locations. Three of the six subareas there—the Colorado Plateau, the Mogollon Highlands, and the Gila River drainage—have long sequences and a secondary development from about 3000 B.C. to the time of Christ. Further, there is the same sort of positive-feedback situation between population increases (element 6, table 12.6), a circumscribed environment (element 8), and adoptions of domesticates from a Center—Mesoamerica—(element 7).

Four other areas also seem to have had secondary developments, but the documentation of the conditions causing these developments is relatively poor. In lowland Mesoamerica the only relatively complete sequence comes from Belize, but subsistence and population data as well as dating are poor. In the Nile region

Table 12.6. Comparison of the traits of regional sequences that may represent the positive-feedback processes of secondary developments, stages I–III

Stage	Area	Region	Cultural phases	Range of dates	Hunting-Collecting Bands	1. Pleistocene faunal extinctions and climatic changes	2. Increased subsistence options	3. Base camps in lush niches	4. Resource specializations	Foraging Villagers	5. Sedentary life	6. Population increases	7. Acceptance of foreign domesticates	Horticultural Villagers	8. Population increases to limits of lush circumscribed ecozones	9. Carrying capacity reduced	10. Local domestications	11. Increasing number of specialists	12. Intensive specialized planting	Agricultural Villagers
	Far East	South China coast	Ching-lien-kang	4000 – 5000 B.C.									?					x	x	x
	Southeastern Asia	India region	Harappan	1600 – 2300 B.C.									x				?	x	x	x
STAGE 3	Africa	Sudan region	L. Khartoum	3000 – 3500 B.C.									x				?	x	x	x
		Egypt-Aegean region	Fayum-Badarian	3700 – 6000 B.C.									x					x	x	x
			Franchthi VI	5300 – 6300 B.C.									x				x	x	x	x
	Near East	Tigris region	Bus Mordeh	6800 – 7200 B.C.									?					x	x	x
		Euphrates region	Ramad	6000 – 6500 B.C.														x	x	x
		Levant region	Jericho PPNB	6500 – 7200 B.C.													?	x	x	x
	U.S. Southwest	Colorado Plateau region	Pueblo I-Basketmaker	1500 – 900 A.D.														x	x	x
		Gila region	Vahki-Pinelawn	300 – 700 A.D.														x	x	x
	Mesoamerica	Belize coast	Swazey	800 – 1100 B.C.														x	x	x
	New World Tropics	Coastal Ecuador region	Manteño Chorrera	1000 – 1500 B.C.														x	x	x
		Far north coast region	Salinas	1000 – 1900 B.C.														x	x	x
	Andes	North coast region	Río Seco	1900 – 2300 B.C.														x	x	x
		North-central coast	Aspero	1750 – 2000 B.C.														x	x	x
		Central coast region	Gaviota-Chuquitanta	1750 – 2000 B.C.														x	x	x
		South coast region																		

Asia-Casavilca		1640 – 1900 B.C.												x	x	x
	Naghar	3000 – 4000 B.C.					x		x	?	x		?	x	?	
	E. Khartoum	5000 – 7000 B.C.					x		x	x	x	x		x	?	
STAGES 2–3	Franchthi V	6300 – 7200 B.C.					x	x	x	?	x	x			x	
	Mureybit	7000 – 8500 B.C.					x	x	x	x	x	x	?	x		
	Jericho PPNA	7250 – 8500 B.C.					x	x	x	x	x	x	?	x		
	Basketmaker II	0 – 500 A.D.					x	?	?	?	?	?				
	Armijo	800 – 1800 B.C.					x	x	x	x	x	x		x		
	San Pedro Cochise	150 A.D. – 1200 B.C.					x	x	x	x	x					
	Progreso	2000 – 3000 B.C.					x	x	x	?	?	?				
	Valdivia	2000 – 2500 B.C.					x	x	x	x	x	?	?	x		
	Padre Abán	2000 – 2300 B.C.					x	x	x	x	x	x	x	x		
	Gavilanes 1-2	2000 – 3100 B.C.					x	x	x	x	x	x	x	x		
	Pampa-Playa Hermosa-Conchas	1900 – 2800 B.C.					x	x	x	x	x	x	x	x		
Otuma-Cabeza Larga		1800 – 3190 B.C.					x	x	x	x	x	?	x	x		
	Hsein-jen-tung	6000 – 7000 B.C.				?										
	Bagor	4000 – 5000 B.C.			x	x	x	x								
STAGE 2	Meso-Khartoum	4500 – 5000 B.C.			x	x	x	x	?	?						
	Argissa	5000 – 8000 B.C.			x	x	x	x	?	?						
	Tell Abu Hureyra	7250 – 8500 B.C.			x	x	x	x								
	Late Natufian	8500 – 9500 B.C.			x	x	x	x								
	San Jose	1800 – 3200 B.C.			x	x	x	x								
	Chiricahua	1200 – 3000 B.C.			x	x	x	x								
	Belize-Melinda	3000 – 5000 B.C.			?	?	?	?								
	Vegas	5000 – 6000 B.C.			x	x	x	x								
	Talera	4000 – 5000 B.C.				x	x									
	Encanto-Chilca-Corbina	2800 – 4200 B.C.			x	x	x	x	?	?						
	Paloma-Canario	4000 – 5500 B.C.			x	x	x	x								
San Nicolas		4000 – 6000 B.C.			?	x	x	?								
	Tseng-pi-yai	8000 – 9000 B.C.	x	x	x											
	India Mesolithic	6000 – 8000 B.C.	x	?	?											
STAGES 1–3	Sebilian	6000 – 9000 B.C.	x	x	x											
	Qadan	6000 – 9000 B.C.	x	x	x											
	Franchthi IV	7200 – 8800 B.C.	x	x	x											
	Early Natufian	9500 –10,000 B.C.	x	x	x											

Table 12.6. (*continued*)

Regions (table column headings): South coast region, Central coast region, North-central coast (Andes); North coast region, Far north coast region, Coastal Ecuador region (New World Tropics); Belize coast (Mesoamerica); Gila region, Colorado Plateau region (U.S. Southwest); Levant region, Euphrates region, Tigris region (Near East); Egypt-Aegian region, Sudan region (Africa); India region (Southeastern Asia); South China coast (Far East)

	Cultural phases	Range of dates	**Hunting-Collecting Bands**	1. Pleistocene faunal extinctions and climatic changes	2. Increased subsistence options	3. Base camps in lush niches	4. Resource specializations	**Foraging Villagers**	5. Sedentary life	6. Population increases	7. Acceptance of foreign domesticates	**Horticultural Villagers**	8. Population increases to limits of lush circumscribed ecozones	9. Carrying capacity reduced	10. Local domestications	11. Increasing number of specialists	12. Intensive specialized planting	**Agricultural Villagers**
	Bajada	3200 – 4000 B.C.			x	x	x											
	Orange Walk-Sand Hill	5000 – 7000 B.C.			x	x	x											
	Manantial	6000 – 7000 B.C.			x	x	x											
	Paijan	5000 – 7000 B.C.			x	x	x											
	Campanario	5000 – 7000 B.C.			x	x	x											
	Luz-Arenal	5500 – 7000 B.C.			x	?	?											
	Playa Chiba	5000 – 7000 B.C.			x	?	?											
	Hoabinhian	9000 –10,000 B.C.	x	x														
STAGE 1	Gujarat	pre – 8000 B.C.	x	x	x													
	Egyptian Paleolithic	pre – 9000 B.C.	x	x	x													
	Franchthi 1-III	pre – 9000 B.C.	x	x	x													
	Kebaran Geometric	pre –11,000 B.C.	x	x	x													
	Cody-Clovis	pre – 6000 B.C.	x	x	x													
	Lowe-ha	pre – 7000 B.C.	x	x														
	Chivateros	pre – 7000 B.C.	x	x														

of Africa, the only relevant data that might fit the pattern come from Khartoum in the Sudan. Another area peripheral to the Near East is the general Harappan region along the Indus River and the area along the Ganges. The poorest data for secondary development—little more than speculation—come from the southern coast of China.

Let us examine the relevant data in terms of a series of positive-feedback processes at sequential stages. All would start with early hunting cultures that were affected by the changing environment at the end of the Pleistocene. Everywhere this seems to have led to increased subsistence options (element 2) and a broad-spectrum procurement system interacting with the settlement into base camps (element 3) and development of various kinds of resource specialization (element 4), resulting in a more sedentary life and concomitant increases in population. Certainly this seems to be what happened from roughly 7000 to 5000 B.C. on the Peruvian coast during Paijan and Campanario times (north coast) and Arenal-Luz (central coast) during the general transition from stage 2 to stage 3. Manantial in coastal Ecuador probably experienced similar processes (Lanning 1967a). In the Near East, Kebaran Geometric and Early Natufian certainly reflect a similar process (Redman 1978), and—though the settlement pattern data are lacking—so did the early Mesolithic of Franchthi Cave stage IV (Jacobsen 1981).

The areas peripheral to the Near East—Egypt and the Sudan with their Sebilian and Qadan complexes (Hassan 1985), as well as the so-called Mesolithic of India (Sankalia 1972) along the Indus and Ganges rivers—seem to show the same elements that could have interacted in a positive-feedback process. The Sand Hill and Orange Walk phases in Belize in lowland Mesoamerica, which have increasing numbers of ground-stone tools, larger sites, and evidence of movement toward the sea, may have been affected by a similar process (MacNeish and Nelken-Terner 1983b). In the Southwest of the United States, the shift from Hunting-Collecting Bands to Sulphur Spring Cochise and Bajada Foragers reflected similar trends (Irwin-Williams 1979).

In stage 2 the result everywhere was Foraging Villagers, upon whom new elements acted as driving forces in a positive-feedback process that led to the adoption of domesticates as the people became Horticultural Villagers. On the Peruvian Coast, Canario and San Nicolas in the south and perhaps Talera to the north and early Vegas in nearby Ecuador saw people settling down in rich circumscribed locales (element 5). This sedentarism seems to have resulted in noticeable increases in population (element 6) with concomitant diminution of natural food supplies, causing the people to import domesticates from the highland areas with which they were interacting (element 7).

Evidence of this feedback pattern—increased sedentarism (element 5), increased population (element 6), and overexploitation of food resources (element 8)—is good for Late Natufian times in the Levant. Almost the same process seems to have occurred in nearby Mureybit and Abu Hureyra on the upper Euphrates. The case is less clear for the Mesolithic at Khartoum in the Sudan and Bagor along the Indus and Ganges (Vishnu-Mittre 1977) and for the preceramic or aceramic villages such as Argissa in the Aegean, but there are hints that the process was similar.

In San Jose and Chiricahua times in the U.S. Southwest the settling down of people in the few lush up-valley or canyon locations led to population increases (element 6) and later to importation of domesticates (element 7—Irwin-Williams 1979). The Belize-Melinda complexes in lowland Mesoamerica (MacNeish and Nelken-Turner 1983b) are not well documented nor are the complexes of Hsien-jen-tung and Tseng-pi-yai on China's south coast (Pearson 1979). In both cases, however, there are hints of people settling down in larger villages before attaining horticulture or agriculture.

The final shift occurred when Foraging Villagers relatively quickly took on imported domesticates (element 7) and/or domesticated local plants (element 10) to meet population increases. Feeding full-time specialists (element 11) and larger populations (element 6), whose needs outgrew the wild foods obtained by foraging (element 8) caused rapid development from Village Horticulturists to Village Agriculturists. This last stage is much better documented in all sixteen or seventeen secondary developments. On the south coast of Peru, the sites of Otuma, Cabeza Larga, and perhaps Pampa de San Diego seem to have been villages planting a few domesticated cucurbits. These villages developed into larger horticultural sites such as Asia and Casavilca, where cotton and other types of domesticates were cultivated. By the initial period of sites with pottery, village agriculture existed. All this took place within about a millennium.

In the south-central Peruvian coast at Pampa, Playa Hermosa, and Conchas there were horticultural villages with increasing amounts of domesticates, and there were populations that become Agricultural Villagers with ceremonialism at Chuquitanta and in the Gaviota phase (Patterson 1973). Gavilanes 1-2 on the north-central coast (Bonavia 1982) developed into huge ceremonial agricultural village sites such as Aspero, while Padre Abán evolved into Río Seco culture on the north coast. Both have a similar set of elements. On the far north Peruvian coast and in Ecuador, where the environments were less dry and more tropical, the same triggering elements may have interacted in the shifts from Talera to Salinas and from Valdivia to Manteño, respectively (Lathrap, Marcos, and Zeidler 1977).

In the Near East, at such sites as Jericho and Beidha (Redman 1978), horticultural villages of PPNA (Pre-Pottery Neolithic A) experienced increasing populations (element 6) and perhaps reduced carrying capacities of their oasis environments (element 9) that led to increased planting of domesticates, perhaps lentils and emmer (element 12), and the development of ceremonialism. These sites became agricultural villages in PPNB (Pre-Pottery Neolithic B) and in the pottery Neolithic.

A similar shift occurred at Mureybit, Abu Hureyra, and Ramad to the north along the Euphrates in Syria, while along the Tigris, Bus Mordeh may represent the final stage of this development, although the early stages are as yet undiscovered. In peripheral Greece the shift from Franchthi phase V with horticulture to phase VI with agriculture and probably the development of villages such as those at Argissa, Sesklo, and Nea Nikomedeia (Jacobson 1979) are well documented. Other elements, however, are more difficult to discern, a problem that also exists with the evolution from Mesolithic and/or early to late Khartoum in the Sudan and the evolution to Harappa in the region of Mohenjo-Daro along the Indus River in Pakistan (Vishnu-Mittre 1977). The developments leading to Fayum in Egypt are unknown, but there are hints of a shift from village horticulture to village agriculture and increasing populations.

While the Centers in the Near East and Peru fit my model well, the other two Centers—the Far East and Mesoamerica—as yet have little data with which to test my hypothesis of secondary development. Hints of development at Ching-lien-kang on the southern China coast, the shift from Progreso to Swazey in Belize, and that from Palo Hueco to Santa Luisa in central coastal Veracruz are more intriguing than convincing.

In the non-Center of the U.S. Southwest, however, developments from San Pedro to Mogollon and/or Hohokam and Medio-Trujillo to Pueblo I-II may fit this pattern, for each experienced increasing populations, increasing adoption of domesticates, and more ceremonialism, suggesting the need to support specialists (perhaps full time). The shift from horticultural villages at about 300 B.C. to agricultural villages had ocurred by roughly A.D. 700 to 900 (Irwin-Williams 1979).

All in all, as a test of my hypothesis of secondary development we have six or seven quite adequate relevant regional sequences from two Centers (the Near East and Peru), at least one from a non-Center (the U.S. Southwest), and one from the Ecuadorian coast in the New World tropics. We also have poor but suggestive data from six other regions in the non-Centers—two in south-central Asia, two in North Africa, one in southeastern Europe, and one in the south-central U.S. Southwest, as well as data from coastal Mesoamerica and China, both in Centers. Although not adequate to prove my hypothesis, these data afford a more extensive testing than was possible for my hypothesis of primary development.

THE TERTIARY DEVELOPMENTAL HYPOTHESIS AS THEORY

Although it was possible to do more tests of my hypothesis of tertiary development, with few exceptions the data are even less reliable. Despite this inadequacy, the data make it clear that tertiary development is markedly different from both secondary and primary, in terms not only of its routes (7, 5, and 4) but also of its necessary and sufficient conditions and its positive-feedback cycles. The general model of tertiary development progresses via route 7 from Hunting-Collecting Bands (system A) to Efficient Foraging Bands (system D). It then moves via route 5 to Semisedentary Bands with Domesticates (system D1) and via route 4 to Horticultural Villagers (system C2). Finally, by way of route 4 and/or 15, the model reaches Agricultural Villagers (system E). These steps differ considerably from those of primary developments, in which agriculture was established before villages. The acquisition of full-time agriculture, however, was a slow process for both primary and tertiary developments. The tertiary model vaguely resembles that for secondary development in that village life generally precedes agriculture, but the process in tertiary development was far slower than in secondary.

Tertiary developments (see table 12.7) shared some necessary conditions with the other models. All had regions with potential domesticates (necessary condi-

Table 12.7. Comparison of the necessary and sufficient conditions of tertiary development with the cultural sequences of various areas that followed the differing routes of development

	Route 1					Route 2				Route 3				
	Far East, Yellow River	Andean highlands	Mesoamerican highlands	Near East highlands	Owens Valley, California	Andean highlands	Mesoamerican highlands	Near East highlands	Far East, Yellow River	Coastal Ecuador	Coastal Peru	Mesoamerican lowlands	Levant (Near East)	Near East riverine
Necessary conditions														
1. Potential domesticates	x	x	x	x	x	x	x	x	x	x	x	x	x	x
11. Casual and/or intermittant interaction with regions that developed domesticates					x									
8. Multiple (food) resources exploitable from strategically located base										x	x	x	x	x
13. Many lush uncircumscribed ecozones with numerous food resources														
14. Seasonality (often without harsh seasons) that rarely limits food resources										x		x		x
Sufficient conditions														
1. End of Pleistocene reduces fauna	x	x	x	x	x	x	x	x	x	x	x	x	x	x
2. Development of new subsistence options	x	x	x	x	x	x	x	x	x	x	x	x	x	x
3. Seasonal scheduling	?	x	x	x	x	x	x	x	?	x	x	x	x	x
11. Intensification of wild food procurement systems and										x	x	x	x	?
resource (wild food) specialization										x	x	?	x	
13. Sedentary life of skilled foragers										x	x	?	x	?
5. Larger populations										x	x	x	x	x
15. Diminished carrying capacity of the exploited ecozones										x	x	x	x	x
19. Development of a primary forager (forest) efficiency														
20. Carrying capacity of the uncircumscribed zones gradually upset														
21. Slow adoption of domesticates as a supplement														
22. Climatic change further limits the carrying capacity														
16. Acceptance of domesticates an/or local domestications that slowly lead to sedentary horticulture														
23. Further reduction of the carrying capacity caused by (a) population increases and														
(b) overexploitation of the land														
24. Breakdown in the ceremonial redistribution system										x				
18. Slow emergence of intensive agriculture														

Route 3					Routes 4-5-12									Rte 9	Rtes 6-10-11			Routes 7-8					Route 13		
Coastal China	Egypt riverine	Sudan riverine	Aegean coastal	India riverine	U.S. Southwest-Jornada	Europe	Eastern U.S.	Arid Africa	Tropical Africa	Tropical New World	Southeastern Asia	Oceania	Japan	Owens Valley, California	Northwest Coast, Canada	California coast	Chile north coast	Monitor Valley, Nevada	Danger Cave, Utah	Eskimo	Araucanians, Chile	Boreal tribes, Canada	Bushman, South Africa	West Australians	Alacaluf, Chile
x	x	x	x	x	x	x	x	x	x	x	x	x	x	x											
-		?	?	?	?	x	x	x	x	x	x	x	?	x			?								
x	x	x	x	x	x	x	x	x	x	x	x	x	x		x	x	x		x						?
						x	x	x	x	x	x	x	x		x	x	x								
x	x	x	x	x		x	x	x	x	x	x	x	x		x	x									
x	x	x	x	x	x	x	x	x	x	x	x	x	x	x	x	x	x	x	x	x	x	x	x	x	x
x	x	x	x	x	x	x	x	x	x	x	x	x	x	x	x	x	x	x	x		x	x	x	x	
x	x	x	x	x	x	x	x	x	?	?	?	x	x	x	x	x	x	x	x	?	x	?	?	?	?
x	x	x	x	x	x	x	x	x	?	?	?				x	x	x			x	x	x			
x	x	x	x	x	x	x	x	x	x		x		x		x	x	x			x					
x	x	x	x	x	x	x	x	x			?	x	x		x	x	x								
x	x	x	x	x	x	x	x	x			x	x	x		x	x	x								
x	x	x	x	x	x	x	x	x																	
					x	x	x	x	?	?	x	x	x		x	x	x								?
					?	x	x	x	?	?	x	x	x												
					?	x	x	x	?	?	x	x	x	?											
					x	x	x	x	?	?	x	x	x												
					x	x	x	x	?	?	x	x	x												
					x	x	x	x	?	?	x	x	x												
					?	x	x	x	?	?	x	x	x												
					?	x	x																		
					x	x	x	x	?	?	x	x	x												

tion 1) and an environment that allowed planting. In tertiary developments, domesticates were often acquired first by interactions with Centers or non-Centers; however, these contacts were relatively casual rather than intense (necessary condition 11). Also, local domestications occurred much later in tertiary developments and did not immediately become important (see table 12.7). The other necessary conditions, which are of a more environmental nature, are quite different for tertiary developments than they are for primary ones, and they are vaguely similar to those for secondary developments in terms of regional diversity and the availability of food in all seasons. A fundamental difference is that, while both secondary and tertiary developments had relatively lush ecozones that could be exploited from a single base (necessary condition 8), the ecozones of tertiary development were not generally circumscribed (necessary condition 13). Further, while there was seasonality in tertiary regions, it usually did not limit the acquisition of wild-food resources in any one season (necessary condition 14).

A similar pattern of similarities and differences is vaguely true of the sufficient conditions. The same three sufficient conditions triggered all models of development: the end of the Pleistocene with its decreasing faunal food resources (sufficient condition 1), the development, in response, of a broad-spectrum subsistence system with many new subsistence options (sufficient condition 2), and a seasonally scheduled settlement and subsistence pattern (sufficient condition 3). As a further response, groups undergoing tertiary development intensified their wild-food procurement system (sufficient condition 11) and often became resource specialists, with a resultant tendency toward sedentary life (sufficient condition 13), which led to population increases (sufficient condition 5). These factors diminished the carrying capacity of the regions occupied (sufficient condition 15). In this regard, tertiary developments were similar to secondary developments as well as to groups who never attained village agriculture. Among the latter were groups who became affluent foragers, such as the peoples of the Northwest and California coasts of the United States and of the Chilean Coast of South America, and groups who became Efficient Foragers, such as the peoples of the Great Plains of the United States, the pampas of Argentina, and the steppes of central Asia.

A comparison of tables 12.4 and 12.7 shows that tertiary responses differed from secondary developments in several ways. Groups undergoing tertiary development evolved further efficient foraging techniques, attaining primary forager efficiency (sufficient condition 19 in table 12.7), like some Affluent Foragers. Because of this efficiency, their interaction with Centers, and their uncircumscribed lush ecozones, areas of tertiary developments adopted domesticates only slowly to supplement their efficient foraging system (sufficient condition 21). This gradual acquisition of domesticates meant that the carrying capacity of their environment was upset only slowly (sufficient condition 20).

In other words, groups with tertiary developments basically remained foragers, using only a few domesticates for a long period. Only as their expanding foraging system and population growth combined with changing climatic conditions that desiccated the carrying capacities of their environments (sufficient condition 22) were they increasingly forced to use their imported domesticates as well as to domesticate some local plants (sufficient condition 16). The increased food supply resulting from village horticulture led to more population increases (sufficient condition 23a). As horticultural practices caused more reduction of natural resources (sufficient condition 23b), breakdowns in the redistribution of foodstuffs occurred (sufficient condition 24), and village agriculture slowly developed. The last six or seven sufficient conditions are unique to tertiary developments and seem to have occurred in Europe, the eastern United States, Japan, arid and tropical Africa, and perhaps in the New World tropics, the Rio Grande and Gila areas of the U.S. Southwest, and the tropics of southeastern Asia.

The conditions of tertiary development may be stated as a positive-feedback process (see table 12.8). When faunal food staples were reduced at the end of the Pleistocene, certain hunter groups living in relatively lush uncircumscribed environments developed a broad-spectrum subsistence system with seasonal scheduling and may have developed resource specialization as well, leading to an efficient foraging system that included some acceptance of domesticates from outside groups with whom there was casual interaction. Population increases and more intensive land use may have reduced the carrying capacity of local environments and occasioned more use of domesticates, both imported and locally cultivated, causing the foragers to settle in sedentary base camps. Sedentarism led to more population increases and further reduction of the environment's carrying capacity; coupled with worsening environmental conditions, these changes caused the foragers to become Horticultural Villagers. A breakdown in the redistribution system led to the planting of more and more domesticates until the horticulturists evolved into Village Agriculturists.

Finding elements in the archaeological record that reflect this positive-feedback process is difficult, yet I discern the following elements that characterize tertiary development (table 12.8):

1. The end of the Pleistocene was accompanied by worsening environmental conditions, as evidenced by pollen profiles and other studies, as well as by the loss of megafauna and a decrease in biomass, as evidenced by paleontological data (equivalent to sufficient condition 1 of table 12.7).
2. A broad-spectrum procurement system with seasonal scheduling developed (equivalent to sufficient conditions 2 and 3).
3. These developments led to primary forager efficiency, as evidenced by ecofact and artifact analysis (equivalent to sufficient condition 19).
4. Resource specialization also occurred, as evidenced by ecofact and artifact analysis (equivalent to sufficient condition 11).
5. Forager efficiency may have included casual adoption of domesticates, as evidenced by plant remains (equivalent to sufficient condition 21).
6. A more sedentary way of life and concomitant population increases resulted, as evidenced by settlement-pattern data (equivalent to sufficient conditions 13 and 5).
7. Sedentarism and population increases may have led to overexploitation and resultant reduction in the environment's carrying capacity, as evidenced

Table 12.8. Necessary and sufficient conditions as a positive-feedback process in tertiary developments

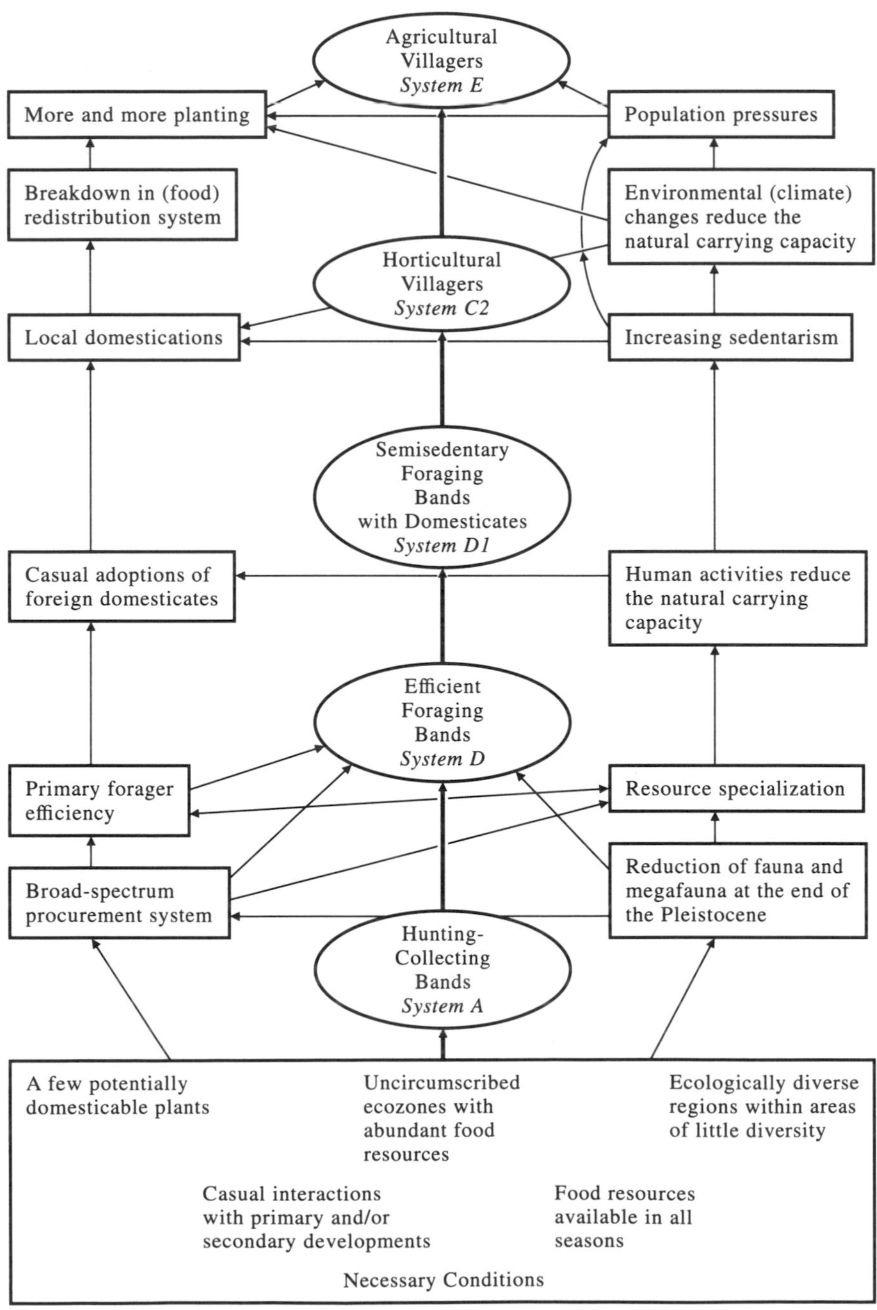

by pollen and other interdisciplinary studies (equivalent to sufficient condition 23).

8. The result may have been more use of domesticates, both imported and locally cultivated, as evidenced by plant remains (equivalent to sufficient condition 16).
9. Natural changes in the environment, as evidenced by paleoecological studies, also limited the carrying capacity (equivalent to sufficient condition 22).
10. The foragers may then have become Horticultural Villagers who further reduced the carrying capacity of the environment through overexploitation, as evidenced by interdisciplinary studies (equivalent to sufficient condition 23b).
11. Population increases, as evidenced by settlement sizes, further reduced the carrying capacity (equivalent to sufficient condition 23a).
12. The above changes, which limited the food supply and broke down the redistribution system (sufficient condition 24), resulted in the slow emergence of village agriculture (equivalent to sufficient condition 18).

Now let us describe how these various features, which are part of the positive-feedback process, operate in causing the tertiary model to shift through the three stages of development (see table 2.9). Stage 1, which began after the hunting stage, was initiated by the end of the Pleistocene with its changing climate and diminishing availability of food resources (element 1, table 12.9), causing people to develop a broad-spectrum procurement system, a number of new subsistence options (element 2), and a seasonally scheduled subsistence system (element 3). Partly because local environments—often temperate or tropical—were relatively lush, people became involved in resource specialization (element 4) and developed into Efficient Foragers (element 3), with a tendency for populations to become more and more sedentary and to increase in size (element 6). As a further subsistence option they adopted domesticates (element 5), in this way becoming Efficient Foragers with Domesticates—system D1.

In Japan at the end of the Pleistocene the Paleolithic peoples became foragers in the period characterized by microblades and tongue-shaped cores (Ikawa-Smith 1982). In Initial Jomon times, some of these groups specialized in exploiting more and more marine resources (elements 3 and 4), which helped make them increasingly sedentary (element 6). As populations grew (element 6), they were receptive to domesticates from mainland China (element 5), although they remained basically Foragers with Domesticates in Early and Middle Jomon times, roughly from 5000 to 2000 B.C. (Kotani 1971).

While the continental environment of western Europe is rather different from the environment of the islands of Japan, the change from the Hunters of the Paleolithic (Magdelenian) to Foragers with a few domesticates in the European Mesolithic or Early Neolithic was also a tertiary development because of similar elements acting in the same kind of positive-feedback process. On the coasts of the Mediterranean, Atlantic, and Baltic, there were Mesolithic groups specializing in exploiting marine resources, while people along the Danube and in interior

Table 12.9. Comparison of the traits of regional sequences that may represent the positive-feedback processes that occurred in tertiary developments, stages I–III

Areas

- Far East: Japan region
- Europe: Danubian region, Med. coast region, Eastern Plains region, Alpine region, Atlantic mainland, Atlantic coast, North and Baltic Sea region
- Eastern U. S.: U.S. Southwest-Jornada region, Caddo-Lower Mississippi region, Middle Mississippi region, Midlands region, Ohio River region, Northeast
- Africa: Western Sahara region, Central Sahara region, Central Eastern region, Guinea coast region, Ethiopia region
- New World Tropics: Panama region, Venezuelan region
- Southeastern Asia: Southeast Asia region, Oceania

	Cultural phases	Range of dates	**Hunting-Collecting Bands**	1. Pleistocene extinctions-climatic changes	2. Increased subsistence options	**Efficient Foraging Bands**	3. Primary forager efficiency	4. Resource specializations	5. Casual adoption of domesticates	6. Population increases, more sedentarism	**Semisedentary Bands with Domesticates**	7. Human reduction of carrying capacity	8. Local domestications-cultivations	**Horticultural Villagers**	9. Natural phenomena reduce carrying capacity	10. Foraging, overexploitation reduce carrying capacity	11. Increased population further reduces carrying capacity	12. Breakdown in (food) redistribution system	13. Intensive specialized planting	**Agricultural Villagers**
STAGE 3	Vaitootia	600 – 1200 A.D.											?	?					x	x
	New Guinea Iron Age	0 – 1500 A.D.											?						x	x
	Lie Siri	0 – 2300 B.C.											?				?		x	x
	Non Nok Tha	2000 – 4000 B.C.											?				?		x	x
	Camruco	400 – 1500 A.D.											?				?		x	x
	Coclé	0 – 1000 A.D.											?				?	?	x	x
	Natchabiet-Axum	0 – 380 B.C.																	x	x
	Kintampo	1400 – 1700 A.D.																	x	x
	Hyrax Hill	0 – 1000 A.D.															?		x	x
	Daima	0 – 500 B.C.																	x	x
	Akjinjeir	400 – 700 B.C.													x	x	x		x	x
	Owasco-Iroquois	1000 – 1600 A.D.													?	x	x		x	x
	Fort Ancient	1000 – 1500 A.D.													?		x		x	x

Cahokia	900 – 1500 A.D.				x	?	x	?	x	x
Kincaid	900 – 1500 A.D.		?		?	?	x		x	x
Coles-Creek Natchez	800 – 1500 A.D.				?		x		x	x
El Paso	1100 – 1300 A.D.		x		?	x	x		x	x
British-Danish Bronze Age	600 – 1200 B.C.		?		?	?	x		x	x
Bell Beaker pottery Neolithic	2000 – 3000 B.C.				x	x	x		x	x
Chassey	2800 – 3800 B.C.		?		x	?	x	?	x	x
Michelsberg	2500 – 3000 B.C.				x	?	x	?	x	x
Langweiler	3000 – 3500 B.C.				x	?	x	?	x	x
Spanish-French Bell Beaker Neolithic	3000 – 4000 B.C.				x	?	x	?	x	x
Vinca Tordos Linear pottery Neolithic	4000 – 4500 B.C.				x		x		x	x
Yayoi	0 – 300 B.C.						x		x	x
Lapita	0 – 1500 B.C.		x	x	x	x	x			
Uai Bobo	2300 – 3000 B.C.		x	x	x	?	x			
STAGES 2–3 Gua Kechil	2000 – 3500 B.C.		x	x	x	?	x			
Corozal I-III	0 – 1000 B.C.	?	x	x	x	?	x			
Monogrillo-Boquete	1000 – 2220 B.C.	?	x	x	x	?	x			
Natchabiet-Lalibela	1000 – 2000 B.C.			x	x					
Naghar	1000 – 2000 B.C.		x	x	x	x	x			
Adrar Bous	2000 – 3000 B.C.	?	x	x	x	?	x			
Chebka-Arriane	700 – 1000 B.C.	x	x	x	x	x	x			
Point Peninsula	0 – 1000 A.D.	?	x	x	x	?	x	?		
Adena-Ohio Hopewell	0 – 900 A.D.	?	x	x	x	x	x	x		
Illinois Hopewell	B.C. 200 – 900 A.D.	x	x	x	x	x	x	x		
Baumer-Lewis	0 – 900 A.D.	?	x	x	x			?		
Marksville-Troyville	0 – 900 A.D.	?	x	x	x		?	?		
Hueco-Mesilla	B.C. 9 – 900 A.D.	x	x	x	x					
British-Danish Megalithic	1500 – 3000 B.C.	x	x	x	x	x	x	x		
French Megalithic	3000 – 4000 B.C.	x	x	x	x	?	x	?		
French Linear pottery	3900 – 4400 B.C.	x		x	x	?	x	?		
Egolzwil	3000 – 4000 B.C.	x		x	x	?	x			
Dneister-Bug	3900 – 4400 B.C.	x		x	x	?	x			
Cardial-Cadium pottery Megalithic	3900 – 4600 B.C.	x	?	x	x	x	x	?		
Koros	4500 – 5500 B.C.	x	?	x	x	x	x			
Late and Final Jomon	300 – 2000 B.C.	x	?	x	x	?	x			

Table 12.9. *(continued)*

Areas

Far East: Japan region
Danubian region
Europe: Med. coast region, Eastern Plains region, Alpine region, Atlantic mainland, Atlantic coast, North and Baltic Sea region
U.S. Southwest-Jornada region
Eastern U. S.: Caddo-Lower Mississippi region, Middle Mississippi region, Midlands region, Ohio River region, Northeast
Africa: Western Sahara region, Central Sahara region, Central Eastern region, Guinea coast region, Ethiopia region
New World Tropics: Panama region, Venezuelan region
Southeastern Asia: Southeast Asia region, Oceania

	Cultural phases	Range of dates	**Hunting-Collecting Bands**	1. Pleistocene extinctions-climatic changes	2. Increased subsistence options	**Efficient Foraging Bands**	3. Primary forager efficiency	4. Resource specializations	5. Casual adoption of domesticates	6. Population increases, more sedentarism	**Semisedentary Bands with Domesticates**	7. Human reduction of carrying capacity	8. Local domestications-cultivations	**Horticultural Villagers**	9. Natural phenomena reduce carrying capacity	10. Foraging, overexploitation reduce carrying capacity	11. Increased population further reduces carrying capacity	12. Breakdown in (food) redistribution system	13. Intensive specialized planting	**Agricultural Villagers**
STAGE 2	Balif	3000 – 4000 B.C.					x	x	x	x	x	?	?							
	Ulu Leang	2000 – 7000 B.C.					?	x	x	x	x	?	?							
	Spirit Cave	5000 – 8000 B.C.					x	x	x	x	x	?	x							
	Ronquin	1000 – 1500 B.C.					x	x	x	?	?	?	?							
	Cueva de los Ladrones	1800 – 4000 B.C.					x	x	x	x	x	?								
	Gobedra	2300 – 2400 B.C.					?	?	x	x	x		?							
	Iwo Eleru	2000 – 3000 B.C.					x	x	?		?									
	Njero Cave	1000 – 1500 B.C.					x	x	x	x	x									
	Meniet	3000 – 4000 B.C.					x	x	x	x	x				x					
	Naghez	1000 – 1100 B.C.					x	x	x	x	x				x					
	Meadowcroft 3-4	500 – 2000 B.C.					x	x	?	x	?									
	Newt Kash Hollow	1000 – 3000 B.C.					x	x	x	x	x	?	x							
	Middle-Late Koster	1000 – 5000 B.C.					x	x	x	x	x		x							

	Dates										
Phillips Spring	1000 – 4000 B.C.				x	x	x	x	x		x
Fresnal	900 – 2600 B.C.				x	x	x	x	x		x
Hembury	2900 – 3900 B.C.				x	x	x	x	x		
Ertebølle	3000 – 3600 B.C.				x	x	x	x	x		
Téviec	3900 – 5000 B.C.				x	x	?	?	?		
Tardenosian 1-3	4000 – 5500 B.C.				x	x	?	?	?	?	
Birsmatten	3500 – 5300 B.C.				x	x	?	?	?	?	
Icoana	4500 – 6000 B.C.				x	x	x	x	x	x	
Baume de l'Abeurador-Fontbregona	5000 – 6000 B.C.				x	x	x	x		x	
Lepinski Vir	5000 – 6000 B.C.				x	x	x	x	x	x	
Early-Middle Jomon	2000 – 5000 B.C.				x	x	x	x	x		
Kuk	4000 – 5000 B.C.	?	x	x	x	x					
Hoabinhian	9000 –12000 B.C.	x	x	x	x						
Cerro Mangote	4000 – 5000 B.C.	?	x	x	x	x					
STAGES 2–3											
LaGruta	2000 – 4000 B.C.		x	x	x	x					
Wilton	2000 – 5000 B.C.	x	x	x	x						
Capsian	4000 – 8000 B.C.	x	x	x	x						
Akreijit-Nkhal	1100 – 2000 B.C.	x	x	x	x						
Kirk-Neville	3000 – 7000 B.C.	x	x	x	x	x					
Modoc-Dalton	5000 – 8000 B.C.	x	x	x	x	x					
Gardner Springs-Keystone	5000 – 2600 B.C.	x	x	x	x	x					
Star Carr-Maglemose	5000 – 8000 B.C.	x	x	x	x	x					
Asturian	5000 – 8000 B.C.	x	x	x	x	x					
Azilian	5000 – 8000 B.C.	x	x	x	x	x					
Furstein	5000 – 8000 B.C.	x	x	x	x	x					
Fatma Koba	7000 – 8000 B.C.	x	x	x	x	x					
Montardian	6000 – 8000 B.C.	x	x	x	x	x					
Odmut	5000 – 8000 B.C.	x	x	x	x	x					
Japanese blade-and-tongue core	9000 –20000 B.C.	?	?	?	?	?					
STAGE 1	pre – 9000 B.C.	x	x	x							
Paleolithic Old World											
Early Man New World											

France and Germany as well as in the South Russian Plain specialized in exploiting riverine resources (elements 3 and 4). Other groups, like those at Star Carr in England, became specialized hunters, while people in the Alps learned to exploit lake resources. This specialization resulted everywhere in groups settling down (element 6), developing a primary forager efficiency (element 3), and adding to their larder domesticates (element 5) such as lentils, vetch, wheat, and barley from the Near East (J. G. D. Clark 1980) to supplement their expanding efficient foraging system. These changes in lifestyle were happening at a time when changing climate increased the forest environments, making hunting more difficult (elements 7, 9, and 11). By Early Neolithic times (7000 B.C.) or Late Mesolithic times (6000–5000 B.C.) these Europeans had become successful Foragers utilizing a few domesticates (Dennell 1983).

Although our data are not as complete for the eastern United States, the shift to Foragers seems to have occurred during the Archaic period. After Clovis and Dalton times (element 1), peoples everywhere developed a broad-spectrum procurement system (element 2), increased subsistence options and resource specialization (element 4), reached a stage of primary forest forager efficiency (element 3), settled down (element 6) into Semisedentary Bands (Caldwell 1958), and acquired gourds and pumpkins (element 5) from Mesoamerica (R. Ford 1981; Watson 1985).

The process of development in the three major tropical areas is less clear, but there are hints that it was similar. In southeastern Asia and Oceania, after an early Hunter stage with a chopping-chopper complex, the Hoabinhian culture developed, characterized by highly efficient jungle Foragers with a variety of resource specializations (elements 3 and 4) and increased subsistence options (element 2). The data from Spirit Cave in Thailand (Solheim 1972) suggest that people there became Efficient Foragers with Domesticates (element 5).

Africa after the Paleolithic saw the development of a microlithic Forager system—Wilton to the south and Capsian to the north. In Mauritania we have a clear picture of the Efficient Forager Akreijit phase (element 3) of about 2000 B.C., which had increasing resource specialization (element 4) and subsistence options (element 2) that gradually caused people to become Semisedentary Bands with Domesticates (element 5) by Naghez times (Munson 1968), with the concomitant sedentarism, population increases (element 6), and further resource specialization.

The picture in the New World tropics is less clear, but everywhere there are indications of a shift from hunting with Fishtail and El Jobo points to Efficient Foraging Bands (element 3) with increased subsistence options (element 2) and marine or riverine resource specializations (element 4)—for example, Cerro Mangote in Panama (Ranere 1980) or La Gruta in Venezuela. Later, at Talamanca-Aguadulce and Correal, peoples who were basically still sedentary Foragers (element 6) were also using maize and perhaps also manioc (element 5), though there is little evidence of the latter (Roosevelt 1980).

The transition from stage 2 to 3 marks the evolution from Semisedentary Bands with Domesticates—system D1—to Horticultural Villagers—system C2. In every area of tertiary development this was a long, slow process, and it is difficult to determine where one stage of development ended and the next one began.

There were population increases (elements 6 and 11), more use of local and imported domesticates (elements 5 and 8), and overexploitation that reduced the local environment's carrying capacity (element 10). These elements of the positive-feedback process that changed system D1 into system C2 are not readily discernable in the archaeological record nor are their feedback cycles well delineated, yet this is what seems to have happened in Japan (Kotani 1971) from Jomon times (2000 B.C.) to Final Jomon (300 B.C.) and from Late Archaic times (1500 B.C.) to Middle Woodland (Hopewell-Point Peninsula) times (A.D. 500) in the eastern United States (Griffin 1978). In Europe the situation seems somewhat more complex, and we have real trouble distinguishing horticulture from agriculture, but everywhere northwest of the Aegean populations were increasing (elements 6 and 11) as was use of domesticates (elements 5 and 8) from Early to Middle Neolithic times. Some pollen studies (Dennell 1983) suggest extensive burning of forests, which could have reduced the land's natural food resources and its carrying capacity (elements 9 and 10).

For the tropics the picture is less clear, but in Mauritania the above process seems well documented from the Naghez phase (1100 B.C.) to Chebka, ending at 700 B.C. (Munson 1968). Whether the same process occurred in similar lagoon situations around Lake Chad in the Sahara, Lake Victoria in Kenya, or Lake Rudolf remains to be seen, but there was a definite shift from Foragers to Village Horticulturists. A similar shift seems to have occurred in southeastern Asia from Spirit Cave times to Gua Kechil. The same sequence may also have occurred in the New World tropics: in Venezuela from Ronquin to Corozal III, and from Talamanca to Boquete-Monagrillo times in Panama.

The final step in tertiary development—from horticultural villages (system C2 of stage 2-3) to agricultural villages (system E of stage 3)—resembles secondary development. Although population pressure was important for both, the tertiary development was much slower (and later) and occurred because of different conditions, feedback processes, and elements. In tertiary developments the key to change and increased use of agriculture was reduction of the carrying capacity of the environment, due to both natural phenomena (element 9)—great droughts in the United States and Africa and at the end of the Atlantic period in Europe—and the overexploitation of food resources (element 10) on land and, in Japan's case, in the sea. Also affecting development in Japan as well as in Central America were changing sea levels. In the final days of Hopewell in the eastern United States as well as in the Megalithic period in Europe, there are hints (from the decrease in material trade items) that an extensive exchange system, which might have included the distribution of food to the growing population, gradually diminished in efficiency or broke down (J. G. D. Clark 1980). All of these changes were slow, making village agriculture for the tertiary development not a great leap forward but a slow crawl onward.

In Japan, this process seems to have occurred from Final Jomon to Yayoi, at the time of Christ, although there is a possibility that migration might have been involved. Certainly all these elements were at work in the shift from Middle Woodland to Mississippian times in the eastern United States (Muller 1983); the slow evolution from Middle Neolithic to Late Neolithic or the Bronze Age of

Europe can similarly be explained by this process (Milisauskas 1978). Explaining the developments in the tropics is more difficult, but the shift from dry-rice to wet-rice agriculture in the period preceding Nok Nok Tha in southeastern Asia could have occurred in this manner as could have the rise of Iron Age village agriculture (Gorman 1977). Equally unclear are the data from the New World tropics, though the development from Boquete to Coclé in Panama (Ranere 1980) and from Corozal to Camoruco in Venezuela (Roosevelt 1980) gives hints that they too were part of this tertiary development. As you can see, this tertiary developmental positive-feedback (or deviation-amplifying) process and its various elements, while having more regional examples, is not as fully documented as the primary and secondary developments, and we need much more data from many regions.

Only one of our regional examples comes from a major Center—the Far East. That region is Japan, whose development has yielded a good test of my hypothesis of tertiary development. We have seen a clear evolution from microblade Hunters through the long Jomon period of Affluent Foraging Bands, who slowly acquired domesticates from the mainland until they became Yayoi Village Agriculturists in the Iron Age at the time of Christ. Although there is little evidence of a late breakdown in redistribution (element 12), most of the conditions for stage 3, including a worsening of the maritime environment, seem to pertain. In addition to more analysis of the Final Jomon period, we obviously need complete sequences from other areas peripheral to the Chinese Center, such as Korea, Okinawa, and the Philippines, to see whether the same tertiary process occurred in these regions.

Perhaps our best testing of the tertiary development hypothesis comes from Europe, which we have divided into several regional sequences, ranging in time from the Paleolithic (9000–8000 B.C.) through the Mesolithic to the Neolithic or Iron Age (4500–2000 B.C.). Although a few regions—the Danubian, western Mediterranean, Alps, and Eastern Plains—have adequate sequential data with some relevant plant remains that conform to my tertiary model, the best data come from the regions of the Atlantic Coast and interior, the British Isles, and the Baltic Sea. The Atlantic Coast regions not only have good data on the later aspects of tertiary development but also tend to confirm my hypothesis. Although further research is needed to derive more adequate information on plant remains, perhaps an even greater need is some sort of reanalysis of the relatively abundant remains we do have from the crucial Mesolithic and Neolithic periods.

The eastern United States has almost as many good regional sequences for the period from Early Man (7000 B.C.) to Mississippian times (A.D. 1000), but the crucial plant remains are much less adequate. Data relevant to hypothesis testing are limited; those that tend to confirm my tertiary development hypothesis come from the regions of the central midlands (Illinois and Missouri), the middle Mississippi, the lower Ohio River, and the eastern prairie-like subareas of Ohio and Indiana; there are also some data from the Iroquois subarea of Pennsylvania and New York. The many other subareas—the Plains, Caddo region, lower Mississippi, southern Appalachians, and so on—have produced tantalizing bits of relevant data but hardly enough to test my hypothesis.

Even more frustrating is the information from the three tropical areas. Africa has yielded the most relevant data for testing my hypothesis, but even in Mauritania the plant remains are not impressive (Munson 1968). Crucial information from other African subareas is even worse. Even poorer are the data from the New World tropics, where only Panama and Venezuela yield any information for hypothesis testing. The least information comes from southeastern Asia, Malaysia, Oceania, Polynesia, and from the Chihuahua Desert subarea of the southwestern United States, where investigation of the problems of the origins of agriculture is in its infancy.

FINAL COMMENTS

In summary, despite the inadequacy of available data, we do have some confirmation of my three hypothetical developments from early Pleistocene Hunters to Village Agriculturists. In light of all the data we have examined since chapter 1, I would modify slightly each development and its routes.

Primary development seems to progress slowly from Hunting-Collecting Bands—system A—via route 1a to Destitute Foraging Bands—system B. It then moves via route 1b to Incipient Agricultural Bands—system B1—proceeding via route 2a to Agricultural Bands—System B2—and then via route 2b to Village Agriculturists—system E. As figure 18 indicates, the process was more one of evolution than of revolution. All the primary developments seem to have taken place only in Centers. Major driving forces in these harsh regions were the search for subsistence security and a reduction in residential mobility. Social interactions, rescheduling, changes in the carrying capacities of the environment, population growth, and new means of production were the results of these forces rather than the causes.

In secondary developments, however, population pressures and new means of production were among the forces causing change. From Hunting-Collecting Bands, people progressed via route 6a to Affluent Foraging Bands—system C—and then via route 6b to Foraging Villagers—system C1. The process continued via route 3 or 12 to Horticultural Villagers—system C2—and finally via route 3 or 15 to Agricultural Villagers—system E. Whereas the primary developmental process was slow, the change in secondary development, from Affluent Foraging Bands to Foraging Villagers to Horticultural Villagers to Agricultural Villagers, was relatively rapid and occurred in both Centers and some non-Centers. Interaction with areas of primary development was crucial to secondary development. The causes were not simple, and the process was complex. As in primary development, change was the result of a variety of tensions, not of affluence or leisure time in which to experiment (Sauer 1952).

The tertiary developments were still slower processes involving peoples who were successful Foragers and adopted agriculture only slowly for reasons often having to do with a reduction in the carrying capacity of an area. Development proceeded from Hunting-Collecting Bands via route 7 to Efficient Foraging Bands—system D—and then via route 5 to Semisedentary (base camp) Bands with Domesticates—system D1. The process continued via route 4 to Horticultural Villagers—system C2—and, finally, via route 4 and/or 15 to Village

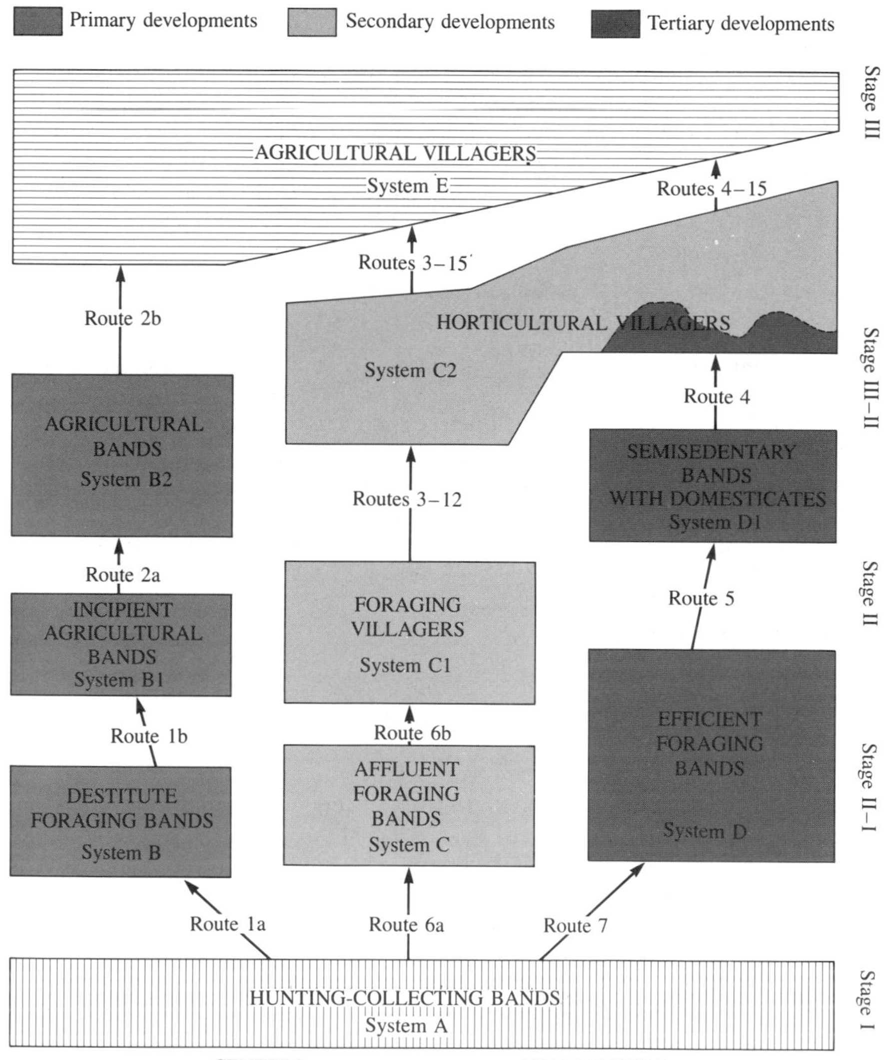

Figure 18. Routes to village agriculture

Agriculturists—system E. The process was long and slow, and there were different tensions at different times.

We have tested all three hypotheses with concrete data, and the information we have accumulated does seem to confirm them. I believe we have moved from speculation and hypothesis toward the ideal state of scientific generalization or laws of cultural change. Obviously, we need more and better data for continued testing of the hypotheses and fleshing out the scientific generalizations. Yet even if we have reached no firm or final conclusions, the pursuit of scientific "truth" has been enjoyable. The search for and collection of relevant data, the examination of archaeological findings, the analysis of the data, the arguments with colleagues, the readings, and the analyses of the relevant comparative data have all been stimulating and fascinating. One might even say it is fortunate we have reached no *final* conclusions—so that this whole satisfying process can continue. I trust that scholars of the future will enjoy their endeavors in this realm as much as I have. I wish them luck and look forward to the day when we really do have the final scientific answers to this intriguing problem on the origin of agriculture and settled life.

Bibliography

Adair, Mary J.
1988 *Prehistoric Agriculture in the Central Plains*. Publications in Anthropology 16. Lawrence, Kansas: University of Kansas.

Adovasio, J. M., J. D. Gunn, J. Donahue, and R. Stuckenrath
1975 "Excavations at Meadowcroft Rockshelter: A Progress Report." Harrisburg, Pennsylvania: *Pennsylvania Archaeology* 45(3): 1–93.

Agogino, George A., and Sherman Feinhandler
1957 "Amaranth Seeds from a San Jose Site in New Mexico." Austin, Texas: *Texas Journal of Science* 9(2): 154–56.

Agogino, George A., and J. Hester
1956 "A Re-evaluation of the San Jose Non-Ceramic Cultures." Santa Fe, New Mexico: *El Palacio* 63(1): 6–21.

Aigner, Jean S.
1972 "Relative Dating of North Chinese Faunal and Cultural Complexes." Madison, Wisconsin: *Arctic Anthropology* 9(2): 36–79.

Akazawa, T. A.
1981 "Maritime Adaptations and Agricultural Transition." In *Affluent Foragers*, edited by Shuzo Koyama and David Hurst Thomas 9:213–60. Osaka, Japan: National Museum of Ethnology.

Alegria, Ricardo
1965 "On Puerto Rican Archaeology." Salt Lake City, Utah: *American Antiquity* 31(2):246–49.

Alegria, Ricardo, H. B. Nicholson, and G. R. Willey
1955 "The Archaic Tradition in Puerto Rico." Salt Lake City, Utah: *American Antiquity* 21(2):113–21.

Allchin, B., A. Goudie, and K. Hegde
1978 *The Prehistory and Paleogeography of the Great Indian Desert*. London: Academic Press.

Allchin, F. R.
1969 "Early Cultivated Plants in India and Pakistan." In *The Domestication and Exploitation of Plants and Animals*, edited by P. Ucko and G. W. Dimbleby, 322–29. London: Duckworth.

Ambrose, Stanley H.
1987 "Chemical and Isotopic Technique of Diet Reconstruction in the Eastern

United States." In *Emergent Horticultural Economics of the Eastern Woodlands,* edited by W. F. Keegan. Center for Archaeological Investigations, Southern Illinois University, Occasional Paper No. 7.:87–108. Carbondale, Illinois.

Andersen, S. H.

1974 "Ringkloster En jysk inlandsboplads med Ertebølle Kultur." Copenhagen: *Kuml* 1973–1974:11–94.

Anderson, Edgar

1952 *Plants, Life, and Man.* Boston: Little, Brown.

1956 "Man as a Maker of New Plants and Plant Communities." In *Man's Role in Changing the Face of the Earth,* edited by W. L. Thomas, Jr., 363–77. Chicago: University of Chicago Press.

Applegarth, J. D.

1977 "Macrofloral Remains in Excavations at Sparks Rockshelter by P. T. Fitzgibbon, J. M. Adovasio, J. Donahue." Harrisburg, Pennsylvania: *Pennsylvania Archaeologist* 47(5):40–48.

Arkell, A. J.

1975 *Prehistory of the Nile Valley.* Vol. 2 of *Handbuch das Orientalistik* 7. Leiden, Netherlands: Brill.

Armenta, Juan

1978 "Vestigios de Labor Humana en Hueso de Animales Extintos de Valsequillo, Puebla, México." Mexico: Consejo Editorial de Gobierno del Estado de Puebla, Mexico 1–123.

Asch, David L., and Nancy B. Asch

1977 "Chenopod as Cultigen: A Re-evaluation of Some Prehistoric Collections from Eastern North America." Chicago: : *Midcontinental Journal of Archaeology* 2(1):3–45.

1985 "Prehistoric Plant Cultivation in West-Central Illinois." In *Prehistoric Food Production in North America,* edited by R. I. Ford, 149–203. Anthropological Papers 75, Museum of Anthropology, University of Michigan. Ann Arbor, Michigan.

Asch, Nancy B., Richard I. Ford, and David L. Asch

1972 *Paleoethnobotany of the Koster Site.* Vol. 6 of Illinois State Museum Research Papers Reports of Investigation 24:1–34. Springfield, Illinois: Illinois State Museum.

Aveleyra, Luis, and M. Maldonado-Koerdell

1953 "Association of Artifacts with Mammoth in the Valley of Mexico." Salt Lake City, Utah: *American Antiquity* 18(4):332–40.

Balout, L.

1981 "The Prehistory of North Africa" In Vol. 1 of *General History of Africa,* edited by J. Ki-Zerbo, 568–81. Berkeley, California: Heinemann.

Barrau, J.

1965 "L'humide et le sec: An Ethnobotanical Adaptation to Contrastive Environments in the Indo-Pacific Area." Honolulu, Hawaii: *Journal of the Polynesia Society* 74:329–46.

1974 "L'Asie du Sud-Est, berceau cultural." Paris: *Etudes Rurales* 53–56:17–39.

Barrau, Lionel, and Roland Portères

1957 *Deux pseudo-indacticos prehistoriques d'Algerie.* Paris.

Bartlett, A. S., E. Barghoorn, and R. Berger

1969 "Fossil Maize from Panama." Washington, D.C.: *Science* 165:389–90.

Bar-Yosef, Ofer
1975 "The Epi-Paleolithic in Palestine and Sinai." In *Problems in Prehistory,* edited by Fred Wendorf and Anthony Marks, 313–71. Dallas: Southern Methodist University Press.
Bar-Yosef, Ofer, and Mordecai Kislev
1988 "Early Farming Communities in the Jordan Valley." In *Foraging and Farming,* edited by D. Harris and C. Hillman, 156–71. London: Allen and Unwin.
Bayard, D. T.
1970 "Excavation at Non Nok Tha, Northeastern Thailand." Madison, Wisconsin: *Asian Perspectives* 13:109–43.
Beckett, Pat
1973 "Cochise culture sites in south-central and north-central New Mexico." MA thesis, Eastern New Mexico University, Portales, New Mexico.
1980 "The AKE Site." Department of Sociology and Anthropology, Cultural Resource Management Division, New Mexico State University, Las Cruces, New Mexico.
Bell, G. D. H.
1965 "The Comparative Phylogeny of the Temperate Cereals." In *Essays on Crop Plant Evolution,* edited by J. Hutchinson, 75–91. Cambridge, England: Cambridge University Press.
Bellwood, Peter
1979 *Man's Conquest of the Pacific.* New York: Oxford University Press.
Bender, Margaret, David A. Baerreis, and Raymond L. Stevenson
1981 "Further Light on Carbon Isotopes and Hopewell Agriculture." Salt Lake City, Utah: *American Antiquity* 46(2):346–53.
Benfer, Robert A.
1983 "The Challenges and Rewards of Sedentarism: The Preceramic Village of Paloma, Peru." In *Paleopathology and the Origin of Agriculture,* edited by M. Cohen and G. Armelagos, 531–58. New York: Seminar Press.
Bennett, Wendell C.
1937 *Excavation at La Malta Maracay, Venezuela.* Anthropological Papers of the American Museum of Natural History 36(2). New York.
1944 *Archaeological Regions of Colombia: A Ceramic Survey.* Yale University Publications in Anthropology 30. New Haven, Connecticut.
Bennett, Wendell C., and Junius Bird
1949 *Andean Culture History.* American Museum of Natural History Handbook Series, 15. New York.
Berry, Michael S.
1985 "The Age of Maize in the Greater Southwest: A Critical Review." In *Prehistoric Food Production in North America,* edited by R. I. Ford. Anthropological Papers 75, Museum of Anthropology, University of Michigan, 279–307. Ann Arbor, Michigan.
Binford, Lewis R.
1962 "Archaeology as Anthropology." Salt Lake City, Utah: *American Antiquity* 28(2):217–25.
1968 "Post Pleistocene Adaptations." In *New Perspectives in Archaeology,* edited by Sally R. and Lewis R. Binford, 313–41. Chicago: Aldine.
Binford, Lewis R., Sally R. Binford, R. Whallen, and M. A. Hardin
1970 "Archaeology at Hatchery West." Salt Lake City, Utah: Society for American Archaeology, *Memoirs* 24.

Bird, Junius B., and Richard Cooke

1978 "The Occurrence in Panama of Two Types of Paleo-Indian Projectile Points." In *Early Man in America,* edited by A. D. Bryan, 263–72. Edmonton, Alberta: University of Alberta.

Bohrer, Vorsila L.

1970 "Ethnobotanical Aspects of Snaketown: A Hohokam Village in Southern Arizona." Salt Lake City, Utah: *American Antiquity* 34(4):413–30.

1972 "Paleoecology of the Hay Hollow site, Arizona." Chicago: *Fieldiana, Anthropology* 63(1):1–30.

Bonavia, Duccio, editor

1982 *Los Gavilanes.* Lima, Peru: Instituto Arqueológico Aleman.

Bonavia, Duccio, and Alexander Grobman

1988 "Andean Maize: Domestication, Racial Evolution, and Spread." In *Foraging and Farming,* edited by D. Harris and C. Hillman, 556–670. London: Allen and Unwin.

Bordaz, Jacques

1966 "Suberde." London: *Anatolian Studies* 16:32–33.

Boserup, Ester

1965 *The Conditions of Agricultural Growth.* Chicago: Aldine.

Braidwood, Linda S., R. J. Braidwood, Bruce Howe, Charles A. Reed, and Patty Jo Watson

1983 *Prehistoric Archaeology along the Zagros Flanks.* Vol. 105 of Oriental Institute Publication. Chicago: University of Chicago Press.

Braidwood, Robert J.

1951 *Prehistoric Men.* Chicago Natural History Museum, Popular Series, *Anthropology,* no. 37. Chicago, Illinois.

Braidwood, Robert J., and Linda S. Braidwood

1950 "Jarmo, A Village of Early Farmers in Iraq." Cambridge, England: *Antiquity* 24:189–95.

Braidwood, Robert J., Halet Gambel, Barbara Lawrence, Charles Redman, and Robert Stewart

1974 "Beginnings of Village-Farming Communities in Southwestern Turkey." In *Proceedings of the National Academy of Sciences* 71(2):568–72.

Braidwood, Robert J., and Bruce Howe

1960 "Prehistoric Investigations in Iraqi Kurdistan." *Studies in Ancient Oriental Civilization,* no. 31. Chicago: University of Chicago Press.

Braidwood, Robert J., Bruce Howe, and Charles A. Reed

1961 "The Iranian Prehistoric Project." Washington, D.C.: *Science* 133:2000–2001.

Braidwood, Robert J., and Gordon R. Willey

1962 *Courses Toward Urban Life.* Chicago: Aldine.

Brown, Kenneth L.

1980 "A Brief Report on Paleoindian Archaic Occupations in the Quiché Basin, Guatemala." Salt Lake City, Utah: *American Antiquity* 45(2):313–24.

Brush, Charles F.

1969 "A Contribution to the Archaeology of Coast Guerrero, Mexico." Ph.D. dissertation, Department of Anthropology, Columbia University, New York.

Bryan, A. L., and M. C. de Beltrao

1978 "An Early Stratified Sequence near Rio Claro, East-Central Sao Paulo State, Brazil." In *Early Man in America,* edited by A. L. Bryan, 303–305. Department of Anthropology Papers 1. Edmonton, Canada: University of Alberta.

Bryan, Kirk, and J. H. Toulouse, Jr.
1943 "The San Jose Non-Ceramic Culture and its Relation to Puebloan Culture in New Mexico." Menasha, Wisconsin: *American Antiquity* 8(3):269–88.

Buikstra, J. E., Jill Bullington, O. F. Charles, D. C. Cook, S. R. Frankenberg, L. W. Kornigberg, J. B. Lambert, and L. Xue
1987 "Diet, Demography, and the Development of Horticulture." In *Emergent Horticultural Economies of the Eastern United States,* edited by W. F. Keegan, 67–86. Center for Archaeological Investigations, Southern Illinois University, Occasional Paper no. 7. Carbondale, Illinois.

Bullen, Ripley P.
1962 "The Preceramic Krum Bay Site: Virgin Islands and Its Relationship to the Peopling of the Caribbean." Vienna, Austria: 34th International Congress of Americanists.

Bullen, Ripley P., and William W. Plowden, Jr.
1963 "Preceramic Archaic Sites in the Highlands of Honduras." Salt Lake City, Utah: *American Antiquity* 28(3):382–85.

Buth, G. M., Faroog A. Lone, and Magsooda Khan
1986 "Plant Husbandry and Ancient Agriculture in North India." Paper presented at World Archaeological Congress, Southampton, England.

Butler, Ann
1988 "Cryptic Anatomical Characters as Evidence of Pre-Agrarian Plant Manipulation." In *Foraging and Farming,* edited by D. Harris and C. Hillman, 390–407. London: Allen and Unwin.

Butzer, Karl W.
1972 *Environment and Archaeology: An Ecological Approach to Prehistory.* Chicago: Aldine.

Butzer, Karl W., and H. B. S. Cooke
1964 *Environment and Archaeology: An Introduction to Pleistocene Geography.* Chicago: Aldine.
1982 "The Palaeo-Ecology of the African Continent—The Physical Environment of Africa from Earliest Geological to Late Stone Age Times." In Vol. 1 of *The Cambridge History of Africa.* Cambridge, England: Cambridge University Press.

Byers, Douglas S, editor
1967a *Environment and Subsistence.* Vol. 1 of *Prehistory of the Tehuacán Valley.* Austin, Texas: University of Texas Press.
1967b *The Non-Ceramic Artifacts.* Vol. 2 of *Prehistory of the Tehuacán Valley.* Austin, Texas: University of Texas Press.

Caldwell, J. R.
1958 "Trend and Tradition in the Prehistory of the Eastern United States." Scientific Papers no. 10. Menasha, Wisconsin: *American Anthropologist,* Memoir 88.

Campbell, J. M., and F. H. Ellis
1952 "The Atriso Sites: Cochise Manifestations in the Middle Rio Grando." Salt Lake City, Utah: *American Antiquity* 17(3): 211–221.

Cardich, Augusto
1964 *Lauricocha, Fundamentos para una Prehistoria de los Andes.* Central Studia Praehistorica, no. 3. Buenos Aires: Centro Argentino de Estudios Prehistoricas.

Carmichael, David L.
1982 *New Radiocarbon Dates and Evidence for Early Cultigens at Fresnal Shelter,*

New Mexico. Urbana, Illinois: Department of Anthropology, University of Illinois.

1986 "Archaeological Survey in the Tularosa Basin of New Mexico," Historic and National Resources Report No. 3. Fort Bliss, Texas.

Carneiro, R. L.

1960 "Slash-and-Burn Agriculture, a Closer Look at its Implications for Settlement Patterns." In *Man and Culture*, edited by A. Wallace, 229–34. Philadelphia, Pennsylvania: University of Pennsylvania Press.

1974 "The Transition from Hunting to Horticulture in the Amazon Basin." In *Man in Adaptation*, edited by Y. A. Cohen, 157–66. Chicago: Aldine.

Carter, George F.

1945 *Plant Geography and Culture History in the American Southwest*. New York: Viking Fund Publication no. 5.

Castetter, Edward F., and Willis H. Bell

1942 *Pima and Papago Indian Agriculture*. Albuquerque, New Mexico: University of New Mexico Press.

Caton-Thompson, G., and E. W. Gardner

1934 *The Desert Fayum*. 2 vols. London: Royal Anthropological Institute.

Chang, K. C.

1967 "The Yale Expedition to Taiwan and the Southeast Asian Horticultural Evolution." New Haven, Connecticut: Yale University Press, *Discovery* 2(2): 3–10.

1969 *Fengpilou, Tapenkeng, and the Prehistory of Taiwan*. Yale University Publications in Anthropology 73. New Haven, Connecticut: Yale University Press.

1970 "The Beginning of Agriculture in the Far East." Cambridge, England: *Antiquity* 44:175–85.

1971 *The Archaic of Ancient China*. New Haven, Connecticut: Yale University Press.

1973 "Radiocarbon Dates from China: Some Initial Interpretations." Chicago: *Current Anthropology* 14(5):525–28.

1979 "The Affluent Foragers in the Coastal Area of China, Extrapolations from Evidence on Transition to Agriculture." In *Affluent Foragers*, edited by Shuzo Koyama and David Hurst Thomas, 101–23. Osaka, Japan: National Museum of Ethnology.

Chang, Te-Tzu

1976 "The Rice Cultures." London: Philosophical Transactions of the Royal Society of London, 275:143–57.

1983 "The Origins and Early Culture of the Cereal Grains and Food Legumes." In *The Origins of Chinese Civilization*, edited by D. V. Keightley, 65–94. Berkeley: University of California Press.

1988 "Domestication and Spread of the Cultivated Rices." In *Foraging and Farming*, edited by D. Harris and C. Hillman, 408–17. London: Allen and Unwin.

Chevalier, A.

1938 "Le Sahara, Centre D'origine des Plantes Cultivée." Paris: *Société de Biogeographie*, VI:307–22.

Childe, V. Gordon

1925 *The Dawn of European Civilization*. New York: Alfred Knopf.

1931 "The Forest Cultures of Northern Europe." Dublin, Ireland: Royal Anthropological Institute of Great Britain and Ireland, *Journal* 61:325–48.

1942 *What Happened in History*. Baltimore, Maryland: Penguin.
1951 *Man Makes Himself*. London: C. A. Watts and Co.
1958 *The Prehistory of European Society*. Harmondsworth, England: Pelican Books.

Chomko, Stephen A., and Gary W. Crawford
1978 "Plant Husbandry in Prehistoric Eastern North America, New Evidence of its Development." Salt Lake City, Utah: *American Antiquity* 43(3):405–407.

Clark, J. Desmond
1968 "The Middle Acheulian Occupation Site at Latamne, Northern Syria." *Quaternaria* 10:1–72.
1970 "The Spread of Food Production in Sub-Saharan Africa." In *Papers in African Prehistory*, edited by J. D. Fage and R. A. Oliver. Cambridge. England: Cambridge University Press.
1981 "Prehistory in South Africa." In Vol. 1 of *General History of Africa*, edited by J. Ki-Zerbo, 523–30. Berkeley, California: Heinemann.

Clark, J. Graham D.
1936 *The Mesolithic Settlement of Northern Europe*. Cambridge, England: Cambridge University Press.
1952 *Prehistoric Europe: The Economic Basis*. London: Methuen.
1957 *Archaeology and Society*. Cambridge, Massachusetts: Harvard University Press.
1972 *Star Carr: A Case Study in Bioarcheology*. Chicago: Addison-Wesley.
1980 *Mesolithic Prelude*. Edinburgh, Scotland: University of Edinburgh Press.

Coe, J. D.
1964 "The Formative Cultures of the Carolina Piedmont." Philadelphia, Pennsylvania: American Philosophical Society, *Transaction* 54(5).

Coe, Michael D., and Kent V. Flannery
1964 "Microenvironments and Mesoamerican Prehistory." Washington, D.C.: *Science* 143:650–54.

Cohen, Mark N.
1977a *The Food Crisis in Prehistory: Overpopulation and the Origins of Agriculture*. New Haven, Connecticut: Yale University Press.
1977b "Population Pressure and the Origin of Agriculture: Peru." In *Origins of Agriculture*, edited by C. Reed, 135–78. The Hague: Mouton Publishers.

Cohen, Morris, and Alfred Nagel
1970 *Logic and the Scientific Method*. New York: University of Columbia Press.

Cole, Fay-Cooper, and Thorne Deuel
1937 *Rediscovering Illinois*. Chicago: University of Chicago Press.

Conklin, H. C.
1957 *Hanunoo Agriculture*. Rome, Italy: Food and Agriculture Organization.

Contenson, Henri de
1971 "Tell Ramad, A Village of Syria of the 7th and 6th Millennia B.C." New York: *Archaeology* 24:278–83.

Coon, Carleton S.
1956 *The Seven Caves*. New York: Knopf.

Cordell, linda S.
1984 *Prehistory of the Southwest*. New York: Academic Press.

Cosgrove, C. B.
1947 *Caves of the Upper Gila and Hueco Areas in New Mexico and Texas*. Papers of the Peabody Museum of American Archaeology and Ethnology 24(2). Cambridge, Massachusetts.

Coursey, D. G.
1976 "The Origins and Domestication of Yams in Africa." In *Origins of African Plant Domestication,* edited by Jack Harlan, John de Wet, and Ann Stemler, 383–407. The Hague: Mouton Publishers.

Cowan, C. Wesley
1985 "Understanding the Evolution of Plant Husbandry in Eastern North America: Lessons from Botany, Ethnography, and Archaeology." In *Prehistoric Food Production in North America,* edited by R. I. Ford, 205–44. Ann Arbor, Michigan. Anthropological Papers 75, Museum of Anthropology, University of Michigan.

Cowan, W., and P. J. Watson
1989 *Origins of Plant Cultivations in World Perspective.* Washington, D.C.: Smithsonian Institution Press.

Crawford, Gary
1983 *Paleoethnobotany of the Kaneda Peninsula Jomon.* Anthropological Paper 73. Museum of Anthropology, University of Michigan. Ann Arbor, Michigan.
1989 "Prehistoric Plant Domestication in East Asia: The Japanese Perspective." In *The Origins of Plant Domestication in World Perspective,* edited by P. J. Watson and C.W. Cowan, 1–37. Washington, D.C.: Smithsonian Institution Press.

Crawford, Gary, William A. Hurley, and Masakuzu Yoshizaki
1976 "Implication of Plant Remains from Early Jomon, Hamonasuno Site." *Asian Perspectives* 19(1): 145–48.

Cressey, George B.
1951 *Asia's Lands and Peoples.* New York: McGraw-Hill.
1960 *Crossroads: Land and Life in Southwest Asia.* Chicago: Lippincott.

Cruxent, J. M.
1967 "El Paleo-Indio en Taima-Taima, Islada Falcon, Venezuela." Caracas, Venezuela: *Acta Científica,* Venezuela 3: 17.

Cutler, Hugh C., and George Agogino
1960 "Analysis of Maize from the Four Bear Site and Two Other Arikara Locations in South Dakota." Albuquerque, New Mexico: *Southwestern Journal of Anthropology* 16(3): 312–16

Cutler, Hugh C., and Leonard W. Blake
1973 *Plants from Archaeological Sites East of the Rockies.* St. Louis, Missouri: Missouri Botanical Gardens.

Cutler, Hugh C., and Thomas W. Whitaker
1967 "Cucurbits from Tehuacán Caves." In *Environment and Subsistence,* vol. 1 of *Prehistory of the Tehuacán Valley,* edited by D. S. Byers, 212–19. Austin, Texas: University of Texas Press.

Damp, J., D. Pearsall, and L. Kaplan
1981 "Beans for Valdivia." Washington, D.C.: *Science* 212: 811–12.

Davidson, Ian
1986 "The Other Side of the First Frontier: Escaped Domesticated Animals, the Introduction of Agriculture to Spain, and the Identification of the Origins of Agriculture." Paper presented at World Archaeological Congress, Southampton, England.

Davies, O.
1967 *West Africa Before the Europeans: Archaeology and Prehistory.* London: Methuen.

De Boer
1975 "The Archaeological Evidence for Manioc Cultivation: A Cautionary Note." Salt Lake City, Utah: *American Antiquity* 40(4):419–33.

DeBono, F.
1981 "Prehistory in the Nile Valley." In Vol. 1 of *General History of Africa*, edited by J. Ki-Zerbo, 634–51. Berkeley, California: Heinemann.

DeCandolle, Alphonse
1884 *Origins of Cultivated Plants*. London: Kegan Paul.

Decker, Deena
1988 "Origins, Evolution, and Systematics of *Cucurbita Pepo*." *Economic Botany* 42:4–15.

Dennell, Robin
1983 *European Economic Prehistory*. New York: Academic Press.
1989 "The Origins and Early Development of European Crop Agriculture: A Summary and Discussion of the Present Evidence." In *The Origins of Plant Domestication in World Perspective*, edited by P. J. Watson and C. Wesley Cowan. Washington, D.C.: Smithsonian Institution Press.

Dick, Herbert W.
1965 *Bat Cave*. Monographs of the School of American Research, no. 17. Santa Fe, New Mexico.

Dimbleby, Geoffrey
1967 *Plants and Archaeology*. London: John Baker Co.

Di Peso, Charles C.
1956 *The Upper Pima of San Cayetano del Tumacacori*. Amerind Foundation Publication 7. Tucson, Arizona.

Dolukhanov, P. M.
1982 "Upper Pleistocene and Holocene Cultures of the Russian Plain and Caucasus." In *Advances in World Archaeology*, edited by Fred Wendorf and Angela Close, 323–58. New York: Academic Press.

Donnan, C.
1964 "An Early House from Chilca, Peru." Salt Lake City, Utah: *American Antiquity* 30(2):137–44.

Drucker, Phillip
1948 "Preliminary Notes on an Archaeological Survey of the Chiapas Coast." In *Middle American Research Records* 1(11):1–16. New Orleans, Louisiana: Tulane University.

Engel, Frederic
1966 *Paracas, Cien Siglos de Cultura Peruana*. Lima, Peru: Juan Mejia Baca.
1972 "La Grotte du Megatherium a Chilca et les Ecologies du Haut-Holocene Peruvien." In *Melangos oflest a Claude Levi-Strauss*, edited by J. Pouillon et P. Maranda, 413–35. The Hague: Mouton Publishers.

Epstein, Jeremiah F.
1969 *The San Isidro Site: An Early Man Campsite in Nuevo Leon, Mexico*. Anthropological Papers of the University of Texas 1:111–23. Austin, Texas.

Epstein, Jeremiah F., T. R. Hester, and Carol Graves, editors
1980. "Papers on the Prehistory of Northeastern Mexico and Adjacent Texas." University of Texas Center for Archaeological Research, Report No. 9. San Antonio, Texas.

Espinosa, Jorge
1976 "Excavaciones Arqueológicas en El Bosque." *Informe* 1:22–55. Managua, Nicaragua: Instituto Geográfico Nacional.

Evans, C., and B. J. Meggers
1960 *Archaeological Investigations in British Guiana.* Bureau of American Ethnology, Bulletin 177. Washington, D.C.
1968 *Archaeological Investigations on the Rio Napo, Eastern Ecuador.* Smithsonian Contributions to Anthropology, vol. 6. Washington, D.C.: Smithsonian Institution Press.

Evans, C., B. J. Meggers, and J. M. Cruxent
1959 "Preliminary Results of Archaeological Investigations along the Orinoco and Ventuari Rivers, Venezuela." San Jose, Costa Rica: 33rd International Congress of Americanists, *Actas* 2:359–69.

Ewing, J. F.
1947 "Preliminary Notes on the Excavation of the Paleolithic Site of Ksar Akil." Cambridge, England: *Antiquity* 21:186–97.

Fay, George E.
1950 "Peralta Complex—A Sonoran Variant of the Cochise Culture." Washington, D.C.: *Science* 124:1029.

Ferdon, Edwin N., Sr.
1946 *An Excavation of Hermit's Cave, New Mexico.* School of American Research Monograph 10. Santa Fe, New Mexico.

Ferguson, Leland
1978 "The Southeast in Current Research." Salt Lake City, Utah: *American Antiquity* 43 (4):756–62.

Fish, Paul R., Suzanne K. Fish, Austin Long, and Charles Mitsicek
1986 "Early Corn Remains from Tumamoc Hill, Southern Arizona." Salt Lake City, Utah: *American Antiquity* 51(3):563–71.

Fisher, W. B.
1968 *The Land of Iraq.* In vol. 1 of *The Cambridge History of Iran.* Cambridge, England: Cambridge University Press.

Flannery, Kent V.
1966 "The Post Glacial Re-Adaptation as Viewed from Mesoamerica." Salt Lake City, Utah: *American Antiquity* 31(6):800–805.
1968 "Archaeological Systems Theory and Early Mesoamerica." In *Anthropological Archaeology in the Americas,* edited by B. J. Meggers. Washington, D.C.: Anthropological Society of Washington.
1970 "Preliminary Archaeological Investigations in the Valley of Oaxaca, Mexico, 1966 through 1969." Report to the Instituto Nacional de Antropología e Historia and the National Science Foundation. MS. Museum of Anthropology, University of Michigan. Ann Arbor, Michigan.
1986 *Guilá Naquitz: Archaic Foraging and Early Agriculture in Oaxaca, Mexico.* New York: Academic Press.

Flannery, Kent V., Joyce Marcus, and Stephen A. Kowalewski
1981 "The Preceramic and Formative in the Valley of Oaxaca." Supplement to vol. 1 of the *Handbook of Middle American Indians,* edited by Victoria R. Bricker and J. A. Sabloff, 48–93. Austin, Texas: University of Texas Press.

Ford, J. A.
1944 *Excavations in the Vicinity of Cali, Colombia.* Yale University Publications in Anthropology 31. New Haven, Connecticut: Yale University.

Ford, J. A., and Gordon R. Willey
1941 "An Interpretation of the Prehistory of the Eastern United States." Menasha, Wisconsin: *American Anthropologist,* 43:325–63.

Ford, Richard I.

1981 "Gardening and Farming before A.D. 1000: Patterns of Prehistoric Cultivation North of Mexico." Ann Arbor, Michigan: *Journal of Ethnobiology* 1(1): 96–127.

Ford, Richard I., editor

1985 *Prehistoric Food Production in North America.* Anthropological Papers 75, Museum of Anthropology, University of Michigan. Ann Arbor, Michigan.

Fowler, Michael L.

1973 "Agriculture and Village Settlement in the North American East—The Central Mississippi Valley Area: A Case History." Seville, Spain: XXXVI Congreso Internacional de Americanistas, Vol 1.

1974 *Cahokia: Ancient Capital of the Midwest.* Menlo Park, California: Addison-Wesley.

Fritz, Gayle

1984 "Identifications of Cultigen Amaranth and Chenopod from Rockshelter Sites in Northwestern Arkansas." Salt Lake City, Utah: *American Antiquity* 49: 555–72.

Galinat, Walton C.

1970 "Maize from the Blaine Site." In *Blaine Village and the Fort Ancient Tradition in Ohio,* edited by O. Prufer and O. J. Shane, 219–26. Kent, Ohio: Kent State University Press.

1980 "The Archaeological Maize Remains from Volcan Panama." In *Adaptation Radiations in Prehistoric Panama,* edited by O. Linares and A. Ranere, 175–81. Peabody Museum monographs 5, Harvard University. Cambridge, Massachusetts.

Gallagher, James P.

1988 "Agriculture Intensification in Temperate Climates." In *Foraging and Farming,* ed. by D. Harris and C. Hillman, 572–84. London: Allen and Unwin.

Gambel, Halct, and R. J. Braidwood

1980 *Prehistoric Research in Southeastern Anatolia.* Istanbul: Istanbul University.

Garcia-Barcena, Joaquín

1981 "Una punta acanalada de la cueva Los Grifos, Ocozocautla, Chiapas." Mexico, D.F.: Depto. de Prehistoria, *Cuadernos* 17:1–19.

Garcia-Barcena, Joaquín, and Diana Santamaria

1982 "La Cueva de Santa Marta, Ocozocoautla, Chiapas." *Colección Científica* 111. Mexico: Instituto Nacional de Antropología e Historia.

Garcia-Barcena, Joaquín, Diana Santamaria, Ticul Alverez, Manuel Reyes, and Fernando Sanchez

1976 *Excavaciones en El Abrigo de Santa Marta, Chiapas.* Mexico: INAH, Depto. de Prehistoria, *Informes* 1:1–21.

Garcia Cook, Angel

1973 *Una Punta acanalada en El Estado de Tlaxcala, Mexico.* Puebla, Mexico: FAIC no. 9, 25–26.

Garcia Moll, Roberto

1977 *Análisis de los materiales arqueológicos de la Cueva del Texcal, Puebla.* Mexico, D.F.: Depto. de Prehistoria, Colección Científica Arqueologia no. 56, 9–89.

Garrod, Dorothy

1930 *The Paleolithic of Southern Kurdistan, Excavations in the caves of Zarsi and Hazar Merd.* Athens: American School of Prehistory Research Bulletin no. 6.

1957 "The Natufian Culture, the Life and Economy of a Mesolithic People in the Near East." London: *Proceedings of the British Academy* 43:55.

Garrod, Dorothy, and J. G. D. Clark

1965 "Primitive Man in Egypt, Western Asia and Europe." In vol. 1 of *Cambridge Ancient History.* Cambridge, England: Cambridge University Press.

Gilmore, Melvin R.

1931 *Vegetable Remains of the Ozark Bluff Dweller Culture.* Papers of Michigan Academy of Sciences, Arts, and Letters, no. 14. Ann Arbor, Michigan: University of Michigan Press.

Glover, I. C.

1979 "Prehistoric Plant Remains from Southwest Asia with Special Reference to Rice." In *South Asian Archaeology,* no. 7, 7–37. Naples, Italy: Papers of 4th International Conference of South Asian Archaeologists.

Golson, Jack, and P. J. Hughes

1976 "The Appearance of Plant and Animal Domestication in New Guinea." In *La Prehistoire Oceanienne,* edited by Jose Granger, 88–100. Paris: *Colloque,* 22, Centre National de la Recherche Cientifique.

Goodyear, Albert C.

1975 *Hecla II and III, an Interpretive Study of Archaeological Remains from the Lake Shore Project, Papago Reservation, South Central Arizona.* Arizona State University Anthropological Research Paper no. 9. University of Arizona Department of Anthropology.

Gorman, Chester F.

1969 "Hoabinhian: A Pebble-tool Complex with Early Plant Association in Southeast Asia." Washington, D.C.: *Science* 163:671–73.

1970 "Excavations at Spirit Cave, North Thailand: Some Interim Interpretations." Madison, Wisconsin: *Asian Perspectives* 13:79–107.

1977 "A Priori Models and Thai Prehistory: A Reconsideration of the Beginnings of Agriculture in Southeast Asia." In *Origins of Agriculture,* edited by C. E. Reed, 321–56. The Hague: Mouton Publishers.

Green, R. C.

1976 "New Sites with Lapita Pottery and Their Implications for an Understanding of the Settlement of the Western Pacific." Working Paper 51, Dept. of Anthropology, University of Auckland, New Zealand.

Griffin, James B.

1965 "Late Quaternary Prehistory in the Northeastern Woodlands." In *Quaternary of the United States,* 663–67. Princeton, New Jersey: Princeton University Press.

1978 "The Midlands." In *Ancient North Americans,* edited by Jesse D. Jennings, 221–80. San Francisco: Freeman Press.

Gruhn, Ruth

1978 "A Note on Excavations at El Bosque, Nicaragua in 1975." In *Early Man in America,* edited by A. D. Bryan, 261–62. Edmonton, Alberta: University of Alberta.

Gruhm, Ruth, and Alan Lyle Bryan

1977 "Los Tapiales: A Paleo-Indian Campsite in the Guatemalan Highlands." Philadelphia: Proceedings of the American Philosophical Society 121:3.

Guernsey, Samuel J., and A. V. Kidder

1921 *Basket-Maker Caves of Northeastern Arizona.* Papers of the Peabody Museum of American Archaeology and Ethnology 8(2). Cambridge, Massachusetts.

Guilaine, Jean

1976 *La Prehistoire Française.* Paris: Centre Nationale de Recherche Cientifique III.

1979 "The Earliest Neolithic in the Western Mediterranean: A New Appraisal." Cambridge, England: *Antiquity* 53:22–30.

1981 *Les Premiers Bergers, et Paysan de l'Occident Mediterranée,* 2d ed. The Hague: Mouton Publishers.

Guilaine, Jean, and M. Barbaza, D. Geddes, J. L. Vernet, M. Llonqueras and M. Hopf

1982 "Prehistoric Human Adaptations in Catalonia." Boston, Massachusetts: *Journal of Field Archaeology* 9:407–16.

Gummerman, George, and R. C. Euler

1976 *Papers on the Archaeology of Black Mesa, Arizona.* Carbondale, Illinois: Southern Illinois University Press.

Hahn, Eduard

1896 *Die Haustierre und ihre Beziehungen zur Wirtschaft des Menschen.* Leipzig: Duncker and Humbell.

Hallam, Sylvia J.

1988 "Plant Usage and Management in Southwest Australian Societies." In *Foraging and Farming,* edited by D. Harris and C. Hillman, 136–51. London: Allen and Unwin.

Harlan, Jack R.

1971 "Agricultural Origins: Centers and Non-Centers." Washington, D.C.: *Science,* 174(4008):465–73.

1977 "The Origins of Cereal Agriculture in the Old World." In *The Origins of Agriculture,* edited by C. Reed, 357–84. The Hague: Mouton Publishers.

1982 "The Origins of Indigenous African Agriculture." In vol. 1 of *Cambridge History of Africa,* 624–57. Cambridge, England: Cambridge University Press.

1986 "Wild Grass-Seed Harvesting in the Sahara and Sub-Sahara of Africa." Paper presented at the World Archaeological Congress, Southampton, England.

1988 "Indigenous African Agriculture." In *Foraging and Farming, The Evolution of Plant Exploitation,* edited by D. Harris and G. C. Hillman, 111–36. London: Allen and Unwin.

Harlan, Jack R., J. M. de Wet, and A. B. L. Stemler

1976 *Origins of African Plant Domestications.* The Hague: Mouton Publishers.

Harlan, Jack R., and Daniel Zohary

1966 "Distribution of Wild Wheats and Barley." Washington, D.C.: *Science* 153(3740):1074–1080.

Harrington, M. R.

1924 "The Ozark Bluff-Dwellers." Menasha, Wisconsin: *American Anthropologist* 26(1):1–21.

Harris, David R.

1969 "Agricultural Systems, Ecosystems and the Origins of Agriculture." In *The Domestication and Exploitation of Plants and Animals,* edited by P. J. Ucko and G. W. Dimbleby, 3–15. London: Duckworth.

1977 "Alternate Pathways Toward Agriculture." In *Origins of Agriculture,* edited by C. Reed, 179–237. The Hague: Mouton Publishers.

Harris, David, and Gordon C. Hillman, editors

1988 *Foraging and Farming: The Evolution of Plant Exploitation.* London: Allen and Unwin.

Harris, Marvin

1975 *Culture, People, and Nature: An Introduction to General Anthropology.* New York: T. Y. Crowell.

1980 *Cultural Materialism.* New York: Vintage.

Hassan, Fekri A.

1977 "The Dynamics of Agricultural Origins in Palestine: A Theoretical Model." In *Origins of Agriculture,* edited by C. Reed. The Hague: Mouton Publishers.

1984 "Mid-Holocene Desertification and Human Responses in the Western Desert of Egypt." Paper presented at the 17th Annual Chacmool Conference, Calgary, Alberta, Canada.

1985 "Holocene Nile Floods and Their Implications for Origins of Egyptian Agriculture." In *Culture and Environment in Late Quaternary of Africa,* Cambridge Monographs in African Archaeology, edited by J. Bower and D. Lubell. Cambridge, England: Cambridge University.

Haudricourt, A. G., and Louis Hedin

1944 *L'Homme et les plantes cultivées.* Paris: Gallimard.

Haury, Emil

1962 "The Greater American Southwest." In *Courses Toward Urban Life,* edited by R. J. Braidwood and G. Willey, 106–31. Chicago: Aldine.

1975 *The Stratigraphy and Archaeology of Ventana Cave, Arizona.* Tucson, Arizona: University of Arizona Press.

Haury, Emil, and E. B. Sayles

1947 *An Early Pit-House Village of the Mogollon Culture, Forestdale Valley, Arizona.* University of Arizona, Bulletin 18(4). Tucson, Arizona: University of Arizona.

Hawkes, J. G.

1988 "The Domestication of South American Roots and Tubers." In *Foraging and Farming,* edited by D. Harris and C. Hillman, 481–503. London: Allen and Unwin.

Hedge, N. T. M.

1978 *Analysis of Ancient Indian deluxe wares,* 401–405. Bonn: Proc. Int. Symp. Archaeometry.

Heer, O.

1866 Treatise on the Plants of the Lake Dwellings. In *The Lake Dwellings of Switzerland and other parts of Europe,* F. Keller, trans. J. E. Lee. London, n.p.

Heiser, Charles B.

1985 "Some Botanical Considerations of the Early Domesticated Plants North of Mexico." In *Prehistoric Food Production in North America,* edited by R. I. Ford. Anthropological Papers 75, Museum of Anthropology, University of Michigan. Ann Arbor, Michigan.

1988 "Domestication of Cucurbitaceae: Cucurbita and Lagenaria." In *Foraging and Farming,* edited by D. Harris and C. Hillman, 471–80. London: Allen and Unwin.

Heiser, Charles B., and David C. Nelson

1974 "On the Origins of the Cultivated Chenopods (Chenopodium)." *Genetics* 78:503–505

Helbaek, Hans

1948 Les empreintes de cereales on Riis. Copenhagen: P. J. Hamu.

1954 "Prehistoric Food Plants and Weeds in Denmark." Copenhagen: *Denmark Geologische Ilders* 2(80):250–61.

1960 "The Paleoethnobotany of the Near East and Europe." In *Prehistoric Investigations in Iraqi Kurdistan,* edited by R. J. Braidwood and B. Howe, 99–118. Chicago: University of Chicago Press.

1962 "Late Cypriot Vegetable Diet at Apliki." Lund, Denmark: *Opuscula Atheniensia* 4:171–83.

1964 "First Impressions of Catal Huyuk Plant Husbandry." Chicago: *Anatolian Studies* 14:121.

1966 "Pre-pottery Neolithic Farming at Beidha." *Palestine Exploration Quarterly* 98:61.

1969 "Plant Collecting, Dry-farming and Irrigation in Prehistoric Deh Luran." In *Prehistory and Human Ecology of the Deh Luran Plain,* Memoir no. 1, edited by Frank Hole, Kent Flannery, and James Neeley, 244–303. Ann Arbor, Michigan: University of Michigan Museum of Anthropology.

Hester, Thomas R., Thomas C. Kelly and Giancarlo Ligabue

1981 "A Fluted Paleo-Indian Projectile Point from Belize, Central America." San Antonio, Texas: University of Texas, Center for Archaeological Research, *Working Papers,* No. 1, 1–16.

Hibben, Frank

1941 "Evidence of Early Occupation of Sandia Cave, New Mexico and other sites in the Sandia-Manzano region." Smithsonian Miscellaneous Collections 99. Washington, D.C.: Smithsonian Institution Press.

Higgs, E. S., and C. Vita-Finzi

1972 "Prehistoric economies: A terratopical approach." In *Papers in economic prehistory,* edited by E. S. Higgs, 1–24. Cambridge, England: Cambridge University Press.

Higham, Charles

1986 "The Archaeology of Mainland Southeast Asia." In *Cambridge World Archaeology.* Cambridge: Cambridge University Press.

1988 "Rice Exploitation in Southeast Asia: New Information from Khok Phanom Di, Central Thailand." In *Foraging and Farming,* edited by D. Harris and C. Hillman, 641–49. London: Allen and Unwin.

Higham, Charles, and Bernard Maloney

1988 "Coastal Adaptations, Sedentism, and Domestication: A Model for Socioeconomic Intensification in Prehistoric Southeast Asia." In *Foraging and Farming,* edited by D. Harris and C. Hillman, 650–60. London: Allen and Unwin.

Hill, Betsy Dupuis

1972 "A New Chronology of the Valdivia Ceramic Complex from the Coastal Zone of Guayas Province, Ecuador." Berkeley, California: *Nawpa Pacha* 10(12): 1–32.

Hillman, Gordon C.

1988 "Plant-Food Economy During the Epi-Paleolithic Period at Tell Abu Hureyra, Syria: Dietary Diversity, Seasonality, and Modes of Exploitation." In *Foraging and Farming,* edited by D. Harris and C. Hillman. London: Allen and Unwin.

Ho, Ping-ti

1977 "The Indigenous Origins of Chinese Agriculture." In *Origins of Agriculture,* edited by C. E. Reed, 413–84. The Hague: Mouton Publishers.

Hole, Frank, and Kent V. Flannery

1967 "The Prehistory of Southwestern Iran, A Preliminary Report." London: *Proceedings of the Prehistoric Society of 1967* 33:147–70.

Hole, Frank, Kent V. Flannery and James A. Neely
1969 *Prehistory and Human Ecology of the Deh Luran Plain*. Museum of Anthropology Memoir no. 1. Ann Arbor, Michigan: University of Michigan.
Hopf, Maria
1965 *Untersuchungen aus dem botanischem Labor am RGZM—Getreidekorn—Abdrucke als schmuckelements in Neolthischer Kerakik*. Mainz, Germany: Jahrbuch Rom. Germanzentralmuseums.
1969 "Plant Remains and Early Farming in Jericho." In *The Domestication and Exploitation of Plants and Animals,* edited by P. J. Ucko and G. W. Dimbleby, 355–59. London: Duckworth.
Horowitz, A., R. E. Gerald, and M. S. Chaiffetz
1981 "Preliminary Paleoenvironmental Implications of Pollen Analysis from Mesilla, El Paso and Historic Sites." El Paso: Centennial Museum, University of Texas.
Howard, George
1943 *Excavations at Ronquin, Venezuela*. Yale University Publications in Anthropology, no. 28. New Haven, Connecticut.
1947 *Prehistoric Ceramic Styles of Lowland South America: Their Distribution and History*. Yale University Publications in Anthropology, no. 37. New Haven, Connecticut.
Huckell, Bruce B., and Lisa W. Huckell
1988 "Crops Come to the Desert: Late Pre-Ceramic Agriculture in Southeastern Arizona." Paper presented at the Society for American Archaeology Meeting, Phoenix, Arizona.
Hugot, H. J.
1981 "The Prehistory of the Sahara. In vol. 1 of *General History of Africa,* edited by J. Ki-Zerbo, 585–603. Berkeley, California: Heinemann.
Huntington, Ellsworth
1915 *Civilization and Climate*. New Haven, Connecticut: Yale University Press.
Huscher, H.
1941 *Continuation of Archaeological Survey of Southern and Western Colorado* (Grant No 557). Yearbook of the American Philosophical Society. Philadelphia, Pennsylvania.
Hutterer, Karl Leopold
1983 "The Natural and Cultural History of Southeast Asian Agriculture." London: *Anthropos* 78:169–212.
Ikawa-Smith, Fumiko
1978 "Introduction: Early Paleolithic Tradition of East Asia." In *Early Paleolithic in South and East Asia,* edited by F. Ikawa-Smith, 1–10. The Hague: Mouton Publishers.
1982 "The Early Prehistory of the Americas as Seen from Northeast Asia." In *Peoples of the New World,* edited by James E. Ericson, R. Ervine Taylor, and Reiner Berger. Los Altos, California: Balena Press.
Irwin-Williams, Cynthia
1967a "Association of Early Man with Horse, Camel, and Mastodon at Hueyatlaco, Valsequillo (Puebla, Mexico)." In *Pleistocene Extinctions,* edited by P. S. Martin, 337–47. New Haven, Connecticut: Yale University Press.
1967b "Picosa: The Elementary Southwestern Culture." Salt Lake City, Utah: *American Antiquity,* 32(4):441–57.
1967c "A Summary of Archaeological Evidence from the Valsequillo Region, Puebla, Mexico." In *Proceedings of the 9th International Congress of Anthro-*

pological and Ethnological Sciences, Chicago, 7–22. The Hague: Mouton Publishers.

1973 *The Oshara Tradition: Origins of Anasazi Culture.* Eastern New Mexico University, Contributions in Anthropology, 5(1). Portales, New Mexico: University of Eastern New Mexico.

1979 "Post-Pleistocene Archaeology, 7000-2000 B.C. in the Southwest." In *Handbook of North American Indians,* 9:31–42. Washington D.C.: Smithsonian Institution Press.

Irwin-Williams, Cynthia, and Patrick Beckett

1973 "Excavations at the Moquis Site: A Cochise Locality in Northern New Mexico." Unpublished manuscript.

Irwin-Williams, Cynthia, and J. H. Irwin

1966 "Excavations at Magic Mountain." In *Proceedings of the Denver Museum of Natural History,* no. 12. Denver, Colorado.

Irwin-Williams, Cynthia, and Phillip H. Shelley

1980 *Investigations at the Salmon Site, The Structure of Chacoan Society in Northern Southwest,* 5 vols. Portales, New Mexico: Eastern New Mexico University Press.

Izumi, Seiichi, and Toshihiko Sono

1963 *Excavations at Kotosh, Peru, 1960.* Tokyo: Kadokawa Publishing.

Jacobsen, Thomas W.

1981 "Franchthi Cave and the Beginning of Settled Life in Greece." *Hesperia* 50(4):6–10.

Jacobson, Jerome

1976 "Evidence for Prehistoric habitation patterns in eastern Malwa." In *Ecological Backgrounds of South Asian Prehistory,* edited by K. A. R. Kennedy and G. L. Possehl. Ithaca, New York: Cornell University.

1979 *Recent Developments in South Asian Prehistory and Protohistory.* Annual Review of Anthropology, vol. 8. Palo Alto, California: Stanford University.

Jarman, M. R., G. N. Bailey, and H. N. Jarman

1982 *Early European Agriculture: Its Foundation and Development.* New York: Cambridge University Press.

Jelinek, Arthur, W. R. Forrand, G. Haes, A. Horowitz, and P. Goldberg

1973 "New Excavations at the Tabun Cave, Mt. Carmel, Israel 1967–1972, A Preliminary Report." Chicago: *Paleo-Orient* 1(2):151–83.

Jennings, Jesse D.

1957 "Danger Cave." *Memoirs of the Society for American Archaeology,* vol. 23. Salt Lake City, Utah.

1978 *Prehistory of Utah and the Eastern Great Basin.* Anthropological Paper, University of Utah, Salt Lake City, Utah.

1979 *The Prehistory of Polynesia.* Cambridge, Massachusetts: Harvard University Press.

Johnson, Alfred E.

1976 "A Model of Kansas City Hopewell Subsistence–Settlement Systems." In *Hopewell Archaeology in the Lower Missouri Valley,* edited by A. E. Johnson. Publications in Anthropology 8:75. University of Kansas Press, Lawrence, Kansas.

Johnson, Frederick

1972 *Chronology and Irrigation.* Vol. 4 of *Prehistory of the Tehuacán Valley.* Austin, Texas: University of Texas Press.

Johnson, LeRoy

1964 *The Devil Mouth Site.* Archaeological Series, Department of Anthropology, no. 6, 46–57. Austin, Texas: University of Texas Department of Anthropology.

Jones, Volney H.

1935 "Ethnobotany." In *Report on the Excavation of Jemez Cave, New Mexico,* edited by H. Alexander and Paul Reiter, 38–49. Santa Fe, New Mexico: School of American Research.

1936 "The Vegetable Remains of Newt Kash Hollow Shelter." In *Rockshelters in Menifee County, Kentucky,* edited by W. S. Webb and W. D. Funkhouse, 147–55. Report on Archaeology and Anthropology, University of Kentucky, Lexington, Kentucky.

1968 "Corn from the McKees Rock Village Site." Harrisburg, Pennsylvania: *Pennsylvania Archaeologist* 38(1):81–86.

Judge, James W.

1979 "The Development of a Complex Cultural Ecosystem in the Chaco Basin, New Mexico." In *Proceedings of the First Conference on Scientific Research in the National Parks,* edited by Robert Limn. Washington, D.C.: U.S. Government Printing Office.

Kajalie, M. D.

1988 "The Exploitation of Wild Plants in the Mesolithic Period During 10,500-8000 B.C. in Sri Lanka: Paleobotanical Study of Cave Sireen Beli." In *Foraging and Farming,* edited by D. Harris and C. Hillman, 269–81. London: Allen and Unwin.

Kaplan, Lawrence

1963 "Archaeoethnobotany of Cordova Cave, New Mexico." *Economic Botany* 17(4):350–59.

1970 "Plant Remains from the Blain Site." In *Blain Village and the Fort Ancient Tradition in Ohio,* edited by O. Prufer and O. J. Shane, 227–31. Kent, Ohio: Kent State University Press.

1980 "Variation in Cultivated Beans." In *Guitarrero Cave,* edited by T. F. Lynch, 145–48. New York: Academic Press.

Kaplan, Lawrence, and R. S. MacNeish

1960 "Prehistoric Bean Remains from Caves in the Ocampo Region of Tamaulipas, Mexico." In *Botanical Museum Leaflets* 19(2):33–56. Cambridge, Massachusetts: Harvard University Press.

Kaplan, Lawrence, and Shirley L. Maina

1977 "Archaological Botany of the Apple Creek Site, Illinois." Washington, D.C.: *Journal of Seed Technology,* 2(2):40–53.

Kaulicke, Peter

n.d. "Beitrage zur Archaeologie des Altiplano von Junin, Peru." PhD dissertation; University of Bonn, Bonn, Germany.

Kautz, Robert R.

1980 "Pollen Analysis and Paleoethnobotany." In *Guitarrero Cave,* edited by T. F. Lynch, 43–64. New York: Academic Press.

Keegan, William F.

1987 *Emergent Horticultural Economies of the Eastern Woodlands.* Center for Archaeological Investigations, 7. Southern Illinois University, Carbondale, Illinois.

Keightley, David N.

1983 *The Origins of Chinese Civilization.* Berkeley: University of California Press.

Kenyon, Kathleen M.
1957 *Digging up Jericho.* New York: Praeger.

Kidder, A. V.
1924 *An Introduction to the Study of Southwestern Archaeology with a Preliminary Account of the Excavations at Pecos.* New Haven, Connecticut: Yale University Press.
1927 "Southwestern Archaeological Conference." Washington, D.C.: *Science* 68: 489–91.

Kidder, Alfred V., II
1944 *Archaeology of Northwestern Venezuela.* In Papers of the Peabody Museum of American Archaeology and Ethnology 26(1). Cambridge, Massachusetts.

King, Frances B.
1985 "Early Cultivated Cucurbits in Eastern North America." In *Prehistoric Food Production in North America,* edited by R. I. Ford, 73–98. Anthropological Paper 75, Museum of Anthropology, University of Michigan, Ann Arbor, Michigan.
1987 "The Evolutionary Effects of Plant Cultivation." In *Emergent Horticultural Economies of the Eastern Woodlands,* edited by W. F. Keegan, 51–66. Center for Archaeological Investigations, 7. Southern Illinois University, Carbondale, Illinois.

King, Frances B., and R. Bruce McMillan
1975 "Plant Remains from a Woodland Storage Pit, Boney Spring, Missouri." Lincoln, Nebraska: *Plains Anthropologist* 20(61): 111–16.

Kirch, Patrick V.
1979 "Subsistence and Ecology." In *The Prehistory of Polynesia,* edited by J. D. Jennings, 286–307. Cambridge, Massachusetts: Harvard University Press.

Kirkbride, Diana
1966 "Beidha, 1965 Campaign." New York: *Archaeology* (19): 268–72.
1968 "Beidha: Early Neolithic Village Life South of the Dead Sea." Cambridge, England: *Antiquity* 42: 263–74.

Kislev, Mordecai
1925 "Early Horsebean from Yiftah'el, Israel." Washington, D.C.: *Science* 228(4697): 319–20.

Kökten, Kilic I.
1955 "Ein allgemeiner Uberblick uber die prahistorischen Forshchungen en Karaim-Hole be Antalya." Bonn, Germany: *Bulletin* 19: 284.

Kotani, Y.
1971 "Evidence of Plant Cultivation in Jomon Japan." In *Affluent Foragers,* edited by S. Koyama and D. H. Thomas. Osaka, Japan: National Museum of Ethnology.

Kozlowski, Janusz K.
1977 *Chipped Flint Industries of Neo-Indian Cultures in the Greater Antilles.* Krakow, Poland: Polish Contributions in New World Archaeology.

Kozlowski, Janusz K., and Andrzej Krzonowski
1977 *New Finds of Leaf Points from the Central Andes of Peru.* Krakow, Poland: Polish Contributions in New World Archaeology.

Kroeber, A. L.
1916 *Zuni Potsherds.* Anthropological Paper 18(1); American Museum of Natural History, New York.

Kunstader, P.
1978 "Subsistence Agricultural Economics of Lua and Karen Hill Farmers, Mae

Sariang District, Northwestern Thailand." In *Farmers in the Forest,* edited by Kunstader, E. C. Chapman, and S. Sabhasri, 74–133. Honolulu, Hawaii: East-West Center.

Kunth, C.

1828 "Examen Botanique in Passalacqua." *Catalogue raisonné et historique de antiquitées decouvertes en Egypte.* Paris.

Ladd, John

1964 "Archaeological Investigations in Parita and Santa Maria Zones of Panama." Washington, D.C.: *Bulletin* 193, Smithsonian Institution Bureau of American Ethnology.

Lanning, Edward P.

1963 "A Pre-Agricultural Occupation on the Central Coast of Peru." Salt Lake City, Utah: *American Antiquity* 28(3):360–71.

1967a "Archaeological Investigations on the Santa Elena Peninsula, Ecuador." Report to the National Science Foundation, ms. Columbia University, New York.

1967b *Peru Before the Inca.* Englewood Cliffs, New Jersey: Prentice-Hall.

Lathrap, Donald W.

1958 "The Cultural Sequence at Yarincocha, Eastern Peru." Salt Lake City, Utah: *American Antiquity* 23(3):379–80.

1967 "Review of Early Formative Period of Coastal Peru: The Valdivia and Machalilla Phases." Menasha, Wisconsin: *American Anthropologist* 69(1): 96–98.

1970 *The Upper Amazon.* London: Thames and Hudson.

1977 "Our Father the Cayman, Our Mother the Gourd: Spinden Revisited or a Unitary Model for the Emergence of Agriculture." In *Origins of Agriculture,* edited by C. E. Reed, 712–91. The Hague: Mouton Publishers.

Lathrap, Donald W., Jorge Marcos, and James Zeidler

1977 "Real Alto: An Ancient Ceremonial Center." New York: *Archaeology* 30(1):3–13.

Lavallée, Danièle

1977 "Telarmachay." Lima, Peru: *Revista del Museo Nacional,* 55–133.

Lawton, Harry W., Philip J. W. Wilke, Mary Decker, and William M. Mason

1976 "Agriculture Among the Paiute of Owens Valley." Berkeley, California: *The Journal of California Anthropology* 3(1):13–50.

Lehmer, Donald J.

1948 "The Jornada Branch of the Mogollon." University of Arizona, Social Science Bulletin no. 17. Tucson, Arizona.

1971 "Introduction to Middle Missouri Archaeology." Anthropological Paper no. 1, National Park Service, Washington, D.C.

Leroi-Gourhan, Arlette

1969 "Pollen Grains of Gramineae and Cerealia from Shanidar and Zawi Chemi." In *The Domestication and Exploitation of Plants and Animals,* edited by P. J. Ucko and C. W. Dimbleby, 143–48. London: Duckworth.

Li, Hui Lin

1983 "The Domestication of Plants in China; Ecogeographical Considerations." In *Origins of Chinese Civilization,* edited by D. W. Keightley, 21–64. Berkeley, California: University of California Press.

Linares, O. F.

1979 "What is Lower Central American Archaeology?" Palo Alto, California: *American Review of Anthropology,* 8:21–43.

Linares, O. F., and Anthony J. Ranere
1980 *Adaptive Radiations in Prehistoric Panama*. Peabody Museum Monographs 5, Harvard University, Cambridge, Massachusetts.

Lindsay, Alexander J., Richard Ambler, Mary Anne Stein, and Phillip Hobler
1968 "Survey and Excavations North and East of Navajo Mountain, Utah, 1959–19—." Museum of N. Arizona Bulletin, no. 43, Glen Canyon Series, no. 8. Flagstaff, Arizona.

Linton, Ralph
1924 "North American Maize Culture." Menasha, Wisconsin: *American Anthropologist* 26:345–99.
1955 *The Tree of Culture*. New York: Alfred A. Knopf.

Lipe, William D.
1978 "The Southwest." In *Ancient North Americans*, edited by J. D. Jennings, 327–402. San Francisco: Freeman and Co.

Lister, R.
1958 *Archaeological Excavations in the Northern Sierra Madre Occidental, Chihuahua and Sonora, Mexico*. Anthropological Series no. 7, University of Colorado State, Boulder, Colorado.

Lone, Faroom A., Magsooda Khan, and G. M. Buth
1986 "Beginnings of Agriculture in India: An Appraisal of the Paleobotanic Evidence." Paper presented at World Archaeological Congress, Southampton, England.
1988 "Beginning of Agriculture in India." In *Foraging and Farming: The Evolution of Plant Exploitation*, edited by D. Harris and C. Hillman. London: Allen and Unwin.

Longacre, William A., and J. J. Reid
1974 "The University of Arizona Field School at Grasshopper, Eleven Years of Multidisciplinary Research and Teaching." Tucson, Arizona: *Kiva* 40:3–38.

Lorenzo, José Luis
1955 "Los Concheres de la costa de Chiapas." *Anales*, no. 7. Mexico, D.F.: Instituto Nacional de Antropología e Historia.
1958 "Un sitio Precerámico en Yanhuitlán, Oaxaca." *Dirección de Prehistoria*, no. 6. Mexico, D.F.: Instituto Nacional de Antropología e Historia.
1961 "Un Buril de la cultura precerámica de Teopisca, Chiapas." *Homenaje a Pablo Martinez del Rio*. Mexico, D.F.: Instituto Nacional de Antropología e Historia.
1967 "La estapa litica en México." Dpto. de Prehistoria, no. 11. Mexico, D.F.: Instituto Nacional de Antropología e Historia.
1975 "Los primeros Pobladores del Nomadismo a los Centros ceremoniales." Dept. of Invest. Hist. no. VI. Mexico, D.F.: Instituto Nacional de Antropología e Historia.
1981 "El Cedral. México, SLP México: un sitio presencio humana de mas de 30,000 B.P." Paper presented at 10th Congress, Internacional de Ciencias Prehistóricas y Protohistoricas.

Lothrop, Samuel K.
1966 "Archaeology of Lower Central America." In vol. 4 of *Handbook of Middle American Indians*, edited by R. Wauchope, 180–208. Austin, Texas: University of Texas Press.

Lumbreras, Luis G.
1974 *The Peoples and Cultures of Ancient Peru*. Washington, D.C.: Smithsonian Institution Press.

Lynch, Tom, editor
1980 *Guitarrero Cave: Early Man in the Andes.* New York: Academic Press.
Lynoff, M. J., T. W. Boutton, J. E. Price, and D. E. Nelson
1986 "Stable Carbon Isotopic Evidence for Maize Agriculture in Southeast Missouri and Northeast Arkansas." Salt Lake City, Utah: *American Antiquity,* 51(3):15–65.
McBurney, C. B. M.
1967 *The Haua Fteah (Cyrenaica) and the Stone Age of the Southwest Mediterranean.* Cambridge, England: Cambridge University Press.
McClung de Tapia, Emily
1989 "Mesoamerica and Central America." In *The Origins of Plant Domestication in World Perspective,* edited by P. J. Watson and C. W. Cowan. Washington, D.C.: Smithsonian Institution Press.
McClurkan, Burney B.
1980 "The Archaeology of la Cueva de la Zona de Derrumbes." In *Papers on the Prehistory of Northeastern Mexico and Adjacent Texas,* edited by Jeremiah Epstein, Thomas Hester and Carol Graves. Center for Archaeological Research, Report 9: 59–69. San Antonio: University of Texas at San Antonio.
McGimsey, Charles R.
1956 "Cerro Mangote: A Preceramic Site in Panama." Salt Lake City, Utah: *American Antiquity* 22(2):151–61.
1958 "Further Data and a Date from Cerro Mangote, Panama." Salt Lake City, Utah: *American Antiquity* 23(4):434.
McGregor, John C.
1941 *Southwestern Archaeology.* Urbana, Illinois: University of Illinois Press.
McKern, C.
1939 "The Midwestern Tazonomic Method as an Aid to Archaeological Study." Menasha, Wisconsin: *American Antiquity* 4(4):310–13.
MacNeish, Richard S.
1947 "A Preliminary Report on Coastal Tamaulipas." Menasha, Wisconsin: Salt Lake City, Utah: *American Antiquity,* 13:1–13.
1948 "The Pre-Pottery Faulkner Site of Southern Illinois." Menasha, Wisconsin: *American Antiquity* 13(3):232–43.
1950 "A Synopsis of the Archaeological Sequence in the Sierra de Tamaulipas." Mexico City: *Revista México de Instituto Antropología* 11:79–96.
1954 *An Early Archaeological Site Near Panuco, Veracruz.* Part 5, vol. 44 of Transactions of the American Philosophical Society, Philadelphia, Pennsylvania.
1957 "The Independent Investigator." In *Identification of Non-Artifactual Archaeological Materials,* edited by W. W. Taylor, 28–29. National Research Council Publication 565, Washington, D.C.
1958 "Preliminary Archaeological Investigations in the Sierra de Tamaulipas, Mexico." Part 6, vol. 48 of Transactions of the American Philosophical Society, Philadelphia, Pennsylvania.
1964 "The Food-Gathering and Incipient Agriculture Stage of Prehistoric Middle America." In *Natural Environment and Early Cultures,* vol. 1 of *Handbook of Middle American Indians,* edited by Robert C. West. Austin, Texas: University of Texas Press.
1965 "The Origin and Dispersal of New World Agriculture." Cambridge, England: *Antiquity* 39:87–94.
1967 "A Summary of the Subsistence." In vol. 1 of *Prehistory of the Tehuacan*

Valley: Environment and Subsistence, edited by D. Byers, 290–309. Austin, Texas: University of Texas Press.

MacNeish, Richard S., editor

1970 "Ceramics." In vol. 3 of *Prehistory of the Tehuacan Valley,* 1–306. Austin, Texas: University of Texas Press.

1973 "The Evolution of Community Patterns in the Tehuacán Valley of Mexico and Speculations about the Cultural Process." In *Ecology and Agricultural Settlements: An Ethnographic and Archaeological Perspective,* edited by R. Trigham, 43–55. Andover, Massachusetts: Warner Modular Publications.

1977 "The Beginning of Agriculture in Central Peru." In *Origins of Agriculture,* edited by C. Reed, 753–810. The Hague: Mouton Publishers.

1978 *The Science of Archaeology?* Belmont, California: Duxbury Press.

1981a "The Transition to Statehood as Seen from the Mouth of a Cave." In *Transition to Statehood in the New World,* edited by Grant D. Jones and Robert R. Kautz, 123–57. Cambridge, England: Cambridge University Press.

1981b "Excavations and Chronology." In vol. 2 of *The Prehistory of the Ayacucho Basin, Peru.* Ann Arbor, Michigan: University of Michigan Press.

1981c "Reviews of 'Zohapilco, Cinco milenios de ocupación humana en un sitio lacustre de la cuenca de México' by Christina Niederberger." Salt Lake City, Utah: *American Antiquity* 46(4):951–52.

1981d *Second Annual Report of the Belize Archaic Archaeological Reconnaissance.* Andover, Massachusetts: Robert S. Peabody Foundation for Archaeology.

1981e "Tehuacan's Accomplishments." Supplement to vol. 1 of the *Handbook of Middle American Indians,* edited by Victoria R. Bricker and J. Sabloff, 31–47. Austin, Texas: University of Texas Press.

1982a "The Ayacucho Preceramic as a Sequence of Cultural Energy Flow Systems." Vol. 4 in *Prehistory of the Ayacucho Basin, Peru,* edited by R. S. MacNeish, 236–80. Ann Arbor, Michigan: University of Michigan Press.

1982b *Third Annual Report of the Belize Archaic Archaeological Reconnaissance.* Andover, Massachusetts: Robert S. Peabody Foundation for Archaeology.

1984 "The Preceramic of Middle America." In *Advances in World Archaeology,* 93–130. New York: Academic Press.

1985 "The Archaeological Record on the Problem of the Domestication of Corn." *Maydica* 30:171–78.

1988 "The Beginning of Agriculture in the American Southwest as Seen from the Chihuahua Tradition." Paper presented at the Society for American Archaeology Meeting, Phoenix, Arizona.

1989 Defining the Archaic Chihuahua Tradition. Annual Report of the Andover Foundation for Archaeological Research. Andover, Massachusetts.

In press "The Peopling of the Antilles: The Mesoamerican Maritime Migration Hypothesis" in William Sears Festschrift.

MacNeish, Richard S., and Patrick H. Beckett

1987 *The Archaic Chihuahua Tradition of South-Central New Mexico and Chihuahua, Mexico.* COAS Monograph 7, Las Cruces, New Mexico.

MacNeish, Richard S., M. L. Fowler, A. Garcia-Cook, F. H. Peterson, A. Nelken-Turner, and J. H. Neely

1975 *Excavation and Reconnaissance, Prehistory of the Tehuacan Valley,* vol. 5. Austin, Texas: University of Texas Press.

MacNeish, Richard S., and C. Irwin

1962 "Preceramic Projectile Point Typology in Mexico." Papers, Congress of Americanists. Mexico City, Mexico.

MacNeish, Richard S., and Antoinette Nelken-Turner
1983a "The Preceramic of Mesoamerica." Boston, Massachusetts: *Journal of Field Archaeology,* 10(1):71–84.
1983b *The Final Annual Report of the Belize Archaic Archaeological Reconnaissance.* Boston, Massachusetts: Center for Archaeological Studies, Boston University.
MacNeish, Richard S., Antoinette Nelken-Turner, and Irmgard Johnson
1967 *Nonceramic Artifacts,* vol. 2 of Prehistory of the Tehuacan Valley, 227–45. Austin, Texas: University of Texas Press.
MacNeish, Richard S., T. Patterson and D. Browman
1975 *The Central Peruvian Prehistoric Interaction Sphere.* Andover, Massachusetts: Robert S. Peabody Foundation.
MacNeish, Richard S., and Frederick A. Peterson
1962 "The Santa Marta Rock Shelter, Ocozocoautla, Chiapas." Papers of the New World Archaeological Foundation 14, 1–46. Provo, Utah: Brigham Young University Press.
MacNeish, Richard S., R. K. Vierra, A. Nelken-Turner, R. Lurie and A. Garcia-Cook
1983 "The Pre-ceramic Way of Life." In vol. 4 of *The Prehistory of the Ayacucho Basin.* Ann Arbor, Michigan: University of Michigan Press.
MacNeish, Richard S., and S. Jeffrey K. Wilkerson, and Antoinette Nelken-Turner
1980 *First Annual Report of the Belize Archaic Reconnaissance.* Andover, Massachusetts: R. S. Peabody Foundation for Archaeology.
Mangelsdorf, Paul C.
1952 *Review of Agricultural Origins and Dispersals.* New York: American Geographical Society.
Mangelsdorf, Paul C., Herbert W. Dick, Julián Cámara-Hernández
1967 "Bat Cave Revisited." Cambridge, Massachusetts: Harvard University, *Botanical Museum Leaflets* 22(1):213–60.
Mangelsdorf, Paul C., and R. H. Lister
1956 "Archaeological Evidence on the Diffusion and Evolution of Maize in Northwestern Mexico." Cambridge, Massachusetts: Harvard University, *Botanical Museum Leaflets* 17, 151–78.
Mangelsdorf, Paul C., Richard S. MacNeish and Walton C. Galinat
1967a "Prehistoric Maize, Teosinte and Tripsacum from Ocampo, Tamaulipas, Mexico." Cambridge, Massachusetts: Harvard University, *Botanical Museum Leaflets* 22(2):33–63.
1967b "Prehistoric Wild and Cultivated Maize." In vol. 1, Prehistory of the Tehuacan Valley, *Environment and Subsistence,* edited by D. S. Byers, 178–200. Austin, Texas: University of Texas Press.
Mangelsdorf, Paul C., R. S. MacNeish, and G. R. Willey
1964 "Origins of Agriculture in Middle America." In *Natural Environments and Early Cultures,* vol. 1 of *Handbook of Middle American Indians,* edited by R. C. West. Austin: University of Texas Press.
Mangelsdorf, Paul C., and C. Earle Smith, Jr.
1949 "New Archaeological Evidence on Evolution in Maize." Cambridge, Massachusetts: Harvard University, *Botanical Museum Leaflets* 13:213–47.
Marcos, Jorge G.
1978 "The Ceremonial Precinct at Real Alto." Ph.D. dissertation, Department of Anthropology, University of Illinois, Urbana, Illinois.
Marino, Bruno, and Richard S. MacNeish
1991 "Carbon 13/12 and Nitrogen 15/14 Isotopic Analysis in Skeletons from the

Jornada Region." In *Preliminary Investigations of the Archaic in the Region of Las Cruces, New Mexico,* edited by R. S. MacNeish. Ft. Bliss, Texas.

Marks, Anthony

1971 "Settlement Patterns and Intrasite Variability in the Central Negev, Israel." Menasha, Wisconsin: *American Anthropologist* 73:1237–44.

Marquardt, William H., and Patty Joe Watson

1977 "Excavation and Recovery of Biological Remains from Two Archaic Shell Middens in Western Kentucky." Tuscaloosa, Alabama: Southeastern Archaeological Conference *Bulletin* 20:112–29.

Martin, Paul

1958 *A Biogeography of Reptiles and Amphibians in the Gomez Farias Region, Tamaulipas, Mexico.* Ann Arbor: Museum of Zoology, University of Michigan.

Martin, Paul S., Elaine Bluhm, Hugh C. Cutler and Roger Grange

1952 "Mogollon Cultural Continuity and Change: The Stratigraphic Analysis of Tularosa and Cordova Caves." Chicago: *Fieldiana: Anthropology* 40:1–528.

Martin, Paul S., J. B. Rinaldo and E. Antevs

1949 "Cochise and Mogollon Sites in the Pine Lawn Valley, Western New Mexico." Chicago: *Fieldiana: Anthropology* 38(1).

Mathyushin, G. I.

1976 *Mezolit Yuzhnogo Urala.* Moscow: Nauka.

Matson, R. C.

1988 "The Adoption of Agriculture in the Southwest." Paper presented at the Society for American Archaeology Meeting, Phoenix, Arizona.

Matsutani, Akiko

1972 "Spodographic Analysis of Ash from the Kotosh Site." Appendix to *Excavations at Kotosh, Peru, 1963 and 1966,* edited by Seiichi Izumi and Kuzuo Terada, 319–23. Tokyo: University of Tokyo Press.

Meacham, William

1983 "Origins and Development of the Yueh Coastal Megalithics." In *Origins of Chinese Civilization,* edited by D. N. Keightley, 147–72. Berkeley, California: University of California Press.

Meggers, Betty J., and C. Evans

1957 "Archaeological Investigations at the Mouth of the Amazon." Washington, D.C.: Bureau of American Ethnology *Bulletin* 167.

1973 "An Interpretation of the Culture of Marajo Island." In *Peoples and Cultures of Native South America,* edited by D. R. Greus, 39–47. Garden City, New York: Doubleday Natural History Press.

1978 "Lowland South America and the Antilles." In *Ancient South Americans,* edited by J. D. Jennings, 543–92. San Francisco, California: Freeman and Co.

Meggers, Betty J., C. Evans and E. Estrada

1965 "Early Formative Period of Coastal Ecuador." Washington, D.C.: *Smithsonian Contributions to Anthropology* 1.

Mellaart, James

1967 *Catal Huyuk: A Neolithic Town in Anatolia.* New York: McGraw-Hill.

1970 *Excavations at Hacilar.* 2 vols. Edinburgh, Scotland: University of Edinburgh Press.

1975 *The Neolithic of the Near East.* New York: Scribner's.

Mellars, P. A.

1978 *The Early Postglacial Settlement of Northern Europe.* London: Duckworth.

Menghin, Oswald
1931 *Weltgeschichte der Steinzeit.* Vienna, Austria: Anton Schroll and Co.
Mera, Henry P.
1938 "Reconnaissance and Excavation in Southeastern New Mexico." Menasha, Wisconsin: *Memoirs of the American Anthropological Association* 15.
Milisauskas, Sarunas
1978 *European Prehistory.* New York: Academic Press.
Miller, Naomi F.
1988 "Origins of Plant Cultivation in the Near East." In *Foraging and Farming,* edited by D. Harris and C. Hillman, 331–343. London: Allen and Unwin.
1989 "Origins of Plant Cultivations in the Near East." In *The Origins of Plant Domestication in World Perspective,* edited by P. J. Watson and C. W. Cowan, 102–46. Washington, D.C.: Smithsonian Institution Press.
Milojeic, V., J. Boessnick and M. Hopf
1962 "Die deutschen Ausgrabungen auf der Argissa Magula." In vol. 1 of *Thessalien.* Bonn, Germany: Habelt.
Minnis, Paul E.
1985 "Domesticating People and Plants in the Greater Southwest." In *Prehistoric Food Production in North America,* edited by R. I. Ford, 309–40. Anthropology Paper 75, Museum of Anthropology, University of Michigan, Ann Arbor, Michigan.
1989 "Earliest Plant Cultivation in Desert North America." In *The Origins of Plant Domestication in World Perspective,* edited by P. J. Watson and C. W. Cowan, 61–83. Washington, D.C.: Smithsonian Institution Press.
Mirambell, Lorena
1973 "El hombre en Tlapacoya desde hace unos 20 mil años." *Boletin INAH* 4, 11–14. Mexico, D.F.: Instituto Nacional de Antropología e Historia.
Mochanov, Y. A.
1978 "Stratigraphy and Absolute Chronology of the Paleolithic in Northeast Asia." In *Yakutia and Her Neighbors in Ancient Times.* Yakut, USSR: Academy of Science.
Moore, Andrew M. T.
1988 "The Transition from Hunting and Gathering to Farming in Southwest Asia: Present Problems and Future Directions." In *Foraging and Farming,* edited by D. Harris and C. Hillman, 620–31. London: Allen and Unwin.
Moorehead, Warren K.
1892 *Primitive Man in Ohio.* New York: Putnam.
Morris, Earl H., and R. F. Burgh
1954 *Basket Maker II Sites Near Durango, Colorado.* Carnegie Institution of Washington, Publication 604. Washington, D.C.
Mortensen, Peder, Jorgen Meldgaard and H. Thane
1964 "Excavations at Tepe Guran, Luristan." Copenhagen: *Acta Archaeologica* 34:97–133.
Moseley, Michael
1978 "The Evolution of Andean Civilization." In *Ancient Native Americans,* edited by J. Jennings. San Francisco: Freeman.
Mountjoy, Joseph B., R. E. Taylor, and Lawrence H. Feldman
1972 "Matanchen Complex: New Radiocarbon Dates on Early Coastal Adaptations in West Mexico." Washington, D.C.: *Science* 175:1242–43.

Muller, Jon
1983 "The Southeast." In *Ancient North Americans,* 2d ed., edited by J. Jennings. San Francisco: Freeman.
1987 "Lower Ohio Valley Emergent Horticulture and Mississippian." In *Emergent Horticultural Economies of the Eastern Woodlands,* edited by W. F. Keegan, 281–326. Southern Illinois Anthropological Papers 7, Southern Illinois University, Carbondale, Illinois.

Munson, P. J.
1968 "Recent Archaeological Researches in the Dhar Tichitt Region of South-Central Mauritania, West Africa." Ibadan, Nigeria: West African *Archaeological Newsletter* 10:6–13.

Murdoch, G. P.
1959 *Africa: Its People and Their Culture History.* New York: McGraw-Hill.

Murphy, James L.
1971 "Maize from an Adena Mound in Athens County, Ohio." Washington, D.C.: *Science* 171(3974):897–98.

Murty, M. L. K.
1979 "Recent Research in the Upper Paleolithic Phase in India." Boston, Massachusetts: *Journal of Field Archaeology* 6(3):301–320.

Nance, C. Roger
1980 "La Calzada and the Prehistoric Sequence in Northeast Mexico." In *Papers on the Prehistory of Northeastern Mexico and Adjacent Texas,* edited by Jeremiah Epstein, Thomas Hester, and Carol Graves. Center for Archaeological Research, Report 9, San Antonio: University of Texas at San Antonio.

Nelson, Nels
1916 "Chronology of the Tano Ruins, New Mexico." Menasha, Wisconsin: *American Anthropologist* 18(2):159–80.
1937 "Notes on Cultural Relations between Asia and America." Salt Lake City, Utah: *American Antiquity* 2(4):367–72.

Niederberger, Christine
1976 "Zohapilco: Cinco Milenios de Ocupación Humana en un sitio Lacustre de la Cuenca de México." Dept. de Prehistoria, *Colección Científica, Arqueología* 30, 23–49. Mexico City.

Norton, Presley
1977 "The Loma Alto Connection." New Orleans, Louisiana: Paper Presented at 42d Annual Meeting of Society for American Archaeology.

O'Laughlin, Thomas C.
1968 "La Cueva," in *The Artifact* 3(2), El Paso Archaeological Society, El Paso, Texas.
1980 *The Keystone Dam Site.* El Paso, Texas: Publications in Anthropology 8, El Paso Centennial Museum, University of Texas at El Paso.

Osgood, Cornelius
1942 "The Ciboney Culture of Ayo Redondo, Cuba." New Haven, Connecticut: *Yale University Publication in Anthropology* 25.
1943 "Excavations at Tocoron, Venezuela." New Haven, Connecticut: *Yale University Publication in Anthropology.*

Parker, Arthur C.
1922 "Archaeological History of New York." Albany: New York State Museum, *Bulletin* 235 (8).

Patterson, Thomas
1973 "Central Peru: Its Population and Economy." Boston: *Archaeology* 21(1): 316–21.
Pearsall, Deborah M.
1978 "Phytolith Analysis of Archaeological Soils—Evidence of Maize Cultivation in Formative Ecuador." Washington, D.C.: *Science* 199:177–78.
1980 "Pachamachay Ethnobotanical Report: Plant Utilization of a Hunting Base Camp." In *Prehistoric Hunters of the High Andes,* edited by J. Rick, 191–232. New York: Academic Press.
1988 "Adaptation of Early Hunter-Gatherers to the Andean Environment." In *Foraging and Farming,* edited by D. Harris and C. Hillman, 318–34. London: Allen and Unwin.
1989 "The Origins of Plant Cultivation in South America." In *The Origins of Plant Domestication in World Perspective,* edited by P. J. Watson and C. W. Cowan, 148–72. Washington, D.C.: Smithsonian Institution Press.
Pearson, Richard
1977 "Paleoenvironment and Human Settlement in Japan and Korea." Washington, D.C.: *Science* 197(4310):1239–46.
1979 "The Neolithic Cultures of the Lower Yangtze River and Coastal China." In *Symposium on the Origins of Agriculture and Technology, West or East Asia,* 1–36. Denmark.
1981 "Social Complexity in Chinese Coastal Neolithic Sites." Washington, D.C.: *Science* 213(4512):1078–86.
Pearson, Richard, and Shyh-Charnq Lo
1983 "The Ch'ing lien-Kang Culture and the Chinese Neolithic." In *Origins of Chinese Civilization,* edited by D. N. Keightley, 121–46. Berkeley: University of California Press.
Pei, Gai
1977 "Preliminary Report on Upper Paleolithic Hutoalcang Sites, Hopei, China." Peking, China: *Vertebrate Paleo Asiatica* 15(4):287–302.
Pendergast, David
1971 "Evidence of Early Teotihuacan-Lowland Maya Contacts at Atun Ha." *American Antiquity,* 36(4):455–60.
1975 "The Church in the Jungle." *Rotunda* 8:6–10.
1981a "Lamanai, Belize: Summary of Excavations, 1974–1980." Boston: *Journal of Field Archaeology* 8:29–53.
1981b "Lamanai." New York: *Archaeology* 30:12–23.
Perkins, Dexter, and Patricia Daley
1968 "A Hunter's Village in Neolithic Turkey." New York: *Scientific American* 219(5):97–106.
Perquart, S. J., M. Perquart, M. Boule and H. Vallois
1937 "Teviec: statim nearopole mesolithique du Morbihan." Paris: Archives de l'Institut de Paleontologie Humaine, 18.
Perrot, Jean
1960 "Excavations at 'Eynan: Preliminary Report of the 1959 Season." *Israel Exploration Journal* 10:14–22.
1966 "La gisement natoufien de Mallaha (Eynan) Israel." Paris: *Anthropologie* 7(5-9):437–84.
1967 "Munhata, un village Prehistorique." Paris: *Bible et terre sainte* 93:4–16.
1968 "La Prehistorie palestinienne." In *Supplement au dictionnaire de la Bible.* Paris: Letouzoy an Ane 8:286–446.

Perry, William J., and A. L. Christensen

1987 *Prehistoric Stone Technology on Northern Black Mesa, Arizona.* Center for Archaeological Investigation 2. Southern Illinois University, Carbondale, Illinois.

Petrie, W. M. F., and J. E. Quibell

1895 *Naquda and Ballas.*

Phillips, James L.

1970 "Old World Domestications." *Ecology* 51(4):752–54.

1986 "Hunters and Gatherers in the Wadi Feiran." *Man* 105:170–76.

Phillipson, D. W.

1977 *The Later Prehistory of Eastern and Southern Africa.* New York: Holmes and Meier.

1982 "Early Food Production in Sub-Saharan Africa." In vol. 1 of *The Cambridge History of Africa,* edited by J. D. Clark, 770–829. Cambridge, England: Cambridge University Press.

Pickerskill, Barbara

1988 "Cytological and Genetic Evidence on Domestication and Diffusion of Crops within the Americas." In *Foraging and Farming,* edited by D. Harris and C. Hillman, 426–39. London: Allen and Unwin.

Piperno, Dolores R.

1984 "A Comparison and Differentiation of Phytoliths from Maize and Wild Grasses: Use of Morphological Criteria." Lawrence, Kansas: *American Antiquity* 49(2):361–83.

1988 "Non-Affluent Foragers: Resource Availability, Seasonal Shortages, and the Emergence of Agriculture in Panamanian Tropical Forests." In *Foraging and Farming,* edited by D. Harris and C. Hillman, 538–54. London: Allen and Unwin.

Plog, Stephen

1980 "Village Autonomy in the American Southwest: An Evaluation of the Evidence, Models and Methods in Regional Exchange," edited by R. E. Frey, 135–46. Lawrence, Kansas: *American Antiquity,* Society for American Archaeology Papers.

Plog, Stephen, Fred Plog and Walter Wait

1978 "Decision Making in Modern Surveys." In *Advances in Archaeological Method and Theory,* edited by M. Schiffer, 383–94. New York: Academic Press.

Portères, Roland

1950 "Vieilles agricultures de l'Afrique intertropicale." Paris: *Agronomie Tropicale* 5:489–507.

1970 "Primary Cradles of Agriculture in the African Continent." In *Papers in African Prehistory,* edited by J. D. Fage and R. A. Oliver, 43–58. Cambridge, England: Cambridge University Press.

1976 "African Cereals: Eleusine, Fonio, Black Fonio, Teff, Brachiaria, Paspalun, Pennisetum and African Rice." In *Origins of African Plant Domestication,* edited by J. Harlan, J. de Wet, and A. Stemler, 410–51. The Hague: Mouton Publishers.

Portères, Roland, and J. Barrau

1981 "Origins, Development and Expansion of Agricultural Techniques." In vol. 1 of *General History of Africa,* edited by J. Ki-Zerbo, 589–703. Berkeley, California: Heinemann.

Pozorski, Sheila, and Thomas Pozorski
1979a "Alto Salavarry: A Peruvian Coastal Pre-ceramic Site." Pittsburgh, Pennsylvania: *Annals of Carnegie Museum,* 48(19):337–75.
1979b "An Early Subsistence Exchange System in the Moche Valley, Peru." Boston, Massachusetts: *Journal of Field Archaeology* 6(4):413–27.
Price, T. Douglas
1983 "The European Mesolithic." Salt Lake City, Utah: *American Antiquity* 48(4):761–78.
Pullar, Judith
1977 "Early Cultivation in the Zagros." London: London University, *Iran* 15: 33–55.
Purseglove, J. W.
1977 "The Origins and Migrations of Crops in Tropical Africa." In *Origins of African Plant Domestication,* edited by J. Harlan, J. de Wet, and A. Stemlar, 291–309. The Hague: Mouton Publishers.
Ranere, A. J.
1980 "Preceramic Shelters in the Talamancan Range." In *Adaptive Radiations in Prehistoric Panama,* edited by D. Linares and A. Ranere. Peabody Museum Monographs 5, Harvard University, Cambridge, Massachusetts.
Ranere, A. J., and P. Hansell
1978 "Early Subsistence Patterns along the Pacific Coast of Central Panama." In *Prehistoric Coastal Adaptations,* edited by Barbara Stark and Barbara Voorhies. New York: Academic Press.
Ratzel, Franz
1882 *Anthropogeographie,* vol. 1. Stuttgart, Germany: J. Engelbonns Nachfolger.
Ravines, Rogger
1965 "Ambo: A New Pre-ceramic Site in Peru." Salt Lake City, Utah: *American Antiquity* 31(1):104.
Redman, Charles L.
1977 "Man, Domestication and Culture in Southwestern Asia." In *Origins of Agriculture,* edited by C. Reed, 523–42. The Hague: Mouton Publishers.
1978 *The Rise of Civilization.* San Francisco, California: Freeman and Company.
Reed, Charles A.
1965 "A Human Frontal Bone from the Late Pleistocene of the Kom Ombo Plain, Upper Egypt." *Man* 95:101–104.
Reed, Charles A., editor
1977 *Origins of Agriculture.* The Hague: Mouton Publishers.
Reichel-Dolmatoff, G.
1961 "Puerto Hormiga: Un complejo prehistórico marginal en Colombia." Bogotá, Colombia: *Revista Colombiana de Antropología* 10:349–54.
1965 *Colombia.* London: Thames and Hudson.
Renfrew, Colin
1981 "The Megalithic Builders of Western Europe." In *Antiquity and Man,* edited by J. D. Evans, B. Cunliff and C. Renfrew, 198–220. London: Thames and Hudson.
Renfrew, Colin, J. E. Dixon, and J. R. Cann
1966 "Obsidian and Early Cultural Contact in the Near East." London: *Proceedings of the Prehistoric Society* 32:30–72.
Renfrew, Jane
1966 "A Report on Recent Finds of Carbonized Cereal Grounds and Seeds from Prehistoric Thessaly." Vales Volas, *Thessalika* 5:21–30.

1973 *Paleoethnobotany: The Prehistoric Food Plants of the Near East and Europe.* New York: Columbia University Press.

Rick, John W.

1980 *Prehistoric Hunters of the High Andes.* New York: Academic Press.

Rindos, David

1980 "Symbiosis, Instability, and the Origins and Spread of Agriculture: A New Model." Chicago: University of Chicago, *Current Anthropology* 21(6):1–14.

Ritchie, William A.

1969 *The Archaeology of New York State.* Garden City, New York: Natural History Press.

Rodden, Robert

1965 "An Early Neolithic Village in Greece." New York: *Scientific American* 212(4):83–91.

Roosevelt, Anna C.

1980 *Parmana: Prehistoric Maize and Manioc Subsistence along the Amazon and Orinoco.* New York: Academic Press.

1989 "Resource Management in Amazonia Before the Conquest." In *Advances in Economic Botany* 7. New York Botanical Gardens, New York.

Roth, H. Ling

1887 "On the Origin of Agriculture." London: Royal Anthropological Institute of Great Britain and Ireland, *Journal* 16:102–36.

Rouse, Irving

1964 "Prehistory of the West Indies." Washington, D.C.: *Science* 144:499–514.

Rouse, Irving, and Jose Cruxent

1963 *Venezuela Archaeology.* New Haven, Connecticut: Yale University Press.

Rue, David J.

1988 "Late Archaic Agriculture and Settlement in Southeastern Mesoamerica and Central America." WAPORA, CA (mimeographed paper).

Sankalia, H. D.

1972 "The Chalcolithic Cultures of India." In *Archaeological Congress and Seminar Papers,* edited by S. D. Dea, 530–44. Nagpur, India: Nagpur University Press.

1974 *Prehistory and Protohistory of India and Pakistan.* Poona, India: Deccan College.

Sanoja, Mario

1963 "Cultural Development in Venezuela." In *Aboriginal Cultural Developments in Latin America,* edited by B. J. Meggers and C. Evans. Smithsonian Miscellaneous Collections 146(1). Washington, D.C.: Smithsonian Institution.

1988 "From Foraging to Food Production in Northeastern Venezuela and the Caribbean." In *Foraging and Farming,* edited by D. Harris and C. Hillman, 523–37. London: Allen and Unwin.

Santamaria, Diana

1981 "Preceramic Occupations at Los Grifos Rock Shelter, Ocozocautla, Chiapas, Mexico." Paper presented at 10th Congress, Unión Internacional de Ciencas Prehistóricas y Protohistóricas, Mexico City, Mexico.

Santamaria, Diana, and Joaquin Garcia-Barcena

1984 Raederas y raspadores de Los Grifos. Mexico, D.F.: Instituto Nacional de Antropología e Historia, Depto. de Prehistoria, No 28.

Satherwaite, Melvin B., and Judy Ehlen

1980 *Vegetation and Terrain Relationships in South-Central New Mexico and Western Texas.* Ft. Belvoir, Virginia: U.S. Army Corps of Engineers.

Sauer, Carl O.
1936 "American Agricultural Origins: A Consideration of Nature and Cultures." In *Essays in Anthropology in Honor of A. L. Kroeber,* 279–97. Berkeley, California: University of California Press.
1952 *Agricultural Origins and Dispersals.* New York: American Geographical Society, Bowman Memorial Lecture 5, Series 2.
Sayles, E. B.
1983 *The Cochise Cultural Sequence in Southeastern Arizona.* Tucson, Arizona: University of Arizona Press.
Sayles, E. B., and Ernest Antevs
1941 "The Cochise Culture." Globe, Arizona: Gila Pueblo Medallion Paper 29.
Scarre, C., editor
1984 *Ancient France.* Edinburgh, Scotland: University of Edinburgh Press.
Schiemann, E.
1951 "New Results on the History of Cultivated Cereals." *Heredity,* no. 5, part 3.
Schiffer, Michael B.
1972 "Archaeological Context and Systemic Context." Salt Lake City, Utah: *American Antiquity* 37(2):156–65.
1978 "Methodological Issues in Ethnoarchaeology." In *Explorations in Ethnoarchaeology,* edited by R. A. Gould, 299–344. Albuquerque, New Mexico: University of New Mexico Press, School of American Research.
Schoenwetter, James
1974 "Pollen Records of Guila Naquitz Cave." Salt Lake City, Utah: *American Antiquity* 39(2):292–302.
Schoenwetter, James, and Landon D. Smith
1986 "Pollen Analysis of the Oaxaca Archaic in Guila Naquitz." In *Guilá Naquitz,* edited by K. Flannery, 179–238. New York: Academic Press.
Schwartz, Douglas, and Richard Lang
1973 *Archaeological Investigations of the Arroyo Hondo Site: Third Field Report.* Santa Fe, New Mexico: School of American Research.
Sears, William
1968 "The State and Settlement Patterns in the New World." In *Settlement Archaeology,* edited by K. C. Chang, 134–53. Palo Alto, California: National Press Books.
Semple, Edward C.
1911 *Influences of Geographic Environment.* New York: Holt.
Shane, Orrin C.
1967 "The Leimbach Site." In *Studies in Ohio Archaeology,* edited by O. Prufer and D. J. McKenzie, 98–120. Cleveland, Ohio: Western Reserve University Press.
Shaw, C. Thurston
1972 *Early Crops in Africa: A Review of Evidence.* Burg Wartenstein Symposium 56. New York: Wenner Gren Foundation.
1981 "The Prehistory of West Africa." In vol. 1 of *General History of Africa,* edited by J. Ki-Zerbo, 611–27. Berkeley, California: Heinemann.
Shinnie, P. L.
1971 *The African Iron Age.* Oxford: Clarendon Press.
Simmons, Alan H.
1986 "New Evidence for the Early Use of Cultigens in the American Southwest." Washington, D.C.: *American Antiquity* 51(1):73–88.

Smiley, F. E.
1985 "The Chronometrics of Early Agriculture Sites in Northeastern Arizona." Ph.D. dissertation, University of Michigan, Ann Arbor, Michigan.
Smith, Bruce D.
1984 "The Role of Chenopodium as a Prehistoric Domesticate in Eastern North America: New Evidence from Russell Cave, Alabama." Washington, D.C.: *Science*(226): 165–67.
1987 "The Independent Domestication of Indigenous Seed-Bearing Plants in Eastern North America." In *Emergent Horticultural Economies of the Eastern Woodlands,* edited by W. F. Keegan, 3–49. Center for Archaeological Investigations, no. 7. Southern Illinois University, Carbondale, Illinois.
1989 "Prehistoric Plant Husbandry in Eastern North America." In *The Origins of Plant Cultivation in World Perspective,* edited by P. J. Watson and C. W. Cowan. Washington, D.C.: Smithsonian Institution Press.
Smith, C. Earle
1967 "Plant Remains." In *Environment and Subsistence,* vol. 1 of *Prehistory of the Tehuacan Valley,* edited by D. S. Byers, 220–555. Austin, Texas: University of Texas Press.
1980a "Plant Remains from Chiriqui Sites and Ancient Vegetational Patterns." In *Adaptive Radiations in Prehistoric Panama,* edited by O. Linares and A. Ranere, 151–74. Peabody Museum Monographs 5, Harvard University, Cambridge, Massachusetts.
1980b "Plant Remains from Guitarrero Cave." In *Guitarrero Cave,* edited by T. Lynch, 121–44. New York: Academic Press.
Smith, C. E., Jr., E. O. Callen, H. C. Cutler, W. C. Galinat, L. Kaplan, T. W. Whitaker and R. A. Yarnell
1966 "Bibliography of American Archaeological Plant Remains." In *Economic Botany* 20(4):446–460.
Smith, Jason
1971 "Socio-Cultural Evolution in China: The Archaeological Evidence." Unpublished manuscript.
Smith, Philip E. L.
1975 "Ganj Dareh Tepe." British Institute of Persian Studies (13): 178–84. London: *Iran.*
Smith, Philip E. L., and T. Cuyler Young
1972 "The Evolution of Early Agriculture and Culture in Greater Mesopotamia: A Trial Model." In *Population Growth: Anthropological Implications,* edited by B. J. Spooner. Cambridge, Massachusetts: M.I.T. Press.
Solecki, Ralph S.
1961 "Prehistory in Shanidar Valley, Northern Iraq." Washington, D.C.: *Science* 139:179–93.
1971 *Shanidar: The First Flower People.* New York: Knopf.
Solecki, Rose L.
1964 "Zawi Chemi, Shanidar, a Post-Pleistocene Village Site in Northern Iraq." Warsaw, Poland: *Report of the Sixth International Congress on the Quaternary,* 1961. 4:405–12.
Solheim, Wilhelm G.
1969 "Reworking Southeast Asia Prehistory." *Paidenma* 15:125–39.
1972 "An Earlier Agricultural Revolution." New York: *Scientific American* 226(4):34–41.

Solheim, Wilhelm G., and Donald Bayard
1973 "Prehistory and Ecological Adaptation in the Phu Wiang Region, Northwestern Thailand." Proposal to the National Science Foundation.

Spier, Leslie
1917 "An Outline for a Chronology of Zuni Ruins." New York: Anthropological Papers, American Museum of Natural History vol. 18, part 4.

Srejovic, D.
1978 "The Odmut Cave: A New Facet of Mesolithic Culture of the Balkan Peninisula." *Archaeologica Jugoslavica* 15:3–12.

Srejovic, D., and Z. Letica-Vlasac
1978 *A Mesolithic Settlement in the Iron Gates*. Belgrade, Yugoslavia: Serbian Academy of Science and Arts, vol. 1

Stark, Barbara
1977 *Prehistoric Ecology at Patarata 52, Veracruz, Mexico: Adaptation to the Mangrove Swamp*. Nashville, Tennessee: Vanderbilt University Publication in Anthropology no. 18.

Stekelis, M., and Tamar Yizraely
1963 "Excavations at Nahel Oren, Preliminary Report." Tel Aviv: *Israel Exploration Journal* 13:1–12.

Steward, Julian H.
1937 "Ecological Aspects of Southwestern Society." Los Angeles: University of California at Los Angeles, *Anthropology* 32.
1946 "Introduction." In *Marginal Tribes,* vol. 1 of *Handbook of South American Indians,* 1–11. Washington, D.C.: Smithsonian Institution Press.
1947 "American Culture History in the Light of South America." Albuquerque, New Mexico: *Southwestern Journal of Anthropology* 3(2).
1955 *The Theory of Culture Change*. Urbana, Illinois: University of Illinois Press.

Stothert, Karen E.
1977 "Preceramic Adaptation and Trade in the Intermediate Area." Houston, Texas: Paper presented at American Anthropological Association.
1985 "The Pre-ceramic Las Vegas Culture of Coastal Ecuador." Salt Lake City, Utah: *American Antiquity* 50(5):613–37.

Stothers, David
1977 *The Princess Point Complex*. no. 58 of the Mercury Series. Ottawa, Canada: National Museum of Man.

Strong, Duncan
1935 *An Introduction to Nebraska Archaeology*. Smithsonian Miscellaneous Collections 93(10). Washington, D.C.: Smithsonian Institution Press.

Struever, Stuart, and Kent D. Vickery
1973 "The Beginnings of Cultivation in the Midwest Riverine Area of the United States." Menasha, Wisconsin: *American Anthropologist* 75(5):1197–1220.

Sutton, J. E.
1981 "The Prehistory of East Africa." In vol. 1 of *General History of Africa,* edited by J. Ki-Zerbo. Berkeley, California: Heinemann.

Taylor, Walter W.
1948 *A Study of Archaeology*. Memoirs of the American Anthropological Association no. 69. Menasha, Wisconsin.
1960 Review of "Preliminary Archaeological Investigation of the Sierra De Tamaulipas by MacNeish." Salt Lake City, Utah: *American Antiquity* 25(3): 591–93.
1966 *Archaic Cultures Adjacent to the Northeastern Frontiers of Mesoamerica,* vol.

4 of *Handbook of Middle American Indians.* Austin, Texas: University of Texas Press.

Terra, Helmut de, J. Romero and T. D. Stewart

1949 *Tepepexpan Man.* New York: Viking Fund Publications in Anthropology no. 11.

Thomas, David Hurst

1969 "Great Basin Hunting Patterns." Salt Lake City, Utah: *American Antiquity* 34(4):392–401

1983 "The Archaeology of Monitor Valley, 1 Epistemology." Anthropology Paper 58, part 1, American Museum of Natural History, New York.

1986 "Contemporary hunter-gatherer archaeology in America." In *American Archaeology Past and Future,* edited by J. Sabloff. Washington, D.C.: Smithsonian Institution Press.

Todd, Ian A.

1966 "Asikli Huyuk: A Proto-Neolithic Site in Central Anatolia." *Anatolian Studies* 16.

Towle, Margaret A.

1961 *The Ethnobotany of Pre-Columbian Peru.* Viking Fund Publication no. 30. Chicago: Aldine.

Treistman, Judith M.

1972 *The Prehistory of China.* New York: Doubleday and Company, American Museum Science Books.

Tringham, Ruth E.

1971 *Hunters, Fishers, and Farmers of Eastern Europe, 6000–3000* B.C. London: Hutchinson.

Turnbull, Priscilla F., and Charles A. Reed

1974 "The Fauna from the Terminal Pleistocene of Palegawra Cave." Chicago: *Fieldiana, Anthropology* 63:81–146.

Upham, Steadman, Richard S. MacNeish, Walton C. Galinat and Christopher M. Stevenson

1987 "Evidence Concerning the Origin of Maiz de Ocho." *American Anthropologist* 89(2):410–19.

Van Loon, Maurits, James H. Skinner and Willem Van Zeist

1968 "The Oriental Institute Excavations at Mureybit, Syria. Preliminary report on the 1961 campaign," part 1, Architecture and General Finds. *Journal of Near Eastern Studies* 27(4):265–90.

1970 "The Oriental Institute Excavations at Mureybit, Syria: Preliminary Report of the 1965 campaign," part 3, Paleobotany. *Journal of Near Eastern Studies* 29:167–96.

Van Zeist, Willem

1967 "Late Quaternary Vegetation History of Western Iran." *Review of Paleobotany and Palynology* 2:310–11.

Van Zeist, Willem, and S. Bottema

1966 "Paleobotanical Investigations at Ramad." Paris: *Annales,* Archaeologiques Arbos Syriennes 16(2):179–80.

1971 "Plant Husbandry in Early Neolithic Nea Nikomedeia, Greece." Netherlands *Acta Botanica* 5(20):31–35.

Van Zeist, Willem, and W. A. Casparie

1968 "Wild Einkorn, Wheat and Barley from Tell Mureybit in Northern Syria," Netherlands *Acta Botanica* 17(1).

Van Zeist, Willem, and H. E. Wright
1963 "Preliminary Pollen Studies at Lake Zeribar, Zagros Mountains, Southwestern Iran." Washington, D.C.: *Science* 140(3562):65–69
Vavilov, N. I.
1935 *Bases des Origines de la Selection des Plantes.* Moscow-Leningrad: Selection General.
1951 "The Origin, Variation, Immunity and Breeding of Cultivated Plants." London: *Chronica Botanica* 13(1/6).
Vishnu-Mittre
1977 "Changing Economy in Ancient India." In *Origins of Agriculture,* edited by C. E. Reed. The Hague: Mouton Publishers.
Vita-Finzi, C., and E. S. Higgs
1970 "Prehistoric Economy in the Mount Carmel Area of Palestine: Site-Catchment Analysis." In *Proceedings of the Prehistoric Society* (34):1–37.
Von Wissmann, H.
1956 "On the Role of Nature and Men in Changing the Face of the Dry Belt of Asia." In *Man's Role in Changing the Face of the Earth,* edited by W. L. Thomas, 278–303. Chicago: University of Chicago Press.
Voorhies, Barbara
1976 "The Chantuto People: An Archaic Period Society of the Chiapas Littoral, Mexico." Provo, Utah: *Papers of the New World Archaeological Foundation* 41:1–147.
Waselkov, G. A.
1975 "A Selected Bibliography for Paleoethnobotany." Tuscaloosa, Alabama: *Journal of Alabama Archaeology* 11(2):171–184.
Waterbolk, H. T.
1968 "Food Production in Prehistoric Europe." Washington, D.C.: *Science* 162: 1093–1102.
1982 "The Spread of Food Production over the European Continent." In *Intruduks jonen av jordbruk i Norden,* edited by T. Sjovold, 19–37. Oslo, Norway: Universitats Forlaget.
Waters, Michael R.
1986 "The Geo-Archaeology of Whitewater Draw, Arizona." Anthropological Paper 45, University of Arizona, Tucson, Arizona.
Watson, Patty Jo
1969 "The Prehistory of Salt Cave, Kentucky." Report 16, Illinois State Museum, Springfield, Illinois.
1974 *Archaeology of the Mammoth Cave Area.* New York: Academic Press.
1985 "The Impact of Early Horticulture in the Upland Drainages of the Midwest and Midsouth." In *Prehistoric Food Production in North America,* edited by R. I. Ford. Anthropological Paper 75, Museum of Anthropology, University of Michigan, Ann Arbor, Michigan.
1988 "Early Plant Cultivation in the Eastern Woodlands of North America." In *Foraging and Farming,* edited by D. Harris and C. Hillman, 555–71. London: Allen and Unwin.
Watson, Patty Jo, and Wesley C. Cowan
1989 *The Origins of Plant Cultivation in World Perspective.* Washington, D.C.: Smithsonian Institution Press.
Wedel, Waldo R.
1961 *Prehistoric Man on the Great Plains.* Norman, Oklahoma: University of Oklahoma Press.

Wendorf, Fred, and F. Thomas
1951 "Early Man Sites Near Concho, Arizona." Salt Lake City, Utah: *American Antiquity* 17(2):107–14

Wendorf, Fred, and J. P. Miller
1959 "Artifacts from High Mountain Site in the Sangre de Cristo Range, New Mexico." Santa Fe, New Mexico: *El Palacio* 66(2):37–52.

Werth, E.
1954 *Grabstock, Hacke, and Pflug*. Ludwigsburg, Germany: Eugen Ulmer.

West, Robert C.
1964 "The Natural Regions of Middle America." In *Natural Environment and Early Cultures*, vol. 1 of *Handbook of Middle American Indians*, edited by Robert C. West. Austin, Texas: University of Texas Press.

Wet, J. M. J. de
1975 "Evolutionary Dynamics of Cereal Domestication." London: *Bulletin of the Torrey Botanical Club* 102:307–17.

Wet, J. M. J. de, and J. R. Harlan
1975 "Weeds and Domesticates: Evolution in the Man-made Habitat." *Economic Botany* 29:99–107.

Whalen, Norman
1971 "Cochise Culture Sites in the Central San Pedro Drainage." Ph.D. dissertation, Department of Anthropology, University of Arizona.
1973 "Agriculture and the Cochise." *Kiva* 39(1):89–96.

Whallen, Robert
1973 "Spatial Analysis of Occupation Floor 1: Application of Dimensional Analysis of Variances." Salt Lake City, Utah: *American Antiquity* 38(3):266–78.

Wheeler, Jane (Pires-Ferreira), Edgardo Pires-Ferreira and Peter Kaulicke
1976 "Preceramic Animal Utilization in the Central Peruvian Andes." Washington, D.C.: *Science* 194:483–90.

Wheeler, Mortimer
1959 *Early India and Pakistan*. London: Thames and Hudson.
1968 *The Indus Civilization*. Cambridge, England: Cambridge University Press.

Whitaker, Thomas W., and Junius B. Bird
1949 "Identification and Significance of the Cucurbit Materials from Huaca Prieta, Peru." New York: American Museum *Novitatas*, no. 1426:1–15.

Whitaker, Thomas W., Hugh C. Cutler, and Richard S. MacNeish
1957 "Cucurbit Materials from Three Caves near Ocampo, Tamaulipas." Salt Lake City, Utah: *American Antiquity* 22(4):352–58.

Whiting, Alfred F.
1944 "The Origin of Corn: An Evaluation of Fact and Theory." Menasha, Wisconsin: *American Anthropologist* 46(4):500–15.

Wilkerson, S. Jeffrey K.
1975 "Pre-agricultural Village Life: The Late Preceramic Period in Veracruz." Berkeley, California: *Contributions of the University of California Archaeological Facility* 27:111–122.

Wilkes, Garrison
1988 "Maize Domestication, Racial Evolution and Spread." In *Foraging and Farming*, edited by D. Harris and C. Hillman, 440–55. London: Allen and Unwin.

Willey, Gordon R.
1971 *An Introduction to American Archaeology: South America*. Vol. 2. Englewood Cliffs, New Jersey: Prentice-Hall.

Willey, Gordon, and C. R. McGimsey
1954 *The Monagrillo Culture of Panama.* Papers of the Museum of American Archaeology and Ethnology 49(2). Cambridge, Massachusetts: Harvard University Press.

Wills, Wirt Henry
1988 *Early Prehistoric Agriculture in the American Southwest.* Santa Fe, New Mexico: School of American Research.

Wimberly, Mark, and Peter Eidenbach
1972 *A Preliminary Analysis of Faunal Material from Fresnal Shelter, New Mexico.* Training Manual, Human Systems Research, Tularosa, New Mexico.

Wissmann, H. Von
1957 *Ursprung und Ausbrectungswege von Pflazen und Tierzucht und thre Abhangigkert von der Klimageschichte, Erdkunde,* 11:81–94.

Wittmack, L.
1880 *Antike Samen aus Troja und Peru.* Monatsschr. ver. Beford. Gartenbau i Preussen.

Woodman, P. O.
1981 "A Mesolithic Camp in Ireland." New York: *Scientific American* 245(2): 1–14.

Woods, William I.
1987 "Maize Agriculture and the Late Prehistoric: A Characterization of Settlement Location Strategies." In *Emergent Horticultural Economies of the Eastern Woodlands,* edited by W. F. Keegan. Center for Archaeological Investigations, no. 7, Southern Illinois University, Carbondale, Illinois.

Wright, Herbert E.
1968 "Natural Environment of Early Food Production North of Mesopotamia." Washington, D.C.: *Science* 161:334–39.
1976 "The Environmental Setting for Plant Domestication in the Near East." Washington, D.C.: *Science* 194:385–89.
1977 "Environmental Change and the Origins of Agriculture in the Old and New Worlds." In *Origins of Agriculture,* edited by C. A. Reed, 281–320. The Hague: Mouton Publishers.
1985 "Climatic Change in the Zagros Mountains Revisited." In *Prehistoric Archaeology Along the Zagros Flanks,* edited by R. Braidwood et al. Oriental Institute, publication 105. Chicago, Illinois: University of Chicago Press.

Wright, James V.
1966 "The Ontario Iroquois Tradition." Publications in Archaeology 3. Ottawa, Canada: National Museum of Man.

Wright, James V., and J. E. Anderson
1963 "The Donaldson Site." Ottawa, Canada: National Museum of Canada, *Bulletin* 184.

Wyss, R.
1979 "Das Mittelsteinzeitliche Hirsch jagerlager von Schotz im Wauwiler moos." Zurich, Switzerland: *Archaeologische Forschungen,* Schweiz Landesmuseum, 7:1–13.

Xia, Nai
1977 "Carbon 14 Dates and the Prehistoric Archaeology of China." *Kaogu* 4.

Yarnell, Richard A.
1964 "Aboriginal Relationships Between Culture and Plant Life in the Upper Great Lakes Region." Anthropological Papers 23, Museum of Anthropology, University of Michigan. Ann Arbor, Michigan.

1972 "Iva annua var. macrocarpa: extinct American cultigens." Menasha, Wisconsin: *American Anthropologist* 74:34–49.

1976 "Early Plant Husbandry in Eastern North America." In *Culture Change and Continuity*, edited by Charles E. Cleland, 265–74. New York: Academic Press.

1977 "Native Plant Husbandry North of Mexico." In *Origins of Agriculture*, edited by C. H. Reed, 861–78. The Hague: Mouton Publishers.

1978 "Domestication of Sunflower and Sumpweed in Eastern North America." In *The Nature and Status of Ethnobotany*, edited by R. I. Ford. Anthropological Papers 67, Museum of Anthropology, University of Michigan. Ann Arbor, Michigan.

Yen, D. E.

1977 "Hoabinhian Horticulture: The Evidence and Questions from Northwestern Thailand." In *Prehistoric Studies in Southeast Asia, Melanesia, and Australia*, edited by J. Allen, J. Galson and R. Jones, 567–700. London: Academic Press.

1980 "The Southeast Asian Foundations of Oceanic Agriculture: A Reassessment." Paris: *Journal de la Société des Oceanistes* 36:140–47.

1982 "Ban Chiang Pottery and Rice." Philadelphia, Pennsylvania: *Expedition* 24(4):51–64.

Zevallos-Menendez, Carlos

1971 *La agricultura en el Formativo temprano del Ecuador (cultura Valdivia).* Guayaquil, Ecuador: Casa de la Cultura Ecuatoriana.

Zhimin, An

1988 "Prehistoric Agriculture in China." In *Foraging and Farming*, edited by D. Harris and C. Hillman, 643–49. London: Allen and Unwin.

Zohary, Daniel

1988 "Domestication of the Southwest Asian Neolithic Crop Assemblage of Cereals, Pulses, and Flaxes: The Evidence from Living Plants." In *Foraging and Farming*, edited by D. Harris and C. Hillman, 358–73. London: Allen and Unwin.

Zucchi, A.

1978 "La variabilidad ecologica y la intensificación de la agricultura en los llanos Venezolanos." In *Ensayos Antropológicos en homenaje a José M. Cruxent*, edited by E. Wagner and H. Zucchi, 349–65. Caracas, Venezuela.

Index